ELECTRICAL ENGINEERING IN CONTEXT

Smart Devices, Robots & Communications

Roman Kuc

Professor of Electrical Engineering
School of Engineering & Applied Science
Yale University

CENGAGE
Learning™

Australia • Brazil • Japan • Korea • Mexico • Singapore • Spain • United Kingdom • United States

Electrical Engineering in Context:
Smart Devices, Robots & Communications

Roman Kuc

Publisher: Timothy Anderson

Senior Developmental Editor: Hilda Gowans

Editorial Assistant: Sam Roth

Senior Content Project Manager:
    Jennifer Ziegler

Production Director: Sharon Smith

Team Assistant: Ashley Kaupert

Intellectual Property:
    Analyst: Christine Myaskovsky
    Project Manager: Amber Hosea

Text and Image Researcher:
    Kristiina Paul

Manufacturing Planner:
    Doug Wilke

Copyeditor: Shelly Gerger Knechtl

Proofreader: Erin Buttner

Indexer: Rose Kernan

Compositor: MPS Limited

Senior Art Director: Michelle Kunkler

Internal Designer: MPS Limited

Cover Designer: Rose Alcorn

Cover Image: © Toria/Shutterstock.com;
    © Romeo1232/Dreamstime.com

For product information and technology assistance, contact us at
Cengage Learning Customer & Sales Support, 1-800-354-9706.
For permission to use material from this text or product,
submit all requests online at **www.cengage.com/permissions.**
Further permissions questions can be emailed to
permissionrequest@cengage.com.

Library of Congress Control Number: 2013957165

ISBN-13: 978-1-285-17918-6

ISBN-10: 1-285-17918-8

Cengage Learning
200 First Stamford Place, Suite 400
Stamford, CT 06902
USA

Cengage Learning is a leading provider of customized learning solutions with office locations around the globe, including Singapore, the United Kingdom, Australia, Mexico, Brazil, and Japan. Locate your local office at: **international.cengage.com/region.**

Cengage Learning products are represented in Canada by Nelson Education Ltd.

For your course and learning solutions, visit
www.cengage.com/engineering.

Purchase any of our products at your local college store or at our preferred online store **www.cengagebrain.com.**

Unless otherwise noted, all items © Cengage Learning.

Unless otherwise noted, all photos © Roman Kuc.

Matlab is a registered trademark of The MathWorks, 3 Apple Hill Road, Natick, MA.

Printed in the United States of America
1 2 3 4 5 6 7   18 17 16 15 14

For my students, ever eager to learn,
For wife Robin, a true companion, and
For Our Father, from Whom all blessings flow.

Photo courtesy of Michael Marsland/Yale University

Roman Kuc ("*Koots*") received the BSEE from the Illinois Institute of Technology, Chicago, IL, and the PhD degree in Electrical Engineering from Columbia University, New York, NY. He started his engineering career as a Member of Technical Staff at Bell Laboratories where he designed electromechanical, analog, and digital systems, and developed efficient digital speech coding techniques. After completing his PhD he was postdoctoral research associate and adjunct professor in the Department of Electrical Engineering at Columbia University and the Radiology Department of St. Luke's Hospital, where he applied digital signal processing techniques to diagnostic ultrasound signals. He then joined the Electrical Engineering faculty at Yale University, where he enjoys teaching courses in system theory, digital signal processing, microcontroller programming, and capstone design projects. He also enjoys describing engineering problem solving to non-science majors, which attracted 780 Yale students during one memorable semester. As Director of the Intelligent Sensors Laboratory, he explores digital signal processing for extracting information from sensor data and implementing brain-based sensor systems. Motivated by biological sensing systems, such as echolocating bats, he has implemented biomimetic and neuromorphic sensing systems for robots.

Prof. Kuc has authored more than 150 technical papers and two books, *Introduction to Digital Signal Processing* in 1988 and *The Digital Information Age* in 1999. He authored the biomimetic sonar chapter in *Echolocation in Bats and Dolphins*, and co-authored the Sonar Sensing chapter with Lindsay Kleeman in the *Springer Handbook of Robotics*.

Prof. Kuc is a member of the Connecticut Academy of Science and Engineering, Fellow of the Shevchenko Scientific Society, Senior Member of the IEEE, and past chairman of the Instrumentation Section of the New York Academy of Sciences. His honors include Yale Engineering's Inaugural Sheffield Distinguished Teaching Award, Yale's Order of the Golden Bulldog, and an honorary doctorate from the Glushkov Institute of Cybernetics in Ukraine.

CONTENTS

3   Electric Circuits   52

8 Digital Signal Processing 202

9 Spectral Analysis 228

10 Detecting Data Signals in Noise 256

11 Designing Signals for Multiple-Access Systems 282

12 Source Coding 304

13 Channel Coding   332

14 Symbology   354

Appendix A Math Details 467

Appendix B Probability 471

Appendix C ASCII Code 491

This text presents an overview of Electrical Engineering suitable for first or second-year undergraduate students majoring in Electrical Engineering. It incorporates topics that are current in the fields of smart devices, robots, and communication systems, and it also includes topics that I found fundamental to and reoccurring in my experience as both a design engineer and adviser of capstone projects. These topics are treated in an introductory manner but in sufficient detail to explain the concepts. Where appropriate, the text indicates extensions that are covered in advanced EE courses. In addition to describing *how* particular systems operate, the book attempts to explain *why* a particular approach has been taken or why a particular design is optimal. Each chapter begins with learning objectives that are used to motivate the student and to provide a concise description of the puzzle part that forms the whole.

Ideas and principles in this book are described with minimal reliance on prerequisites in mathematics and science. All that are required is a good grasp of algebra and a knowledge of physics at the high school level. In contrast to a traditional science text, which starts with theory and then illustrates it with applications, this book starts with the application and then presents the physical theory and mathematical analysis required to understand the application. The applications illustrate both how ideas can be expressed mathematically and how mathematical models are used in practice.

This text is intended to be used for a one-semester course, although any particular course will probably not cover all chapters. The chapter length was intentionally kept small to allow each instructor to structure a course that fits into their program's particular curriculum needs or instructor's expertise. Each chapter ends with problems and projects. The problems are intended for paper-and-pencil solutions, and the projects are intended to be solved by writing a Matlab® script. Many of the projects are extensions of Matlab programs that are included in the *Matlab Best Practices* chapter, which help the student get started. Programs that acquire and process real data, such as microphone speech and jpeg camera images, illustrate the application of the theory. One easily can imagine a two-semester course that includes all of the chapters and also teaches programming techniques with Matlab to probe different designs to optimize the performance.

This book tries to tell a story that describes digital information systems, presenting them in a context that provides intuition and understanding of the technological world students currently inhabit. The hope is that students gain an appreciation of where we are and can form an educated guess about where we are going. The text tries to present the *big picture*, using broad strokes to form the relationships among various aspects of digital information in three main application areas of smart devices, robots, and communications. The main topics are illustrated with many examples and sample Matlab code for the investigation of extensions in the projects. The main topics include:

- *Electrical engineering basics*—Starting from basic resistor circuits to illustrate Ohm's Law and Kirchoff's Voltage and Current Laws, these basic analytic tools are then extended to complex-valued impedances of capacitor and inductor circuits. The operational amplifier forms the basic building block in common analog circuits, while the field-effect transistor is presented as the basic digital switch. A similar approach describes ideas in digital signal processing and spectral analysis with the fast Fourier transform in order to provide a fundamental understanding of the important concepts of bandwidth and signal-to-noise ratio.

- *Binary data generation*—The creation of digital data is considered from simple switch closures that occur on touch pads that form the basic informational binary digits or bits. The second source of digital data occurs through analog-to-digital conversion, which transforms signal waveforms into their digital counterparts for applying digital signal processing. The third source of digital data is produced by scanning digital symbols, such as bar codes with digital cameras.

- *Digital signal processing*—Digital data are processed initially with basic logic gates, which are then configured into logic circuits that form the building blocks found in any computer. Digital signal processing describes how matched filters robustly extract data from transmission signals. Complementary and orthogonal signals illustrate how transmission signals can be designed to minimize the error probability when these signals are detected in the presence of noise. For electrical engineering (EE) majors, this approach provides a fundamental understanding of the methods and motivates discussion in follow-on systems and signals courses.

- *Data coding*—Examples of coding techniques illustrate the important and unique features of digital communications. Digital data can be coded by the source for compression and encryption. Information is modeled using elementary ideas in probability that permits measuring its quantity in the form of entropy, with units of bits per symbol. The difference between information and data then becomes clear, and a simple method to compress the data, the Huffman code, achieves a value that is close to the entropy value. The allure of this section is that software projects exemplify the generation, encoding and compression, that illustrate entropy calculations. Redundant data can be compressed to allow efficient transmission of video and audio signals. Secrecy is obtained through encryption. Channel encoders provide error detection and correction for robust transmission. A single-error assumption allows simple block-code error correction to be understood at an intuitive level. The Shannon-Hartley Capacity equation and the inverse square law illustrate why the transmission rate decreases with range. Symbology illustrates how machine-readable codes, in credit-card magnetic strips and in bar codes, are designed for robust scanning and reliable data extraction, while steganography inserts secret data into large data files that result in almost imperceptible changes to the original. Machine *unreadable* codes illustrate one method to thwart artificial intelligence from hacking Web sites.

- *Networks*—Computers connect to networks through wired and wireless channels. Data packets and protocols are common to both. Packet collisions are discussed in wired and wireless networks and simulated with software to illustrate the effects of packet sizes. The trace route utility, which is available on both PCs and Macs, explores the structure of the Internet, determines the speed of data packets traveling over the Internet, and is very popular with students.

An Instructor's Solutions Manual will be available in print or online for registered instructors of the course.

In addition to the print version, this textbook also will be available online through MindTap, a personalized learning program. Students who purchase the MindTap version will have access to the book's, MindTap Reader and will be able to complete homework and assessment material online, through their desktop, laptop, or iPad. If your class is using a Learning Management System (such as Blackboard, Moodle, or Angel) for tracking course content, assignments, and grading, you can seamlessly access the MindTap suite of content and assessments for this course.

In MindTap, instructors can:

- Personalize the Learning Path to match the course syllabus by rearranging content, hiding sections, or appending original material to the textbook content

- Connect a Learning Management System portal to the online course and Reader
- Customize online assessments and assignments
- Track student progress and comprehension with the Progress app
- Promote student engagement through interactivity and exercises

Additionally, students can listen to the text through ReadSpeaker, take notes and highlight content for easy reference, and check their understanding of the material.

ACKNOWLEDGMENTS

I gratefully acknowledge the assistance and encouragement of my Development Editor, Hilda Gowans, of Cengage Learning, whose guidance and advice eased the author's burden. The following reviewers generously contributed their pedagogical experience to improve early versions of the manuscript. These include Pelin Aksoy, George Mason University, Steve P. Chadwick, Embry-Riddle Aeronautical University, Michael Chelian, California State University, Long Beach; Simon Y. Foo and Bruce A Harvey, FAMU-FSU, Roland Priemer, University of Illinois at Chicago, Michael Reed, University of Virginia, Christopher Schmitz, University of Illinois, and Andrew Tubesing, University of St Thomas.

Over the thirty years of teaching at Yale I have had conversations with numerous students at various stages of their academic careers: from high school students taking a college summer course, college first-years thinking about majoring in Electrical Engineering, seniors completing an electrical engineering program, and to recent graduates who were employed as electrical engineers. This book attempts to address the concerns former students had: *What will I learn in an electrical engineering program?* and include the advice that the latter students offered: *What do electrical engineers do?* This material has been classroom tested with Yale and non-Yale students. Their experience ranges from bright high school seniors and freshmen intending to major in engineering to non-science majors interested in technology.

ROMAN KUC
New Haven, CT

ELECTRICAL ENGINEERING IN CONTEXT

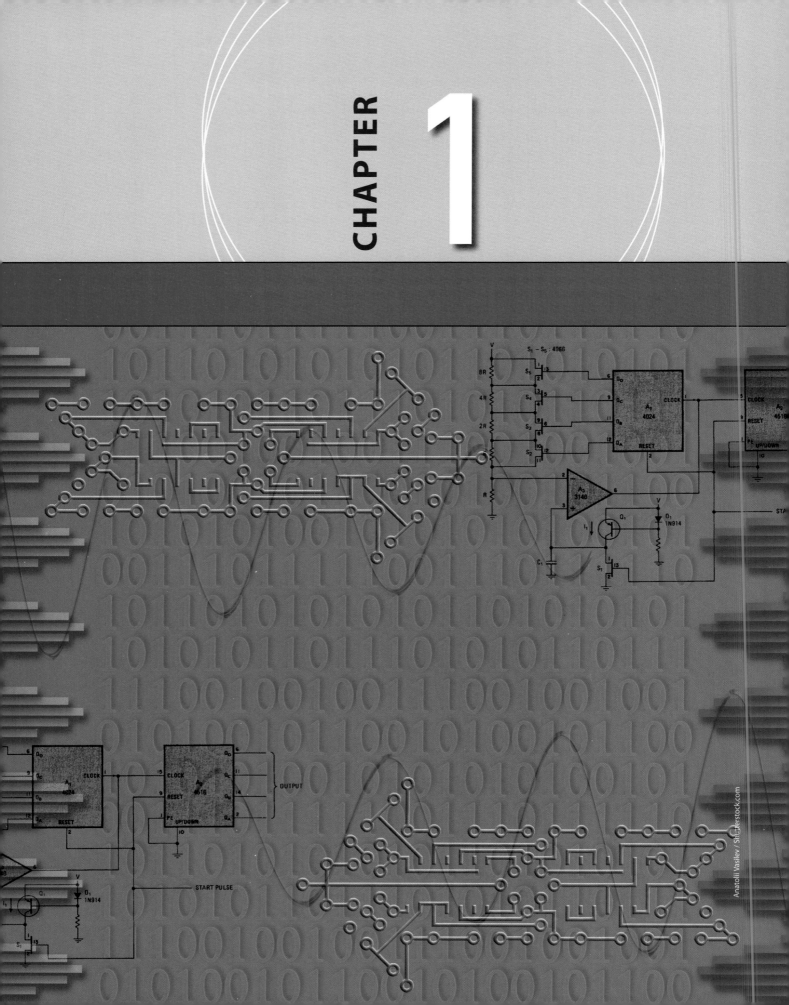

INTRODUCTION

After completing this chapter, the reader should be able to:

- Understand the design process in the electrical engineering profession.
- Understand the importance of computers in smart systems.
- Appreciate the types of systems and signals that carry information.
- Know the magnitudes of relevant time and frequency scales in information systems.

1.1 INTRODUCTION

Electrical engineering is a broad profession that encompasses a wide variety of systems that deal with the movement of electrons to communicate information, control operation, and provide power. The scale of hardware varies from nanodevices to trans-world power and communication networks. The associated problem-solving activity is firmly based in science and mathematics, making an electrical engineering education a desirable and sought-after preparation for many careers.

This chapter provides a general introduction to electrical engineering and the topics discussed in this book in the following manner:

Problem descriptions—Electrical engineers formulate problems in terms of block diagrams that indicate the main component of a system and then apply their tools to design a solution.

Analog and digital signals—Electrical systems communicate through waveforms that typically start and end in analog waveforms but more often much of the processing is performed by computers on digital values. Digital signals form the basis for communications because of their ability to form perfect copies.

Future trends—Digital systems follow Moore's law to indicate improvements that double every 18 months, which bodes a bright future for digital devices.

 ## 1.2 ELECTRICAL ENGINEERING FOR THE DIGITAL AGE

This is the age of computers, smart systems, and the communication networks that link them. Computers are becoming cheaper and more powerful, and smart devices that use computers easily connect to networks and each other. Electrical engineers play a major role is designing the devices, components, systems, and processes that result in the operations that meet specified requirements.

1.2.1 Block Diagrams

Electrical engineers describe system operation using *block diagrams*. A block diagram uses arrows to indicate the direction of signal flow and blocks to represent specific operations. It is a convenient way to summarize the role of components in a system, to show information flow, and to present the *big picture*.

In this book, *information is the quantity of data that is required to complete a task.* Information in its most basic form is transmitted by some form of energy.

- We see things because light energy is scattered from objects.

- We hear sounds because of acoustic energy caused by pressure variations uttered by a speaker, produced by an instrument, or generated by a loud speaker.

- We feel mechanical energy in the vibration caused by a small motor in our cell phone to indicate an incoming call.

- We exert mechanical energy to press a switch to answer a call.

- Our cell phone communicates wirelessly using antennas that transmit and receive electromagnetic energy.

- A thermometer in our computer senses thermal energy, and when it is too high, the operating frequency or voltage is reduced to prevent thermal damage.

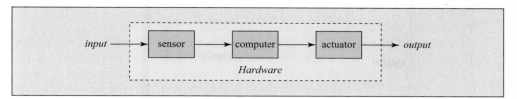

Figure 1.1

Block diagram that summarizes the operation of many smart systems.

Figure 1.1 summarizes the systems described in this book using a simple block diagram that illustrates informational flow through the basic components. Smart systems at a variety of sizes (from smartphones, robots, to communication networks) have the following common features.

Input: Information is embodied in some form of energy that occurs naturally, such as light reflected from objects, or is produced by humans in the sounds generated by speaking or button presses. This is the human communicating with the smart system.

Hardware: Sensors convert the energy into electrical signals that are processed by computers to generate electrical signals that drive actuators.

Output: Actuators convert electrical signals into perceptual energy, such as the light generated by displays and mechanical vibrations produced by motors. This is the smart system communicating with the human.

Cell phones and Smartphones　EXAMPLE　1.1

Cell phones and smartphones are two of the most common digital devices in society. They are differentiated as follows:

Cell phones are devices that are designed primarily for wireless voice communication over a cellular system defined by antennas and transmission rules. A cell phone digitizes speech signals, encodes them for efficient transmission, and stores user data, such as the phone numbers of contacts, to simplify operations required to complete a connection.

Smartphones are cell phones that also connect to the Internet and are implemented with enhanced computers and displays to provide video features.

Both cell phone and smartphone operations, circuits, and capabilities are described in this book.

To illustrate the application of the techniques discussed in this book, Figure 1.2 shows three systems that represent the current trends in electrical engineering.

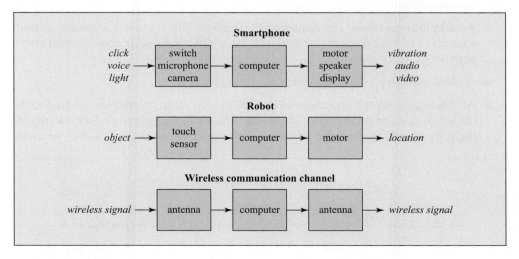

Figure 1.2

Block diagrams of three systems considered in this book.

1. **Smartphones** form the interface between the human user and a variety of networks, from your cell phone system to the Internet. The input includes user clicks on screen icons, speech commands, or pictures of two-dimensional bar codes. This data speeds to its intended destination, which then provides requested information, such as the location of the closest restaurant or entertainment in audio or video forms. This digital information is then provided to the human through speakers, color displays, or vibrating motors.

2. **Robots** perceive their environments by acquiring sensor information, for example, a touch sensor that detects the presence of an object. This information is processed, and aided with data stored in memory, actuators are energized to produce manipulation or movement. In complex environments that produce ambiguous sensor reading, robots accept instructions through teleoperation, that is aided by sensory information transmitted back to a base station to accomplish a desired task. Robots on Mars present particular challenges to teleoperation because their signals are very weak and the travel time delay is long.

3. **Wireless communication channels** acquire and transmit signals through antennas. Detected data are processed, stored, and encrypted to communicate data from smart device to smart device or from computers located across the globe. Various applications present different challenges, from fast download times to real-time viewing without annoying pauses.

1.2.2 Describing System Operation

How do these devices work? What are their limitations? Such questions are in the domain of Electrical Engineering, which is a profession of analysis, design, and problem solving. Engineers use mathematical tools and scientific principles to design devices that improve life and solve problems that occur in society. This book focuses on *electrical engineering* (EE) and shows how engineers control the movement of electrons to implement the amazing devices we encounter in our daily lives.

A typical system consists of the following parts, as illustrated in Figure 1.3, which shows a block diagram of cell phone operation for transmitting digitized speech over a wireless channel.

- An input, typically some form of energy $e(t)$ that varies over time t.

- The sensor, or transducer, converts $e(t)$ into an electrical analog waveform $x(t)$.

- An ADC converts $x(t)$ into a sequence of numbers x_i, which indexed with i. Typically, we are interested in the data x_i for $i = 0, 1, 2, \ldots, n_x - 1$, that encodes the information.

- A computer processes the sequence of numbers into a second sequence y_i that enhances the information (for example, by removing noise or non-relevant components).

- A DAC converts y_i back into an analog waveform $y(t)$.

- An actuator converts the electrical waveform $y(t)$ into an output $e(t)$ that exits the block diagram on the right. In this text, the output is typically a form of energy (light or sound) that can be perceived by human senses. In the case of wireless

Figure 1.3

Block diagram of cell phone operation for transmitting digitized speech over a wireless channel.

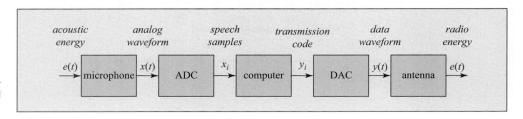

Figure 1.4

Smartphones have cameras that capture images of two-dimensional bar codes.

systems, the energy is in the form of radio waves, or electromagnetic radiation, that communicates with other systems that are described with their own block diagrams.

Smartphone camera   EXAMPLE 1.2

Figure 1.4 shows a smartphone camera taking a picture of a two-dimensional bar code. The camera generates an image array of pixel values. A small computer within the smartphone decodes the bar code image into digital data and transmits the data wirelessly to the Internet to obtain additional information. Similar functions occur in intelligent robots.

The following technical questions occur in the design of a system that interprets the bar code:

- The camera image contains the entire symbol, but the symbol does not occupy the entire image. What processing steps are required for the cell phone computer to decode the image?

- How will your camera know that it interpreted the symbol correctly and there is no error?

- The distance between your camera and the symbol increases as you move away from the symbol. Under what conditions will your camera no longer be able to decode the symbol correctly?

Teleoperation of a robot on Mars   EXAMPLE 1.3

Mobile robots are equipped with sensors to navigate around obstacles. Figure 1.5 shows a base station on Earth transmitting data to and from a camera-wielding mobile robot on Mars through a satellite relay orbiting Mars. The size of the arrows indicates the capacity of the various data channels to display an image from Mars. To prevent accidents on Mars, teleoperation by an operator viewing camera images overrides autonomous operation when the situation is questionable. The signals are transmitted between antennas with a strength that is limited by the available power. The vast distance between Earth and Mars causes signals from Mars to be very weak, and signal processing is required to extract data from noise. A teleoperated robot system raises technical questions including:

- What are the data-carrying capacities of the various channels?

- What transmitted signals produce the best images in a reasonable amount of time?

- With the travel delay between Earth and Mars and available power, how should the processing tasks be divided between the robot and the base station on Earth?

Figure 1.5

A base station on Earth transmits data to and from a camera-wielding mobile robot on Mars through a satellite relay orbiting Mars. The size of the arrows indicates the capacity of the various data channels to display an image from Mars.

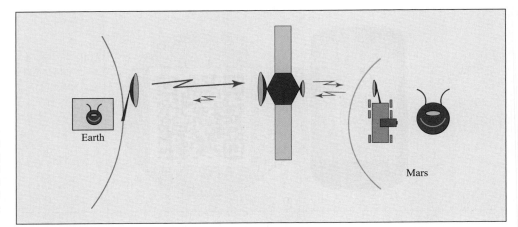

EXAMPLE 1.4

Digital cell phone system

Figure 1.6 shows the transmission activity in a cell phone system, in which many users transmit and receive simultaneously. A computer sorts the routing information and data from different sources and directs them to their respective destinations.

Technical questions that occur in the design of a digital cell phone system include the following.

- The telephone system operates in a particular audio frequency range, typically up to 3,500 Hz. The recording microphone produces sound waveforms having frequency components up to 10,000 Hz. A low-pass filter typically connects to the microphone output to condition the signal for digitization. The filter connects to an analog-to-digital converter (ADC) that samples the audio waveform and uses an 8-bit quantizer to store the samples into a digital memory. What values of sampling period (T_s) and quantizer resolution affect the audio quality?

- To make the system secure, the data is encrypted using a random number generator. How much encryption provides sufficient security?

- Your data is placed on a communication channel shared by other users. How can data transmission signal design affect the number of users within a particular cell?

- How does the spacing of antennas affect system performance?

- On the receiving end, digital data are decrypted and are applied to a digital-to-analog converter (DAC) that reconstructs the original speech waveform. How does the DAC design affect the audio quality?

Figure 1.6

In a cell phone system many users transmit and receive digital data simultaneously. A computer sorts the routing information and directs digital speech data from sources to their respective destinations.

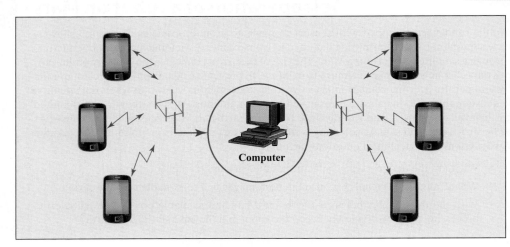

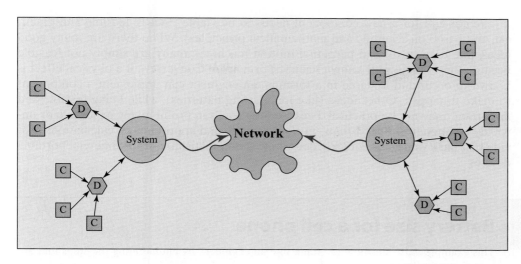

Figure 1.7

Electrical engineers design at various levels: components (C), devices (D), systems, and networks.

1.2.3 What Electrical Engineers Do

Current smart devices contain a mix of analog electronics and digital system components. Electrical engineers, although familiar with the basic principles, typically design at one of four levels of size and complexity: components, devices, systems, and networks, as illustrated in Figure 1.7. We examine each level separately here.

Components

Electrical components are the basic electronic elements that exploit physical principles to produce the electrical signals at the *front end* of devices. Components in a smartphone include the camera, microphone, and processor. A robot has sensor components to indicate where it is and actuator components to move it to where it should be.

Even components can be subdivided into more fundamental parts. Physical interactions at the atomic level convert electrical, optical, thermal, or mechanical energy into electron flow that is then captured and amplified by other electronic devices, such as transistors. Such interactions occur at the micro-scale with dimensions measured in micrometers (μm) or nano-scale, measured in nanometers (nm). At these scales, engineering scientists or applied physicists devise clever techniques to detect individual ions, electrons, or light photons or to construct novel materials that may contain carbon nanotubes to have larger surface areas for a given volume, which is important for high-capacity battery fabrication. Components are designed to be manufactured inexpensively, to operate under a specified temperature and other environmental range, and to require a simple power supply, such as a single battery.

Smartphone camera EXAMPLE 1.5

The camera in a smartphone uses photo-detector components to convert optical energy into electrical signals and amplifiers to generate the signals that can be stored or transmitted. Electrical engineers do not need to design a camera each time they need one in a system; rather, they simply specify an existing camera component and design its integration into a smartphone. The design challenge is to select the *best* camera, which usually involves a trade-off of cost, size, weight, power consumption, availability, reliability, and resolution.

The concept for a new component needs to be viable (that is, feasible and practical, and based on scientific and mathematical principles). While there are many good ideas, like time travel and perpetual motion machines, many are simply not feasible because they violate well-known scientific principles. Conversely, if a physical effect is known to occur and is stated in a formula, an engineer can implement a component to make it happen. Other ideas (like long-lasting batteries), while being feasible and important, may not be practical, because they result in prohibitively expensive or unwieldy components. *Back-of-the-envelope* (simple and approximate) calculations provide ball-park values that quickly determine their practicality for commercial portable devices.

EXAMPLE 1.6

Battery size for a cell phone

This example illustrates how to "think like an engineer" by enumerating the steps that are part of the design process.

An electrical engineer determines the practicality of a long-lasting cell phone battery using the following steps.

STEP 1. Required power. A typical cell phone is meant to fit in the hand and uses one Watt (1 W) of power while it is operating.

STEP 2. Available power. A lithium-ion battery (LIB) has an energy density of 200 Watt-hour (W-hr) per kilogram of mass or 200 W-hr/kg.

STEP 3. Back-of-the-envelope calculation. An LIB weighing 0.1 kg (1/4 pound) provides 20 W-hr of energy. Such a LIB powers the 1 W cell phone for 20 hours, is easy to hold and carry, but requires a nightly recharge.

STEP 4. Conclusion. A battery weighing 0.5 kg (1.1 lb) provides 100 W-hr of energy and will last for about four days, but would make the cell phone five times larger and heavier. Such a cell phone would not fit comfortably in the hand. Therefore, a long-lasting battery using LIB technology is not practical.

> *Factoid:* Scientific and mathematical principles provide the fundamentals that govern the basic operation and form the limitations of component performance.

Devices

Electrical devices are products that are designed by interconnecting components and have a particular purpose. A smartphone is a stand-alone device that performs all of the functions needed to communicate digital data. A device is typically given a model number, which indicates the particular list of components and features that are included.

While engineers typically design individual components, an engineering *manager* coordinates their designs in order to meet a set of constraints. The design of a device typically balances or *trades-off* different criteria, such as cost, size, weight, speed, reliability, and accuracy, as well as marketing ideas, such as new features and ease of use. New models are introduced when the features they offer are significantly better, usually by a factor of two or more, than the current or competitor's model.

Mobile robot   EXAMPLE   1.7

A mobile robot is a device that has sensor components, including a GPS for global localization, a detection sensor for obstacle avoidance, a computer that analyzes the data to determine a path that leads to the destination, and motors that propel the robot in the desired direction.

Systems

An electrical system is an assembly of devices that allows them to operate through communication links. These links may be physically wired connections or wireless connections using radio waves. The devices connecting to a system may have different capabilities, such as the variety of smartphones that can connect to a specific system (AT&T or Verizon). A system contains an organizational management that sets the rules for operation of interconnecting devices. Even systems have constraints, such as the frequency spectrum (within which it can transmit) and environmental factors (size and location of transmission towers). System control becomes complicated because devices can connect or disconnect at random times and locations.

Challenges for a cellular telephone system   EXAMPLE   1.8

Before 1982, the telephone network in the U.S. was a Bell System monopoly. With the support of Bell Labs, the telephone quality and reliability were very good, because all of the devices connected to the network were manufactured by the Bell System. Telephone connections were made more reliable by implementing digital switching technology to replace electromechanical switches. Enlarging an infrastructure mainly involved laying more wires and radio antennas to address the increased demand for telephone service. Planning for future expansion was easier, because there were no competitors and the incoming revenue was reliable.

In the current cellular telephone market, there are several major competitors that vie for customers by providing adequate service, offering differentiating features for a reasonable fee. The smartphones are no longer made by a cell phone company but by high-tech companies. This leads to negotiations about what features can be made available in the smartphone computer and yet be supported by the transmission capabilities of the cell phone company network. To provide the features that consumers demand, generations of improved communication networks have evolved and are currently in the fourth generation (4G). These new generations are made possible through ever-improving electronic and nanotechnology devices that operate at faster speeds, lower power, and increased complexity.

Networks

In today's digital information age, there are many systems that communicate with each other over the most complex network that has ever been implemented: the Internet. Electrical engineers face many challenges in trying to make a network function quickly, robustly, and securely under a wide variety of traffic conditions. Networks need to accommodate ever-increasing data communication needs, such as increased data throughput, or providing large quantities of data and minimum delay so that the video on your smartphone does not pause. Also, these need to be accomplished without errors and in a secure manner to guarantee privacy.

What makes a network particularly complex is that each system operates at its own pace and uses its own data format. Hence, this makes it difficult (even impossible) for a network to have a central control structure. At best, the central organization managing the network specifies guidelines in the form of a network architecture that needs to be followed to allow systems to communicate.

EXAMPLE **1.9**

Internet of Things

Not only smartphone users communicate over the Internet. Researchers are planning for the *Internet of Things* (IoT) that involves putting wirelessly connected computers into things, such as cars, traffic monitoring stations, household appliances, etc. In the IoT, any device that interacts with humans or affects their activity will have a presence on the Internet.

The most ambitious forecast talks about "*6A connectivity*:" Anything, Anytime, Anyone, Any place, Any service, and Any network. Clearly, advances in computer cost, power, connectivity, and flexibility are needed to realize this vision.

1.3 ANALOG AND DIGITAL SIGNALS

Signals come in two basic flavors: *analog* and *digital*. Figure 1.8 shows these signals viewed on an *oscilloscope*, which is an instrument that displays waveforms.

- *Analog signals* are waveforms produced by sensors and typically vary smoothly over time, such as sinusoidal waveforms or signals that encode music. Analog audio signals, such as speech or music, vary on the millisecond time scale. The prototypical analog waveforms are sinusoids. Sinusoidal waveforms at different frequencies are used for analyzing the frequency behavior of systems and for the spectral decomposition of more complex signals.

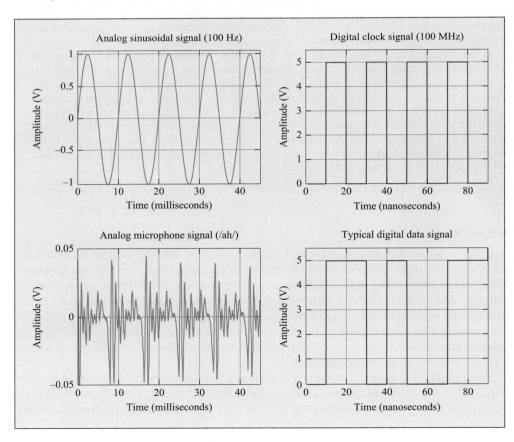

Figure 1.8

Analog and digital signals. Analog signals typically vary smoothly over time, while digital signals change abruptly between specific levels.

- *Digital signals* occur in logic circuits and computers and are produced by *detectors* that signal the occurrence of an event. Digital signals typically take on only discrete values, usually only two, such as 0 and 5 V (Volts) that encode binary logic levels (0/1) in a computer. The constraint in digital waveforms allows computers to quickly, reliably, and flexibly process digital data.

1.3.1 Converting Between Analog and Digital Signals

The device that transforms an analog waveform, such as speech, into a digital form consisting of a sequence of numbers is called an *analog-to-digital converter* (*ADC*). Once in digital form, the digital samples can be processed, stored, or transmitted using digital technology. In the digital audio case, the acoustic signal is converted to an electrical analog waveform by a microphone. This analog signal is applied to an ADC that then produces a sequence of numbers in binary (0/1) form. This data representing audio information is stored in a computer memory on your mp3 player or smartphone. This digital data is then converted back into an electrical waveform using a digital-to-analog converter (DAC). This electrical signal is fed into an amplifier that drives a speaker that reproduces the original acoustic signal.

Transforming an analog waveform into a sequence of numbers must be done carefully to avoid information loss and be undistinguishable from the original. Specifically, two questions must be answered to perform analog-to-digital conversion correctly:

- How often do we need to sample the analog signal in order to retain all the information that it contains?

- How accurately must the resulting digital samples be represented in order to form a satisfactory reproduction of the original analog signal?

To answer the first question we must describe a signal in terms of its frequency content (how many different frequency components are present in the analog signal). The important parameter is the highest frequency component present. To answer the second question, we investigate the process called *quantization*, which approximates a numerical value using a specified number of bits that provides a fixed precision. Transforming analog signals to digital representation comes at a price: While an analog signal can take on any value (within limits), its digital representation needs to be expressed in a finite number of bits. Figure 1.9 shows that analog-to-digital conversion results in a finite-precision approximation called *quantization*. The resulting error is viewed as

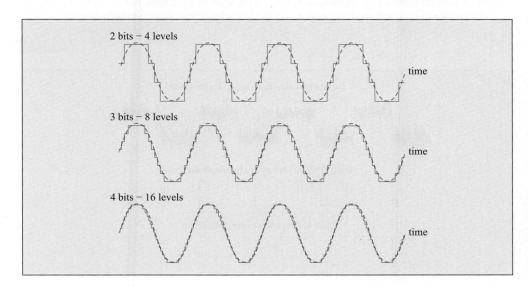

Figure 1.9

Representing an analog waveform using a finite number of bits results in a finite-precision approximation called quantization.

noise that has been added to the analog signal. A large number of bits, typically eight or more, is used to make this noise tolerable so that the performance of a particular system is not degraded by becoming digital.

1.3.2 Advantages of Digital Signals

Modern communication systems transmit data in digital form because digital data are *robust*. Many signals originate as analog waveforms, such as those produced by a microphone that converts acoustic energy produced by your utterances into electrical waveforms. While such waveforms can be transformed back into acoustic energy by driving a speaker, there are advantages in first converting analog waveforms into digital signals using analog-to-digital conversion.

Data Restoration: Digital values can be restored exactly. Unlike analog signals that can take on any value within limits, digital signals are constrained to have one of two allowed values. Small deviations produced by noise can then be restored to their original values. This operation is not possible for analog signals, because their waveforms can have any value between limits.

Figure 1.10 shows that digital signals can be restored to their original values even after the signals have been corrupted with noise. While all analog signals degrade with time, restoring digital values is possible, because digital signals are allowed to have only certain values, with the figure showing -1 V and $+1$ V. Signals corrupted with small amounts of noise typically will be close to these values. Data signals that are between 0.8 and 1.2 can then be reset to 1.0.

EXAMPLE 1.10   Restoring corrupted digital signals

Consider digital waveforms that can be either -1 V or 1 V. The following digital signals were corrupted by additive noise during transmission to produce the detected signal values:

$$-1.1,\ 0.9,\ -0.95,\ 1.1,\ -.75,\ 0.85$$

Recognizing that these values correspond to digital levels at either -1 or $+1$, the original values can be restored by setting them by using threshold $\tau = 0$, with values $< \tau \to -1$ and $\geq \tau \to +1$. Thus,

$$-1.1 \to -1,\ 0.9 \to 1,\ -0.95 \to -1,\ 1.1 \to 1,\ -0.75 \to -1,\ 0.85 \to 1$$

If the noise causes a wrong decision to be made by threshold detection, various error correction techniques can be employed, depending on the severity of the noise, as discussed next.

Figure 1.10

Processing digital data waveforms corrupted with noise to retrieve data.

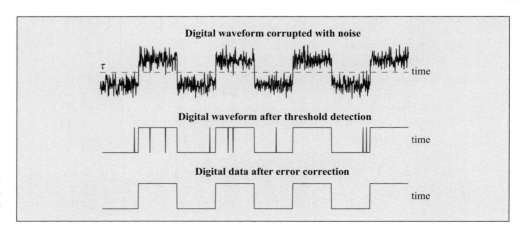

Error Detection and Correction: Occasionally, the corruption is sufficiently severe to cause an erroneous restoration. Unlike analog signals, digital data also have the capability of detecting and even correcting errors. Adding some additional digital values to the original data provides error detection, and adding even more provides error correction. Of course, the penalty for these capabilities is that additional data must be transmitted, increasing both data transmission times and memory requirements. The following example illustrates this process using the simplest method of error detection and correction.

Error detection and error correction in transmitted data

EXAMPLE 1.11

To detect transmission errors, we can transmit each digital value twice: an original 0 is transmitted as 00 and an original 1 as 11. If the detected sequence of data pairs is

$$00\ 11\ 01\ 10\ 11\ 00$$

it is readily apparent that the third and fourth digital pairs contain errors. In the important data-entry application, a digital device can easily detect such a mismatch and instruct the user to re-enter the data.

To correct transmission errors, we can transmit each digital value three times: an original 0 is transmitted as 000 and an original 1 as 111. If the detected sequence of such data triplets is

$$000\ 111\ 010\ 110\ 111\ 000$$

it is readily apparent that the third and fourth digital triples contain errors. A majority count process allows these errors to be corrected:

$$010 \rightarrow 000 \quad \text{(original value} = 0) \qquad 110 \rightarrow 111 \quad \text{(original value} = 1)$$

Factoid: The ability to restore digital signals and to achieve error correction provides a *data permanence* that was not possible with analog signals. So, while *ALL* forms of analog signals degrade with time, digital data can theoretically last forever.

Compression: Digital *data* is different from digital *information.* A completely blue screen may require one million bytes of data to generate but contains very little information. Typically, the quantity of data is much larger than its information content, which is measured using *entropy* with units of bits/symbol. Often, the data size is increased intentionally to achieve *error correction,* which is the correcting of errors that may have occurred in transmission caused by noise.

Unlike analog waveforms, it is possible to exploit the structure of digital data to reduce the number of digital values that must be transmitted or stored. The following example illustrates a simple method of reducing the number of values that represent digital data.

EXAMPLE 1.12

Data compression using run-length coding

Consider the following binary sequence that may be typical of that produced by a particular sensor:

$$00001111110000000001100000011111000 \ldots$$

Notice that binary values occur in groups. Rather that transmitting each binary value separately, note that the binary values repeat and (of course) a string of 0's is always followed by a 1 and a string of 1's is always followed by a 0. These observations lead to the *run-length code* that provides the numbers of consecutive 0's and 1's in the data. For our sequence, the run-length code would be

$$4\ 6\ 9\ 2\ 6\ 5\ 3 \ldots$$

The convention is to assume that the sequence starts with binary value 0. If a data sequence starts with a 1, the run-length code starts with a 0, indicating that there are no 0's. Techniques for handling runs greater than 9 are described later.

1.4 WHERE IS EE GOING?

Computers, and the devices and systems that use them, are constantly becoming more powerful and less expensive with each new generation. The computational power and connectivity of computers has increased to the point that many useful and desirable tasks can be accomplished. Small and portable hand-held devices containing computers called *microprocessors* are capable of performing tasks using *apps* that were once deemed "smart" or intelligent. Once the basic sensors and actuators are available to a large number of *app* designers, a variety of popular applications termed *killer apps* will be developed.

The measure of computer power most often cited is *Moore's law*, named after Gordon Moore, a co-founder of Intel Corporation, who observed that the number of transistors on a computer chip doubles approximately every 18 months (or 1.5 years). For example, if computer circuits contained $N_o = 2{,}500$ transistors in year $t_o = 1971$, the mathematical relation that indicates that the number of transistors in year t, which is denoted as $N(t)$, is given by

$$N(t) = N_o\,2^{\frac{t-t_o}{1.5}} = 2{,}500\,2^{\frac{t-1971}{1.5}} \tag{1.1}$$

Figure 1.11 shows Moore's law on a semi-logarithmic scale indicating the trend in the number of transistors in computer chips over time. The following example illustrates the technological impact.

EXAMPLE 1.13

Moore's law

The original small computer or *microprocessor* developed around 1971 had approximately 2,500 transistors. Eighteen months later, microprocessors had about 5,000 transistors. By 2013, only 42 years or twenty-eight 1.5-year cycles later, there are

$$2^{28} \approx 256{,}000{,}000$$

TIMES MORE transistors in a typical multi-core microprocessor. This computes to 640 BILLION transistors.

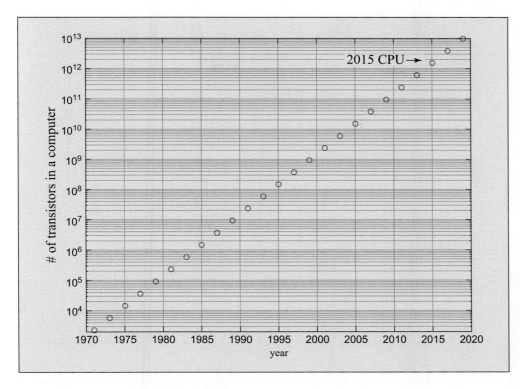

Figure 1.11

Plot of Moore's law on a semilogarithmic scale displaying a linear trend in the number of transistors in computer chips over time. This trend also predicts other features in computing power.

Factoid: A useful approximation is $2^{10} \approx 1,000$. With $2^8 = 256$, the value $2^{28} = 2^8 \times 2^{10} \times 2^{10} \approx 256 \times 10^3 \times 10^3 = 256$ million.

Additional computer features that tend to follow Moore's law include the following:

- the speed of computer computations increases mainly through the introduction of additional computer *cores* on a single chip.

- the size of computer memories is evident in the development of *solid-state drives* (*SSDs*) that are more reliable and faster than conventional mechanical hard drives, thus reducing the nuisance of long boot-up times when your start your computer.

Other technological improvements that scale similarly to Moore's law include

- Internet data transfer speeds

- Battery power storage capacities

- Size and weight reductions of portable devices

- The number of devices connected on a network

Factoid: Computer speed and size are coupled by the laws of physics: Signals cannot travel faster than the speed of light (1 ft/ns), so smaller computers operate more quickly because they communicate signals faster over smaller distances.

Moore's law often guides companies on deciding whether it is time to upgrade to newer computers. The same trend also applies to personal computing devices, such as the smartphone and laptop: In 18 months, your smartphone will be twice as fast, twice as light, twice as intelligent, and will only need to be recharged half as often.

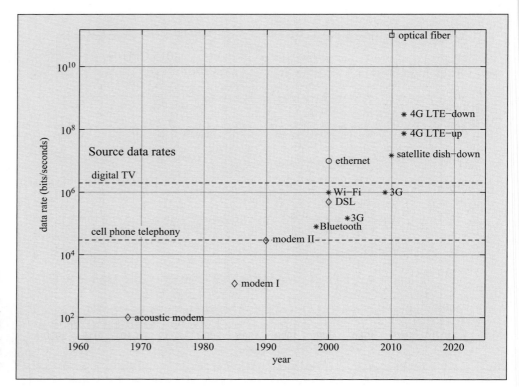

Figure 1.12

Transmission rates provided by different data channels: (◇ - standard telephone land lines, ○ - wired cable, ∗ - wireless, □ - optical fiber). Source data rates required for digital systems shown on left.

Data Transmission Capabilities and Requirements

Accompanying the increase in computing power is the increase in digital transmission rates provided by digital data channels, as well as the digital data rates required for popular digital services. Figure 1.12 shows how digital data communication rates evolved over time. The salient features of digital data transmission over wired channels include the following:

Acoustic modem (modulator/demodulator)—In an initial attempt to transmit data over telephone wires, different acoustic tones in the standard voice bandwidth were used to transmit 1's and 0's in both directions simultaneously, yielding a paltry 100 bps (bits/s).

Modem I—Using data signals rather than acoustic tones increased the data rate to 1200 bps.

Modem II—More clever coding that varied both frequencies and amplitudes increased the data rate to 28, 800 bps.

DSL (Digital subscriber loop)—Digital signals transmitted over tuned telephone wires increased the data rate to 0.5×10^6 bps = 0.5 Mbps.

Coaxial cable—The increased bandwidth provided by coaxial cables increased the data transmission rate to 500×10^6 bps = 500 Mbps.

Optical fiber—Transmitting data using light signals over optical fibers offers extremely high capabilities of up to 100×10^9 bps = 100 Gbps (100 Gigabits/s).

Transmitting data over wireless channels is now commonplace. One differentiation is the range of operation, over short ranges, typically less than 20 m, and over longer ranges. Short-range data transmission channels include

Bluetooth—Bluetooth is a wireless standard for securely exchanging data over short distances between mobile devices at a 80 kbps transmission rate. It was originally designed to replace cables that connected computers to peripheral devices, such as printers. Smartphones and other devices equipped with Bluetooth can form *piconets*, networks of data-exchanging devices that are within close range of each other.

Wi-Fi—Wi-Fi is a wireless standard that was designed to replace wired ethernet connections to the Internet by forming a wireless local area network (WLAN). Higher transmission powers permit larger transmission rates and longer ranges.

Longer-range data channels are

Cellular service—Cellular services that communicate with your smart device, have undergone improvements classified as *generations*, such as 3G and the ten times more powerful 4G LTE (*Long Term Evolution*). The first generation, 1G, transmitted analog signals, and digital data began with 2G. Cellular systems operate with a dense network of transmitting antennas that maintain connectivity with your mobile smart device. The more powerful antenna network transmits data *down* to your smart phone at rates up to 300 Mbps, while your less powerful smart device (for battery life and safety reasons) transmits data back *up* to the network at 75 Mbps. A proposed 5G system will have increased data transmission rates by steering a narrow beam using antenna arrays at both ends.

Satellite dish—Geostationary satellites located 24,000 miles above the earth can beam down data to satellite dish antennas at rates up to 15 Mbps. The 0.25 second delay due to the speed of light (data signal speed) to bounce the data off the satellite is not significant, especially for users in remote locations, who would otherwise have no access at all.

Data transmission channels are sources of digital information that send content to your smart device. Some information, such as Web pages and texting, is not time sensitive and is displayed as it is received. Delays vary according to the congestion of data on the various networks that connect the source to your smart device. In contrast, other sources must transmit data at the rate it is produced for the system to operate successfully. These include

Digital speech—When you speak into your smartphone, your speech waveform is converted into digital data at a rate of 30 kbps. For your speech to be transmitted without interruption, the cell phone system must be able to maintain this 30 kbps average transmission rate *continuously*. Actually, your speech digital data is stored in short, sub-second durations that are transmitted quickly as separate data packets, which are then recombined at the receiving end, with no perceptual distortion. Modern data channels easily transmit at this data rate, and use the excess channel capacity to service additional users.

Digital TV—Similarly, each standard digital TV channel transmits 8 Mbps to display the program you are watching. Note that Wi-Fi and 3G will not support even a single channel. You can watch programs on your laptop on a Wi-Fi network because either the smaller data rate produces a lower-quality video, or a smaller-sized video, both relaxing the data transmission requirement.

1.5 OVERVIEW OF CHAPTERS

This book attempts to tell a story that provides a coherent framework for understanding digital systems. The following list presents a road map through the chapters.

Chapter 2 describes sensors and actuators that are present in smart devices and robots. Sensors produce informational waveforms and indicate events that form the human-to-computer interface, while actuators produce sensory excitations that form the computer-to-human interface.

Chapter 3 describes analog signal physical parameters and passive components that implement electrical circuits. Electrical waveforms that sensors produce and that drive actuators are typically analog, and these are processed by analog circuits and analog electronics. The basic laws that are used to analyze analog circuits are presented.

Chapter 4 extends the analysis from passive electrical circuits to analog electronics that provide amplification and filtering. The operational amplifier is the simple-to-use and simple-to-analyze device that performs these and many other functions and illustrates the ground-breaking idea of *negative feedback*. The field-effect transistor (FET) acts like an electrically controlled switch and is the most common device in digital circuits. Devices that convert between electrical and light signals are also described.

Chapter 5 describes the basic combinational logic operations and their corresponding elemental logic gates. Elementary gates are interconnected to form logic circuits that implement a truth table to perform a useful task, such as controlling digital displays and doing arithmetic operations.

Chapter 6 introduces sequential logic circuits that perform memory and counting operations that occur within smart devices.

Chapter 7 describes analog-to-digital converters (ADCs) that transform analog waveforms into digital data for processing by computer and digital-to-analog converters (DACs) that transform digital data back to analog waveforms.

Chapter 8 describes *how* digital samples of waveforms are processed in the time domain using numerical techniques called *digital filters*.

Chapter 9 describes processing digital samples to determine their properties in the frequency domain using numerical techniques made popular by the efficient *fast Fourier transform* algorithm.

Chapter 10 considers the special case of data signals, whose values are known by the transmitter and receiver and how to use this knowledge to extract the data values. A common and important problem is the detection of data signals that have been corrupted by random noise, as in typical cell phone transmission signals.

Chapter 11 designs signals that obey the orthogonality condition to serve multiple simultaneous users or to allow one user to transmit multiple bits during a single transmission interval. Signals are described that are orthogonal in time, in frequency, and in code, where the latter is used in cellular networks.

Chapter 12 describes the fundamental ideas in *source coding* that perform data compression to contain the same information in a smaller quantity of data and encryption to make data secure. The amount of information produced by a source or contained within a data file is measured by computing its entropy. A simple method to encrypt data illustrates the steps involved in the process.

Chapter 13 describes the fundamental ideas motivating *channel coding* methods that implement error detection and correction. The capacity of a channel to transmit data is computed from the bandwidth and signal-to-noise ratio of the transmitted signals.

Chapter 14 describes the design of machine-readable codes that facilitate the transfer of data from commonly used devices, including credit cards and bar codes.

Chapter 15 describes how networks, such as the Internet, cope with data that occur at random times and that have various sizes by employing data packets with additional information. Solving problems that occur when data packets interfere, called *collisions*, are described in both wired and wireless transmission networks.

Chapter 16 provides a primer in the Matlab software and presents examples of programs that illustrate the calculations performed in the previous chapters.

The Appendices provide background material used to explain the results presented in the chapters.

1.6 Further Reading

Most books include references for further investigations of particular topics. One disadvantage of this traditional approach is that references quickly become outdated as new technology becomes available, and hence are more of historical value. A second disadvantage is that references are often difficult and expensive to obtain, especially textbooks and papers that are provided by professional organizations.

Modern telecommunication technology offers an alternative: Internet search engines offer a means of accessing current information by searching for key words and terms. Such searches provide results almost immediately and at minimal cost. The only caveat to this on-line approach is that one often is deluged with *hits* from each query that include vendor sites and useless, or even worse—wrong, information. The task is then to separate the wheat from the chaff. This book identifies useful search terms by using *italic font* in the text. For example, consider learning more about Moore's Law. Searching for *Moore's Law*, using Google or another search engine, produces over one million hits. One useful resource is *Wikipedia*, which is an evolving on-line encyclopedia that is edited and updated continuously by people knowledgable in their particular field. Such search results, and the links they provide, allow you to learn the correct jargon or terminology, to find the latest features, or to probe as deeply as your needs require.

1.7 Summary

This chapter introduced the basic ideas that are explored in smart devices and communication systems described in this book. While analog signals are often the originating and terminal waveforms in a system, current *smart* systems convert these analog waveforms into digital data to take advantage of the flexibility and robustness inherent in the communication and manipulation of digital data. Electrical engineers having various specialties design systems that perform the tasks needed to make such systems operate successfully and reliably, including:

- Transmitting data from one point to another (Communications)
- Storing data in memories that are optimized for size and power consumption (Microelectronics)
- Manipulating data to accomplish a desired computational task (Computer engineering)
- Extracting waveform informational features for reliable operation (Signal processing)
- Operating a system in a stable fashion in the presence of unexpected disturbances and device variations (Control systems)
- Implementing systems to assist humans by combining different forms of sensory data and stored information (Robotics)
- Encoding data for reliable communications more quickly and over farther distances (Information theory)

1.8 Problems

1.1 Illuminated mouse. You often power your laptop with battery while you are traveling. You need to buy a new mouse, but want to maximize the battery life. Explain why buying the illuminated mouse is not a wise choice.

1.2 Threshold detection. Digital signals that occur within your computer are designed to be either 0 V or 5 V. Additive noise produced the following detected values:

$$-0.1, \; 3.9, \; 0.9, \; 5.1, \; 0.7, \; 4.85$$

What threshold value would you use to restore the values? Explain why. Restore these detected values to their designed values.

1.3 Error correction. Threshold detection converted signal values 0 V and 5 V into binary logic values 0 and 1 respectively. For transmission over a noisy channel, each binary value is transmitted five times. A threshold detector produces the following binary sequence:

$$00100 \; 11001 \; 01000 \; 10110 \; 10001$$

- **a.** Assuming at most 2 errors occur per 5-bit code word, estimate the probability of error in the channel as the number of errors in the sequence divided by the number of data bits in the code words.
- **b.** What rule would you apply to try to correct the errors?
- **c.** Write your corrected binary sequence.

1.4 Prediction with Moore's law. Using the current year's performance as the base, how much more powerful will your computer be in 6 years?

1.5 **Prediction with Moore's law.** How long will you need to wait for your next computer to be 100 times more powerful than your current computer?

1.6 **Simultaneous users on a 4G LTE network.** How many digital speech signals can a 100 Mbps 4G LTE service simultaneously?

1.7 **Simultaneous TV channels on an optical fiber.** Assuming an HDTV program requires a data rate of 15 Mbps, how many channels can an optical fiber provide simultaneously.

1.9 Matlab Projects

1.1 **Simulating noisy digital signals.** Using Example 16.5 as a guide, compose a Matlab script to specify amplitude A and noise level σ_N to form one 100-element array of noisy binary data, each value having a duration of 20 samples.

1.2 **Thresholding noisy digital signals.** Extend Example 16.6 to specify amplitude A and noise level σ_N to form a 100-element array of noisy data and apply the data to a threshold detector.

1.3 **Exploring signal-to-noise ratio.** Extend Example 16.7 to specify signal-to-noise ratio A/σ_N to form a 100-element array of noisy data. Display the value of the A/σ_N ratio that produces between 5 to 10 errors in the 100 data values.

1.4 **Moore's law.** Extend Example 16.8 to plot Moore's law from 1980 to 2020 in two-year increments, comparing linear and logarithmic plots of the y values.

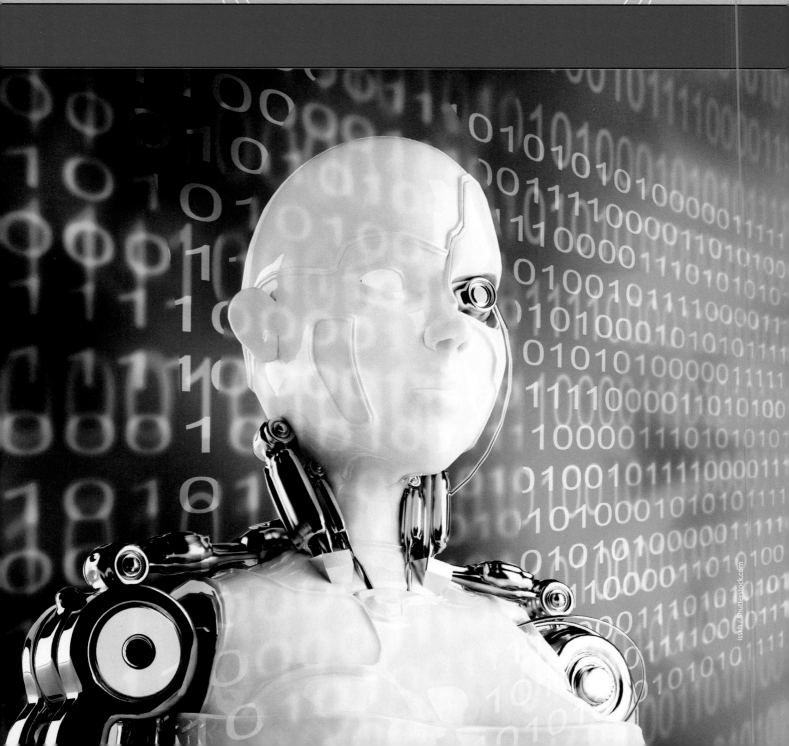

SENSORS AND ACTUATORS

LEARNING OBJECTIVES

After completing this chapter, the reader should be able to:

■ Understand the importance of sensors and actuators in smart devices and robots.

■ Differentiate sensors that produce waveforms from detectors that produce events.

■ Appreciate human-made sensing system as analogs of biological systems.

Sensors and actuators are devices that let the smartphone or tablet interact with the human user, permit a robot to navigate an environment, and allow devices to communicate over wireless networks. Sensors let the smartphone know what you want to do, such as when you operate the touch screen. Actuators let the smartphone indicate to you that it wants your attention through vibrations or sounds. Sensors and actuators perform dual roles: Whereas sensors convert energy that includes mechanical, optical or electrical forms into electrical signals, actuators convert electrical signals into energy, which is something you can see, hear, or feel. Robots offer analogs to biological sensing and actuation through sensors that perceive the environment and actuators that accomplish locomotion and grasping tasks.

2.1 INTRODUCTION

This chapter provides a general introduction to sensors and actuators in the following manner:

Sensors—In many smart systems sensors form the front end by sensing the environment, monitoring system operation, and interpreting human intensions.

Mechanical sensors and actuators—Switches in the form of push buttons, keyboards, and tactile displays are a basic communication channel with smart devices, which often respond through vibrational cues.

Acoustic sensors and actuators—Acoustic sensors detect human speech and actuators play audio and sounds to form a richer communication interface between humans and computers.

Optical sensors and actuators—Optical sensors interpret the environment through images and actuators display video to form the richest communication interface between humans and computers.

Proprioception—Accelerometers and GPS localize smart devices to enhance their capabilities to assist humans in a variety of tasks.

Active sensing—By probing the environment with optical or acoustic energy, active sensors are used to automate many tasks and guide robots and cars to prevent collisions.

2.2 ANALOG AND DIGITAL SENSORS

A sensor is a device that converts physical energy into an electrical signal, which information processing systems typically convert into digital form for their interpretation. Sensors form the *front end* of information systems because they indicate what you want the device to do. This chapter describes sensors that provide information in a numerical or digital form.

We classify a sensor by the type of signal it produces, as Figure 2.1 illustrates.

Analog sensors produce *waveforms* (or signals) that vary continuously in time. One example of an analog sensor is a mercury thermometer whose column height rises to a height proportional to the temperature.

Digital sensors produce discrete values. For example, a digital thermometer produces a number that indicates the temperature to the nearest 0.1° Fahrenheit, *e.g.*, 98.7° F. A digital sensor is typically composed of an analog sensor that converts energy into a waveform, which is followed by a device that converts the analog value into a digital approximation.

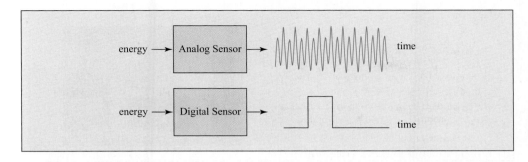

Figure 2.1

Analog and digital sensors.

Analog and digital sensors each have their own advantages. Although an analog sensor seems more accurate than a digital sensor, a digital sensor provides sufficient accuracy for most tasks. For example, an analog thermometer that reads 104.5472867329°F is not really more useful than the digital thermometer that reads 104.5°. Both indicate a fever. However, a digital thermometer that produces readings in 0.5° increments has the potential for error detection: If a nervous parent using a digital thermometer writes down the temperature of a sick child as 104.3°, health care workers would detect the error because 104.3° is not a valid reading.

Digital Events

The simplest digital sensors are *binary sensors* that produce *binary values* that take on only one of two possible values. Typically, for reliable operation, these two values are the extreme voltages in the systems: zero volts (0 V or *ground*) and the maximum value V_{max}, corresponding to the battery voltage. To normalize variable system voltages, *logical* levels of either 0 or 1 are used. A detector that changes logic levels in response to some action indicates the occurrence of an *event*. Events considered in this book include:

- A switch closure produced by pressing a key pad or tactile display.

- The occurrence of a data signal on a communication channel.

- A significant change of an accelerometer indicating that a cell phone display is rotated to a new orientation.

The following sections describe *sensors* according to the energy form that actuates them and *actuators* that convert electrical signals into that energy.

2.3 MECHANICAL SENSORS AND ACTUATORS

2.3.1 Switches

The simplest sensor is a mechanical switch that produces a two-level or binary (0 or 1) output. For example, the open switch can represent a 0, and the closed switch can represent a 1. This type of sensor is well suited to digital logic and computer systems that store and process data in binary form. We encounter these sensors in many applications.

The computer can implement a sense of touch with a *mechanical switch*. Figure 2.2 shows the most flexible mechanical switch, which has three terminals: *contact (c)*, *normally open (no) contact*, and *normally closed (nc) contact*. The *no* contacts complete an electrical circuit when the switch is activated, while the *nc* contacts open an electrical circuit. A computer interprets the closed or open pair of contacts as a binary logical value.

The most common switch on an electrical device, such as a smartphone, computer, or robot, is the *on/off* switch. Figure 2.3 shows a switch that allows current I to flow to the device by connecting the battery.

Figure 2.2

The most flexible mechanical switch has three terminals: contact (c), normally open contact (no), and normally closed contact (nc). The small black square measures 1 mm × 1 mm.

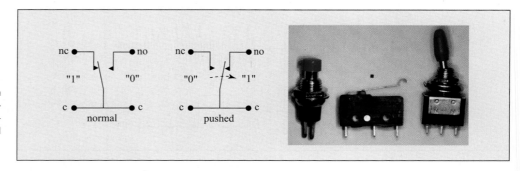

Figure 2.3

An electrical switch connects a battery to an electrical device to cause current *I* to flow.

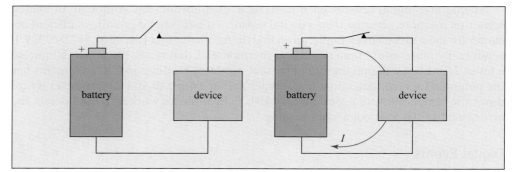

Figure 2.4

A switch closure provides a binary (0/1) indication to a digital device.

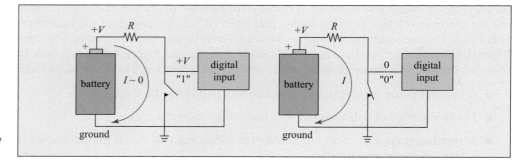

Figure 2.4 shows a switch closure that provides a binary (0/1) indication to a digital device. Resistor *R* is an electrical component shown in the connection from the battery to limit the current *I* that can flow. When the switch is open, $I \approx 0$ and the voltage at the device digital input is the battery voltage $+V$. The digital device typically interprets the $+V$ voltage as as a logical 1. There is negligible current that flows into the digital device.

When the switch is closed, the digital input is connected to *ground*, which is the negative side of the battery corresponding to 0 V. The digital device then interprets a 0 V as a logical 0.

EXAMPLE 2.1   Robot bumper

A pair of switches can provide additional information, such as direction. Figure 2.5 shows an application of a pair of switches for mobile robot navigation and obstacle avoidance. A pair of bumper contact switches mounted on the front of the robot—one to each side—normally in the open state. Each switch connects to a digital input on the robot controller, and both inputs are normally at logic 1. When the robot bumps into an obstacle, the switch on that side closes, converting that input to a logic 0. With two switches and their corresponding inputs, two logic value indicate to a robot when the obstacle is to the left or right. The robot is then programmed to stop, reverse a little, and then turn to avoid the object in its path.

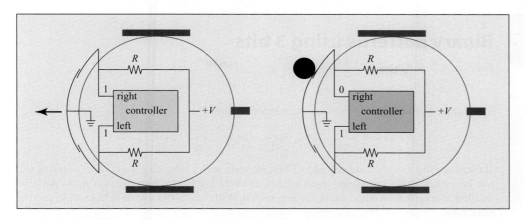

Figure 2.5

A pair of bumper contact switches indicate to a mobile robot when an obstacle is to the left or right. Right figure shows contact with object on the right side of the robot.

2.3.2 Switch Arrays

Each key of a computer keyboard is actually a switch. When you type, each key you press closes a switch that corresponds to a particular character. You could connect each switch to a pair of wires and then a sensing circuit, but this would be wasteful. To reduce the number of wires, a keyboard is structured as a two-dimensional array of switches formed by the rows and columns of keys (typically 5 rows and 13 columns). Figure 2.6 shows a partial keyboard matrix.

For computer consumption, it is convenient to employ binary addresses to identify switches. If b bits are used to specify the address, 2^b unique addresses can be specified, although not all 2^b addresses need to be used. The conventional method for assigning addresses to switches uses the binary counting sequence. Consider a binary pattern with the left-most bit being the *most significant bit (MSB)* and the right-most bit the *least significant bit (LSB)*:

$$\overbrace{0/1}^{MSB} \quad 0/1 \quad 0/1 \quad \cdots \quad 0/1 \quad \overbrace{0/1}^{LSB}$$

The counting value of a b-bit pattern, V, equals

$$V = \overbrace{2^{b-1} \times \overbrace{0/1}^{MSB} + 2^{b-2} \times 0/1 + \cdots + 2^1 \times 0/1 + 2^0 \times \overbrace{0/1}^{LSB}}^{b \text{ bits}} \tag{2.1}$$

$V = 0$ when all bits values equal 0, and $V = 2^b - 1$ when all equal 1.

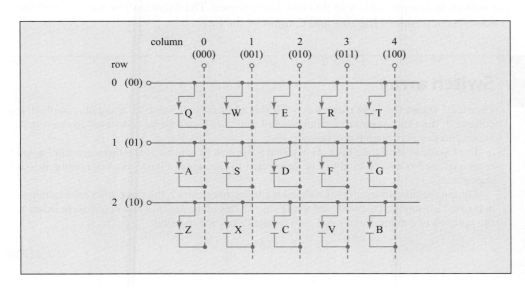

Figure 2.6

Keyboard switch array consists of a two-dimensional array of switches formed by the rows and columns of keys. Binary addresses are shown in parentheses.

EXAMPLE **2.2**

Binary patterns using 3 bits

For $b = 3$, $2^3 = 8$, and

$$V = 2^2 \times 0/1 + 2^1 \times 0/1 + 2^0 \times 0/1$$

The unique patterns and their counting equivalents are

$$\underbrace{000}_{=0}, \underbrace{001}_{=1}, \underbrace{010}_{=2}, \underbrace{011}_{=3}, \underbrace{100}_{=4}, \underbrace{101}_{=5}, \underbrace{110}_{=6}, \underbrace{111}_{=7}$$

If there are five switches, we still need 3 bits, because with 2 bits, $2^2 = 4$, and four patterns will not be sufficient for specifying five switches. If there are fewer than the maximum patterns needed, it is typical to count them starting with 0. With $b = 3$, the five switches would use the patterns

$$\underbrace{000}_{=0}, \underbrace{001}_{=1}, \underbrace{010}_{=2}, \underbrace{011}_{=3}, \underbrace{100}_{=4}$$

EXAMPLE **2.3**

Telephone key pad

A standard telephone key pad has 12 buttons arranged as a 4-row-by-3-column switch matrix. The four rows have 2-bit row addresses 00, 01, 10, and 11. The three columns with addresses require 2-bit column addresses, normally using the first three of the four possibilities. Hence, a 4-bit row/column address is required to determine the switch that was pressed. The code 0110 is interpreted as

$$\underbrace{01}_{\text{row}} \quad \underbrace{10}_{\text{column}}$$

to indicate the switch in row 1 and column 2 was pressed.

A microcomputer chip in the keyboard interprets this row-to-column connection and transmits the corresponding code to the computer. A voltage is applied to each column—one at a time—for a short time interval, typically 1 millisecond (1 ms). During this interval, each row is examined—one at a time—for the presence of the voltage. If the voltage is detected, a key in that row is depressed. The examination time the voltage is detected determines the row and column of the key.

EXAMPLE **2.4**

Switch array

Figure 2.7 shows the D key depressed. During the first time interval, a signal is applied to column 0, the three rows are examined, and none are found to be connected to the signal. Hence, no key in column 0 is depressed.

During the second time interval, a signal is applied to column 1, the three rows are again examined, and none are found to be connected to the signal. Hence, no key in column 1 is depressed.

During the third time interval, a signal is applied to column 2, the three rows are examined, and row 1 is found to be connected to the signal. This identifies that a switch closure exists at the junction of column 2 and row 1, or that the D key is depressed.

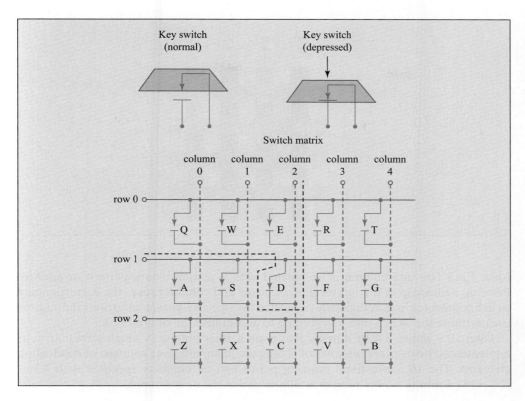

Figure 2.7

Pressing a key closes the switch that makes the connection shown by the blue dashed line.

2.3.3 Touch Screens

Switches can be arranged as a two-dimensional (2D) array for graphical user interface (GUI, pronounced *gooey*) applications. The switch in a touch screen display consists of a pair of contacts separated by small compressible non-conducting spacers, as illustrated in Figure 2.8. Fingertip pressure on the screen surface compresses the insulators, and the two conductors make contact.

Other touch screens work through *capacitive coupling* where only finger proximity to the screen is sufficient to register a reading. However, these are prone to false readings caused by water droplets on the screen.

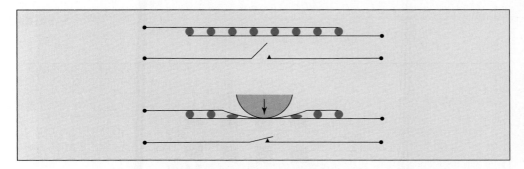

Figure 2.8

The switch in a touch screen display consists of a pair of contacts separated by the small, compressible non-conducting spacers shown in blue.

> *Factoid:* Such pressure sensors are used in robot manipulators to pick up both delicate and heavy objects.

The 2D switch array is constructed by overlapping rows of conducting elements over columns of conduction elements. Figure 2.9 shows four columns overlapped by four

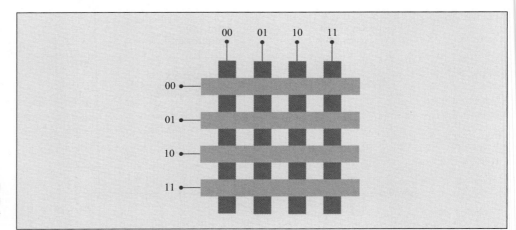

Figure 2.9

A touch screen array is implemented by overlapping rows of conducting elements over columns of conduction elements.

rows. Each row and column is designated with a unique binary pattern or *address*. Pressing a fingertip at the intersection of a row and column causes the corresponding switch contact to close. A digital device called the *switch controller* interprets the address to determine the switch that was pressed to accomplish the desired task.

Figure 2.9 shows a 4 by 4 array that is on a smartphone. A touch screen array is implemented by overlapping rows of conducting elements over columns of conduction elements. The 16 row-column crossing points can be uniquely specified with 4 bits ($2^4 = 16$), forming a 4-bit address, as illustrated in the next example.

EXAMPLE 2.5 Smartphone touch screens

A smartphone home screen display shows icons with one centered over each cross-over point. Figure 2.10 shows a friends icon (smiley face) and a music icon. When we depress the friends icon, the controller decodes the row-column address as 01-11, while the music icon corresponds to the row-column address as 10-01.

To differentiate this icon set from that on another screen, the controller may append a screen address to the row-column address. If the smartphone has eight screens, it assigns a 3-bit address with the home screen address as 000. Then the friends icon has the screen-row-column address of 000-10-11.

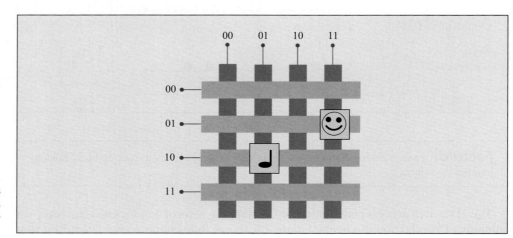

Figure 2.10

Pressing the music icon produces the row-column bit pattern 10-01, and pressing the friends icon produces 01-11.

High-Resolution Touch Screens

The resolution of a touch screen R_s is measured in terms of the number of switches per unit length, to accommodate separate row and column calculations. A switch located every 1 mm results in a linear switch resolution equal to $R_s = 1$ switch/mm. As technology improves, the resolution increases. A typical finger tip closes multiple switches on a high-resolution display. To determine the finger tip position, the smartphone computer senses all the switches that are closed, using the same addressing method described above, and then computes the row and column switch addresses closest to the midpoint locations. To simplify the calculations, the finger tip that closes row switches from r_{min} to r_{max} the finger location is approximated the row midpoint r_{mid} value

$$r_{mid} = \text{floor}\left(\frac{r_{min} + r_{max}}{2}\right) \qquad (2.2)$$

where the floor operation ignores any remainder, which is the normal result that occurs with integer division. A similar result occurs for the column midpoint c_{mid}.

High-resolution touch screen EXAMPLE 2.6

This example illustrates a high-resolution touch screen by considering one-dimensional (row) switch array having a resolution $R_s = 1$ switch/mm. The array is 6.3 cm long, contains 64 switches, and is encoded using a 6-bit address starting at leftmost 0 (= 000000) to rightmost 63(= 111111).

A fingertip width measuring approximately 1 cm depresses 10 switches simultaneously. If the finger closes switches $r_{min} = 40$ to $r_{max} = 49$, then the midpoint is computed as

$$r_{mid} = \text{floor}\left(\frac{r_{min} + r_{max}}{2}\right) = \text{floor}\left(\frac{40 + 49}{2}\right) = 44$$

Gestures, such as finger swipes, are determined by sampling the midpoint locations very often and computing the velocity of the drag motion from the changes in the midpoint locations during the sample time. If T_s is the sample period, R_s is the linear switch resolution in switches/mm and δx_m is the minimum change in midpoint position during this sample period that forms a detectable change, the minimum speed of a finger swipe is

$$v_{min} = \frac{\delta x_m}{R_s T_s} \qquad (2.3)$$

This threshold speed would ignore slow, unintended finger movements.

Finger swipes EXAMPLE 2.7

As a finger moves across the array described in Example 2.6, the computer determines the finger midpoint positions every $T_s = 0.05$ seconds (s). A swipe is sensed when a midpoint position changes by more than four switch locations between position determinations, making $x_m = 4$ switches. With $R_s = 1$ switch/mm, the minimum speed of a finger swipe is then

$$v_{min} = \frac{\delta x_m}{R_s T_s} = \frac{4 \text{ switches}}{1 \text{ switch/mm} \times 0.05 \text{ s}} = 80 \text{ mm/s}$$

Multiple finger presses are computed in the same manner, but require a more powerful processor. Luckily, the power of newer processors increases, while the processor cost remains almost constant, allowing additional features to be included in new models for approximately the same price as the previous model.

EXAMPLE 2.8

Multiple finger gestures

A computer monitors a linear (row) array with $R_s = 1$ switch/mm every $T_s = 0.05$ s. Multiple finger presses exhibit themselves as *regions* of switch closures with one region for each finger. The smartphone computer determines the row and column midpoints of each region, as shown in Example 2.6, and the velocity of each finger as shown in Example 2.7. Two fingers press the switch array with finger $F1$ closing switches 15 to 23 and the smaller finger $F2$ closing 40 to 44. The computer determines midpoint positions using integer (rounding down) arithmetic

$$P1 = \text{floor}\left(\frac{23 + 15}{2}\right) = 19$$

and

$$P2 = \text{floor}\left(\frac{40 + 44}{2}\right) = 42$$

These finger positions produce finger separation at

$$S = P2 - P1 = 42 - 19 = 23 \quad (= 23\,\text{mm})$$

At time T_s later, $F1$ closes switches 11 through 20, producing midpoint $P1' = 15$, and $F2$ closes 48 through 52, producing $P2' = 50$. The new finger separation is

$$S' = P2' - P1' = 50 - 15 = 35 \quad (= 35\,\text{mm})$$

Comparing S' to S indicates an widening gesture. If $T_s = 0.05$ s, then the finger separation speed is

$$v_F = \frac{S' - S}{T_s} = \frac{35\,\text{mm} - 23\,\text{mm}}{0.05\,\text{s}} = 240\,\text{mm/s}$$

2.3.4 Vibrating Motor

A common actuator in a smartphone is a vibrating motor that provides a tactile sensation to indicate an incoming call. Such a motor is shown in Figure 2.11. Connecting a motor to a power supply with a switch causes the motor to turn. An eccentric (off-center) weight connected to the motor shaft causes a smartphone to vibrate to provide a silent

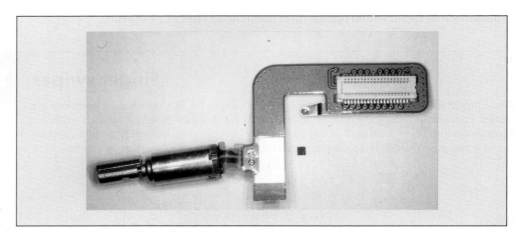

Figure 2.11

Vibrating motor in a cell phone. The black square is 1 mm × 1 mm in size.

indication of an incoming call or an alarm condition. Clever system designers implement other functions with a vibrating motor by varying its speed or on/off pattern.

2.4 ACOUSTIC SENSORS AND ACTUATORS

Acoustic energy carries the sounds we hear, including speech generated by humans, music generated by instruments, and alarms generated by loud speakers.

2.4.1 Microphone

Audio signals are pressure variations in air that contain information, although we perceive the world mostly through visual images with our eyes. Microphones are sensors that convert pressure variations into electrical signals, while a speaker and beeper convert electrical waveforms into sounds. Figure 2.12 shows a microphone, speaker, and beeper in a cell phone. The microphone waveform is typically converted to digital data with an *analog-to-digital converter (ADC)*, described in Chapter 7.

Microphone waveforms   EXAMPLE 2.9

Figure 2.13 shows three displays of the waveform of the utterance /ah/ (as in f/ah/ther) produced by laptop computer microphone. The ADC in the laptop used a sampling rate $f_s = 8$ kHz, sampling period $T_s = 125$ μs, and encoded the sample values using 8 bits/sample. The number of different levels encoded with 8 bits is $2^8 = 256$.

The three displayed waveforms show:

1. Two seconds of microphone output that includes the silent period before the speech, the speech waveform, and the silence after the speech. This "silent period" indicates the background noise that is present.

2. The speech waveform is extracted from the 2-second microphone output by applying a threshold to find the beginning and end points of the speech.

3. A 500-sample section of the extracted waveform displays details in the speech waveform. The time window duration corresponds to 62.5 ms. Note that the waveform appears to be approximately periodic with the *pitch period* equal to 60 samples (7.5 ms).

Figure 2.12

Microphone, speaker, and beeper (from left to right) in a cell phone. The black square is 1 mm × 1 mm in size.

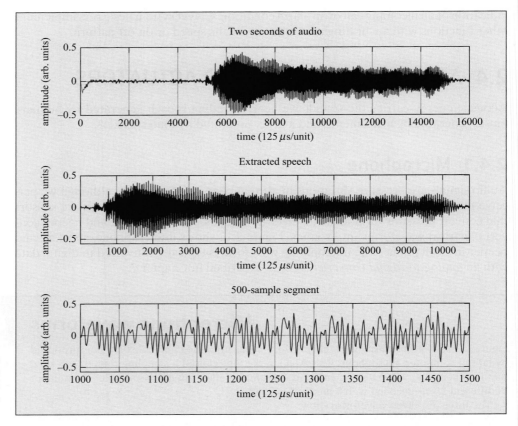

Figure 2.13

Speech waveform of the utterance /ah/ (as in f/ah/ther) produced by laptop computer microphone and sampled using sampling rate $f_s = 8$ kHz.

EXAMPLE 2.10

High-quality digital audio

High-quality audio is represented with digital values each having 16 bits. The number of steps representing the analog level is then

$$N_{steps} = 2^{16} = 65,536$$

If your digital player is powered with 3 V supplied by two AAA batteries, the step size Δ is

$$\Delta = \frac{3\text{ V}}{65,536} = 4.58 \times 10^{-5}\text{ V} = 45.8\,\mu\text{V}$$

2.4.2 Loud Speaker

A common actuator in audio systems is the familiar loud speaker. The one in your smartphone allows you to hear a telephone conversation or listen to music that is either stored in memory or transmitted over a wireless connection. A pair of speakers in your headset emits stereo sound that is slightly different to your left and right ears to provide directionality to the sound sources.

A speaker works by transforming the electric current into a coil of wire into a magnetic field that varies with current amplitude. This field interacts with a permanent magnet that is attached to a diaphragm. The magnet moves, vibrating the speaker diaphragm, which varies the adjacent air pressure that your ears sense as sound.

The audio signals in your cell phone start in digital form either stored in a digital memory or transmitted over a digital data channel. A *digital-to-analog converter*

(DAC) converts these digital samples into analog waveforms that produce the speaker current.

A second acoustic actuator found in a smartphone is the *beeper* that signals the arrival of an incoming call. The beeper uses a *piezoelectric transducer* containing a disc that vibrates like a drum surface when driven by an electrical voltage at a particular frequency (about 1 kHz) to produce the tone you hear. This device works only at this *resonant* frequency, making it very efficient (that is, producing a loud sound with very little electrical power).

2.5 OPTICAL SENSORS AND ACTUATORS

Optical energy carries the images we see, including those illuminated by sunlight or by an artificial source of light, such as a flashlight. The camera is the optical sensor in a smartphone, while LED lamps and color displays are the actuators.

2.5.1 Camera

Figure 2.14 shows a typical smartphone camera. Cameras are composed of picture elements or *pixels* that determine how accurately they record a scene. The resulting gray-scale (black-and-white) image can be considered to be a two-dimensional array where each pixel has an x and y location, and the value at that location is the pixel intensity. An intensity value equal to zero displays as black and a maximum value as white. The maximum value depends on the number of levels that are stored in the camera. Typical maximum values include 255 (8-bit camera) and 1023 (10-bit camera).

A color camera stores the intensity in a three-dimensional array with two dimensions representing (x, y) position and the third dimension having three levels, corresponding to the energy in red, green, and blue light conventionally referred to as *RGB*.

Cameras are typically specified in terms of number of pixels. For example, a 1 megapixel camera would have one million pixels that may be structured as one thousand rows and one thousand columns, representing a square image. Rectangular image formats are more common, having more columns than rows with the number of pixels being equal to the number of rows times the number of columns.

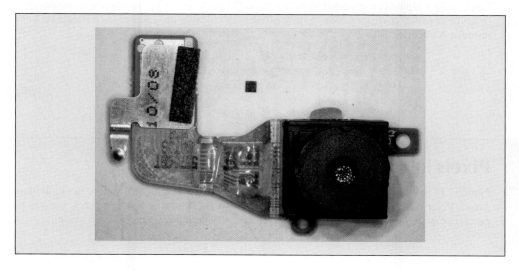

Figure 2.14

The camera circuit in a smartphone. The black square measures 1 mm × 1 mm.

EXAMPLE 2.11

Smartphone cameras

Most current smartphone camera sensors have between 5 and 13 megapixels and easily rival stand-alone digital cameras. While some camera sensors are considering > 40 megapixel sensors, other sensor systems are exploring advanced features by including additional cameras—beyond the front and back facing cameras used for video/audio communications. These additional cameras could gather environmental data, perform face tracking, and produce other useful information.

2.5.2 Display

Optical actuation occurs by displaying binary digits stored in memory as an image on a digital display, which range from a monitor measuring 24 inches or more to a 5-inch smartphone display. In any case, light sensors are replaced by tiny light emitters, such as *light emitting diodes (LEDs)*, that produce red, blue and green (RGB) light in each picture element (pixel) in the display. Controlling the intensity of each R-G-B LED pixel triplet in a display produces the colors that we see. Newer organic LEDs (OLEDs) are smaller, brighter, and more efficient than standard LEDs. Liquid crystal displays (LCDs), commonly used in inexpensive electronics, do not produce light; instead, they block light from passing through them. Hence, LCDs are the most energy efficient but have limited applications. Display sizes vary from small (5 inch smartphones) to very large (highway billboards or *jumbotrons* in sports venues).

EXAMPLE 2.12

Number of LEDs in a display

Consider a 15-inch laptop monitor that has a 2D array of 1440 (horizontal) by 900 (vertical) pixels with each pixel containing a set of R, G, and B LEDs. Hence, there are

$$1440 \times 900 \times 3 = 3,888,000 \text{ LEDs}$$

Such large numbers are usually expressed using scientific notation ($x.xx \times 10^y$). In this case,

$$3,888,000 \text{ LEDs} = 3.89 \times 10^6 \text{ LEDs}$$

Each LED is typically controlled with 256 levels (8 bits, $2^8 = 256$) to produce a corresponding brightness with 0 corresponding to fully off and 255 to fully on. Hence, each image on your monitor is encoded with

$$3,888,000 \text{ LEDs} \times 8 \text{ bits/LED} = 31,104,000 \text{ bits} = 3.11 \times 10^7 \text{ bits}$$

EXAMPLE 2.13

Pixels in an image

Figure 2.15 shows a gray-scale digital camera image that is zoomed in to show the individual pixels. Pixel sizes in high-resolution cameras are so small that individual pixels cannot be perceived with the eye. When viewed from closeup (magnified) pixels become evident.

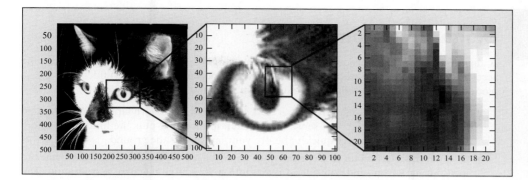

Figure 2.15

Digital camera images are composed of square pixels that have constant values and exhibit sharp transitions at the edges.

Digital image processing applications

EXAMPLE 2.14

With digital images represented by a matrix of numbers, digital signal processing techniques allow interesting and clever features to be implemented with fast computers, allowing the results to be produced quickly. Two examples come to mind.

- Camera images can be processed to locate faces. Face recognition techniques rely on the expected positions of two eyes, a nose, and a mouth. These features can be quickly detected in the image data.

- Modern cameras watch you and recognize your face and eyes when you are watching a video. This allows smartphones to pause the video when you are not looking at the screen.

2.6 PROPRIOCEPTION

Proprioceptor sensors indicate the current positional state of the system—either a smartphone or robot. Just like nerves in your arm tell you where your arm is located (such as when you reach for a glass), smart devices have similar self-location sensors. The two that we consider here are accelerometers and global positioning system (GPS).

2.6.1 Accelerometers

An accelerometer is a sensor that measures acceleration, which typically comes in one of two forms.

1. Acceleration is a change in velocity either in its magnitude (known commonly as speed) or in its direction (as when your car makes a turn). The accelerometer in your smartphone can count your footsteps, because each step causes abrupt accelerations.

2. Accelerometers are also sensitive to the force of gravity. This *always-downward* component of acceleration tells your smartphone what its orientation is when you are viewing the display.

> ***Factoid:*** Researchers are investigating accelerometer waveforms produced by walking to identify a person's physical condition and identity.

EXAMPLE 2.15 Digital accelerometer

A digital accelerometer produces an indication (similar to a switch closure) of when the acceleration exceeds a threshold value. A *digital* accelerometer detects sudden changes in motion that are greater than those encountered in normal usage.

- If your laptop were to fall from your desk to the floor, a modern laptop would contain a digital accelerometer that senses the sudden motion and puts the hard drive into a collision-safe mode (very quickly) before your laptop hits the ground.

- A digital accelerometer in a car activates an air bag in an accident. As a car comes to rest during a front-end collision, it undergoes a very large negative acceleration. The accelerometer has a built-in threshold so that relatively small accelerations caused by speeding up, braking, driving over potholes, and bumping another car while parking are not sufficient to activate the air bag.

EXAMPLE 2.16 Smartphone accelerometer

An accelerometer in your smartphone also measures g, which is the gravitational acceleration. Figure 2.16 shows how a pair of accelerometers detect the g value to determine how to orient the display on your smartphone.

2.6.2 GPS

Global positioning systems (GPS) use time-stamped information transmitted from a set of satellites at known locations to compute their positions. While only time differences from additional satellites are sufficient for localization, consider here the simpler case in which the travel times from satellites to the GPS themselves are known. For example,

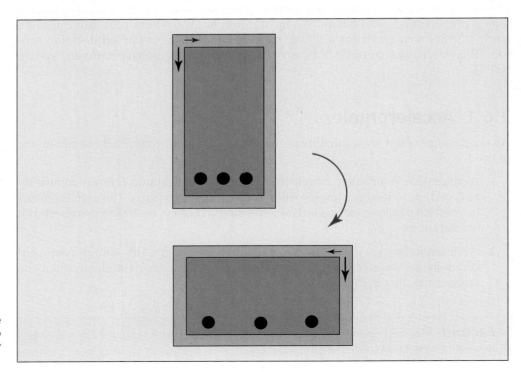

Figure 2.16

A pair of accelerometers measure the gravitational acceleration g to determine how to orient the display on your smartphone.

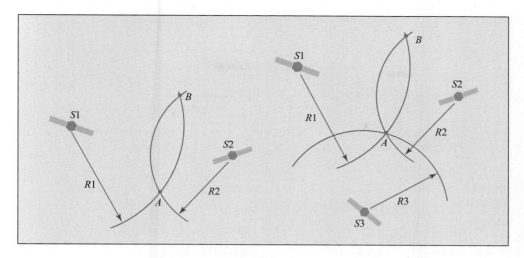

Figure 2.17

GPS localization. Three satellites are needed for localization in two dimensions.

Figure 2.17 shows two satellites $S1$ and $S2$ transmitting the current time signal simultaneously. The GPS measures the travel times from the delays in the received signals. The signal delay from $S1$ determines its range $R1$ and that from $S2$ determines $R2$. With two satellites, there are two locations on a planar surface A and B that correspond to these ranges. By adding another satellite $S3$, the GPS can differentiate A from B to determine a unique location in the plane. A fourth satellite provides data to locate the GPS in three dimensions.

GPS calculations   EXAMPLE 2.17

To illustrate GPS localization, consider the simple two-dimensional case shown in Figure 2.18. Geometry indicates that we can simplify the problem by defining an axis that passes through $S1$ and $S2$, and then centering the coordinate system on $S1$, $S2$ is at $(d, 0)$ and

$$x_G^2 + y_G^2 = R1^2$$

where $R1$ is the distance from $S1$ to the GPS derived from the travel time of the signal from $S1$ to GPS. Consider knowing the distance $R2$ derived from the signal travel time from $S2$ to GPS. Then that GPS lies somewhere on a circle with the $S2$ located at its center. Analytically,

$$(x_G - d)^2 + y_G^2 = R2^2$$

Equating both to y_G^2

$$y_G^2 = R1^2 - x_G^2 = R2^2 - (x_G - d)^2$$

Expanding the right equation, we get

$$R1^2 - x_G^2 = R2^2 - x_G^2 + 2dx_G - d^2$$

yielding the unique solution for x_G

$$x_G = \frac{R1^2 - R2^2 + d^2}{2d}$$

Inserting this value into the $R1$ equation yields the two solutions for y_G as

$$y_G = \pm\sqrt{R1^2 - x_G^2}$$

Solving similar equations for $R1$ or $R2$ and $R3$ leads to a unique location for the GPS.

   This analysis assumes that the range values are exact and correct. Figure 2.18b shows the case when there are errors in the $R1$ and $R2$ values. While particular $R1$ and $R2$ values do yield solutions, errors cause the true GPS location to lie within a region whose dimensions increase with the error magnitudes. Signal processing techniques are applied to the satellite signals to reduce this region size to one that is useful in practice.

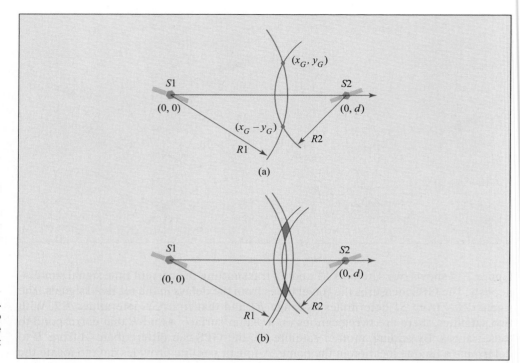

Figure 2.18

GPS localization in two dimensions. (a) Exact range values provide two locations. (b) Range errors reduce the location resolution from a point to a region.

GPS-equipped smart devices provide position and velocity data for the family car as well as autonomous mobile robots. This information is superimposed on maps stored within the smart device memory to indicate road locations and their speed limits. If you add your trip destination, the smart device uses the distance traveled along a desired path (either shortest, quickest, or toll-free) and their speed limits to estimate your arrival time. The time automatically updates as you encounter traffic delays and detours.

EXAMPLE 2.18 GPS on smartphones

GPS-equipped smartphones connected through wireless networks to the Internet are providing increasingly information-rich utilities that include the following.

Personal travel guide: As long as your smartphone can access the satellites, your location is known, and the device plans a route to your destination that includes public transportation.

Points of interest: If you are traveling and searching for a restaurant, a smartphone provides the locations of nearby dining facilities as well as their menus.

Traffic monitoring: If every vehicle on the road was equipped with a smart device that communicates with a computer, traffic flow could be monitored and alternate routes suggested to relieve congestion.

Golf caddy: When on a golf course, a smartphone tells you the distance to the center of the green, keeps track of your hitting distances, and (after learning your golfing ability) suggests which club to use on the next shot.

Sorting pictures: When you take a picture, the smartphone also records its GPS location. Then, if you want to view pictures taken when you visited your grandparents, you could sort them according to the GPS location of their home.

GPS for autonomous vehicles: While a GPS provides location information, it does not provide the location of other nearby (non-GPS equipped) vehicles or obstacles—at least currently. For this information, we need active ranging sensors for obstacle detection and object localization, which are discussed in the next section.

2.7 ACTIVE SENSORS

The sensors discussed previously are *passive* in that they detect energy produced by another source. *Active* sensors produce their own energy that interacts with the environment and is then sensed. Such sensors detect the presence of objects, as in people counters at store entrances or bar code readers. Other sensors determine the range of an object, such as infrared ranging systems found in auto-focus cameras and sonar systems commonly used in robotics.

2.7.1 Optical Detection

An active optical detector consists of two components: a transmitter and a receiver. Figure 2.19 shows an infrared (IR) light-emitting diode (LED) in a clear enclosure and an IR detector in a dark enclosure. The detector enclosure is dark because visible light does not pass through it, but it is transparent to IR light. When current is passed through an LED, it generates optical power.

Figure 2.20 shows these two components positioned so that the transmitter directs the IR light beam directly into the receiver. When no object interrupts the beam, the receiver detects the light energy and produces an output signal interpreted by the system that no object is present. It may be helpful to think of the receiver operation in terms of a switch: the receiver acts as an open switch when a beam is detected. When an object interrupts the beam, the receiver produces a signal interpreted by the system that an object is present. In this case, the receiver acts as a closed switch to indicate the presence of the object.

The transmitter and receiver are separated by distances from centimeters for counting small objects, such as pills entering a bottle, to meters for counting large objects, such

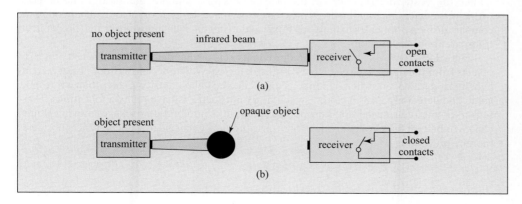

Figure 2.19

Infrared (IR) light-emitting diode (LED on left) and IR phototransistor (PT). The square measures a 1 mm × 1 mm. The LED is located in the remote control and the PT in the TV.

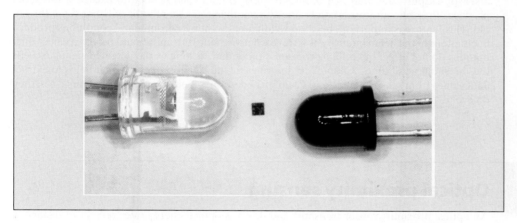

Figure 2.20

IR beam interrupt sensor. (a) When no object is present, the receiver detects the IR beam, producing an open pair of contacts. (b) When an object interrupts the IR beam, the receiver causes the contacts to close.

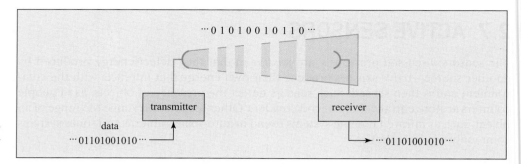

Figure 2.21

Pulses of electromagnetic energy send binary data from a transmitter to a receiver.

as people entering a store or boxes moving down a conveyer in a manufacturing plant. Such systems are used in automatic garage doors to detect the presence of cars or people beneath the door and in home burglar alarms to detect the presence of a burglar.

EXAMPLE 2.19 — Binary data transmission

The basic idea of the beam interrupt sensor also applies to sensors that detect radio waves used in digital TV and cellular telephone systems. A similar transmitter/receiver system communicates data many hundreds of kilometers (radio links) or thousands of kilometers (satellite links). Instead of an object blocking the optical beam, the transmitter *modulates* the beam, or turns it on and off to transmit binary data. For example, the beam is switched on to transmit a 1 and switched off to transmit a 0. A remote receiver senses this modulated beam to detect the data, as shown in Figure 2.21.

One desirable feature of a digital data transmission system is that the actual beam intensity is not important as long as it causes the receiver to reliably detect the signal. For example, a sensor output value that is a thousand times greater than needed to exceed a threshold produces the same interpretation by the system, than the value that is only twice the threshold. This feature is important, because weather and other environmental conditions produce fluctuations in the beam intensity as it travels between the transmitter and receiver. These fluctuations do not affect the reliable transmission of data as long as the detected signal exceeds the threshold at all times. Chapter 10 describes digital signal processing that helps to extract data values from sensor signals, and Chapter 13 describes methods to correct errors when they do occur.

EXAMPLE 2.20 — Optical proximity sensing

An optical proximity sensor transmits and receives optical energy over very short distances. Unlike the beam interrupt sensor, the proximity sensor contains the emitter and detector within one device, making the proximity sensor compact. The sensors that read bar codes and compact disks are optical proximity sensors that detect non-reflecting objects that are typically transparent or black in color.

Figure 2.22 shows the operation of a proximity sensor that detects the presence of nearby objects from the light they reflect back to the detector. When the sensor does not detect any light, it acts as an open switch. Upon sensing a reflected light intensity greater than a threshold value, the detector acts as a closed switch. Proximity sensors have many applications. These sensors count items, such as pills dropping from a chute into bottles, packages moving along a conveyor belt, or bars in a bar code. Robots use proximity sensors to detect physical obstacles, thereby preventing collisions.

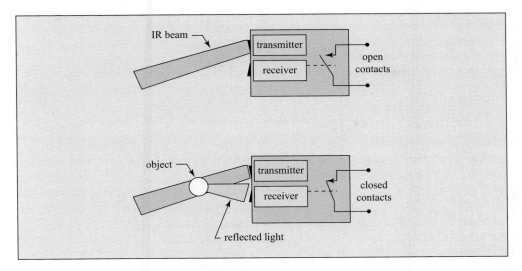

Figure 2.22

An infrared (IR) proximity sensor provides an switch closure when a reflecting object is present.

A wide variety of detectors use optical energy to function. Some employ the red neon-helium laser, such as the UPC symbol scanners in supermarkets.

Bar code scanner **EXAMPLE 2.21**

Figure 2.23 shows a laser scanning a black-and-white bar code. When the laser spot shines on a white region, light bounces back to the detector, and when it shines on a black region, the light is absorbed and no light returns to the detector. This figure shows a black bar produces a binary 1 (a high level in the waveform) and a white space produces a binary 0 (a low level).

2.7.2 Infrared Ranging

Not all optical energy is in the visible range. Many sensors use infrared (IR) light, because being in the low (below red) end of the light spectrum, it can be reliably detected under various lighting conditions.

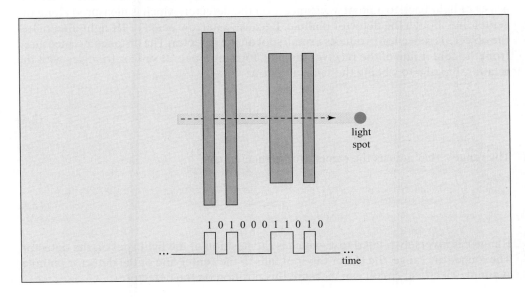

Figure 2.23

Laser scanning a bar code.

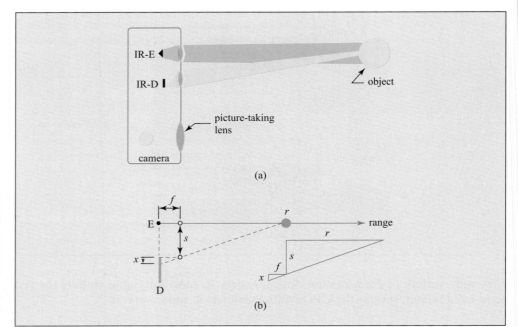

Figure 2.24

Infrared (IR) range sensor. (a) The emitter (E) produces a narrow IR beam that reflects from a object at range r. The reflected light produces a small spot on the detector (D) at a location that determines the range. (b) A pinhole model used for the calculation of range. The pair of similar triangles leads to the formula for range of $r = sf/x$.

An autofocus camera uses a variation of the proximity sensor to determine the range of an object for setting the focus. Figure 2.24a shows an auto-focus camera that includes an infrared (IR) emitter and an IR detector, similar to the proximity sensor. The emitter sends a thin beam of infrared light using a lens. When the beam encounters a reflecting object, the light scatters. The detector lens collects some of the scattered light. The detector determines the object's range from the location of the light spot on the detector.

To illustrate the operation more clearly, the two lenses are modeled as pinholes, as shown in Figure 2.24. These pinholes enable us to trace the path of the light emitted from the transmitter and reflected back to the detector. Lenses concentrate and collect the light rays, making the system more sensitive. The distance from the receiver pinhole to the focal plane is f, and the separation between the transmitter and receiver pinholes is s. The values of f and s are fixed by the camera design. An object at range r reflects a spot of light back to a point a distance x on the detector, which is measured along the center line though the detector pinhole. Because a narrow beam of IR light illuminates the object, its reflection produces a small spot on the detector. The distance x is measured from the center line of the receiver pinhole, forming a pair of similar triangles with the relationship of r to s being the same as f to x:

$$\frac{r}{s} = \frac{f}{x} \tag{2.4}$$

The range r that adjusts the camera focus then equals

$$r = \frac{sf}{x} \tag{2.5}$$

Range r is inversely related to x, which is the position of the light spot on the detector. The longer the range, the closer the spot falls to the center line of the detector pinhole. A simple electrical circuit can measure this position on the detector.

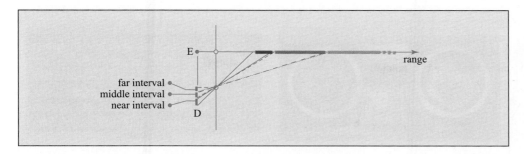

Figure 2.25

Digital IR ranging system. Multiple detector elements determine the range interval of an object rather than the exact range value. The detector array (D) sets the camera's focus without any calculations.

IR ranging   EXAMPLE 2.22

Consider a camera with an IR emitter and detector spaced 2 cm apart ($s = 2$ cm) and the detector positioned 0.5 cm behind the lens ($f = 0.5$ cm). An object casts a spot image 0.1 mm from the center line of the detector lens ($x = 0.1$ mm). The range of the object equals

$$r = \frac{sf}{x} = \frac{0.02 \text{ m} \times 0.005 \text{ m}}{0.0001 \text{ m}} = \frac{10^{-4} \text{ m}^2}{10^{-4} \text{ m}} = 1 \text{ m}$$

We can also digitally determine the distance between the camera and object without doing any calculations. To create a simple digital IR ranging system, the IR range sensor's single detector is divided into an array of smaller detector elements, as shown in Figure 2.25. Each detector element corresponds to a particular range interval. When a detector element senses the light spot reflected from the object, it sends a signal that adjusts the lens focus to the appropriate range interval. Note that this digital sensor uses a *qualitative method*, which requires no calculations, to determine the range interval. The detector element that senses the reflected light sets the focus.

Digital IR range sensor   EXAMPLE 2.23

Consider a camera that has an IR emitter and detector spaced 2 cm apart ($s = 2$ cm) and an array of detector elements positioned 0.5 cm behind the lens ($f = 0.5$ cm). Each detector element has a width of 0.1 mm. The light reflected from an object produces a spot on the second detector element from the detector pinhole centerline.

The second detector element spans $x = 0.1$ mm to $x = 0.2$ mm. From these two limits we can find the range interval that extends from the near range limit r_N (corresponding to $x_N = 0.2$ mm) to the far range limit r_F (corresponding to $x_F = 0.1$ mm). When we convert all measurements to meters, these two limits equal

$$r_N = \frac{sf}{x_N} = \frac{0.02 \text{ m} \times 0.005 \text{ m}}{0.0002 \text{ m}} = \frac{10^{-4} \text{ m}^2}{2 \times 10^{-4} \text{ m}} = 0.5 \text{ m}$$

$$r_F = \frac{sf}{x_F} = \frac{0.02 \text{ m} \times 0.005 \text{ m}}{0.0001 \text{ m}} = \frac{10^{-4} \text{ m}^2}{10^{-4} \text{ m}} = 1 \text{ m}$$

Any object 0.5 m to 1 m in range casts a light spot onto the second detector element. When the second detector element senses a spot anywhere along its width, it focuses the camera at an intermediate range, such as 0.7 m.

Figure 2.26

Sonar ranging system that operates in air at 40 kHz. (a) Transmitting (T) and receiving (R) transducers. (b) Integrated circuit chips and components. The square is 1 mm × 1 mm.

2.7.3 Acoustic Ranging

Sonar is a common sensor for mobile robots because it is inexpensive, works in dark environments, and provides the location of obstacles. Figure 2.26 shows a popular inexpensive sonar. A sonar transmitter emits an acoustic pulse having a frequency around 40 kHz, as shown in Figure 2.27a. The acoustic energy is contained within a beam, which is similar in shape to the beam produced by a flashlight. The sound pulse travels at the speed of sound at $c = 343$ m/s in air. A reflection is produced when the acoustic pulse encounters an object. If the reflection is detected by the sonar receiver, usually located near the transmitter, the echo is detected. Most sonars detect echo arrival times with the first echo determining the range to the nearest obstacle.

Figure 2.27a shows the sonar echo waveform applied to a threshold τ. Thresholding eliminates artifacts caused by noise. The time at which the echo amplitude first exceeds the threshold defines the echo *time-of-flight* (TOF), which is the time the acoustic pulse travels the round-trip distance between the sonar and closest object. The range to the closest object is computed from the sound speed c and the TOF using

$$r = \frac{c \times \text{TOF}}{2} \tag{2.6}$$

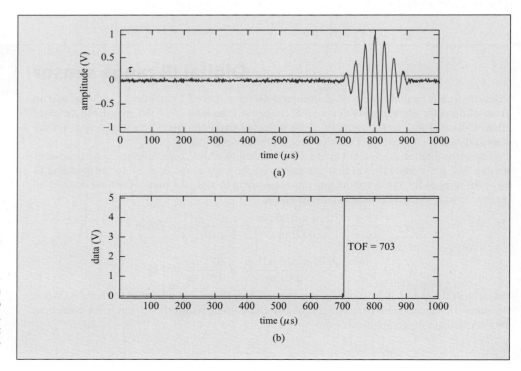

Figure 2.27

Waveform produced by echo from object at 0.12 m range. (a) Echo waveform. (b) Digital TOF indication. Acoustic pulse was transmitted at time $t = 0$, and the echo exceeds the threshold τ at time $t = 703\,\mu s$.

A sonar transmits an acoustic pulse, thus defining time $t = 0$. Figure 2.27 shows the echo waveform exceeds threshold τ at time TOF = 703 μs. The closest object range detected by sonar is then

$$r = \frac{c \times \text{TOF}}{2} = \frac{(343\,\text{m/s}) \times (703 \times 10^{-6}\,\text{s})}{2} = 0.121\,\text{m}$$

Radar Ranging

Sonar ranging works well for short ranges, less than 5 m. Obstacle detection for a car traveling on a highway requires detection for ranges that are ten times farther. Radar (radio detection and ranging) systems provide this extra range, and operate using time-of-flight similar to sonar—except that the speed is that of light ($c = 3 \times 10^8$ m/s, or 1 foot per nanosecond). Fast electronics determine the range to the nearest 0.1 m, which is sufficiently accurate for long-distance detection.

A radar transmits a pulse, thus defining time $t = 0$. The echo waveform exceeds threshold τ at time TOF = 30 ns. The closest object range detected by radar is then

$$r = \frac{c \times \text{TOF}}{2} = \frac{(3 \times 10^8\,\text{m/s}) \times (30 \times 10^{-9}\,\text{s})}{2} = 4.5\,\text{m}$$

2.8 Research Challenges

2.8.1 Computer Perception

The current bottleneck in robotics is the lack of suitable sensing for understanding unstructured environments. One active area of sensor research is in *perception*—the assignment of sensor waveforms into meaningful symbols. These symbols are then useful as landmarks in robot navigation and as biometric features for human identification.

2.8.2 Human/Computer Interaction

A bottleneck in human/computer interaction is the slow communication speed from a human to a computer.

While computers communicate millions of bytes per second in videos, a human pecking away at a touch pad is (at most) 100 bits per second. The progression of human-to-computer interfaces includes keyboard, mouse with graphical user interface (GUI), 2D touch screens with multi-finger gesture commands, and current attempts at 3D and non-touch pen and hand gesture interpretation. Research to further increase the human-to-computer communication speed involves reliable speech recognition, identifying repetitive patterns by users to infer their intent (such as word completion), and using camera vision to identify user gestures for 3D mouse manipulation. More ambitious projects explore processing brain EEG signals for determining user intent.

2.9 Summary

This chapter describes the inputs to and outputs from smart digital systems. These are motivated by tasks that these systems serve to perform. The following chapters indicate the processing performed by computer and communication systems to acquire the sensor outputs and process these data to accomplish desired tasks.

2.10 Problems

2.1 Music icon address. What screen-row-column address would the controller assign to the music icon shown in Figure 2.10 if the icon is located on the third screen of 16 possible screens?

2.2 Calculator switch array. A scientific calculator has 50 keys for digits and logarithmic and trigonometric functions arranged in five rows and ten columns. Specify a binary address code to indicate which key was pressed.

2.3 Forming a touch screen switch array. A touch screen array has a count of rows and columns that sums to 10. What is the structure of the array that accommodates the maximum number of keys?

2.4 Finger swipe along a switch array. Extending Example 2.7, a linear switch array is 10 cm long and has a resolution $R_s = 2$ switches/mm. A swipe motion is detected if the mid-point location changes by more than 8 switches. If the sampling period $T_s = 0.1$ s, what is the minimum finger swipe speed along the linear array that indicates a swipe motion?

2.5 Multiple finger gesture. Extending Example 2.8, a linear switch array is 10 cm long and has a resolution $R_s = 2$ switches/mm, and sampling period $T_s = 0.1$ s. If $P1 = 20$, $P2 = 40$, $P1' = 22$, and $P2' = 36$ is sensed as a gesture, what is the finger swipe speed? Is it widening or spreading?

2.6 Number of bits in a large color LED display. A large color billboard is a two-dimensional array of $2^{10} \times 2^{10}$ pixels, with each pixel containing red, green and blue LEDs. (Single LED packages contain separate R, G, and B LEDs inside.) Assuming that each LED is controlled to shine at one of 256 levels, how many bits are needed to specify a color image on the billboard? How many different colors can each 3-LED pixel display?

2.7 Number of possible images in a large color LED display. A large color billboard is a two-dimensional array of $2^{10} \times 2^{10}$ pixels, with each pixel containing red, green and blue LEDs. Assuming that each LED is controlled to shine at one of 256 levels, what is the number of different images that can be displayed? Express answer as a power of 10.

2.8 Bit rate to generate a full-screen movie. A video game displays images on your laptop monitor having a resolution of 1680×1050 pixels. Each pixel contains a red, green, and blue LEDs, and each LED is controlled to shine at one of 256 levels. The game produces a new image on the screen 60 times per second. How many bits per second are being sent to your monitor while you are playing your game? Give answer in scientific notation ($x.xx \times 10^y$).

2.9 Smartphone location from two range measurements. This problem considers the location information using the range values measured by two antennas. Let antennas $A1$ and $A2$ be located 5 km apart. Determine the two possible locations for the smartphone relative to antenna $A1$ when the smartphone range from $A1$ is 3 km and from $A2$ is 3.5 km.

2.10 Smartphone location region caused by range errors. Sketch and determine the four points defining the region that contains your smartphone when the range measured from antenna $A1$ is (3 ± 0.1) km and that from $A2$ is (3.5 ± 0.1) km.

2.11 Pulse time for a bar code scan. In Example 2.21, if a laser spot moves across the bar code at 10 m/s, and the width of the thinnest bar is 1 mm, what is the duration of the shortest pulse produced by the scanner? Give answer in μs (10^{-6} s).

2.12 IR range sensor. In an IR autofocus camera, the emitter and detector are separated by 1 cm and positioned 1 cm behind the lenses, which are modeled as pinholes. The light reflected from an object produces a spot 1 mm from the centerline of the detector pinhole. What is the range of the object from the camera in meters (m)?

2.13 Digital IR range sensor. In a digital IR autofocus camera, the emitter and detector are 1 cm apart and the detector array is 1 cm behind the lens. An IR detector element has near and far limits $x_F = 0.01$ mm and $x_N = 0.02$ mm that senses light reflected from an object located from r_N to r_F in range. Determine the values of r_N and r_F in m.

2.14 Digital IR range sensor dimensions. In a digital IR autofocus camera, the emitter and detector are 1 cm apart and the detector array is 1 cm behind the lens. What are the detector element's near and far limits (x_F and x_N) that senses light reflected from an object located 1 m to 4 m away? Give answer in millimeters (mm).

2.15 Sonar ranging–range to TOF. A sonar system operates in air up to a maximum range of 4 m. What is the maximum TOF? Give answer in ms (10^{-3} s)?

2.16 Sonar ranging–TOF to range. A sonar system observes a TOF = 10 ms. What is the object range in meters (m)?

2.17 Sonar ranging resolution. A sonar system experiences a jitter in the echo arrival time because of dynamic temperature variations in air, which limits the TOF resolution to ΔTOF $= \pm 50$ μs. What is the corresponding sonar range resolution Δr in mm?

2.18 Radar ranging–range to TOF. A radar system operates up to a maximum range of 100 m. What is the maximum TOF?

2.19 Radar ranging–TOF to range. A radar system observes a TOF = 0.1 μs. What is the object range in meters (m)?

2.20 Radar ranging resolution. A radar system is specified to have a range resolution of ± 0.1 m. What is the corresponding resolution in the radar TOF?

2.11 Matlab Projects

2.1 Acquire microphone speech signal. Using the Matlab script in Example 16.9 as a guide, acquire speech data from the microphone on your laptop and display 100-sample and 1,000-sample waveforms.

2.2 Having fun with speech. Write a Matlab program that plays acquired microphone speech normally and after a one second pause *backwards*, that is, in time-reversed order.

2.3 Transform a jpeg image file into 3D matrix. Using the Matlab script in Example 16.12 as a guide, acquire a jpeg image file on your laptop, transform it into 3D matrix, and display it in image format.

2.4 Transform Matlab color image. The 3D matrix produced by a jpeg displays the x, y spacial location in the first two dimensions and the third dimension defining the red, blue, and green (RGB) values at each spacial location. Modify an acquired jpeg image to display its R, G, and B components as separate images.

2.5 Convert a color jpeg image into a gray-scale image. Using the Matlab script in Example 16.12, generate a gray-scale image that is a 2D matrix of numbers that vary from 0 to 255.

ELECTRIC CIRCUITS

LEARNING OBJECTIVES

After completing this chapter, the reader should be able to:

- Appreciate the issues related to energy storage and power consumption in electrical devices.

- Apply the fundamental laws governing currents and voltages in basic circuits.

- Analyze circuits consisting of series and parallel connections of resistors.

- Appreciate frequency-variable impedances produced by capacitors and inductors.

- Design impedances having specified frequency properties.

This chapter describes basic electrical circuits and the batteries that power portable devices. The components that implement circuits include resistors, capacitors, and inductors. The classical laws that govern circuit behavior include Ohm's law and Kirchoff's Current and Voltage laws. The chapter starts by considering resistor circuits to illustrate the basic concepts and to provide understanding governing the operation of many digital devices. Circuits containing capacitors and inductors are more complicated because their effects vary with frequency. Their operation forms the basis for the implementation of frequency-selective filters.

3.1 INTRODUCTION

This chapter provides a general introduction to electric circuits by describing the following topics:

Electrical quantities and components—Electric circuits control charge and its flow by using three basic components of resistors, capacitors, and inductors.

Batteries—Electric circuits are powered by voltage supplies. Understanding the advantages and disadvantages of batteries is important for this important source of power for portable devices, ranging from smart phones, and robots to automobiles.

Analysis of voltages and currents—The measurable quantities in electric circuits are voltages and currents that are governed fundamentally by Kirchoff and Ohm laws.

Resistor circuits—The basics of electric circuit operation are illustrated with simple resistor circuits.

Capacitor and inductor circuits—Capacitors and inductors add frequency-dependence to circuits that is useful for filtering operations that are described using two-dimensional or complex numbers.

3.2 ELECTRICAL QUANTITIES

Electric circuits, like those shown in Figure 3.1, use components described in this chapter. These circuits operate by controlling the following physical quantities.

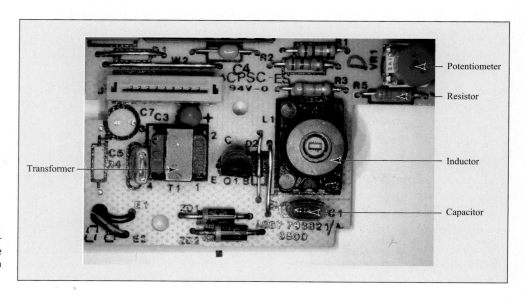

Figure 3.1

Circuit board with electrical components described in this chapter. The small black square at the bottom measures 1 mm × 1 mm.

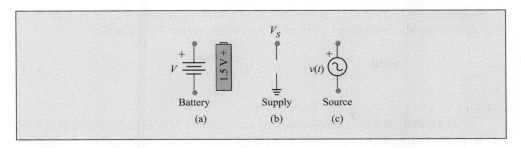

Figure 3.2

Voltage sources: (a) Battery, (b) power supply providing V_S Volts relative to *ground* (the three-line triangle), and (c) a source generating a time-varying voltage waveform.

Charge (Q): This refers to the quantity of excess electrons that are present in a device or material measured in units of Coulombs (C). Early engineers thought charge was carried by positive particles rather than electrons, so they assigned a negative charge to electrons.

Current (I): The flow of positive charge from one point in an electrical circuit to another is the current. This flow can generate heat, produce light or a magnetic field, and also cause injury. The average current is measured as the change of charge (ΔQ) divided by the time interval of the flow (Δt)

$$I = \frac{\Delta Q}{\Delta t} \tag{3.1}$$

Current is measured in units of *Ampere* or amp (A), where 1 A equals a flow of 1 Coulomb (C) in 1 second (s). By convention, current is indicated in an electrical circuit with an arrow pointing in the direction of positive charge flow, although in reality negatively-charged electrons are flowing in the opposite direction.

Voltage (V): This is the electrical potential difference between two points measured in units of *Volts* (V). One volt is the amount of work, Joules (J), required to move 1 C of charge from one point to another, making 1 V = 1 Joule/Coulomb (J/C). Once this work is completed, it can be extracted again for useful purposes, as in a battery. Figure 3.2 shows three typical voltage sources:

Battery: The symbol of a battery contains four horizontal bars: two wide and two narrow. An ideal battery provides a constant voltage V with the positive voltage being supplied at the wide bar end.

Power supply: A power supply is a component that produces a constant voltage V_S that is derived from a primary source of energy, such an electrical outlet, a solar panel, or wind turbine. One example is the *power brick* that connects to the electrical outlet to power your computer or recharge your smartphone. The power supply has two terminals: the supply terminal designated V_S and the return terminal designated by the triangular symbol that is also called *ground* (GND).

Signal Source: Sensors such as microphones convert mechanical, acoustic, or optical energy into electrical voltage waveforms that vary in time and are denoted $v(t)$. Signal sources are designated by a small circle containing a sinusoidal symbol with + indicating the positive reference point.

Energy ($\mathcal{E}$): Energy is the total amount of work that can be delivered by a voltage source, such as a battery or capacitor. The basic unit of work is a Joule (J), while electrical energy is typically measured in Watt-hours (W-hr), with 1 J = 1 W-hr.

Energy ($\mathcal{E}_S$) measures the strength of a signal that is transmitted over a channel and is important for determining the reliability of its detection in the presence of noise.

Power ($\mathcal{P}$): Electrical power measures the rate at which energy is used in units of Watts (W), where 1 W = 1 Joule/second (J/s). The power rating of a device indicates its capability to perform a task reliably, as in a 60 W light bulb or 1,000 W hair dryer. A power supply providing V_S V and I_S A delivers power $\mathcal{P}_S$ equal to

$$\mathcal{P}_S = V_S\, I_S \tag{3.2}$$

with 1 V-A = 1 W.

Power is *delivered* by a supply, *used* by an electrical device, and *dissipated* as heat. The power delivered by a power supply produces useful work, typically as mechanical power in the motion of a motor, as optical power in the display of an LED monitor, and as heat either intended for generating warmth or wasted as a by-product (the heat you feel in your laptop). Electrical engineers strive to minimize the wasted energy by designing more efficient devices, such as the light-emitting diodes (LEDs) that replace hot incandescent bulbs.

EXAMPLE 3.1 Flashlight

Consider a flashlight that is powered by 2 AA nickel-cadmium (NiCd) rechargeable batteries. Each battery supplies 1.2 V, and the two are connected in series to supply 2.4 V.

Let us compare a flashlight containing an incandescent bulb with a newer LED type.

Bulb: A typical bulb consumes 3 W of power that must be supplied by the batteries. We can compute the current supplied by the batteries as

$$I_S = \frac{\mathcal{P}_S}{V_S} = \frac{3\,\text{W}}{2.4\,\text{V}} = 1.25\,\text{A}$$

The energy that a battery can deliver is determined from its *Ampere-hour (A-hr) rating*, usually printed on the battery case. For example, an AA NiCd battery is rated at 600 mA-hr. Roughly speaking, the battery can deliver 600 mA (0.6 A) for one hour. Because our flashlight bulb requires 1.25 A, the NiCd batteries last for

$$\frac{0.6\,\text{A-hr}}{1.25\,\text{A}} = 0.48\,\text{hr}$$

Incandescent bulbs convert 3% of their consumed power into visible light, producing 0.09 W of light, with the remaining 97% or 2.91 W generating heat, making them hot to the touch.

LED: An LED converts 15% of the consumed power to light, being a factor of five more efficient than an incandescent bulb. Hence, LEDs produce the same light using only 20% of the power or

$$\mathcal{P}_{\text{LED}} = 0.2 \times 3\,\text{W} = 0.6\,\text{W}$$

To produce this power requires a current equal to

$$I_S = \frac{\mathcal{P}_S}{V_S} = \frac{0.6\,\text{W}}{2.4\,\text{V}} = 0.25\,\text{A}$$

to be delivered by the batteries. The 600 mA-hr battery will then last

$$\frac{0.6\,\text{A-hr}}{0.25\,\text{A}} = 2.4\,\text{hr}$$

This comparison indicates why LEDs have replaced bulbs in many applications. High-intensity LEDs also get hot, but not in the same way that incandescent bulbs do, providing design challenges for heat removal in automobiles.

3.3 ELECTRICAL COMPONENTS

Electric circuits are implemented using three components: resistors, capacitors, and inductors. These components control the charge, current, and voltage in a circuit to perform a particular task. Each component has an electrical value that is determined in part by its physical dimensions and by electrical properties of its material.

Resistors

A resistor is an electrical component that impedes (resists) the flow of electrons. By analogy, consider electron flow (current) to be modeled as a flow of water in a hose. In this case, voltage is the water pressure in a faucet and charge is the volume of water in the bucket we are filling. A resistor is then a hose with a narrow diameter (large resistance) restricting the flow, while a large diameter hose (small resistance) allows greater flow.

Figure 3.3(a) shows the construction and electrical symbol for a resistor. Individual resistors are constructed by inserting a piece of carbon between two leads, while resistors used in integrated circuits are much smaller in size and use silicon semiconductor material. The resistance value R of a resistor is measured in units of Ohms (Ω) and is determined by

$$R = \rho \frac{\ell}{A} \tag{3.3}$$

where ℓ is the length of the material, A is its cross-sectional area, and ρ is its *resistivity*. Resistors are constructed using carbon having resistivity $\rho \approx 10^{-5}$ Ω-m. In contrast, wires are made of copper or aluminum, which have $\rho \approx 10^{-8}$ Ω-m.

Figure 3.4 shows two generations of resistors, one in a older computer circuit and a modern cell phone.

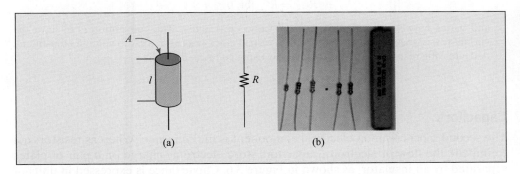

(a)   (b)

Figure 3.3

(a) A resistor R is constructed using carbon, and its electrical symbol indicates the circuitous path electrons must take to flow through it. (b) Resistance values of devices shown are from left to right: 1$k\Omega$ (1/8 W), 1 $k\Omega$ (1/4 W), 10 $k\Omega$ (1/4 W), 100 $k\Omega$ (1/4 W), 1 $M\Omega$ (1/4 W), and 10 Ω (10 W). The small square measures 1 mm $\times$ 1 mm.

Figure 3.4

Resistors in older circuits typically have $\ell = 10$ mm while surface-mount resistors in a smartphone have $\ell = 1$ mm.

EXAMPLE 3.2

EXAMPLE 3.2 Resistance values

Resistors in electrical circuits typically vary between 1 Ω and $10^6 \Omega$ ("1 meg-Ohm"= 1 $M\Omega$). The physical size of a resistor depends on the amount of power it can handle without generating a destructive amount of heat. To simplify electronic device manufacturing, resistors for a particular power rating all have the same size, no matter what the resistance value. For example, 1/4 Watt (W) resistors have 6.4 mm length and 2.5 mm in diameter, while 1/8 W resistors are 3 mm in length and 1.8 mm diameter. The resistance value of resistors is indicated by colored bands or a numerical designation.

Inside the resistor, the carbon length ℓ is typically held constant and the area A is adjusted to set the resistance value. For example, a 1,000 Ω (= 1 $k\Omega$) 1/4 W carbon film resistor has $\ell = 6.4$ mm and area

$$A_{1\,k\Omega} = \rho\frac{\ell}{R} = \left(10^{-5}\,\Omega\text{-m}\right)\frac{6.4 \times 10^{-3}\,\text{m}}{1{,}000\,\Omega} = 6.4 \times 10^{-11}\,\text{m}^2$$

while a 1 $M\Omega$ resistor has area

$$A_{1\,M\Omega} = \rho\frac{\ell}{R} = \left(10^{-5}\,\Omega\text{-m}\right)\frac{6.4 \times 10^{-3}\,\text{m}}{10^6\,\Omega} = 6.4 \times 10^{-14}\,\text{m}^2$$

EXAMPLE 3.3 Potentiometer

Some circuits require a change in a resistor value during normal operation, such as when adjusting the volume on a radio. The device that accomplishes this is a *potentiometer*, or *pot*, shown in Figure 3.5. The pot has a constant resistance value between contacts 1 and 2, such as 10 $k\Omega$, that is distributed over length ℓ. A sliding contact connected to a mechanical device, such as a knob, electrically connects to the distributed resistor at point y, which can be adjusted to be $0 \le y \le \ell$. The resistance between terminals 2 and 3 is then

$$R_{2-3} = R\frac{y}{\ell} \quad \text{for } 0 \le y \le \ell$$

and varies between 0 and R as y varies between 0 and ℓ. The convention is to have the variable resistance increase as the knob is turned clockwise, as in a radio volume adjustment. A decreasing resistance can be achieved by using contacts 1 and 2.

Capacitors

The second most common electrical component is the *capacitor*. Whereas resistors act to impede the flow of electrons, capacitors store electronic charge on a pair of plates separated by an insulator, as shown in Figure 3.6. Capacitance is expressed in units of *Farads* (F). If A is the overlapping area between two plates separated by distance d, the capacitance value is given by

$$C = \frac{\epsilon A}{d} \tag{3.4}$$

Figure 3.5

A potentiometer increases the resistance between terminals 1 and 2 by adjusting a knob in the clockwise direction. The devices shown have a 10 $k\Omega$ resistance between terminals 1 and 3.

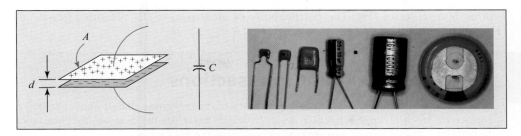

Figure 3.6

A capacitor is fabricated with two charge-storing plates having area A and separated by distance d. Capacitance values of devices shown are from left to right: 0.001 μF, 0.01 μF, 0.1 μF, 10 μF, 10000 μF, and 1 F ultracapacitor. The small square measures 1 mm × 1 mm.

where ϵ is the *dielectric constant*, which in air $\epsilon \approx 10^{-11}$ F/m. Early capacitors were fabricated using a pair of thin plastic (insulator) films that were coated with metal on one surface. These two films were rolled into a cylinder and a wire was connected to each plate.

Capacitors in electrical circuits are typically in the micro-Farad (μF $= 10^{-6}$ F) range and less. Advanced processing techniques have developed *ultracapacitors* that are measured in Farads and are made possible by new materials and fabrication techniques.

Making a 1 F capacitor   EXAMPLE 3.4

Integrated circuit fabrication techniques allow insulators made of silicon dioxide to have a thickness $d = 100$ nm (10^{-7} m) and sustain a voltage of 5 V before they break down. Assuming $\epsilon = 10^{-11}$ F/m and to make a 1 F capacitor, the area that would be needed is

$$A = \frac{dC}{\epsilon} = \frac{10^{-7}\,\text{m} \times 1\,\text{F}}{10^{-11}\,\text{F/m}} = 10^{4}\,\text{m}^2 = 10^{8}\,\text{cm}^2$$

Assuming that the thickness of the metal is negligible compared to d, then stacking 10^8 1-cm^2 sheets each having a thickness $d = 10^{-7}$ m results in a stack 1 m high, forming a volume of 100 cm^3. Modern ultracapacitors achieve volumes of approximately 10 cm^3 by using a carbon film *sponge* that produces the required area and yet fits into a volume that becomes practical.

Inductors

The third electrical component is an *inductor*, which is implemented with a coil of wire that converts a current into a magnetic field B. The resulting magnetic field can then turn a motor, operate an electrically-controlled switch, called a relay, magnetically couple with another inductor to form a *transformer*, or sense the presence of an automobile passing over the coil.

Figure 3.7 shows a coil of wire having N turns, cross-sectional area A, and length ℓ. The coil core contains a material that has a property called the magnetic permeability μ. Then, the coil inductance value is given by

$$L = \frac{\mu N^2 A}{\ell} \tag{3.5}$$

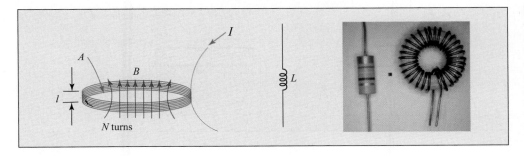

Figure 3.7

An inductor is fabricated by winding a coil of wire in N turns having area A and length l. The magnetic permeability of the material within the vicinity the coil is μ. Current I produces magnetic field B. Inductance values of devices show are from left to right: 100 μH and 50 μH. The small square measures 1 mm × 1 mm.

with units of Henries (H). The permeability of air is $\mu_{\text{air}} \approx 10^{-6}$ Henries per meter (H/m), while that of steel is $\mu_{\text{steel}} \approx 10^{-3}$ H/m.

EXAMPLE 3.5 — Sensor coils at traffic intersections

When a car drives over a coil, the coil inductance changes because the air above the coil is replaced by steel, having a different μ. An electronic circuit senses the change in the inductance to determine the presence of an automobile and turns the light green.

For example, with $\mu_{\text{air}} = 10^{-6}$ H/m, the $\mu_{\text{steel}} = 10^{-3}$ H/m is a thousand times greater than that of air. A coil is constructed to have an inductance of $10\,\mu$H in air. When a car passes over the coil, the inductance can change by a factor of up to 1,000 or up to 10 mH, depending on how much of the coil area is occupied by steel. Such changes in inductance are easy to measure with inexpensive circuits leading to their common use in traffic applications.

EXAMPLE 3.6 — Reed relay

Reed relays are electrically activated switches. As shown in Figure 3.8, the switch is constructed by placing two long *reed* contacts in a glass envelope, which is situated within a coil having many turns. When no current is applied, there is no magnetic field, and the reed contacts are open. Applying a current into the coil generates a magnetic field that causes the reeds to touch, thus completing an electrical contact.

EXAMPLE 3.7 — Two inductors form a transformer

Two inductors in close proximity form a magnetic circuit in which a magnetic field couples the two electrical circuits. A current that varies sinusoidally flowing in one coil produces a sinusoidally varying magnetic field that interacts with the second coil. This second coil responds by *generating* sinusoidally alternating current and voltage in the second circuit. These two inductors form a *transformer*.

One use for such a transformer is to charge your smartphone *wirelessly*. That is, unlike a power brick that connects to the smartphone through a cable, wireless charging is accomplished by merely laying the smartphone down on a surface close to the active inductor coil. Figure 3.9 shows a wireless charging arrangement. A coil in a charging pad is driven by an AC power source. A coil within the smartphone couples to the magnetic field B causing a current to flow in the smartphone that acts to charge its battery. Because the magnetic coupling reduces as the cube of the separation between the two coils, wireless charging is efficient only for small separations.

Figure 3.8

The magnetic field produced by passing current *I* through a coil closes the contacts in a reed relay.

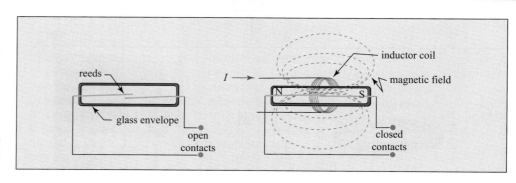

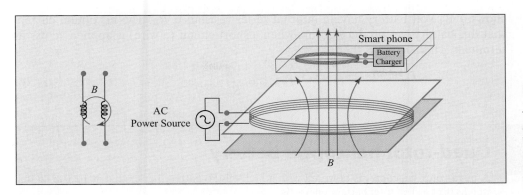

Figure 3.9

Two inductor coils form a magnetic coupling through magnetic field B. When one coil is located in a base and the other in a smartphone, this magnetic coupling acts to charge the smartphone wirelessly.

3.3.1 Batteries as Power Sources

Batteries are important components in portable smart devices and robots as a means of storing and providing electrical energy. Batteries come in two basic forms:

Non-rechargeable (single use) batteries include non-reversible alkaline chemistry processes that exhaust their ability with usage and over time. The associated chemical reaction produces voltages in increments of 1.5 V with the simplest being a single AA battery.

Rechargeable batteries can have their energy restored through a chemical process. These include nickel-cadmium (NiCd) and nickel-metal hydride (NiMH) batteries, which produce voltages in increments of 1.2 V per cell. Lithium polymer (LiPo) batteries are less massive and are currently popular for powering electric vehicles, including automobiles and radio-controlled (RC) flying vehicles, such as helicopters and quad-rotors.

If a battery having voltage V_B produces current I_B, the power it is providing equals

$$\mathcal{P}_B = I_B \times V_B \tag{3.6}$$

If the battery operates for a maximum of T_B seconds before its voltage reduces to an unacceptably low level, the total energy the battery supplies is $\mathcal{E}_B$, also called its *capacity*, which equals

$$\mathcal{E}_B = \mathcal{P}_B \times T_B = I_B \times V_B \times T_B \tag{3.7}$$

The capacity is proportional to battery *mass*. Roughly speaking, a battery using a given technology that provides twice the energy will also weigh twice as much.

Batteries do not have unlimited current capabilities. For example, a battery rated at 2.5 A-hr *cannot* supply 1,000 A for 0.0025 hrs, because there is a *current limit* imposed by the chemical process and construction. In addition to the A-hr rating, batteries also have a maximum current rating, which is determined by its *internal resistance* R_B. Figure 3.10

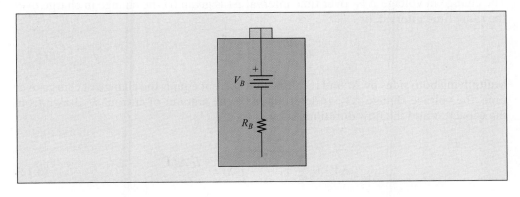

Figure 3.10

All batteries have an internal resistance R_B that limits its maximum current.

shows a V_B volt battery having internal series resistance R_B. The maximum current that this battery can produce occurs when a short-circuit (a wire) is applied across its terminals. This current is then

$$I_{B,\text{max}} = \frac{V_B}{R_B} \tag{3.8}$$

EXAMPLE 3.8

Quad-rotor helicopter battery

A quad-rotor helicopter is powered with a 11.1 V LiPo battery pack rated at $1000\,mAh$ that weighs 0.1 kg. The total battery energy is

$$\mathcal{E}_B = V_B \times I_B \times T_B = 11.1\,\text{V} \times 1\,\text{A-hr} = 11.1\,\text{W-hr}$$

The LiPo battery can be charged with a maximum current of 1 A and can be discharged with a maximum current of 10 A.

The helicopter can fly for about 15 minutes with a fully charged battery. Assuming a constant current is supplied by the battery, $\mathcal{E}_B$ is expended in 0.25 hours, and the power supplied by the battery is

$$\mathcal{P}_B = \frac{\mathcal{E}_B}{T_B} = \frac{11.1\,\text{W-hr}}{0.25\,\text{hr}} = 44.4\,\text{W}$$

The current supplied by the battery during flight is

$$I_B = \frac{\mathcal{P}_B}{V_B} = \frac{44.4\,\text{W}}{11.1\,\text{V}} = 4\,\text{A}$$

which is below the rated maximum value of 10 A.

3.3.2 Capacitor as a Battery Alternative

Because capacitors store charge, they can power a circuit by dispensing the stored charge as a current. If the capacitor voltage is V_C, the energy stored in the capacitor equals

$$\mathcal{E}_C = \frac{1}{2} C V_C^2 \tag{3.9}$$

As this energy is supplied to a circuit, V_C will decay, eventually reaching zero when all of the stored energy is expended. The voltage V_C is related to the charge Q_C stored in the capacitor and its capacitance value C by

$$V_C = \frac{Q_C}{C} \tag{3.10}$$

The change in voltage ΔV_C over time interval Δt is given by the change in charge over the same time interval, or

$$\frac{\Delta V_C}{\Delta t} = \frac{1}{C} \frac{\Delta Q_C}{\Delta t} \tag{3.11}$$

Multiplying both sides by Δt and recalling that current equals the change of charge over time, the voltage change ΔV_C is determined by the amount of current I_C drawn from the capacitor and the flow duration Δt by

$$\Delta V_C = \frac{1}{C} \overbrace{\left(\frac{\Delta Q_C}{\Delta t} \right)}^{=I_C} \Delta t = \frac{I_C \Delta t}{C} \tag{3.12}$$

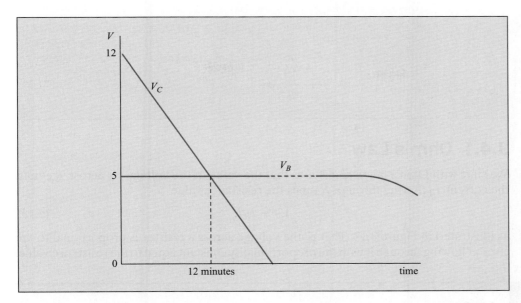

Figure 3.11

Battery voltage V_B and capacitor voltage V_C when both deliver a constant current to a circuit.

Capacitor as a power source for your smartphone

EXAMPLE 3.9

Say your smartphone requires a 5 V voltage source that supplies a constant 0.01 A current to operate. Assume we have a capacitor initially charged to $V_C = 12$ V. The capacitor needs to supply $I_C = 0.01$ A to operate the circuit, causing V_C to decrease linearly with time. A device called a *voltage regulator* converts this decreasing V_C to a constant 5 V, by taking up the excess voltage as needed, as long as $V_C > 5$ V. When $C = 1$ F, V_C drops from 12 V to 5 V in Δt seconds where

$$\Delta t = \frac{C \Delta V_C}{I_C} = \frac{(1\,\text{F})(12\,\text{V} - 5\,\text{V})}{0.01\,\text{A}} = \frac{(1\,\text{Coulomb/V})(12\,\text{V} - 5\,\text{V})}{0.01\,\text{Coulomb/s}} = 700\,\text{s}$$

or approximately 12 minutes.

Figure 3.11 compares battery voltage V_B with capacitor voltage V_C when both deliver a constant current to a circuit.

In contrast, the chemical process that generates the battery voltage maintains an almost constant level until it exhausts itself, causing the voltage to drop quickly afterwards. Clearly, a battery that produces voltage through chemical means has its advantages.

> *Factoid:* Ultracapacitors are competing with chemical batteries for powering electric automobiles. Compared to batteries, ultracapacitors are desirable because they charge up more quickly and can endure more charge/discharge cycles.

3.4 CIRCUIT ANALYSIS LAWS

An *electric circuit* is the interconnection of voltage or current sources and electrical components. The electrons are provided by *sources*, such as batteries, and electron flow is directed by the electrical components. To design a circuit or to determine what a particular circuit does, the analytic tools described in this section are employed.

Figure 3.12

Ohm's law relates the voltage across a component to the current flowing through it.

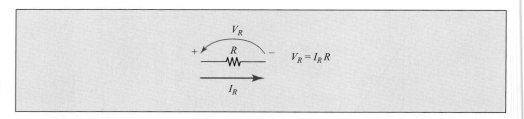

$$V_R = I_R R$$

3.4.1 Ohm's Law

We start with Ohm's law, which states that the value of the voltage V_R across R equals the current I_R flowing through R times the resistance value

$$V_R = I_R R \tag{3.13}$$

as illustrated in Figure 3.12. That is, the voltage across a resistor is proportional to the current flowing through it with the proportionality constant equal to the resistance value.

EXAMPLE 3.10 **Voltage across resistor *R***

A typical current value observed in consumer electronics is one-one thousandths of an Ampere or milli-ampere ($= 10^{-3}$ A, which is denoted mA). Typical values of resistance are in the thousand Ω (Ohm) range or kilo-Ω ($= 10^3\,\Omega$, which is denoted $k\Omega$ and pronounced "kill-Ohm") but can be in the million Ohm (meg-Ohm $M\Omega$) range.

When $I_R = 1$ mA flows through $R = 1\,k\Omega$, the voltage measured across the terminals of the resistor is

$$V_R = I_R \times R = 1\,\text{mA} \times 1\,k\Omega = 10^{-3}\text{A} \times 10^3\,\Omega = 1\,\text{V}$$

When the current $I_R = 0$, the voltage drop across the resistor is $V_R = 0$. This case commonly occurs in digital circuits.

EXAMPLE 3.11 **Current flowing through resistor**

The voltage supplied by a pair of AA batteries is $V_B = 3$ V. When connected across a $2\,k\Omega$ resistor, $V_R = V_B$, and the current I_R flowing through R is

$$I_R = \frac{V_R}{R} = \frac{3\,\text{V}}{2\,k\Omega} = \frac{3\,\text{V}}{2{,}000\,\Omega} = 1.5 \times 10^{-3}\,\text{A} = 1.5\,\text{mA}$$

Power Dissipation in a Resistor

Resistors convert electrical energy into heat. Unless the device is a heater, the heat dissipated by resistors is a nuisance, causing computers and smartphones to *run hot* and batteries to *run out*. This heat must be controlled by devices such as heat sinks and fans. The power $\mathcal{P}_R$ dissipated by resistor R equals the product of the voltage across R and the current flowing through R:

$$\mathcal{P}_R = V_R I_R \tag{3.14}$$

Applying Ohm's law ($V_R = I_R R$), describes the power dissipated by the resistor in terms of the current flowing through R:

$$\mathcal{P}_R = (V_R)\,I_R = (I_R R)\,I_R = I_R^2 R \tag{3.15}$$

The power in terms of the voltage across R is then

$$\mathcal{P}_R = V_R (I_R) = V_R \left(\frac{V_R}{R} \right) = \frac{V_R^2}{R} \tag{3.16}$$

Hand dryer power — EXAMPLE 3.12

Next time you are drying your hands using an electric hand dryer, you can read the label that indicates that it uses 1,000 W, or 1 kW ($= \mathcal{P}_R$). Such dryers are typically connected to the 110 V supply ($= V_R$). Hence, the current flowing through the heating coils (resistors) equals

$$I_R = \frac{\mathcal{P}_R}{V_R} = \frac{1,000\,\text{W}}{110\,\text{V}} = 9.1\,\text{A}$$

The coil resistance equals

$$R = \frac{V_R^2}{\mathcal{P}_R} = \frac{(110\,\text{V})^2}{1,000\,\text{W}} = 12.1\,\Omega$$

If a calculator is not handy, engineers approximate these values by using $V = 100\,\text{V}$, producing $I_R = 10\,\text{A}$ and $R = 10\,\Omega$.

3.4.2 Nodes and Loops

Electric circuits are implemented by interconnecting components such as resistors, capacitors, and inductors to voltage supplies. To assist in circuit analysis, we first define the *topology* or circuit configuration in terms of nodes and loops.

Nodes: A *node* in an electrical circuit is the junction of two or more wires. Figure 3.13 shows that a simple node connects two components and may also connect to voltage sources as well.

General nodes — EXAMPLE 3.13

The supply voltage and ground connections also form nodes that typically connect to many components. Because the resulting connecting lines produce messy circuit drawings, it is conventional to designate the connections to the ground node by triangular symbols. Similarly, connections to the power supply are designated using the letter V. Figure 3.14 shows the same circuit drawn using these two conventions. One explicitly shows lines connecting components to the supply voltage and to ground, while the other shows the symbolic connection through supply and ground symbols. It is easy to forget that the supply and ground connections form their respective nodes.

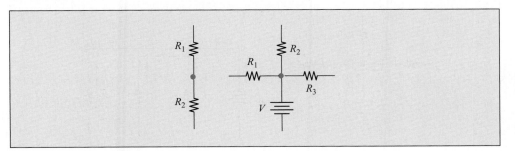

Figure 3.13

Nodes in an electrical circuit are indicated as blue circles and may connect two or more components and voltage sources.

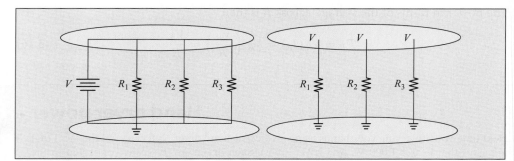

Figure 3.14

Connections to voltage supplies and ground also form nodes. Both circuits are equivalent.

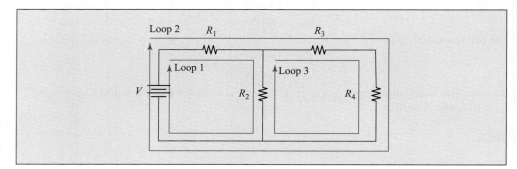

Figure 3.15

Loops in a circuit are indicated with arrows that start and terminate at the same node.

Loops: A *loop* is a path through components and supplies in an electrical circuit that starts and terminates at the same node. Figure 3.15 shows that a loop can pass through only components or through a voltage supply as well.

3.4.3 Kirchoff's Current and Voltage Laws

The analysis of electric circuits is simplified by applying two laws governing the voltages and currents.

Kirchoff's Current Law (KCL) The sum of currents flowing into a node equals zero. We define a current flowing into a node as positive and a current flowing out of a node as negative. Figure 3.16 illustrates Kirchoff's Current Law: If n wire indexed with $i = 1, 2, \ldots, n$ connect at a node, with current I_i flowing in wire i, the sum of the currents flowing in all the wires sums to zero, as

$$\sum_{i=1}^{n} I_i = 0 \tag{3.17}$$

Figure 3.16

Kirchoff's Current Law states that the sum of the currents flowing into a node equals zero. Left figure shows a node contains all wires connected to it.

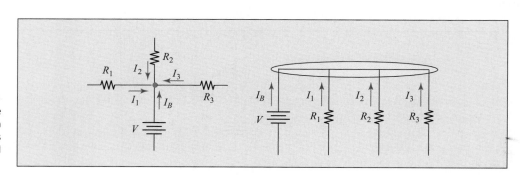

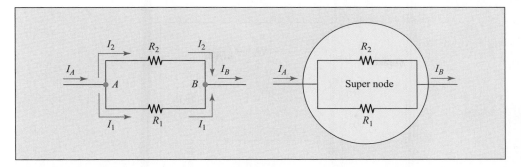

Figure 3.17

Kirchoff's Current Law applied at nodes A and B and a super node.

Kirchoff's Current Law (KCL) $\quad$ **EXAMPLE 3.14**

Figure 3.17 shows a simple circuit containing two resistors, R_1 and R_2 connected at nodes A and B. Kirchoff's Current Law applied at node A yields

$$I_A - I_1 - I_2 = 0$$

Bringing I_1 and I_2 to the left side of the equation yields

$$I_A = I_1 + I_2$$

that is, the sum of the currents flowing into a node equals the sum flowing out of a node.
Applying the KCL to node B, we get

$$-I_B + I_1 + I_2 = 0$$

or equivalently

$$I_1 + I_2 = I_B$$

We can combine nodes A and B to form a *super node* that clearly illustrates

$$I_A = I_B$$

This result, although trivial, shows that the current supplied to any device by a battery is equal to current returning to the battery.

Kirchoff's Voltage Law (KVL) The sum of voltages along any closed loop in a circuit equals zero. If a closed loop consists of n sources or components with voltage V_i across source or component i,

$$\sum_{i=1}^{n} V_i = 0 \qquad\qquad (3.18)$$

The *polarity* (direction) of the voltage is indicated by $+$ and $-$ signs. The $+$ side of a voltage source supplies current for the circuit (that is, the current arrow points out of the $+$ terminal of a supply). Current flowing into a passive component (R, L, or C) defines the $+$ side for the voltage across the component. If while traveling along the loop one encounters the $+$ sign first, the corresponding voltage value is positive, otherwise the voltage value is negative.

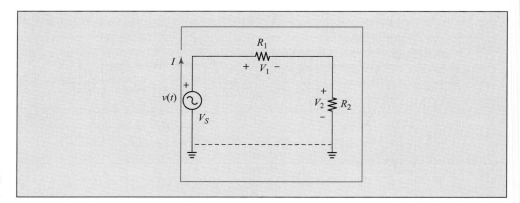

Figure 3.18

Kirchoff's Voltage Law (KVL) applied along loop defined by current flow I.

 EXAMPLE **3.15**

Kirchoff's Voltage Law (KVL)

Figure 3.18 shows a simple circuit having a voltage source V_S, which may be a battery supplying voltage to a device or the voltage waveform produced by a microphone. This voltage source causes a current I to flow into the circuit with the flow coming out of the voltage source terminal marked with +. The current flow through R_1 causes voltage V_1 with the indicated polarity, and current flow through R_2 causes voltage V_2.

The current flow forms a convenient loop through all the components in the circuit. Following this loop and starting with V_1, we have the following voltage progression around the loop that forms Kirchoff's voltage law:

$$+V_1 + V_2 - V_S = 0$$

or equivalently

$$V_S = V_1 + V_2$$

That is, the voltage provided by the source is divided among the components along the loop.

▶ 3.5 RESISTOR CIRCUITS

We first analyze circuits implemented only with resistors because they are simpler yet still illustrate the fundamental ideas. We then consider circuits that also contain capacitors and inductors. Two resistors R_1 and R_2 can be connected in series or in parallel, and we examine each case separately.

3.5.1 Series Resistor Circuits

Figure 3.19 shows two resistors R_1 and R_2 connected in series. Applying KVL and following the current loop,

$$V_{R_1} + V_{R_2} - V_S = 0 \tag{3.19}$$

Hence,

$$V_S = V_{R_1} + V_{R_2} \tag{3.20}$$

Applying Ohm's law to compute the voltage across each resistor, we get

$$V_S = I\,R_1 + I\,R_2 = I(R_1 + R_2) \tag{3.21}$$

The equivalent resistance is the value of the resistor that would appear across the same two terminals. The equivalent resistance that draws the same current from supply V_S is

$$R_S = \frac{V_s}{I} = R_1 + R_2 \tag{3.22}$$

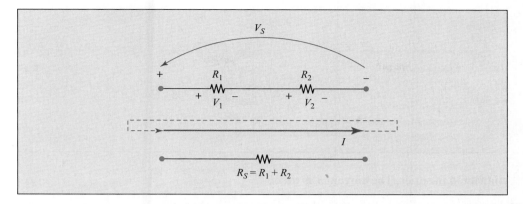

Figure 3.19

Resistors R_1 and R_2 connected in series produce the equivalent resistance R_S. The dashed line passes through V_S and completes the loop.

Series connection of resistors

EXAMPLE **3.16**

Let both resistors have equal values, $R_1 = R_2 = R$. Then the series connection produces

$$R_S = R_1 + R_2 = 2R$$

or twice the resistance value. By extension, connecting n resistors, each having resistance R, in series results in a total resistance R_S between the ends and equals

$$R_S = nR$$

Series connection of three resistors

EXAMPLE **3.17**

Consider connecting three resistors R_1, R_2 and R_3 in series. Extending the circuit shown in Figure 3.19, three resistors would produce the KVL result

$$V_{R_1} + V_{R_2} + V_{R_3} - V_S = 0$$

Hence,

$$V_S = V_{R_1} + V_{R_2} + V_{R_3}$$

Applying Ohm's law to compute the voltage across each resistor, we get

$$V_S = I R_1 + I R_2 + I R_3 = I(R_1 + R_2 + R_3)$$

The equivalent resistance that draws the same current from supply V_S is

$$R_S = \frac{V_S}{I} = R_1 + R_2 + R_3$$

Extending the result in the previous example to the series connection of n resistors, connecting $R_1, R_2, \ldots, R_n$ yields an equivalent resistance R_S equal to the sum of the individual resistances

$$R_S = \sum_{i=1}^{n} R_i \qquad (3.23)$$

Voltage Divider

The series connection of two resistors leads to a common and important circuit, called the *voltage divider*. Consider the output voltage V_{OUT} to be the value at the node connecting the resistors. Figure 3.20 shows an input voltage V_{IN} applied to R_1 and R_2

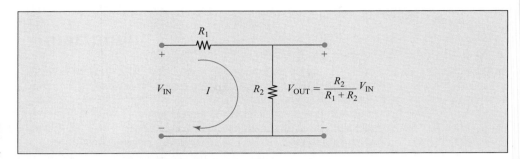

Figure 3.20

Two resistors form a voltage divider.

connected in series. The current I is then

$$I = \frac{V_{IN}}{R_S} = \frac{V_{IN}}{R_1 + R_2}$$

The output voltage V_{OUT} is

$$V_{OUT} = I\, R_2 = \frac{R_2}{R_1 + R_2} V_{IN}$$

Substituting the value of I, we get

$$V_{OUT} = \frac{R_2}{R_1 + R_2} V_{IN}$$

Hence, the resistors divide the input voltage by the ratio of R_2 to the sum of R_1 and R_2 to produce an output voltage.

EXAMPLE 3.18 Voltage divider with a switch

Consider R_2 to be a push-button switch, as shown in Figure 3.21. This behavior is important in digital circuits where a digital switch can produce either very high resistance or very low resistance.

When the switch contacts are open, $R_2 = \infty$. Then,

$$R_1 + R_2 = R_1 + \infty = \infty(= R_2)$$

The output voltage then equals

$$V_{OUT} = \frac{R_2}{R_1 + R_2} V_{IN} = \frac{R_2}{R_2} V_{IN} = V_{IN}$$

That is, the input voltage appears at the the output when the switch is open.

When the switch contacts are closed, $R_2 = 0$. Then,

$$R_1 + R_2 = R_1$$

The output voltage then equals

$$V_{OUT} = \frac{0}{R_1 + 0} V_{IN} = 0$$

That is, the output voltage is zero when the switch is closed.

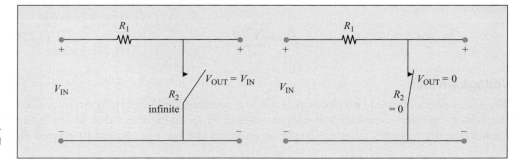

Figure 3.21

A switch and a resistor form a voltage divider that is common in digital circuits.

Automobile battery and starter motor circuit

EXAMPLE 3.19

Most conventional gasoline-powered automobiles are equipped with a $V_B = 12\,\text{V}$ lead acid battery. The engine is started using a key that operates an ignition relay switch that connects the battery to the starter motor, which is a strong electrical motor that turns the gasoline engine. Figure 3.22 shows the battery/starter motor circuit that has the following components.

Battery provides 12 V and has an internal resistance R_B that is initially small but increases with battery age. The $+$ terminal connects through the ignition switch and a wire directly to the starter motor. The $-$ terminal (*ground*) connects to the metal car chassis, so that all the metal parts in a car are connected to ground. The R_B value limits the maximum current I_{max} that the battery can provide:

$$I_{max} = \frac{V_B}{R_B}$$

Every battery has such an internal resistance that limits I_{max}, even AA cells. A 12 V car battery that can produce $I_{max} = 600\,\text{A}$ when new has internal resistance as

$$R_{B,new} = \frac{V_B}{I_{max}} = \frac{12\,\text{V}}{600\,\text{A}} = 0.02\,\Omega$$

Wire resistance R_W is the resistance in the wire and the connections from the battery to the starter motor. With the large currents involved, this resistance becomes important and is often the cause of failures.

Starter motor has internal resistance R_S, which limits the initial current, which occurs before the motor starts turning. Once a motor begins to turn, the current is reduced by a reverse voltage produced by the motor that is proportional to the motor speed.

The initial current flowing from the battery to the starter motor I_S then equals

$$I_S = \frac{V_B}{R_B + R_W + R_S}$$

To operate the starter motor requires an initial minimum current $I_S > 100\,\text{A}$. The resistance of the starter motor R_S must be sufficiently small to allow such a current to flow. The maximum resistance is then

$$R_{S,max} < \frac{V_B}{I_S} = \frac{12\,\text{V}}{100\,\text{A}} = 0.12\,\Omega$$

To allow a safety margin, starter motors are designed to have half this resistance value, as

$$R_S = \frac{R_{S,max}}{2} = 0.06\,\Omega$$

Thick wires connect the battery to the starter motor because their resistance should be $R_W \ll R_S$. Engineers take $\ll$ to mean a factor of 10 smaller, making

$$R_W = \frac{R_S}{10} = 0.006\,\Omega$$

Hence, when everything is working properly, the current produced by a new battery in a new car is computed by inserting the nominal values, so

$$I_S = \frac{V_B}{R_B + R_W + R_S} = \frac{12\,\text{V}}{0.02\,\Omega + 0.006\,\Omega + 0.06\,\Omega} = \frac{12\,\text{V}}{0.086\,\Omega} = 140\,\text{A}$$

Hence, when new, the starter motor starts quickly and reliably.

As the battery ages, R_B increases. As a car ages, R_W increases through resistive corrosion at the connections to the battery. This increased resistance then limits the current to the starter motor. For example, let the combined resistance of the connections and/or an old battery equal $R_{old} = 1\,\Omega$, then the starter motor current is

$$I_{old} = \frac{V_B}{R_{old} + R_S} = \frac{12\,\text{V}}{1.06\,\Omega} \approx 12\,\text{A}$$

which is not sufficient to start the car. It is time to check the battery connections for corrosion or buy a new battery.

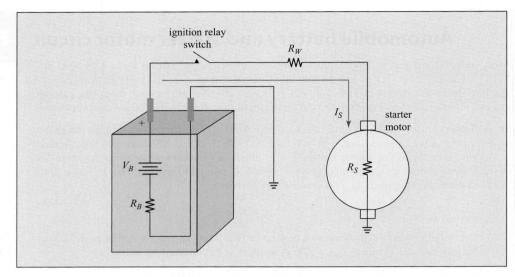

Figure 3.22

The battery/starter motor circuit in a car.

3.5.2 Parallel Resistor Circuits

When resistors R_1 and R_2 are connected in parallel, as shown in Figure 3.23, the same voltage V_S appears across each resistor. By Kirchoff's Current Law (KCL), the current I_S flowing into node A equals the sum of the currents flowing out of the node and through the resistors. The value of this current then equals

$$I_S = I_1 + I_2 = \frac{V_S}{R_1} + \frac{V_S}{R_2} \tag{3.24}$$

The equivalent resistance R_P draws the same current I_S from the source V_S, so

$$I_S = \frac{V_S}{R_P} \tag{3.25}$$

Equating these last two equations, we have

$$\frac{V_S}{R_P} = \frac{V_S}{R_1} + \frac{V_S}{R_2} \tag{3.26}$$

or

$$\frac{1}{R_P} = \frac{1}{R_1} + \frac{1}{R_2} \tag{3.27}$$

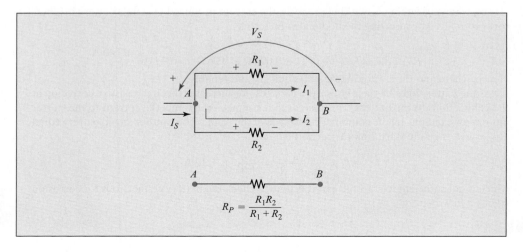

Figure 3.23

Resistors R_1 and R_2 connected in parallel produce the equivalent resistance R_P.

Multiplying both sides by $R_1 R_2$, we get

$$\frac{R_1 R_2}{R_P} = \frac{R_1 R_2}{R_1} + \frac{R_1 R_2}{R_2} = R_2 + R_1 \tag{3.28}$$

Solving for R_P, we have the result

$$R_P = \frac{R_1 R_2}{R_1 + R_2} \tag{3.29}$$

or the value of the equivalent resistance for the parallel connection equals the product of the resistances over their sum

Parallel connection of two resistors   EXAMPLE 3.20

Let both resistors have equal values or $R_1 = R_2 = R$. The parallel connection then produces

$$R_P = \frac{R_1 R_2}{R_1 + R_2} = \frac{R^2}{2R} = \frac{R}{2}$$

That is, connecting identical resistors in parallel produces an equivalent resistor having half the resistance.

Extending the result in the previous example, connecting n equal resistors having resistance R in parallel produces equivalent resistance R_P, as

$$R_P = \frac{R}{n} \tag{3.30}$$

Parallel connection of three resistors   EXAMPLE 3.21

Consider connecting three resistors R_1, R_2 and R_3 in parallel. Extending the circuit shown in Figure 3.24 and applying KCL, we get

$$I_S = I_1 + I_2 + I_3 = \frac{V_S}{R_1} + \frac{V_S}{R_2} + \frac{V_S}{R_3}$$

The equivalent resistance R_P draws the same current I_S from the source V_S, so

$$I_S = \frac{V_S}{R_P}$$

Equating these last two equations, we have

$$\frac{V_S}{R_P} = \frac{V_S}{R_1} + \frac{V_S}{R_2} + \frac{V_S}{R_3}$$

or

$$\frac{1}{R_P} = \frac{1}{R_1} + \frac{1}{R_2} + \frac{1}{R_3}$$

Multiplying both sides by $R_1 R_2 R_3$, we get

$$\frac{R_1 R_2 R_3}{R_P} = \frac{R_1 R_2 R_3}{R_1} + \frac{R_1 R_2 R_3}{R_2} + \frac{R_1 R_2 R_3}{R_3} = R_2 R_3 + R_1 R_3 + R_1 R_2$$

Solving for R_P,

$$R_P = \frac{R_1 R_2 R_3}{R_1 R_2 + R_1 R_3 + R_2 R_3}$$

As a verification, let all resistors have the same value $R_1 = R_2 = R_3 = R$

$$R_P = \frac{R^3}{R^2 + R^2 + R^2} = \frac{R}{3}$$

which is the expected result.

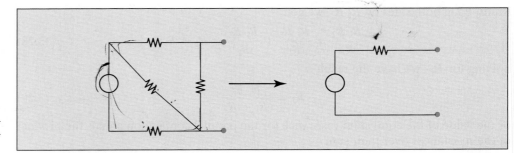

Figure 3.24

Any complex circuit can be expressed as a simpler Thevenin equivalent circuit.

3.5.3 Thevenin Equivalent Circuit

Electric circuits can become very complicated when they include a large number of components. When one circuit (the *source*, say a microphone) connects to another (the *load*, which can be an amplifier), electrical engineers employ a simplifying concept called the *Thevenin equivalent circuit* (TEC). Figure 3.24 shows that a complex circuit can be expressed as a simpler TEC having a single voltage source V_{TH} and a single resistance R_{TH}.

Associated with the TEC are three useful quantities illustrated in Figure 3.25:

1. V_{OC} is the *open-circuit voltage*, which is the voltage that appears across the output terminals under *open-circuit conditions* (that is, when there is nothing connected to the output terminals). Figure 3.25 shows that because $I = 0$, there is no voltage drop across R_{TH}, making

$$V_{OC} = V_{TH} \tag{3.31}$$

2. I_{SC} is the *short-circuit current*, which is the current that flows out of (and back into) the TEC when the output terminals are *short circuited* (that is, a wire connects the two output nodes). Figure 3.25 shows that R_{TH} limits the current, making

$$I_{SC} = \frac{V_{TH}}{R_{TH}} \tag{3.32}$$

3. Thus, the *Thevenin Equivalent resistance* is the open-circuit voltage divided by the short-circuit current:

$$R_{TH} = \frac{V_{OC}}{I_{SC}} \tag{3.33}$$

When connecting two circuits, the load is represented as an equivalent *load resistance* (R_L) that connects to the TEC of the source, as shown in Figure 3.26. It is then a simple matter to show the effect of connecting a load to a source:

1. I_L is the current delivered to the load and is

$$I_L = \frac{V_{TH}}{R_{TH} + R_L} \tag{3.34}$$

Note, $R_L > 0$ makes $I_L < I_{SC}$.

Figure 3.25

Open-circuit voltage and short-circuit current in a Thevenin equivalent circuit.

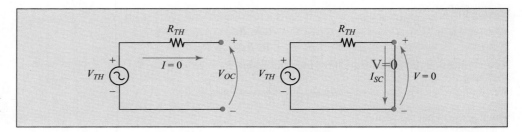

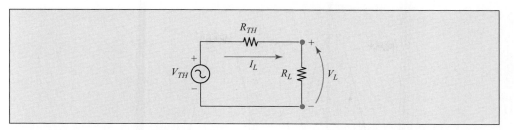

Figure 3.26

Thevenin equivalent circuit driving a load being represented by equivalent input resistance R_L.

2. V_L is the voltage across the load and equals

$$V_L = I_L R_L = \frac{V_{TH}}{R_{TH} + R_L} R_L = \frac{R_L}{R_{TH} + R_L} V_{TH} \qquad (3.35)$$

Note, $R_L < \infty$ makes $V_L < V_{TH}$.

Determining the TEC when a short-circuit is ill-advised   EXAMPLE 3.22

While the open-circuit voltage is a simple matter to measure, thus determining V_{TH}, determining R_{TH} with the short-circuit current is totally different. Applying a short circuit across terminals is sometimes a bad idea. For example, in testing a battery, connecting the terminals with a wire, called *shorting*, would result in large (sometimes *dangerously large*) currents.

In cases when a short-circuit is ill-advised, a known R_L is attached, V_{TH} and V_L are measured, and R_{TH} can be computed. Starting with

$$V_L = \frac{R_L}{R_{TH} + R_L} V_{TH}$$

we solve for R_{TH} to find

$$R_{TH} = \left(\frac{V_{TH} - V_L}{V_L} \right) R_L$$

Let us reduce R_L while measuring V_L. When $V_L = V_{TH}/2$, we get

$$R_{TH} = \left(\frac{V_{TH} - V_{TH}/2}{V_{TH}/2} \right) R_L = \left(\frac{V_{TH}/2}{V_{TH}/2} \right) R_L = R_L$$

That is, the value of $V_L \approx V_{TH}$ when $R_L \gg R_{TH}$. As R_L is reduced, V_L begins to drop. When $V_L = V_{TH}/2$, then $R_L = R_{TH}$.

3.6   FREQUENCY-DEPENDENT IMPEDANCES

In describing resistor circuits previously, consider non-time varying voltage (V) and current (I). This is appropriate for DC circuits or time scales over which these values can be approximated by constants. We now turn our attention to time-varying voltages denoted $v(t)$ and currents $i(t)$, which typically vary sinusoidally in time. For example, Figure 3.27 shows a sinusoidal voltage waveform described by

$$v(t) = A \sin(2\pi f_o t + \phi) \qquad (3.36)$$

where A is the amplitude, f_o is the frequency with units of cycles per second or Hertz (Hz), and ϕ is the phase with units of radians.

The *period* is the time interval over which the waveform repeats its values

$$T_o = \frac{1}{f_o} \qquad (3.37)$$

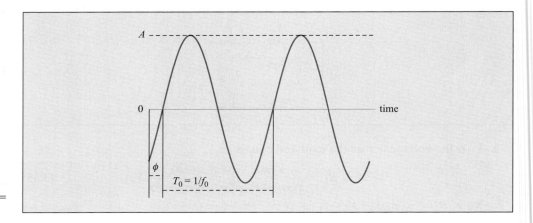

Figure 3.27

Sinusoidal voltage waveform $v(t) = A \sin(\omega_0 t + \phi)$ with $\phi = -\pi/4$.

Another convenient symbol for frequency is the Greek lower-case *omega*

$$\omega = 2\pi f \tag{3.38}$$

with units of *radians/second* (*rad/s*). Multiplying ω by time t (seconds) produces the dimensionless quantity *radian*. Hence, (ωt) is commonly used as the argument of the sine function, as in

$$v(t) = A \sin(\omega_o t + \phi) \tag{3.39}$$

EXAMPLE 3.23 Sine and cosine waveforms

The common sinusoidal waveforms start at $t = 0$ at 0 $(\sin(0) = 0)$, or at 1 $(\cos(0) = 1)$. More general sinusoids include the phase to indicate the value at $t = 0$. Figure 3.28 shows sine and cosine waveforms having the same frequency. Note that

$$\cos(\omega t) = \sin\left(\omega t + \frac{\pi}{2}\right)$$

This is most easily seen at $t = 0$, at which

$$\cos(0) = 1 = \sin\left(0 + \frac{\pi}{2}\right)$$

Also, a waveform that is constant over time can be considered to be a cosine waveform with $\omega_o = 0$, or

$$V_S = A\cos(0t) = A \quad \text{for all } t$$

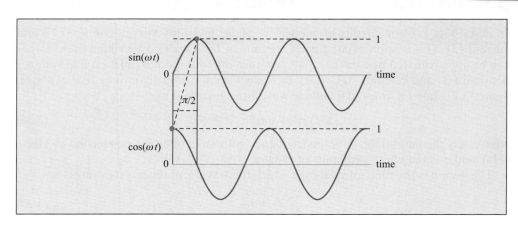

Figure 3.28

Sine and cosine waveforms having the same frequency $\cos(\omega_0 t) = \sin\left(\omega_0 t + \frac{\pi}{2}\right)$.

> **Factoid:** Electrical circuits that are driven with sinusoidal supply voltage $v(t)$ having frequency ω will have a sinusoidal current $i(t)$ at the same frequency ω that will typically differ in amplitude and phase.

3.6.1 Complex Impedance

The general term that relates the voltage across an electrical component to the current flowing through it is called the *impedance*, which is denoted by Z and expressed in Ohms (Ω). Previously, we considered resistors whose impedance did not vary with frequency. We now extend the analysis of electric circuits to include capacitors and inductors, whose impedance varies with frequency.

Ohm's law relates the concepts of voltage, current, and impedance. When the value of the time-varying current $i(t)$ flowing through frequency-dependent impedance $Z(\omega)$ is known, the voltage $v(t)$ produced across the impedance equals

$$v(t) = i(t)\, Z(\omega) \tag{3.40}$$

Figure 3.29 shows the voltage $v(t)$ of a sinusoidal waveform component having frequency ω measured across the impedance $Z(\omega)$ with current $i(t)$ flowing through it.

An impedance is a complex number that can be expressed as

$$Z(\omega) = \mathcal{R}e\, Z(\omega) + j\mathcal{I}m Z(\omega) \tag{3.41}$$

where $\mathcal{R}e\, Z(\omega)$ is the *real* part, $\mathcal{I}m Z(\omega)$ is the *imaginary part*, and $j = \sqrt{-1}$. A complex number also can be expressed as the vector

$$Z(\omega) = |Z(\omega)|e^{j\angle Z(\omega)} \tag{3.42}$$

where $|Z(\omega)|$ is the magnitude and $\angle Z(\omega)$ is the phase. To compute the magnitude, use the result for the magnitude of complex number c as

$$|c| = \sqrt{c\, c^*} \tag{3.43}$$

where c^* is the *complex conjugate* of c obtained by replacing j by $-j$ where ever it occurs in c.

We display $Z(\omega)$ as a vector in the complex $(\mathcal{R}, j\mathcal{I})$ plane, where $\mathcal{R}$ is the real abscissa axis, $j\mathcal{I}$ is the imaginary ordinate axis. Then, $|Z(\omega)|$ is the length of the vector, and $\angle Z(\omega)$ is the angle with respect to the $\mathcal{R}$ axis in the counter-clockwise direction.

This section examines the impedance of the resistor, capacitor, and inductor and then considers how circuits implemented with these components affect sinusoidal waveforms.

Resistive Impedance

The impedance associated with resistors $Z_R(\omega)$ is the same at all frequencies. Hence, $Z_R(\omega)$ is not a function of frequency, and it it is simply the resistance value of R, or

$$Z_R(\omega) = R \tag{3.44}$$

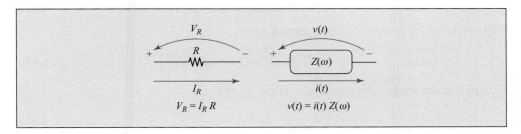

Figure 3.29

Ohm's law extends from resistor R to complex impedance $Z(\omega)$ that changes with frequency.

Comparing the real and imaginary terms of a complex number, we see

$$Z_R = \mathcal{R}e\, Z_R(\omega) + j\mathcal{I}m\, Z_R(\omega) = R \tag{3.45}$$

Hence, Z_R is purely real,

$$R = \mathcal{R}e\, Z_R(\omega) \tag{3.46}$$

with the zero imaginary part as

$$\mathcal{I}m\, Z_R(\omega) = 0 \tag{3.47}$$

Expressing Z_R in vector form, we have

$$Z_R = |Z_R|\, e^{j\angle Z_R} = R \quad (= Re^{j0}) \tag{3.48}$$

Comparing the magnitude and angle terms, we see

$$|Z_R| = R \quad \text{and} \quad \angle Z_R = 0 \tag{3.49}$$

Applying Ohm's law to relate the current flowing through the resistor to the voltage across it, we have

$$i_R(t) = \frac{v_R(t)}{R} \tag{3.50}$$

Figure 3.30 shows that the sinusoidal current waveform has the same frequency and phase as that of the voltage waveform, that is, the current through a resistor is *in phase* with the voltage across it.

Capacitive Impedance

With capacitors, the rate at which its charge can change determines is capacitive impedance, which is frequency dependent. No DC current can flow through a capacitor, no matter what V_C is, making the impedance of a capacitor infinite at $\omega = 0$. As ω increase, the impedance decreases, having the analytic form

$$Z_C = \frac{1}{j\omega C} \tag{3.51}$$

The conventional form of a complex number has a j term in the numerator. We obtain this form by multiplying both the numerator and denominator by $-j$ to get

$$Z_C = \left(\frac{-j}{-j}\right)\frac{1}{j\,\omega C} = \frac{-j}{\underbrace{(-j^2)}_{=1}\,\omega C} = \frac{-j}{\omega C} \tag{3.52}$$

Comparing the real and imaginary terms of a complex number, we see

$$Z_C = \mathcal{R}e\, Z_C(\omega) + j\mathcal{I}m\, Z_C(\omega) = \frac{-j}{\omega C} \tag{3.53}$$

Hence, Z_C is purely imaginary,

$$Z_C = \mathcal{I}m\, Z_C(\omega) = \frac{-1}{\omega C} \tag{3.54}$$

with zero real part

$$\mathcal{R}e\, Z_C(\omega) = 0 \tag{3.55}$$

Expressing Z_C as a vector in the complex plane

$$Z_C = |Z_C|\, e^{j\angle Z_C} = \left(\frac{1}{\omega C}\right) e^{-j\pi/2} \tag{3.56}$$

Comparing the magnitude and angle terms, we see

$$|Z_C| = \frac{1}{\omega C} \quad \text{and} \quad \angle Z_C = -\frac{\pi}{2} \tag{3.57}$$

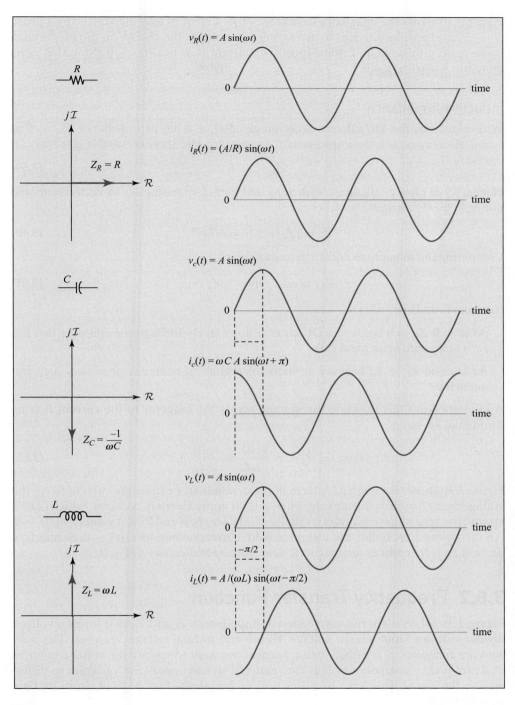

Figure 3.30

Comparing the impedances and sinusoidal voltage and current waveforms in a resistor, capacitor, and inductor.

Note:

At $\omega = 0$ $Z_C = \infty$, because a constant (DC) current is blocked by the insolating gap between the capacitor plates.

At $\omega = \infty$ $Z_C = 0$, because a capacitor acts as a short circuit at very high frequencies.

Applying Ohm's law to relate the current flowing through the capacitor to the voltage across it gives

$$i_C(t) = \frac{v_C(t)}{Z_C} = \overbrace{j}^{e^{j\pi/2}} \omega C \, v_C(t) = \omega C \, v_C(t) \, e^{j\pi/2} \qquad (3.58)$$

Figure 3.30 shows the imaginary term causes the current to *lead* the voltage by a phase $\phi = \pi/2$ radians. Intuitively, an increasing $v_C(t)$ means the charge on the plates must be increasing through an inward (positive) current flow or $i_C(t) > 0$. Note that $v_C(t)$ increases when $i_C(t) > 0$.

Inductive Impedance

With inductors, the DC current flows unimpeded, making its impedance $Z_L = 0$ at $\omega = 0$. As ω increases, the impedance of an inductor also increases and is given by

$$Z_L = j\omega L \tag{3.59}$$

Hence, Z_L is purely imaginary with zero real part. Expressing Z_L in vector form and noting $j = e^{j\pi/2}$, we get

$$Z_L = |Z_L| e^{j\angle Z_L} = \omega L e^{j\pi/2} \tag{3.60}$$

Comparing the magnitude and angle terms gives

$$|Z_L| = \omega L \quad \text{and} \quad \angle Z_L = \frac{\pi}{2} \tag{3.61}$$

Note the following:

At $\omega = 0$ $Z_L = 0$ because a DC current flows freely through the coil wire that has zero resistance in an ideal coil.

At $\omega = \infty$ $Z_L = \infty$ because an inductor acts as an open circuit at very high frequencies.

Applying Ohm's law to relate the voltage across the inductor to the current flowing through it, we have

$$i_L(t) = \frac{v_L(t)}{Z_L} = \frac{v_L(t)}{j\omega L} = \frac{v_L(t)}{\omega L} e^{-j\pi/2} \tag{3.62}$$

Figure 3.30 shows the imaginary term in the denominator causes the current to *lag* the voltage by $\pi/2$ radians. Intuitively, when $i_L(t)$ is approximately constant, which occurs around the zero-slope regions of its sinusoid (at its highest and lowest values), $v_L(t) \approx 0$. An increasing $i_L(t)$ builds the magnetic field, corresponding to $v_L(t) > 0$. Similarly, a decreasing $i_L(t)$ reduces the magnetic field, corresponding to $v_L(t) < 0$.

3.6.2 Frequency Transfer Function

Figure 3.31 shows a a series connection of impedances Z_i and Z_2 that forms a voltage divider with an input source voltage $V_{IN}(\omega)$ and output voltage $V_{OUT}(\omega)$. The voltages are functions of ω rather than t, because we want to specify the voltage in terms of its frequency components. It is conventional to use upper-case variables to differentiate the signal waveform as a function of frequency, that is $V(\omega)$, from its time waveform $v(t)$.

Figure 3.31

A voltage divider configuration of Z_1 and Z_2 forms a filter having a frequency transfer function $H(\omega) = V_{OUT}(\omega)/V_{IN}(\omega)$.

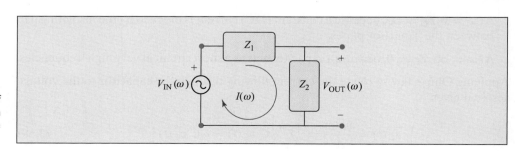

Because the same current $I(\omega)$ flows through Z_1 and Z_2, Ohm's law produces two formulas:

$$I(\omega) = \frac{V_{IN}(\omega)}{Z_1 + Z_2} \tag{3.63}$$

and

$$V_{OUT}(\omega) = I(\omega)\, Z_2 \tag{3.64}$$

Substituting $I(\omega)$ from the first equation into the second, the voltage-divider formula is

$$V_{OUT}(\omega) = \frac{Z_2(\omega)}{Z_1(\omega) + Z_2(\omega)}\, V_{IN}(\omega) \tag{3.65}$$

Because the impedances vary on frequency, the *frequency transfer function $H(\omega)$* that converts $V_{IN}(\omega)$ into $V_{OUT}(\omega)$ is defined as

$$V_{OUT}(\omega) = H(\omega)\, V_{IN}(\omega) \tag{3.66}$$

or the equation

$$H(\omega) = \frac{V_{OUT}(\omega)}{V_{IN}(\omega)} = \frac{Z_2(\omega)}{Z_1(\omega) + Z_2(\omega)} \tag{3.67}$$

As with any complex-valued quantity, $H(\omega)$ is expressed in terms of its real and imaginary components

$$H(\omega) = \mathcal{R}e\, H(\omega) + j\mathcal{I}m\, H(\omega) \tag{3.68}$$

Its magnitude and phase are

$$H(\omega) = |H(\omega)|e^{j\angle H(\omega)} \tag{3.69}$$

The magnitude is computed most easily by using

$$|H(\omega)|^2 = H(\omega)H^*(\omega) \tag{3.70}$$

where $H^*(\omega)$ is the complex conjugate of $H(\omega)$ found by replacing each occurrence of j by $-j$.

The phase is computed using the arctangent function

$$\angle H(\omega) = \arctan\left(\frac{\mathcal{I}m\, H(\omega)}{\mathcal{R}e\, H(\omega)}\right) \tag{3.71}$$

The next section illustrates these caluclations for an *RC* circuit that acts as a low-pass filter that commonly models the frequency characteristics of communication channels that have limited frequency bandwidth.

3.7 *RC* CIRCUITS

This section describes circuits that contain resistors and capacitors. We start with *RC* circuits because they are the most common in practice. The extension to *RL* circuits is straightforward, and because *RL* circuits are rarely used, we then proceed directly to *LC* circuits.

This section computes the equivalent impedance that occurs when a resistor and capacitor are combined in one circuit. Figure 3.32 shows that this combination can be implemented either in parallel or in series.

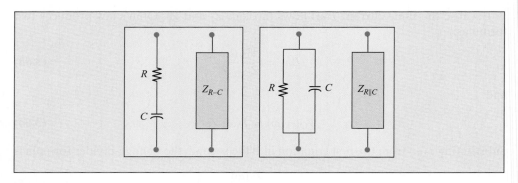

Figure 3.32

A resistor and capacitor can be combined either in series ($R-C$) or in parallel ($R||C$) to produce the corresponding impedances Z_{R-C} and $Z_{R||C}$.

3.7.1 Series *RC* Circuits

Similar to resistors in series, combining R and C in series causes their impedances (vectors) to add and produce the series combination impedance Z_{R-C}, which is equal to

$$Z_{R-C} = Z_R + Z_C = R - \frac{j}{\omega C} \tag{3.72}$$

The analysis simplifies if we express Z_{R-C} in terms of its real and imaginary parts. In the $R-C$ case, this is straightforward, but it helps in the transition to more complicated cases.

$$Z_{R-C} = \mathcal{Re}\,(Z_{R-C}) + j\mathcal{Im}\,(Z_{R-C}) \tag{3.73}$$

Equating the real and imaginary parts in the previous two equations, we have

$$\mathcal{Re}\,(Z_{R-C}) = R \quad \text{and} \quad \mathcal{Im}\,(Z_{R-C}) = -\frac{1}{\omega C} \tag{3.74}$$

Impedance Magnitude

The impedance magnitude is found by applying Pythagoras' theorem to the right triangle formed by R and Z_C:

$$|Z_{R-C}|^2 = \mathcal{Re}\,(Z_{R-C})^2 + \mathcal{Im}\,(Z_{R-C})^2 \tag{3.75}$$

Substituting the values for the real and imaginary parts gives

$$|Z_{R-C}|^2 = R^2 + \left(\frac{-1}{\omega C}\right)^2 \tag{3.76}$$

Taking the square root and factoring out R, the magnitude equals

$$|Z_{R-C}| = R\sqrt{1 + \frac{1}{(\omega RC)^2}} \tag{3.77}$$

To simplify the analysis, we define the frequency ω_x at which the two impedance magnitudes are equal, called the *cross-over frequency*. We find frequency ω_x such that $|Z_R| = |Z_C|$ or

$$R = \frac{1}{\omega_x C} \quad \rightarrow \quad \omega_x = \frac{1}{RC} \tag{3.78}$$

The corresponding cross-over frequency in units of Hz is

$$f_x = \frac{\omega_x}{2\pi} = \frac{1}{2\pi RC} \tag{3.79}$$

Substituting ω_x into $|Z_{R-C}|$, the simpler form is

$$|Z_{R-C}| = R\sqrt{1 + \frac{1}{(\omega/\omega_x)^2}} = R\sqrt{1 + \frac{1}{(f/f_x)^2}} \tag{3.80}$$

Note the change in $|Z_{R-C}|$ as the frequency f varies.
   As $f \to 0$

$$|Z_{R-C}| = R\sqrt{1 + \frac{1}{(0/f_x)^2}} = \infty \qquad (3.81)$$

that is, Z_C dominates, because $|Z_C| \gg R$ at low frequencies.
   As $f \to \infty$,

$$|Z_{R-C}| = R\sqrt{1 + \frac{1}{(\infty/f_x)^2}} = R \qquad (3.82)$$

that is, R dominates, because $R \gg |Z_C|$ at high frequencies.
   When $f = f_x$ (the *cross-over frequency*),

$$|Z_{R-C}| = R\sqrt{1 + \frac{1}{(f_x/f_x)^2}} = \sqrt{2}\,R \qquad (3.83)$$

Note that $|Z_{R-C}| \neq 2R$ when the two impedances are equal, because it is a vector sum. (Recall that a right triangle with equal sides has a hypotenuse that is $\sqrt{2}$ larger.)

Impedance Angle

Figure 3.33 shows $\angle Z_{R-C}$ is found from the arctangent of the imaginary part divided by the real part

$$\angle Z_{R-C} = \arctan\left(\frac{\mathcal{I}m\,(Z_{R-C})}{\mathcal{R}e\,(Z_{R-C})}\right) \qquad (3.84)$$

Substituting the real and imaginary components gives

$$\angle Z_{R-C} = \arctan\left(\frac{-1/\omega C}{R}\right) = -\arctan\left(\frac{1}{\omega RC}\right) \qquad (3.85)$$

Incorporating cross-over frequencies ω_x and f_x results in the simple form

$$\angle Z_{R-C} = -\arctan\left(\frac{\omega_x}{\omega}\right) = -\arctan\left(\frac{f_x}{f}\right) \qquad (3.86)$$

Note the change in $\angle Z_{R-C}$ as the frequency f varies.
   As $f \to 0$,

$$\angle Z_{R-C} = -\arctan\left(\frac{f_x}{f}\right) = -\arctan\left(\frac{f_x}{0}\right) = -\arctan(\infty) = -\frac{\pi}{2} \qquad (3.87)$$

That is, the angle points in the purely Z_C direction.

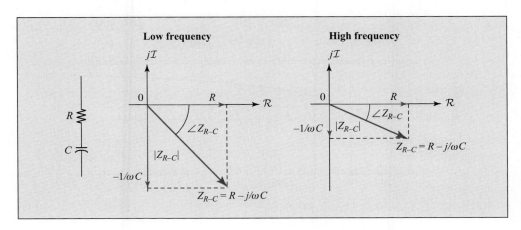

Figure 3.33

Impedance of series R and C circuit at low and high frequencies.

As $f \to \infty$,

$$\angle Z_{R-C} = -\arctan\left(\frac{f_x}{f}\right) = -\arctan\left(\frac{f_x}{\infty}\right) = -\arctan(0) = 0 \qquad \text{(3.88)}$$

That is, the angle points in the purely Z_R direction.
When $f = f_x$ (the *cross-over frequency*),

$$\angle Z_{R-C} = -\arctan\left(\frac{f_x}{f}\right) = -\arctan\left(\frac{f_x}{f_x}\right) = -\arctan(1) = -\frac{\pi}{4} \qquad \text{(3.89)}$$

which would be expected from $|Z_R| = |Z_C|$.

EXAMPLE 3.24   Variation of Z_{R-C} with frequency

Figure 3.34 shows $|Z_{R-C}|$ and $\angle Z_{R-C}$ plotted as a function of frequency for $R = 1,592\,\Omega$ and $C = 1\,\mu\text{F}$.
The cross-over frequency is

$$f_x = \frac{1}{2\pi RC} = \frac{1}{2\pi(1,592)(10^{-6})} = 100\,\text{Hz}$$

The magnitude of the impedance at $f = f_x$ is

$$|Z_{R-C}(f_x)| = \sqrt{2}R = 2,251\,\Omega \quad \text{(dotted line)}$$

The smooth curve was obtained by computing $|Z_{R-C}|$ and $\angle Z_{R-C}$ at a dense set of frequency points, whihc is every 10 Hz for $0 \le f \le 500$ Hz.
Variations in $|Z_C|$ and R are shown in dashed curves. At low frequencies, $|Z_{R-C}| \approx |Z_C|$, and at high frequencies, $|Z_{R-C}| \approx R$.

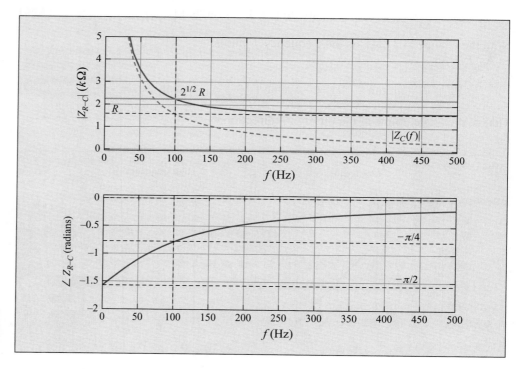

Figure 3.34

$Z_{R-C}(f)$ (heavy curves) for $R = 1.67\,k\Omega$ and $C = 1\,\mu\text{F}$, producing $f_x = 100$ Hz. Variations in $|Z_C|$ and R are shown in dashed curves.

RC low-pass filter frequency transfer function   EXAMPLE   3.25

Figure 3.35 shows an RC low-pass filter with $Z_1 = R$ and $Z_2 = 1/(j\omega C)$. The low-pass frequency transfer function is

$$H_{LP}(\omega) = \frac{Z_2}{Z_1 + Z_2} = \frac{\frac{1}{j\omega C}}{R + \frac{1}{j\omega C}}$$

Multiplying numerator and denominator by $j\omega C$, we get

$$H_{LP}(\omega) = \frac{1}{1 + j\omega RC}$$

The magnitude is computed most easily from

$$|H_{LP}(\omega)|^2 = H(\omega)H^*(\omega)$$

where

$$H_{LP}^*(\omega) = \left(\frac{1}{1 - j\omega RC}\right)$$

Then

$$|H_{LP}(\omega)|^2 = \left(\frac{1}{1 + j\omega RC}\right)\left(\frac{1}{1 - j\omega RC}\right) = \frac{1}{1 + (\omega RC)^2}$$

Hence,

$$|H_{LP}(\omega)| = \sqrt{\frac{1}{1 + (\omega RC)^2}}$$

The real and imaginary parts are found by multiplying the numerator and denominator of $H(\omega)$ by $H^*(\omega)$, yielding

$$H_{LP}(\omega) = \frac{1}{1 + j\omega RC}\frac{1 - j\omega RC}{1 - j\omega RC} = \frac{1 - j\omega RC}{1 + (\omega RC)^2}$$

This separates into the real and imaginary parts, with

$$\mathcal{R}e\,H_{LP}(\omega)) = \frac{1}{1 + (\omega RC)^2}$$

and

$$\mathcal{I}m\,H_{LP}(\omega)) = -\frac{\omega RC}{1 + (\omega RC)^2}$$

The phase then equals

$$\angle H_{LP}(\omega) = \arctan\left(\frac{\mathcal{I}m\,H_{LP}(\omega)}{\mathcal{R}e\,H_{LP}(\omega)}\right)$$

$$= \arctan\left(\frac{-\dfrac{\omega RC}{1 + (\omega RC)^2}}{\dfrac{1}{1 + (\omega RC)^2}}\right)$$

$$= \arctan(-\omega RC)$$

Figure 3.35 shows plots of $|H_{LP}(\omega)|$ and $\angle H_{LP}(\omega)$ for $R = 1\,k\Omega$ and $C = 1\,\mu F$.

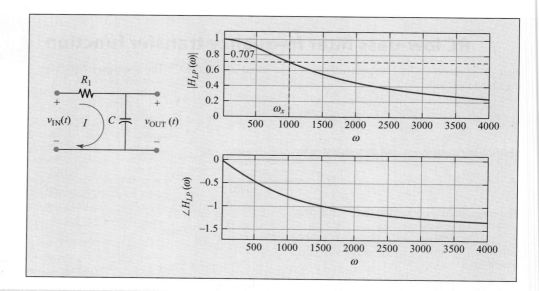

Figure 3.35

An RC low-pass filter.

Note that the high frequencies are attenuated more than the low frequencies, hence the name *low-pass filter*. Specifically,

- For $\omega = 0$,

$$|H_{LP}(0)| = 1 \quad \text{and} \quad \angle H_{LP}(0) = 0$$

because $|Z_C(0)| = 1/\omega C = \infty$, all of the voltage from the input appears at the output.

- For $\omega \to \infty$,

$$|H_{LP}(\infty)| = 0 \quad \text{and} \quad \angle H_{LP}(\infty) = -\pi/2$$

because $|Z_C(\infty)| = 0$, the capacitor acts as a short circuit.

- For $\omega = (RC)^{-1}$,

$$|H_{LP}((RC)^{-1})| = \sqrt{\frac{1}{1 + ((RC)^{-1} \times RC)^2}} = \sqrt{\frac{1}{2}} = 0.707$$

and

$$\angle H_{LP}((RC)^{-1}) = -\pi/4$$

because $|Z_C| = R$, but R and Z_C add vectorially.

EXAMPLE 3.26   Voltage divider for an *RC* filter

The $R-C$ filter shown in Figure 3.36 has $R = 1\,k\Omega$ and $C = 1\,\mu F$. The input $v_i(t) = A_i \sin(\omega t)$, where $A_i = 10$ V and $\omega = 2\pi f$. Determine the output $v_{OUT}(t) = A_o \sin(\omega t + \phi)$ by computing A_o and ϕ when $f = 1\,kHz$.

We find the answer using the following three steps.

1. **Voltage divider formula:** The voltage divider formula for the $R-C$ circuit yields the ratio of $v_{OUT}(t)$ to $v_i(t)$

$$\frac{V_{OUT}(\omega)}{V_{IN}(\omega)} = \frac{Z_C}{R + Z_C} = \frac{\dfrac{1}{j\omega C}}{R + \dfrac{1}{j\omega C}}$$

Multiplying numerator and denominator by $j\omega C$, we get

$$\frac{V_{OUT}(\omega)}{V_{IN}(\omega)} = \frac{1}{1 + j\omega RC}$$

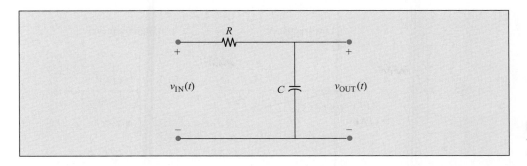

Figure 3.36

RC voltage divider circuit.

2. **Magnitude response:** To compute A_o/A_i, we need the magnitude response. We obtain the square of the magnitude response by multiplying by the complex conjugate to get

$$\left|\frac{A_o}{A_i}\right|^2 = \left|\frac{V_{OUT}(\omega)}{V_{IN}(\omega)}\right|^2 = \left(\frac{1}{1+j\omega RC}\right)\left(\frac{1}{1-j\omega RC}\right) = \frac{1}{1+(\omega RC)^2}$$

Inserting the values for ω, R, and C, we get

$$\left|\frac{A_o}{A_i}\right|^2 = \frac{1}{1+[(2\pi)(10^3\,\text{Hz})(10^3\,\Omega)(10^{-6}\,\text{F})]^2} = \frac{1}{1+[2\pi]^2} = 0.0247$$

Hence,

$$A_o = \sqrt{0.0247}\,A_i = 0.157(10\,\text{V}) = 1.57\,\text{V}$$

3. **Phase response:** To compute ϕ, we need the phase response. We obtain the real and imaginary parts by multiplying numerator and denominator by the complex conjugate of the denominator to get

$$\frac{V_{OUT}(\omega)}{V_{IN}(\omega)} = \left(\frac{1}{1+j\omega RC}\right)\left(\frac{1-j\omega RC}{1-j\omega RC}\right) = \frac{1}{1+(\omega RC)^2} - j\frac{\omega RC}{1+(\omega RC)^2}$$

The phase is then the arctangent of the ratio of the imaginary part and real part

$$\phi = \arctan\left[\frac{-\dfrac{\omega RC}{1+(\omega RC)^2}}{\dfrac{1}{1+(\omega RC)^2}}\right] = -\arctan(\omega RC)$$

Inserting the values for ω, R and C we get

$$\phi = -\arctan\left((2\pi)(10^3\,\text{Hz})(10^3\,\Omega)(10^{-6}\,\text{F})\right) = -\arctan(2\pi) = -1.41\,\text{radians}$$

Hence,

$$v_{OUT}(t) = 1.57\sin(\omega t - 1.41)$$

3.7.2 Parallel *RC* Circuits

The parallel combination of impedances obeys the same product over sum law as for resistors above. Applying this to impedances with $Z_R = R$ and $Z_C = -j/\omega C$, we get

$$Z_{R||C} = \frac{Z_R\,Z_C}{Z_R + Z_C} \tag{3.90}$$

Inserting these values, we can simplify by removing the denominator in Z_C and setting $-j^2 = 1$:

$$Z_{R||C} = \frac{R\left(-j\frac{1}{\omega C}\right)}{R - j\frac{1}{\omega C}}\left(\frac{j\omega C}{j\omega C}\right) = \frac{R}{1+j\omega RC} \tag{3.91}$$

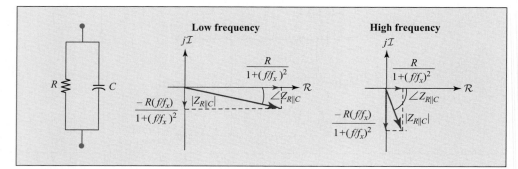

Figure 3.37

Impedance of parallel R and C circuit at low and high frequencies.

Finally, substituting $\omega_x = 1/(RC)$, we get

$$Z_{R\|C} = \frac{R}{1 + j(\omega/\omega_x)} = \frac{R}{1 + j(f/f_x)} \tag{3.92}$$

Figure 3.37 shows how $Z_{R\|C}$ appears in the complex plane.

Impedance Magnitude

The magnitude of $Z_{R\|C}$ is

$$\left| Z_{R\|C} \right| = \sqrt{\left(\frac{R}{1 + j(f/f_x)} \right) \left(\frac{R}{1 - j(f/f_x)} \right)} = \frac{R}{\sqrt{1 + (f/f_x)^2}} \tag{3.93}$$

Note how $\left| Z_{R\|C} \right|$ changes as the frequency f varies.

As $f \to 0$,

$$\left| Z_{R\|C} \right| = \frac{R}{\sqrt{1 + (f/f_x)^2}} = \frac{R}{\sqrt{1 + (0/f_x)^2}} = R \tag{3.94}$$

As $f \to \infty$,

$$\left| Z_{R\|C} \right| = \frac{R}{\sqrt{1 + (f/f_x)^2}} = \frac{R}{\sqrt{1 + (\infty/f_x)^2}} = 0 \tag{3.95}$$

Thus C dominates, because it forms a short-circuit across R.

When $f = f_x$ (the *cross-over frequency*),

$$\left| Z_{R\|C} \right| = \frac{R}{\sqrt{1 + (f/f_x)^2}} = \frac{R}{\sqrt{1 + (f_x/f_x)^2}} = \frac{R}{\sqrt{2}} \tag{3.96}$$

Impedance Angle

To find the real and imaginary parts of $Z_{R\|C}$, multiply the numerator and denominator by the complex conjugate of the denominator to get

$$Z_{R\|C} = \frac{R}{1 + j(f/f_x)} \left(\frac{1 - j(f/f_x)}{1 - j(f/f_x)} \right) = \frac{R(1 - j(f/f_x))}{1 + (f/f_x)^2} \tag{3.97}$$

Then the real part of $Z_{R\|C}$ is

$$\mathcal{R}e\left(Z_{R\|C} \right) = \frac{R}{1 + (f/f_x)^2} \tag{3.98}$$

and the imaginary part of $Z_{R\|C}$ is

$$\mathcal{I}m\left(Z_{R\|C} \right) = -\frac{R(f/f_x)}{1 + (f/f_x)^2} \tag{3.99}$$

After we cancel the common denominator term, the phase is then

$$\angle Z_{R||P} = \arctan\left(\frac{\mathcal{I}m\left(Z_{R||C}\right)}{\mathcal{R}e\left(Z_{R||C}\right)}\right) = \arctan\left(-\frac{R(f/f_x)}{R}\right) = -\arctan\left(\frac{f}{f_x}\right) \quad \textbf{(3.100)}$$

Note how $\angle Z_{R-C}$ changes as the frequency f varies.

As $f \to 0$,

$$\angle Z_{R||P} = -\arctan\left(\frac{f}{f_x}\right) = -\arctan\left(\frac{0}{f_x}\right) = 0 \quad \textbf{(3.101)}$$

As $f \to \infty$,

$$\angle Z_{R||P} = -\arctan\left(\frac{f}{f_x}\right) = -\arctan\left(\frac{\infty}{f_x}\right) = -\frac{\pi}{2} \quad \textbf{(3.102)}$$

When $f = f_x$ (the *cross-over frequency*),

$$\angle Z_{R||P} = -\arctan\left(\frac{f}{f_x}\right) = -\arctan\left(\frac{f_x}{f_x}\right) = -\frac{\pi}{4} \quad \textbf{(3.103)}$$

Variation of $Z_{R||C}$ with frequency   EXAMPLE 3.27

Figure 3.38 shows $\left|Z_{R||C}\right|$ and $\angle Z_{R||C}$ plotted as a function of frequency for $R = 1,592\,\Omega$ and $C = 1\,\mu\text{F}$. Variations in $\left|Z_C\right|$ and R (no variation) are shown in dashed curves. The cross-over frequency is

$$f_x = \frac{1}{2\pi\,RC} = \frac{1}{2\pi(1,592)(10^{-6})} = 100\,\text{Hz}$$

The magnitude of the impedance at $f = f_x$ is

$$\left|Z_{R||C}(f_x)\right| = \frac{R}{\sqrt{2}} = 1,126\,\Omega \quad \text{(dotted line)}$$

At low frequencies, $\left|Z_{R||C}\right| \approx R$, and at high frequencies, $\left|Z_{R||C}\right| \approx \left|Z_C\right|$, which approaches zero as f increases.

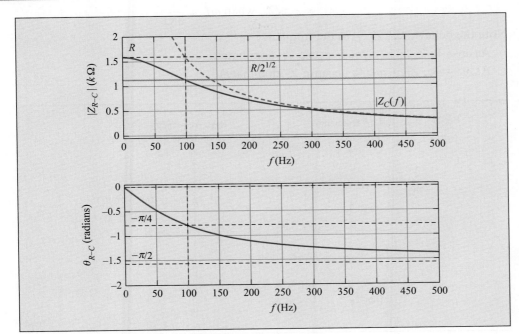

Figure 3.38

$Z_{R||C}(f)$ (heavy curves) for $R = 1.67\,k\Omega$ and $C = 1\,\mu\text{F}$, producing $f_x = 100\,\text{Hz}$. Variations in $\left|Z_C\right|$ and R are shown in dashed curves.

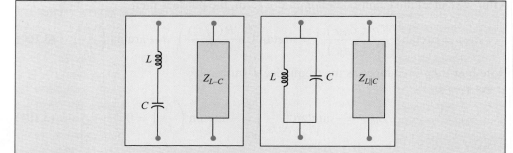

Figure 3.39

An inductor and capacitor can be connected either in series (L-C) or in parallel ($L\,||C$) to produce the corresponding impedances $Z_{L\text{-}C}$ and $Z_{L\,||C}$.

3.8 *LC* CIRCUITS

This section computes the equivalent impedance that occurs when we combine a inductor and capacitor. Figure 3.39 shows that this combination can be implemented either in parallel or in series.

3.8.1 Series *LC* Circuits

Connecting L and C in series shown in Figure 3.40 causes their impedances (vectors) to add and produce the series combination impedance $Z_{R\text{-}C}$, which is equal to

$$Z_{L\text{-}C} = Z_L + Z_C = j\omega L + \frac{1}{j\omega C} = j\left(\omega L - \frac{1}{\omega C}\right) \tag{3.104}$$

Note that $Z_{L\text{-}C}$ is purely imaginary, with the magnitude

$$|Z_{L\text{-}C}| = \left|\omega L - \frac{1}{\omega C}\right| \tag{3.105}$$

and phase

$$\angle Z_{L\text{-}C} = \pi/2, \quad \text{when } \omega L > \frac{1}{\omega C} \tag{3.106}$$

$$= -\pi/2, \quad \text{when } \omega L < \frac{1}{\omega C}$$

Note the behavior of $Z_{L\text{-}C}$ as the frequency varies.
 As $\omega \to 0$, $Z_{L\text{-}C} = -j\infty$ because Z_C dominates.
 As $\omega \to \infty$, $Z_{L\text{-}C} = j\infty$ because Z_L dominates.

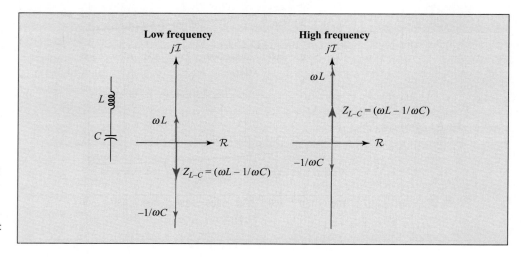

Figure 3.40

Impedance of series L and C circuit at low and high frequencies.

When $\omega = \omega_o$ (the *resonant frequency*),

$$Z_{L-C} = j\left(\omega_o L - \frac{1}{\omega_o C}\right) = 0 \tag{3.107}$$

Solving for ω_o, we obtain

$$\omega_o L = \frac{1}{\omega_o C} \quad \rightarrow \quad \omega_o = \sqrt{\frac{1}{LC}} \tag{3.108}$$

For frequencies $\omega < \omega_o$, Z_{L-C} acts as a capacitor, and for $\omega > \omega_o$, Z_{L-C} acts as an inductor.

With L and C connected in series, the currents flowing through both components must be equal, or

$$i_L(t) = i_C(t) = i(t) \tag{3.109}$$

By Ohm's law, the voltages across each component equal

$$v_L(t) = Z_L i_L(t) = Z_L i(t) \quad \text{and} \quad v_C(t) = Z_C i_C(t) = Z_C i(t) \tag{3.110}$$

Because at $\omega = \omega_o$, $Z_L = -Z_C$ we have

$$v_L(t) = -v_C(t) \tag{3.111}$$

and the sum of voltages across both L and C gives

$$v_L(t) + v_C(t) = 0 \tag{3.112}$$

Thus, the current out of the capacitor flows directly into the inductor in such a way that the voltages cancel out. This, at $\omega = \omega_o$, the series L–C circuit acts like a wire.

Figure 3.41 shows the magnitude and phase of Z_{L-C} for $L = 2.7$ mH and $C = 1\,\mu$F.

3.8.2 Parallel *LC* Circuit

The parallel combination of L and C shown in Figure 3.42 obeys the same product over sum law as for resistors previously. Applying this to impedances with $Z_L = j\omega L$ and

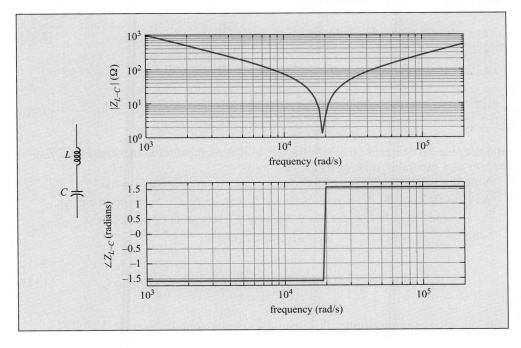

Figure 3.41

Impedance of series L and C circuit.

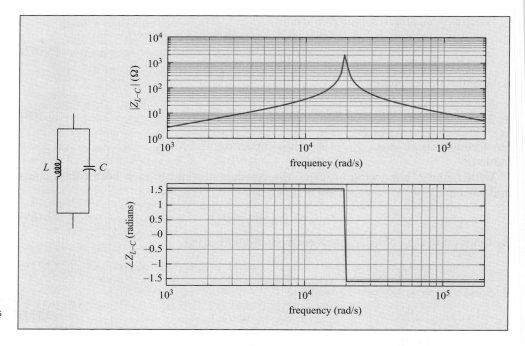

Figure 3.42

Parallel L and C circuit and its impedance.

$Z_C = 1/(j\omega C)$ we get

$$Z_{L\|C} = \frac{Z_L\, Z_C}{Z_L + Z_C} \tag{3.113}$$

$$= \frac{j\omega L\, \frac{1}{j\omega C}}{j\omega L + \frac{1}{j\omega C,}} \tag{3.114}$$

$$= -j\frac{L/C}{\omega L - \frac{1}{\omega C}}$$

Multiplying by $\omega C/\omega C$ to simplify, gives

$$Z_{L\|C} = -j\frac{\omega L}{\omega^2 LC - 1} \tag{3.115}$$

Note that $Z_{L\|C}$ is purely imaginary. The magnitude is

$$|Z_{L\|C}| = \left|\frac{L/C}{\omega L - \frac{1}{\omega C}}\right| \tag{3.116}$$

and the phase is determined by whether $Z_{L\|C}$ acts as an inductor, in which case $\angle Z_{L\|C} = \pi/2$, or as a capacitor, making $\angle Z_{L\|C} = -\pi/2$.

Note the behavior of $Z_{L\|C}$ as frequency varies from $\omega = 0$ to $\omega = \infty$.

As $\omega \to 0$, $Z_{L\|C} \to Z_L = j\omega L$, because the capacitor acts as an open circuit.

As $\omega \to \infty$, $Z_{L\|C} \to Z_C = 1/j\omega C$, because Z_L acts as an open circuit.

When $\omega = \omega_o = (LC)^{-1/2}$ (the *resonant frequency*),

$$|Z_{L\|C}(\omega_o)| = \left|\frac{L/C}{\frac{L}{\sqrt{LC}} - \frac{\sqrt{LC}}{C}}\right| = \left|\frac{L/C}{\sqrt{L/C} - \sqrt{L/C}}\right| = \infty \tag{3.117}$$

The following example applies this result to a circuit containing an R, L, and C.

Transfer function for an *R L C* circuit — EXAMPLE 3.28

The *RLC* filter in Figure 3.43 has $R = 1\,k\Omega$, $L = 1\,mH$, and $C = 1\,\mu F$. Determine $|H(\omega)|^2$ and $\phi(\omega)$.

We find the answer using the following three steps:

1. **Voltage divider formula:** The voltage divider formula for the circuit yields the ratio of $V_{OUT}(\omega)$ to $V_{IN}(\omega)$

$$H(\omega) = \frac{V_{OUT}(\omega)}{V_{IN}(\omega)} = \frac{Z_{L||C}}{R + Z_{L||C}}$$

The impedance of the *LC* parallel circuit is given by the product over sum formula

$$Z_{L||C} = \frac{Z_L Z_C}{Z_L + Z_C} = \frac{(j\omega L)(\frac{1}{j\omega C})}{j\omega L + \frac{1}{j\omega C}} = \frac{L/C}{j\omega L + \frac{1}{j\omega C}}$$

Multiplying the numerator and denominator by $j\omega C$ and recalling $j^2 = -1$ gives

$$Z_{L||C} = \frac{j\omega L}{1 - \omega^2 LC}$$

The frequency transfer function is then

$$H(\omega) = \frac{Z_{L||C}}{R + Z_{L||C}} = \frac{j\frac{\omega L}{1 - \omega^2 LC}}{R + j\frac{\omega L}{1 - \omega^2 LC}}$$

Dividing numerator and denominator by the numerator gives

$$H(\omega) = \frac{1}{1 + \frac{R(1 - \omega^2 LC)}{j\omega L}} = \frac{1}{1 - j(R/(\omega L) - \omega RC)}$$

Inserting the break frequencies $\omega_{RL} = R/L$ and $\omega_{RC} = 1/(RC)$ gives the final frequency transfer function result as

$$H(\omega) = \frac{1}{1 - j(\omega_{RL}/\omega - \omega/\omega_{RC})}$$

2. **Magnitude response:** We obtain the square of the magnitude response by multiplying the frequency transfer function by its complex conjugate to give

$$|H(\omega)|^2 = \left(\frac{1}{1 - j(\omega_{RL}/\omega - \omega/\omega_{RC})}\right)\left(\frac{1}{1 + j(\omega_{RL}/\omega - \omega/\omega_{RC})}\right)$$

Multiplying the denominator terms gives the desired frequency transfer function as

$$|H(\omega)|^2 = \frac{1}{1 + (\omega_{RL}/\omega - \omega/\omega_{RC})^2}$$

At $\omega = 0$, $|H(0)|^2 = 0$, because $\omega_{RL}/\omega = \infty$. The circuit shows this occurs because $|Z_L| = \omega L = 0$ at $\omega = 0$. Thus, the inductor acts as a short-circuit at DC.

At $\omega = \infty$, $|H(\infty)|^2 = 0$, because $\omega/\omega_{RC} = \infty$. This occurs because $|Z_C| = 1/(\omega C) = 0$ at $\omega = \infty$. Thus, the capacitor acts as a short-circuit at high frequencies.

At resonant frequency $\omega = \omega_o = (LC)^{-1/2}$, substituting $\omega_{RL} = R/L$ and $\omega_{RC} = 1/(RC)$ gives

$$|H(\omega_o)|^2 = \frac{1}{1 + (R\sqrt{C/L} - R\sqrt{C/L})^2} = 1$$

This occurs because Eq. 3.117 showed $|Z_{L||C}(\omega_o)| = \infty$.

3. **Phase response:** To compute ϕ we need the real and imaginary parts of $H(\omega)$. Multiplying the numerator and denominator of $H(\omega)$ by the complex conjugate of the denominator gives

$$H(\omega) = \left(\frac{1}{1 - j(\omega_{RL}/\omega - \omega/\omega_{RC})}\right)\left(\frac{1 + j(\omega_{RL}/\omega - \omega/\omega_{RC})}{1 + j(\omega_{RL}/\omega - \omega/\omega_{RC})}\right)$$

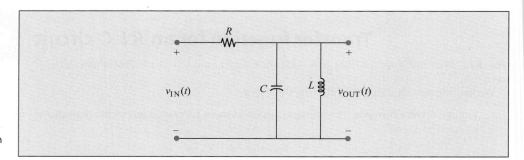

Figure 3.43

Circuit containing R in series with parallel connection of L and C.

Separating the real and imaginary parts gives

$$H(\omega) = \overbrace{\frac{1}{1 + (\omega_{RL}/\omega - \omega/\omega_{RC})^2}}^{\text{Real } H(\omega)} + j \overbrace{\frac{\omega_{RL}/\omega - \omega/\omega_{RC}}{1 + (\omega_{RL}/\omega - \omega/\omega_{RC})^2}}^{\text{Imag } H(\omega)}$$

The phase equals the arctangent of the ratio of the imaginary to real parts. Canceling the common denominator term gives the desired phase response

$$\phi(\omega) = \arctan(\omega_{RL}/\omega - \omega/\omega_{RC})$$

At $\omega = 0$, $\phi(0) = \arctan(\infty) = \pi/2$, because $Z_{L\|C}$ acts like an inductor when the capacitor is an open-circuit at DC.

At $\omega = \infty$, $\phi(\infty) = \arctan(-\infty) = -\pi/2$, because $Z_{L\|C}$ acts like an capacitor when the inductor is an open-circuit at high frequencies.

At $\omega = \omega_o = (LC)^{-1/2}$ (the resonant frequency), inserting $\omega_{RL} = R/L$ and $\omega_{RC} = 1/(RC)$ gives

$$\phi(\omega_o) = \arctan\left((R/L)\sqrt{LC} - RC/\sqrt{LC}\right) = \arctan\left(R\sqrt{LC} - R\sqrt{LC}\right) = 0$$

3.9 Research Challenges

3.9.1 New Materials—Graphene

Graphene is a layer of interconnected carbon atoms that is single atomic layer thick that offers novel electrical properties. Graphene can fold in upon itself to form Buckey balls, cylinders (nanotubes), or be used in sheet form. The electrical properties of graphene are being investigated for use as digital memories and ultracapacitors.

3.9.2 New Components— Memristors

Some describe the memristor as the missing component that completes the passive component set R, L, and C.

The memristor was discovered in 1971 by Leon Chua, who proposed that it relates electric charge to the magnetic field. Its curious property is that the resistance increases with current flow in one direction and decreases with opposite current flow, and when the current stops, the memristor *remembers* the last resistance value it had. This property makes it a candidate for high-density, nonvolatile memories or for those that do not forget when the voltage supply is removed.

3.10 Summary

This chapter analyzed basic circuits implemented using resistors, capacitors, and inductors. Purely resistive circuits are the simplest to analyze and are used in implementing voltage dividers and understanding the behavior of digital circuits. Circuits containing capacitors and inductors are more complicated to analyze because their impedance are complex-valued to describe amplitude and phase variations with frequency. This variation is useful for implementing filters and understanding the frequency limitations of data transmission channels.

3.11 Problems

3.1 Hand dryer power. An electric hand dryer consumes 1 kW. Models in the US operate on 110 V and those in Europe on 220 V. The voltages are AC (alternating current) values, but they *effectively* produce the same power as if they were DC values. What is the current the hand dryer uses in the United States? What is the current it uses in Europe?

3.2 Resistance value of an incandescent light bulb. Incandescent light bulbs use tungsten filaments made of wires with circular cross section. The resistivity of tungsten is $\rho = 5.6 \times 10^{-8}$ Ω-m at room temperature. If the wire diameter is $d = 0.1$ mm, what is the length ℓ of a tungsten filament that has a resistance of 10 Ω at room temperature?

3.3 Resistance value of a cold 100 W incandescent light bulb. An incandescent light bulb consumes 100 W when connected to a 100 V supply. The resistivity of tungsten becomes 15 times greater when it is glowing hot than the value at room temperature. What is the tungsten filament resistance of a 100 W bulb at room temperature? What is the initial value of current that the cold bulb draws when it is turned on?

3.4 Potentiometer value. Your radio volume is controlled by a $10\,k\Omega$ potentiometer that rotates over a 270° angle. Fully counterclockwise rotation defines 0°, making $R_{2-3} = 0$. You turn the potentiometer clockwise to 45°. What are the resistance values of R_{2-3}, R_{1-2}, and R_{1-3}?

3.5 Capacitance value. Two aluminum foil plates having an area $A = 40$ cm^2 are separated by a thin insulator plastic film of thickness $d = 0.01$ mm. If the plastic has value $\epsilon = 10^{-11}$ F/m, what is the capacitance value of the structure? What A is needed to form a 10 μF capacitor?

3.6 Inductor value. A coil located below the road surface senses the presence of an automobile that changes the permeability value when it is situated over the coil. Let the coil have $N = 100$ turns, area $A = \pi$ m^2, and length $\ell = 10$ cm. What is the coil inductance in air? If the presence of a car increases the effective permeability by a factor of 100, what is the inductance value then?

3.7 Electric car battery. The battery in an electric vehicle is made up by connecting many smaller batteries, similar to those used in a cell phone. A car travelling at 60 mph uses 12 kW of power. Assuming an LIB energy density of 200 W-hr/kg, what is the mass of a LIB that allows a 120 mile trip in two hours?

3.8 Capacitor as a power source. A capacitor technology implements an ultracapacitor that can sustain a voltage of 20 V without breaking down. What capacitance value when fully charged supplies a continuous current of 1 A and still has $V_C \geq 5$ V after one hour?

3.9 Ohm's law—AA batteries. Four AA batteries, each 1.5 V, are connected in series to supply voltage to a lamp circuit. The lamp has a resistance $R = 12\,\Omega$. What is the current flowing through the lamp?

3.10 Ohm's law—Battery types. Although AA batteries are nominally 1.5 V, the chemical process within the battery determines the actual voltage. For example, the terminal voltage on an alkaline battery is 1.6 V, while that of a nickel-metal hydride (NiMH) battery is 1.4 V, and nickel-cadmium (NiCd) measure 1.2 V. Your electric toothbrush takes 2 AA batteries. Which battery type would you use in your toothbrush and why?

3.11 Ohm's Law—Automobile starter motor. Most conventional gasoline-powered automobiles are equipped with a 12 V (nominal) battery. The starter motor requires an initial current of at least 100 A to turn the engine. What is the maximum resistance of the starter motor for proper operation?

3.12 Hair dryer power. Assume your hair dryer consumes 2,000 W of power. A switch adjusts the resistance to allow its use either in the US (110 V) or in Europe (220 V). The voltages are AC (alternating current) values, but they *effectively* produce the same power as if they were DC values.

a. You are in the US and mistakenly switched it to the 220 V setting. How much power is your hair dryer producing?

b. You are in Europe and mistakenly switched it to the 110 V setting. How much power is your hair dryer producing?

c. In terms of the reliability (or warranty) of your hair dryer, which of the two mistakes is more dire, and why?

3.13 Kirchoff's Current Law—LED display. A 5 V supply provides up to 30 A to drive a color light-emitting diode (LED) display that has 480 rows and 640 columns, corresponding to 307, 200 picture elements (*pixels*). Each pixel contains a red, green, and blue LED. When the display shows all white, all the LEDs have maximum current. Assuming that all LEDs are electrically identical, what current I_{LED} flows through each LED?

3.14 Kirchoff's Voltage Law—Holiday lamps. A string of holiday lamps connects to a 100 V supply. If each lamp requires 2.2 V to operate, how many lamps can be on a string if they are all connected in series, as shown in Figure 3.18 with $V_i = V_2 = \cdots = V_n$?

3.15 Series connection of four resistors. Find the equivalent resistance of the serial connection of resistors R_1, R_2, R_3 and R_4 by applying KVL and Ohm's law.

3.16 Voltage divider. Compute the output voltage of the circuit shown in Figure 3.44.

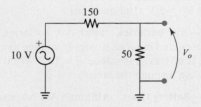

Figure 3.44

Voltage divider circuit. Resistance values are in Ω.

3.17 Parallel connection of four resistors. Find the equivalent resistance of the parallel connection of resistors R_1, R_2, R_3 and R_4, by applying KCL and Ohm's law. Verify the result by letting all resistors have the same value R.

3.18 Thevenin equivalent circuit. Compute and draw the TEC of the circuit in Figure 3.45 at the output terminals on the left.

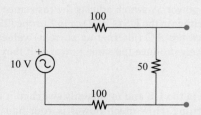

Figure 3.45

Circuit with output terminals on left. Resistance values are in Ω.

3.19 Thevenin equivalent circuit of a new battery. A 12 V car battery is advertised to provide 650 A of cranking current. Assuming that this value is the short-circuit current, what is the Thevenin resistance of this battery?

3.20 Thevenin equivalent circuit of an old battery. A 12 V battery degrades by increasing its R_{TH}. To test the battery, apply a load resistor $R_L = 12\,\Omega$ across the terminals and measure the voltage drop from the open-circuit value $V_{OC} = 12$ V to $V_L = 11.5$ V. Find R_{TH}.

3.21 Impedance diagram of R-C circuit. Compute and plot Z_R, Z_C, and $Z_{R\text{-}C}$ of an R-C circuit with $R = 10\,k\Omega$ and $C = 1\,\mu F$ at $f = f_x/2$, f_x, and $2f_x$.

3.22 R value for cross-over frequency of R-C circuit. Compute the value of R that produces $f_x = 1$ kHz when $C = 0.01\,\mu F$.

3.23 Impedance diagram of $R\|C$ circuit. Compute and plot $\mathcal{R}e\,Z_{R\|C}$, $\mathcal{I}m\,Z_{R\|C}$, and $Z_{R\|C}$ of an $R\|C$ circuit with $R = 10\,k\Omega$ and $C = 1\,\mu F$ at $f = f_x/2$, f_x, and $2f_x$.

3.24 Cross-over frequency of $R\|C$ circuit. Compute the value of R for $f_x = 1$ kHz when $C = 0.01\,\mu F$ and sketch the impedance in the complex plane. Sketch the impedance magnitude as a function of frequency and indicate the values at $f = f_x/2$, f_x, and $2f_x$.

3.25 Voltage divider for an RC filter. The RC low-pass filter shown in Figure 3.46 has $R = 10\,k\Omega$ and $C = 1\,\mu F$. The input voltage is $v_i(t) = 2\sin(4, 000\pi t)$. Determine $v_o(t)$.

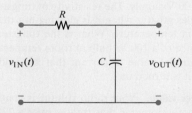

Figure 3.46

Voltage divider for an RC filter.

3.26 RC low-pass filter design. The RC low-pass filter shown in Figure 3.46 has $C = 1\,\mu F$. Determine the value of R that makes $|v_o| = |v_i/10|$ at $f_o = 10^4$ Hz .

3.27 Multistage RC low-pass filter design. One stage of an RC low-pass filter is shown in Figure 3.46 has cross-over frequency f_x, where $|H(f_x)| = 1/\sqrt{2}$. If n stages connect in series and act independently, what is n, the minimum number of RC stages that are connected in series so that $|H(f_x)|^n \leq 0.1$.

3.28 LC circuit. Compute the resonant frequency f_o of an LC series circuit when $L = 50\,\mu H$ and $C = 0.1\,\mu F$.

3.12 Matlab Projects

3.1 *R-C* **circuit.** Using Example 16.13 as a guide, compose a Matlab script to compute f_x and plot the frequency dependence of the impedance magnitude and phase for a series RC circuit with $R = 10\,k\Omega$ and $C = 2\,\mu F$ for $f_x/10 \leq f \leq 5f_x$.

3.2 **Voltage and current in a series** *R-C* **circuit.** Using Example 16.14 as a guide, compose a Matlab script that computes the amplitude and phase relationship between the voltage and current in a series RC impedance with $R = 10\,k\Omega$ and $C = 2\,\mu F$ at $f = f_x$.

3.3 **Parallel** *RC* **circuit.** Using Example 16.15 as a guide, compose a Matlab script that inputs the values of R and C and plots the impedance magnitude and phase for $f_x/10 \leq f \leq 5f_x$.

3.4 *R$\|$C* **circuits.** Using Example 16.15 as a guide, compose a Matlab script to compute and plot the impedance magnitude and phase for the parallel RC circuit with $R = 10\,k\Omega$ and $C = 2\,\mu F$ for $f_x/10 \leq f \leq 5f_x$.

3.5 **Low-pass** *RC* **filter.** Using Example 16.16 as a guide, compose a Matlab script to design an $R - C$ low-pass filter using $C = 0.1\,\mu F$ to produce $f_x = 10\,kHz$. Compute and plot the frequency dependence of RC filter for $f_x/10 \leq f \leq 5f_x$.

3.6 **Multistage** *RC* **low-pass filter design.** Using Example 16.17 as a guide, compose a Matlab script to design an n-stage RC low-pass filter with $C = 1\,\mu F$ to produce $f_x = 10\,kHz$. Plot the frequency response magnitude comparison as the number of stages increases until you observe the n value that has $|H_{LP}(f_x)|^n < 0.4$.

3.7 **LRC band-pass filter design.** Using Example 16.18 as a guide, compose a Matlab script to plot the magnitude response for $f_o/10 \leq f \leq 10f_o$ of the RLC filter given in Example 3.28.

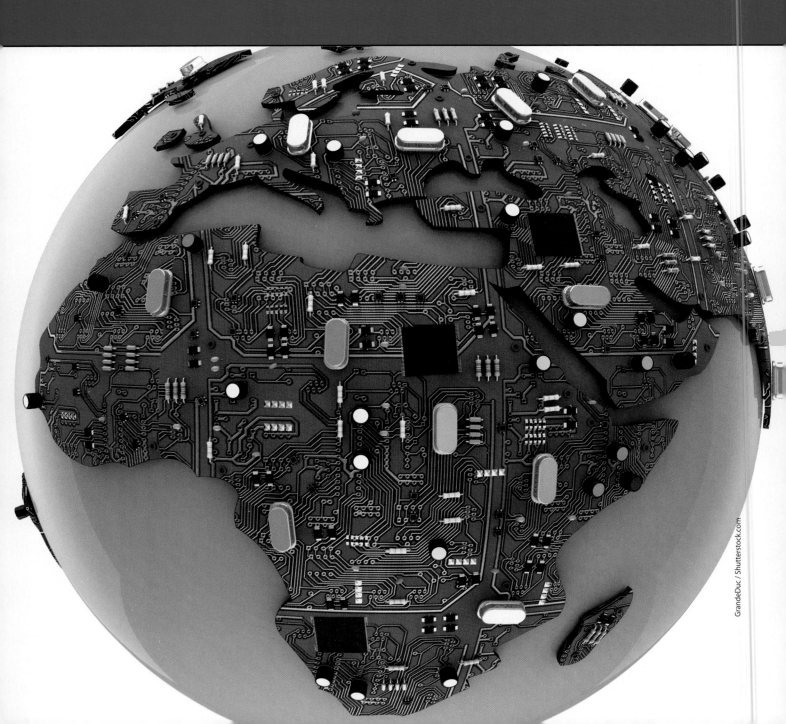

ELECTRONICS

LEARNING OBJECTIVES

After completing this chapter, the reader should be able to:

■ Understand operation of operational amplifiers in several useful circuits.

■ Design low-pass filters for modeling data transmission channels and band-pass filters for modeling communication channels.

■ Understand how field-effect transistors act as digital switches.

■ Use diodes for rectification and conversion of electrical signals into light.

Electronic circuits transform electrical waveforms using both *passive* electrical components, such as resistors and capacitors, and *active* devices, such as transistors and amplifiers, that require a power supply. Computers and smartphones contain active circuits to amplify signals to drive actuators and filter waveforms to reduce noise. The most versatile electronic device is the *operational amplifier* (op amp), which is an integral component in circuits that amplify and filter waveforms, set thresholds, and sum waveforms. Negative-feedback amplifiers convert op amps with variable gains into a stable amplifier with a gain that is determined by the ratio of resistors. Adding capacitors to an op-amp circuit produces a *filter* whose gain changes with frequency. Such filters reduce noise in signals and limit their frequency content for conditioning waveforms to accommodate communication channels and for preventing aliasing in analog-to-digital conversion. Field-effect transistors act as voltage-controlled switches that implement digital electronics. This chapter also describes diodes that convert current into light power and sensors that convert light power back to electrical current.

4.1 INTRODUCTION

This chapter provides a general introduction to electronic circuits by describing the following sequence of topics.

Operational amplifiers—The op amp is a very versatile circuit for transforming analog waveforms, and is simple to analyze.

Field effect transistors—An FET is a simple semiconductor device that forms an electrically activated switch that is the basis of digital logic circuits.

Diodes—Semiconductor physics has developed interesting devices called diodes that let current flow in only one direction and convert current into light power to form displays and efficient sources of illumination.

4.2 OPERATIONAL AMPLIFIER

The operational amplifier (op amp) is one of the simplest and most useful devices for designing electronic systems. The op amp is powered by two connections to voltage supplies: the positive supply voltage V_{CC} and the negative supply voltage V_{EE}. When there is only one supply—as in the case of a single battery—$V_{EE} = 0$ V, which is also called *ground (GRD)*. These ever-present voltage connections are typically not shown in circuits to clarify the operations on the signals. Figure 4.1 shows an op amp having three signal connections:

V^- is the inverting input.

V^+ is the non-inverting input.

V_{OUT} is the output.

Figure 4.1

Operational amplifier (op amp). (a) Circuit diagram. (b) Connections on a integrated circuit (IC) chip. (c) Image of an 8-pin IC package. Image grid has 1 mm spacing.

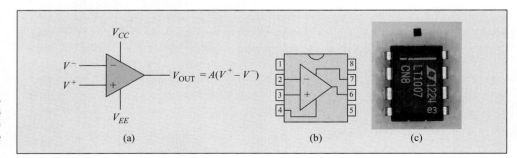

The op amp amplifies the voltage difference ($V^+ - V^-$) by the op amp *gain A* to produce the output voltage:

$$V_{OUT} = A(V^+ - V^-) \tag{4.1}$$

The gain A is typically very large but can vary between one thousand to one million, depending on the manufacturing conditions producing that particular integrated circuit device. A clever design procedure called *negative feedback* converts this large variation into an amplifier with a smaller, but more precise, gain.

The next sections describe op amp applications that are useful for both analog and digital systems.

4.2.1 Comparator

A voltage comparator, also known as a *threshold-detector*, is a circuit that compares input voltage V_{IN} with a voltage V_τ, called a *threshold*. Figure 4.2 shows that this comparison produces two easily distinguishable output values that indicate when $V_{IN} < V_\tau$ and when $V_{IN} \geq V_\tau$. Such a circuit is commonly found in data receivers and at the input of digital logic circuits that operate with small or slowly varying sensor voltages. Figure 4.3 shows comparator outputs for restoring noise-corrupted digital signals and conditioning sensor waveforms as input for digital circuits.

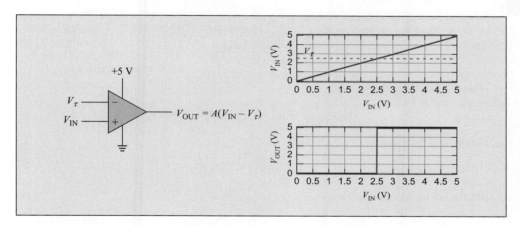

Figure 4.2

Op amp comparator circuit with $V_{CC} = 5$ V and $V_{EE} = 0$ V.

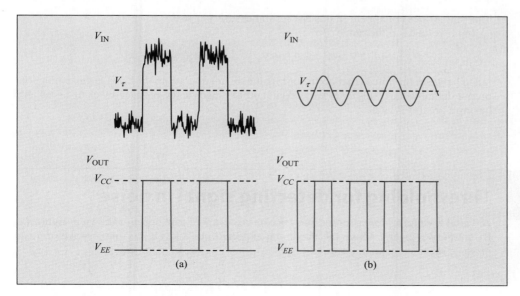

Figure 4.3

Op-amp comparator outputs for (a) digital signal restoration and (b) sensor signal conditioning for connecting to digital circuits.

To perform threshold detection, the input voltage V_{IN} connects to the non-inverting (+) input, while a threshold voltage V_τ is applied to the non-inverting (−) input. The output voltage V_{OUT} multiplies the voltage difference $(V_{IN} - V_\tau)$ by a large gain A for

$$V_{OUT} = A(V^+ - V^-) = A(V_{IN} - V_\tau) \qquad (4.2)$$

The V_{OUT} value is limited by the op-amp supply voltage values V_{CC} and V_{EE}. When powered by a single power supply such as $V_{CC} = +5$ V and $V_{EE} = 0$ V, V_{OUT} is limited to the range from 0 V to +5 V. The large op-amp gain A quickly drives V_{OUT} between these two extreme values

$$V_{IN} - V_\tau \geq \frac{V_{CC}}{A} \rightarrow V_{OUT} = V_{CC} \qquad (4.3)$$

$$V_{IN} - V_\tau < \frac{V_{EE}}{A} \rightarrow V_{OUT} = V_{EE}$$

A large A causes the threshold detector to produce either V_{CC} or V_{EE} with $V_{CC} = 5$ V and $V_{EE} = 0$ V commonly associated with digital circuits.

EXAMPLE 4.1

Threshold detection

A comparator is implemented using an op amp with gain $A = 10^3$, a single 5 V supply, and $V_\tau = 2$ V. Then the op amp produces $V_{OUT} = 5$ V when

$$V_{IN} - V_\tau \geq \frac{V_{CC}}{A}$$

This occurs for V_{IN} values, as

$$V_{IN} \geq \frac{V_{CC}}{A} + V_\tau = \frac{5}{10^3} + 2\,V = 2.005\,V$$

The op amp produces $V_{OUT} = 0$ V when

$$V_{IN} - V_\tau < \frac{V_{EE}}{A} = 0$$

This occurs for V_{IN} values, as

$$V_{IN} < V_\tau = 2\,V$$

The gain A does not appear in this last result because $V_{EE} = 0$.
Input voltages in the range

$$2\,V < V_{IN} < 2.005\,V \rightarrow 0 < V_{OUT} < 5\,V$$

fall in between the values that occur for reliable digital operation: thus, producing ambiguous output digital values. This *ambiguous* digital input range can be reduced by using an op amp having a greater A value.

EXAMPLE 4.2

Thresholding for detecting signal in noise

A digital waveform from a sensor often becomes corrupted with noise, as shown in Figure 4.4. By setting V_τ slightly above the maximum expected noise level, the sensor waveform can produce reliable digital indications.

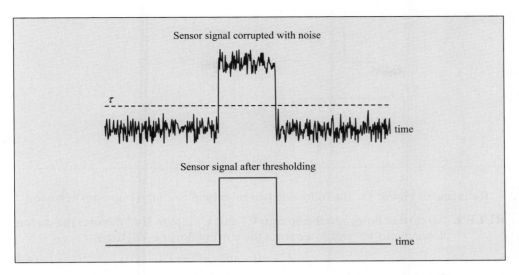

Figure 4.4

Setting τ slightly above the maximum noise level allows detecting a signal in noise.

Comparator for digital data waveform restoration

EXAMPLE 4.3

A digital data waveform often becomes corrupted with noise during transmission. Figure 4.5 shows the data values in black are corrupted with noise to produce blue waveform. By setting V_τ equal to the midpoint between the two expected voltage levels, the original digital waveform values can be restored perfectly when the noise is not too large, as shown on the left side of the figure. When the noise becomes too large, errors occur, as shown in red in the right side of the figure.

4.2.2 Feedback Amplifier

The op-amp gain A can vary widely between devices. A stable amplifier is designed to have gain $G \ll A$ for a smaller but more reliable value that is insensitive to variations in A through the use of *negative feedback*, which is one of the great ideas in electrical engineering. By *feeding back* a portion of V_{OUT} to the V^- input, the op-amp gain becomes insensitive to variations in the op-amp's internal components.

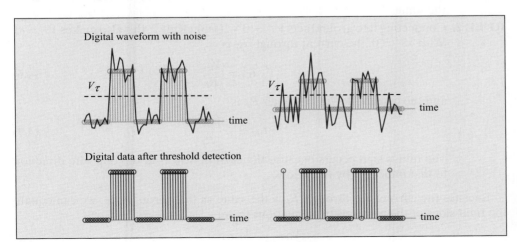

Figure 4.5

Restoring digital data waveform with op-amp comparator. Errors are shown in red.

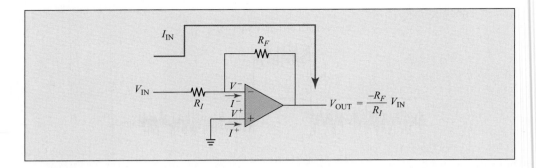

Figure 4.6

Op amp employing negative feedback.

Referring to Figure 4.6, the following two rules help to analyze op-amp behavior:

RULE 1. No current flows into the op amp V^+ and V^- inputs. If I^+ denotes the current flowing into V^+ and I^- denotes the current flowing V^-, then

$$I^+ = 0 \quad \text{and} \quad I^- = 0 \tag{4.4}$$

This rule occurs because the op amp is designed so that the inputs V^+ and V^- have infinite (very, very large) resistance values.

RULE 2. The voltage difference under normal operation between the op-amp inputs is effectively zero, or

$$V^+ - V^- \approx 0 \quad \rightarrow V^+ \approx V^- \tag{4.5}$$

This zero voltage difference times the nearly infinite gain A in Eq. (4.1) produces the finite value of the V_{OUT} voltage.

These two rules allow a simple analysis of an op-amp negative-feedback amplifier.

Figure 4.6 shows the following signal connections that implement the op-amp feedback amplifier.

- The input voltage V_{IN} (for example, a microphone output) connects to the inverting input through input resistor R_I.

- The input current I_{IN} flows through input resistor R_I.

- The resistor R_F connects (or feeds back) part of the op-amp output voltage V_{OUT} to the inverting input.

- The non-inverting input connects to ground, making $V^+ = 0$.

Now apply the two rules to analyze the feedback amplifier.

RULE 1. Interpreted as $I^- = 0$. Hence, the current I_{IN} flowing through R_I and R_F is the same.

RULE 2. Connecting it to ground sets $V^+ = 0$ V. Using Rule 2, this also makes $V^- = 0$. With $V^- = 0$, the current through R_I is

$$I_{\text{IN}} = \frac{V_{\text{IN}}}{R_I} \tag{4.6}$$

and the current flowing through R_F is

$$I_{\text{IN}} = -\frac{V_{\text{OUT}}}{R_F} \tag{4.7}$$

The minus sign occurs because the current flow is in the opposite direction to that indicated by the arrow.

Because the current I_{IN} through R_I is the same as that through R_F, we can equate the right sides of the previous two equations to get

$$\frac{V_{\text{IN}}}{R_I} = -\frac{V_{\text{OUT}}}{R_F} \tag{4.8}$$

Solving for the output voltage V_{OUT} we obtain the simple amplifier equation

$$V_{OUT} = -\frac{R_F}{R_I} V_{IN} \tag{4.9}$$

The minus sign means V_{OUT} is an inverted form of V_{IN} (that is, if $V_{IN} > 0$ then $V_{OUT} < 0$). The gain of the op-amp circuit G is defined as the ratio of the output amplitude to the input amplitude

$$G = \frac{V_{OUT}}{V_{IN}} = -\frac{R_F}{R_I} \tag{4.10}$$

Note that G is not a function of op-amp gain A, when $|A| \gg |G|$. This is an important consequence of the negative-feedback amplifier, stable gain: The gain A of the op amp itself is typically very large but can vary depending on manufacturing variables (often $10^3 < A < 10^6$). The negative-feedback amplifier design converts this large, but un-controlled variable gain into a stable gain $G = -R_F/R_I$ that is determined solely by the ratio of resistance values, which can be manufactured more accurately. (Typical R values are within 5% of the nominal values, and 1% accuracy is only a bit more expen-sive.) As long as the internal op-amp gain A is much greater than the designed gain G or $A > 10G$, the op-amp feedback amplifier provides the accurate gain G.

> **Factoid:** Negative feedback is one of the great ideas in electrical engineering. It trades off a stable, lower gain at the expense of a large, but uncontrolled, variable gain.

Op-amp feedback amplifier having $G = -10$ EXAMPLE 4.4

To implement an op-amp feedback amplifier with a gain $G = -10$, set

$$G = \frac{V_{OUT}}{V_{IN}} = -\frac{R_F}{R_I} = -10$$

A reasonable value for $R_I = 1\,k\Omega$, making $R_F = 10\,k\Omega$. Figure 4.7 shows the input and the resulting output sinusoidal waveforms. Note that the negative gain makes V_{OUT} go negative when V_{IN} goes positive.

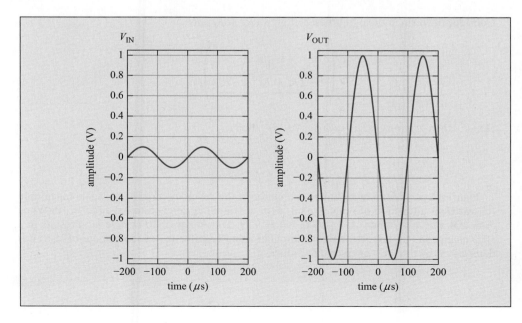

Figure 4.7

Input and output waveforms in an op-amp feedback amplifier.

EXAMPLE **4.5**

Stable *G* with variable *A*

We analyze the negative feedback amplifier by starting with the op-amp gain equation

$$V_{OUT} = A(V^+ - V^-)$$

By its connection to ground, $V^+ = 0$, making

$$V_{OUT} = -AV^-$$

Applying the voltage divider rule, the voltage V^- is

$$V^- = \frac{R_F}{R_I + R_F} V_{IN} + \frac{R_I}{R_I + R_F} V_{OUT}$$

Substituting $V^- = -V_{OUT}/A$, multiplying through by $-A$

$$V_{OUT} = -\frac{A R_F}{R_I + R_F} V_{IN} - \frac{A R_I}{R_I + R_F} V_{OUT}$$

Putting the terms containing V_{OUT} on the left side of the equation yields

$$V_{OUT} + \frac{A R_I}{R_I + R_F} V_{OUT} = -\frac{A R_F}{R_I + R_F} V_{IN}$$

Multiplying by $(R_I + R_F)$

$$V_{OUT}(R_I + R_F) + A R_I V_{OUT} = -A R_F V_{IN}$$

Solving for $G = V_{OUT}/V_{IN}$, we get

$$G = \frac{V_{OUT}}{V_{IN}} = -\frac{A R_F}{(R_I + R_F) + A R_I}$$

Dividing numerator and denominator by $A R_1$, we get

$$G = \frac{V_{OUT}}{V_{IN}} = -\frac{R_F}{R_I} \left(\frac{1}{1 + (R_I + R_F)/(A R_I)} \right)$$

Consider the term in the parentheses. If $A \gg (R_F + R_I)/R_I$

$$\left(\frac{1}{1 + \underbrace{(R_I + R_F)/(A R_I)}_{\to 0}} \right) \to 1$$

Hence, when A is very large, such as $A > 10^3$, then

$$G = -\frac{R_F}{R_I}$$

Figure 4.8 provides a quantitative evaluation of this result by showing how gain magnitude $|G|$ varies as a function of op-amp gain A for an amplifier deigned to have $|G| = 10$. When $A = 100$, $|G| = 9$. When $A = 1,000$, $|G| = 9.99$. This factor of 10 change in A results in a 10% change in $|G|$. For $A > 10^3$, the amplifier gain G is effectively independent of A, as it is determined by the ratio of resistance values.

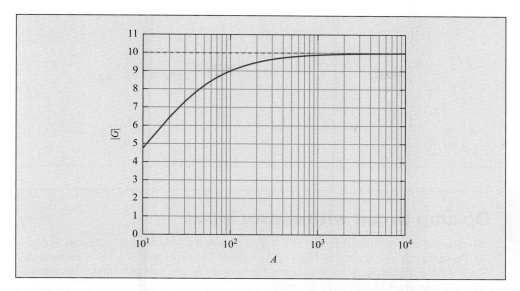

Figure 4.8

Variation in $|G|$ as A varies.

Varying *G* with a potentiometer   EXAMPLE 4.6

The gain of a negative feedback amplifier G can be varied using an adjustable resistor, called a *potentiometer*, in place of a fixed-value for R_F.

Figure 4.9 shows the schematic of potentiometer and its use in the feedback path of the op-amp negative-feedback amplifier. The potentiometer has three terminals with the resistance between terminals 1 and 3 (denoted R_{1-3}) being a fixed resistance, such as $10\,k\Omega$. Terminal 2 contacts the surface of this resistor at a point that can varied mechanically, such as twisting a knob or sliding a lever. Varying the location of the contact varies the resistance between terminals 1 and 2 (R_{1-2}) and between terminals 2 and 3 (R_{2-3}). A circuit connects to these terminals so that the perceptual effect increases (that is, the volume get louder or a display becomes brighter) when the knob is turned clockwise or a lever is shifted upward. For example, the knob you turn to increase the volume on your radio connects to a potentiometer, and you turn the knob clockwise to increase the volume. Potentiometers are often used to set the gain of circuits within devices and are typically made inaccessible and sealed to make the setting permanent.

The figure shows the potentiometer adjusting the gain of an operational amplifier. By connecting terminals 1 and 2 to the op amp and leaving terminal 3 unconnected, the value of the feedback resistor $R_F = R_{1-2}$. When the knob is turned fully counterclockwise, terminal 2 connects to terminal 1 making $R_{1-2} = 0$. Turning the knob clockwise increases R_{1-2} and op-amp gain magnitude $|G| = R_F/R_I$ until the knob is turned fully, making $R_{1-2} = R_{1-3}$ and producing the maximum gain

$$|G|_{\max} = \frac{R_{1-3}}{R_I}$$

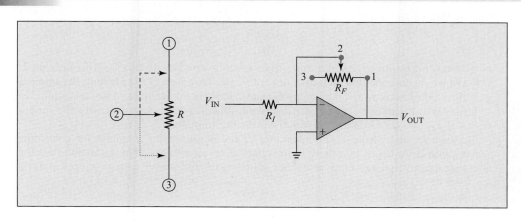

Figure 4.9

A potentiometer is a device whose resistance between terminals is varied by mechanically adjusting the position where terminal 2 contacts the resistor surface. The gain of a negative feedback amplifier can be adjusted using a potentiometer.

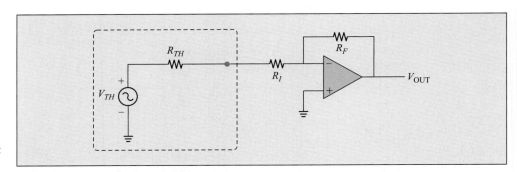

Figure 4.10

Thevenin equivalent circuit of input source affects op amp gain.

EXAMPLE **4.7**

Op amp circuit with sensor input

The effect of connecting an op amp to a sensor is computed most easily by first determining the Thevenin equivalent circuit (TEC) of the sensor. The resistance of a TEC can be seen in Figure 4.10 to connect in series to the amplifier resistor R_I, and this then affects the gain on a non-inverting negative feedback op amp by

$$G = -\frac{V_{OUT}}{V_{TH}} = -\frac{R_F}{R_{TH} + R_I}$$

Some sensors have large R_{TH}, causing the gain to have a smaller magnitude than that calculated from R_F/R_I.

EXAMPLE **4.8**

Op amp feedback amplifier connecting to non-inverting input

An input signal can be applied to the V^+ input of an op amp, as shown in Figure 4.11. Because negative feedback equalizes the input terminal voltages

$$V^- = V^+ = V_{IN}$$

The current I_I flowing in the direction shown through R_I equals

$$I_1 = -\frac{V^-}{R_I} = -\frac{V_{IN}}{R_I}$$

With no current flowing into the V^- input, $I^- = 0$, the same current (I_I) flows through R_F

$$I_1 = -\frac{V_{OUT} - V_{IN}}{R_F} = -\frac{V_{OUT}}{R_F} + \frac{V_{IN}}{R_F}$$

Equating the previous two equations for I_I and solving for V_{OUT} gives

$$V_{OUT} = V_{IN}\left(1 + \frac{R_F}{R_I}\right)$$

The amplifier gain is then

$$G = \frac{V_{OUT}}{V_{IN}} = 1 + \frac{R_F}{R_I}$$

Note that G is a positive value, indicating V_{OUT} is not inverted relative to V_{IN}. This additional 1 in the gain is inconsequential for large values of R_F/R_I. However, $G > 1$ prevents the amplifier gain from being zero, which may be problematic in some applications. Also, this additional 1 term makes frequency-dependent gains more difficult to compute than with the simpler $G = -R_F/R_I = -Z_F/Z_I$ obtained with the inverting amplifier.

Because $I^+ = 0$ (Rule 1) the advantage of the non-inverting amplifier for connecting sensors with large R_{TH} is that there is no voltage drop across R_{TH}. That is, the non-inverting amplifier has a very large input resistance, allowing the full sensor voltage to appear at the amplifier input.

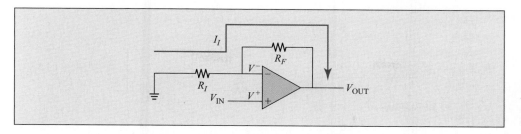

Figure 4.11

Op amp gain when input connects to non-inverting input.

4.2.3 Summing Amplifier

An op amp is easily configured to add two signals, as shown in Figure 4.12. Input voltage V_1 connects through R_1 to the inverting op-amp input, and V_2 connects though R_2. Because the non-inverting op-amp input connects to ground $V^+ = 0$, the inverting input voltage $V^- = 0$ by the negative feedback. Hence, the current flowing through R_1 is

$$I_1 = \frac{V_1}{R_1} \tag{4.11}$$

and that through R_2 is

$$I_2 = \frac{V_2}{R_2} \tag{4.12}$$

The current flowing through R_F in the direction shown is

$$I_F = -\frac{V_{OUT}}{R_F} \tag{4.13}$$

Because $I^- = 0$, no current flows into the op-amp input, making

$$I_F = I_1 + I_2 \tag{4.14}$$

Substituting these current values, we get

$$-\frac{V_{OUT}}{R_F} = \frac{V_1}{R_1} + \frac{V_2}{R_2} \tag{4.15}$$

Solving for V_{OUT}, we get

$$V_{OUT} = -\left(\frac{R_F}{R_1} V_1 + \frac{R_F}{R_2} V_2 \right) \tag{4.16}$$

demonstrating that the output voltage equals the (negative or inverted) sum of the input voltages. The following examples illustrate two useful applications of the summing op-amp circuit to implement functions discussed in later chapters.

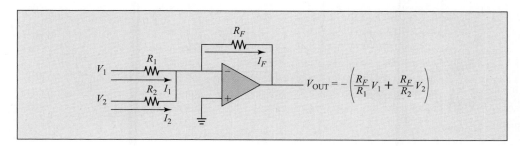

Figure 4.12

The op amp designed as a summing circuit.

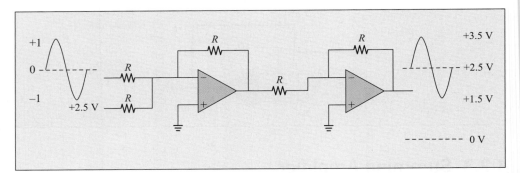

Figure 4.13

The op-amp summing circuit adds an offset to make a bipolar waveform strictly positive.

EXAMPLE 4.9

Adding an offset to a zero-mean waveform for ADC

Figure 4.13 shows a op-amp summing circuit that is important for an analog-to-digital converter (ADC), which is discussed in Chapter 7. Analog waveforms that have both positive and negative voltages with a zero average are typically converted to positive-only voltage waveforms before being applied to an ADC. Many ADCs have an input range from $0\,V$ to $5\,V$, so it is desirable to shift the original waveform average of $0\,V$ to $2.5\,V$, by adding $2.5\,V$ to the analog waveform. Before this shift in average value, an amplifier having an appropriate gain limits the original analog waveform to lie within $\pm 2.5\,V$.

To compensate for the negative gain of the inverting negative feedback amplifier, a second amplifier with $G = -1$, by making $R_F = R_I$, connects to the output of the first to obtain an overall positive gain. Figure 4.13 shows this configuration by making all resistors equal to value R, thus providing $G = -1$ for both the first stage and the second stage to obtain an overall gain of $G = +1$.

EXAMPLE 4.10

Forming a digital-to-analog converter

Figure 4.14 shows an op-amp summing circuit that performs quantization using a staircase pattern and forms the digital-to-analog converter (DAC) discussed in Chapter 7. As in the previous example, two inverting op amps are connected in series to form a positive gain.

To illustrate how binary data can be transformed into analog voltage levels, let digital data voltages D_0 and D_1 be either $0\,V$ or $+1\,V$. Note that, unlike logic levels that can be 0 or 1, we set these digital signals to have voltages equal to $0\,V$ and $1\,V$. These inputs correspond to the binary representation of an analog signal amplitude.

D_0 connects to the inverting input through resistor $2R$, providing an overall gain G_{D_0} equal to

$$G_{D_0} = \overbrace{\left(-\frac{R_F}{R_I}\right)}^{\text{first stage}} \times \overbrace{\left(-\frac{R_F}{R_I}\right)}^{\text{second stage}} = \overbrace{\left(-\frac{R}{2R}\right)}^{=-0.5} \times \overbrace{\left(-\frac{R}{R}\right)}^{=-1} = 0.5$$

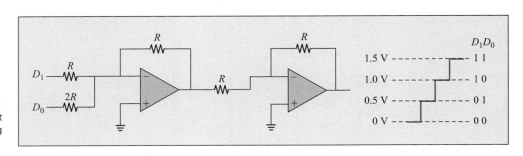

Figure 4.14

The op-amp summing circuit converts digital data into analog voltages.

Hence, $D_0 = 1\,\text{V}$ increases the output voltage by 0.5 V. This input corresponds to the least significant bit in a binary representation of an analog signal amplitude.

D_1 connects to the inverting input through resistor R, providing an overall gain G_{D_1} equal to

$$G_{D_1} = \overbrace{\left(-\frac{R_F}{R_I}\right)}^{\text{first stage}} \times \overbrace{\left(-\frac{R_F}{R_I}\right)}^{\text{second stage}} = \overbrace{\left(-\frac{R}{R}\right)}^{=-1} \times \overbrace{\left(-\frac{R}{R}\right)}^{=-1} = 1$$

Hence, $D_1 = 1\,\text{V}$ increases the output voltage by 1 V.

Then, the digital values convert to analog values:

$$D_1 D_0 = 00 \rightarrow V_{\text{OUT}} = 0\,\text{V}$$
$$D_1 D_0 = 01 \rightarrow V_{\text{OUT}} = 0.5\,\text{V}$$
$$D_1 D_0 = 10 \rightarrow V_{\text{OUT}} = 1.0\,\text{V}$$
$$D_1 D_0 = 11 \rightarrow V_{\text{OUT}} = 1.5\,\text{V}$$

Additional digital input signals can be added by reducing their input resistances by additional factors of 2 in the first stage. For example, input D_2 would connect through $R/2$ to increase the output voltage by 2 V. The supply voltages of the op amp need to accommodate the maximum value produced by the DAC.

4.2.4 Filters

A filter is a circuit whose gain varies with frequency. Filters extract the signal in the informational bandwidth by attenuating frequency components outside this bandwidth. For example, the antenna in a cell phone system transmits many simultaneous signals over a set of pre-defined bandwidths. A particular cell phone is assigned to a specific frequency band, and a filter within the cell phone extracts the waveforms in this frequency band while attenuating all other signals used by other channels.

An op-amp filter is designed by including frequency-selective impedances in the input and feedback paths. Figure 4.15 shows the impedance Z_I in the input path and Z_F in the feedback path, so the op-amp negative feedback amplifier gain is given by

$$G(\omega) = -\frac{Z_F(\omega)}{Z_I(\omega)} \tag{4.17}$$

where $\omega = 2\pi f$ is the frequency in units of *radians/second*. Using ω rather than f simplifies the results, because it is easier to write ω rather than $2\pi f$ and the dimensionless quantity (ωt) rather than $(2\pi f t)$.

If $Z_F(\omega)$ and $Z_I(\omega)$ are complex-valued impedances that vary with frequency, the gain also will be complex-valued and vary with frequency. The filter *frequency response* is then expressed in the form

$$G(\omega) = |G(\omega)| e^{j\angle G(\omega)} \tag{4.18}$$

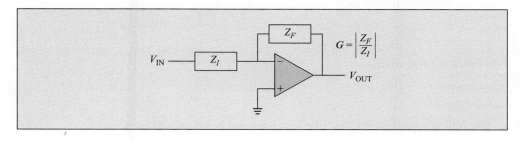

Figure 4.15

Op amp filter with frequency-selective feedback and/or input impedances.

where the *magnitude response* $|G(\omega)|$ is given by

$$|G(\omega)| = \frac{A_{OUT}(\omega)}{A_{IN}(\omega)} \qquad (4.19)$$

where $A_{OUT}(\omega)$ and $A_{IN}(\omega)$ are the amplitudes of the sinusoids at frequency ω at the output and input respectively.

The *phase response* $\angle G(\omega)$ equals the phase difference between the output and input sinusoids, as

$$\angle G(\omega) = \angle V_{OUT}(\omega) - \angle V_{IN}(\omega) \qquad (4.20)$$

The phase is computed from the arctangent of the ratio of the imaginary to the real parts of $G(\omega)$. Thus,

$$\angle G(\omega) = \arctan\left(\frac{\mathcal{I}m\, G(\omega)}{\mathcal{R}e\, G(\omega)}\right) \qquad (4.21)$$

where $\mathcal{I}m\, G(\omega)$ is the imaginary part of $G(\omega)$ and $\mathcal{R}e\, G(\omega)$ is the real part.

EXAMPLE 4.11 · Gain and phase response

Consider a filter with the input sinusoid at frequency $\omega = \omega_1 = 2\pi f_1$. The frequency values are

$$f_1 = 1\,\text{kHz} \rightarrow \omega_i = 2\pi \times 10^3 \text{ radians/second}$$

If

$$V_{IN}(\omega_1) = 2\sin(\omega_1 t)$$

produces output

$$V_{OUT}(\omega_1) = 10\sin\left(\omega_1 t - \frac{\pi}{5}\right)$$

the magnitude response at $\omega = \omega_1$ equals

$$|G(\omega_1)| = \frac{A_{OUT}(\omega_1)}{A_{IN}(\omega_1)} = \frac{10}{2} = 5$$

and the phase response equals

$$\angle G(\omega_1) = \angle V_{OUT}(\omega_1) - \angle V_{IN}(\omega_1) = -\frac{\pi}{5} - 0 = -\frac{\pi}{5}$$

The input sinusoid at higher frequency $\omega = \omega_2 = 10\omega_1$ ($f_2 = 10\,\text{kHz}$) is

$$V_{IN}(\omega_2) = 2\sin(\omega_2 t)$$

and produces output

$$V_{OUT}(\omega_2) = \sin\left(\omega_2 t - \frac{\pi}{3}\right)$$

The magnitude response at $\omega = \omega_2$ equals

$$|G(\omega_2)| = \frac{A_{OUT}(\omega_2)}{A_{IN}(\omega_2)} = \frac{1}{2} = 0.5$$

and the phase response equals

$$\angle G(\omega_2) = \angle V_{OUT}(\omega_2) - \angle V_{IN}(\omega_2) = -\frac{\pi}{3} - 0 = -\frac{\pi}{3}$$

Because $|G(\omega_1)| > |G(\omega_2)|$, this filter has a gain that decreases with increasing frequency and is called a *low-pass filter*.

The following examples describe low-pass and band-pass filter designs.

Op-amp low-pass filter   EXAMPLE 4.12

Consider the op-amp circuit shown in Figure 4.16. The filter gain is given by

$$G(\omega) = -\frac{-Z_F(\omega)}{Z_I(\omega)} = -\frac{-Z_F(\omega)}{R_I}$$

where the impedance of the $R\|C$ circuit is given by the product-over-sum rule

$$Z_{R\|C}(\omega) = \frac{R_F\left(\dfrac{-j}{\omega C_F}\right)}{R_F + \left(\dfrac{-j}{\omega C_F}\right)} = \frac{R_F}{1 + j\omega R_F C_F}$$

The magnitude of the $R\|C$ impedance is given by

$$|Z_{R\|C}(\omega)| = \frac{R_F}{\sqrt{1 + (\omega R_F C_F)^2}} \quad (= |Z_F(\omega)|)$$

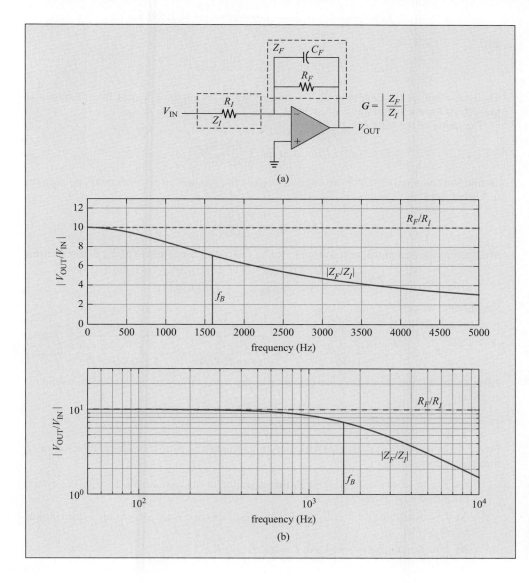

(a)

(b)

Figure 4.16

Op-amp low-pass filter with R in input path and $R\|C$ circuit in feedback path: (a) circuit, (b) frequency response.

The gain $G(\omega)$ is

$$G(\omega) = \frac{-Z_F(\omega)}{Z_I(\omega)} = \frac{-R_F/R_I}{1 + j\omega R_F C_F}$$

Note that $|G(0)| = R_F/R_I$ is modified by a term that varies with frequency.
The magnitude response is

$$|G(\omega)| = \frac{|Z_F(\omega)|}{|Z_I(\omega)|} = \frac{|Z_F(\omega)|}{R_I}$$

Define the *break frequency* as

$$\omega_B = \frac{1}{R_F C_F} \qquad \left(f_B = \frac{\omega_B}{2\pi} = \frac{1}{2\pi R_F C_F} \right)$$

Then write the gain in the following more simple form

$$|G(\omega)| = \frac{R_F/R_I}{\sqrt{1 + (\omega/\omega_B)^2}} \qquad \left(|G(f)| = \frac{R_F/R_I}{\sqrt{1 + (f/f_B)^2}} \right)$$

When $R_I = 1 \, k\Omega$, with $R_F = 10 \, k\Omega$, and $C_F = 0.01 \, \mu F$

$$f_B = \frac{1}{2\pi R_F C_F} = \frac{1}{2\pi (10^4)(10^{-8})} = \frac{10^4}{2\pi} = 1,592 \text{ Hz}$$

and

$$|G(f)| = \frac{10}{\sqrt{1 + 3.9 \times 10^{-7} f^2}}$$

The magnitude of $|G|$ decreases as f increases, which forms the low-pass filter. Figure 4.16 plots the $|G|$ using linear and log-log scale.
The phase of the filter is

$$\angle G(\omega) = \arctan \left(\frac{\mathcal{I}m \, G(\omega)}{\mathcal{R}e \, G(\omega)} \right)$$

To find $\mathcal{I}m \, G(\omega)$ and $\mathcal{R}e \, G(\omega)$, we express $G(\omega)$ as the sum of real and imaginary parts in the following steps

$$G(\omega) = \frac{R_F/R_I}{1 + j\omega R_F C_F}$$

Multiplying the numerator and denominator by the complex conjugate of the denominator

$$G(\omega) = \frac{R_F/R_I}{1 + j\omega R_F C_F} \times \frac{1 - j\omega R_F C_F}{1 - j\omega R_F C_F}$$

Simplifying, we get

$$G(\omega) = \frac{(R_F/R_I)(1 - j\omega R_F C_F)}{1 + (\omega R_F C_F)^2} = \overbrace{\frac{R_F/R_I}{1 + (\omega R_F C_F)^2}}^{\mathcal{R}e \, G} + j \overbrace{\frac{-(R_F/R_I)\omega R_F C_F}{1 + (\omega R_F C_F)^2}}^{\mathcal{I}m \, G}$$

After canceling the terms that are common in the numerator and denominator, we finally arrive at

$$\angle G(\omega) = \arctan \left(\frac{\mathcal{I}m \, G(\omega)}{\mathcal{R}e \, G(\omega)} \right) = \arctan \left(-\omega R_F C_F \right)$$

Figure 4.17 shows $\angle G(\omega)$ decreases with increasing frequency. Note the values at the following three frequency values.

At $f = 0$: $\angle G(0) = 0$. At $f = 0$, the capacitor is an open circuit, leaving the effects produced by only R_F.

At $f = \infty$: $\angle G(\infty) = -\pi/2$. At $f = \infty$, the capacitor is a short circuit, removing the effect of R_F.

At $f = f_B$: $\angle G(f_B) = -\pi/4$. At $f = f_B$, $Z_{C_F} = R_F$.

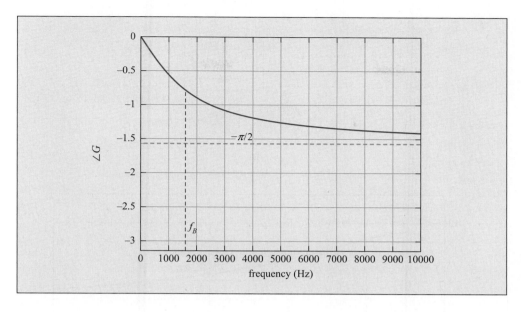

Figure 4.17

Phase response of an op-amp low-pass filter.

Band-Pass Filter

While low-pass filters remove frequency components that are higher than a specified break frequency, *band-pass filters* allow components to pass that lie between two specified frequencies: f_L at the low end and f_H at the high end. Such band-pass filters are important for extracting desired signals from all those that are detected.

- When you turn to a particular channel on your TV set, you are actually adjusting a band-pass filter to select the signals corresponding to that channel and direct them to the display. The total signals that are transmitted to your TV either by cable or satellite actually contain all of the channels.

- When your cell phone receives a call, the cell phone network assigns a particular frequency band over which you will communicate. Your cell phone contains a band-pass filter that isolates the signals meant for you and ignores the signals intended for other users.

Using RC impedances in the input and feedback paths, op amps become band-pass filters for such applications, as described in the following examples.

Band-pass filter EXAMPLE 4.13

Consider the op-amp circuit shown in Figure 4.18. The filter gain is given by

$$G(\omega) = \frac{-V_{\text{OUT}}(\omega)}{V_{\text{IN}}(\omega)} = \frac{-Z_F(\omega)}{Z_I(\omega)}$$

Only the magnitude response is considered here, which equals

$$|G(\omega)| = \frac{|V_{\text{OUT}}(\omega)|}{|V_{\text{IN}}(\omega)|} = \frac{|Z_F(\omega)|}{|Z_I(\omega)|}$$

The impedance magnitude of the $R\|C$ circuit on the feedback side is given by

$$|Z_F(\omega)| = |Z_{R\|C}(\omega)| = \frac{R_F}{\sqrt{1 + (\omega R_F C_F)^2}}$$

The impedance magnitude of the R-C circuit on the input side is given by

$$|Z_I(\omega)| = |Z_{R-C}(\omega)| = R_I \sqrt{1 + (\omega R_I C_I)^{-2}}$$

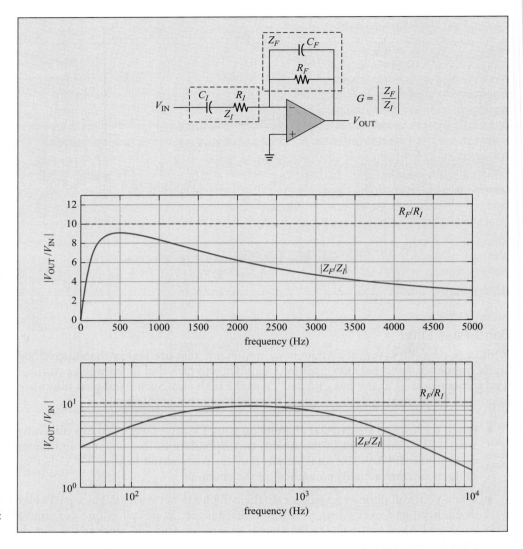

Figure 4.18

Op-amp band-pass filter with R-C circuit in input path and $R \| C$ circuit in feedback path.

Substituting these values into the gain equation gives

$$|G(\omega)| = \frac{\dfrac{R_F}{\sqrt{1+(\omega R_F C_F)^2}}}{R_I \sqrt{1+(\omega R_I C_I)^{-2}}}$$

Simplifying gives

$$|G(\omega)| = \frac{R_F/R_I}{\sqrt{1+(\omega R_F C_F)^2}\sqrt{1+(\omega R_I C_I)^{-2}}}$$

Note that $|G(0)| = R_F/R_I$ is modified by terms that vary with frequency. In this case, there are two break frequencies: ω_H (for *high*) and ω_L (for low), which are equal to

$$\omega_H = \frac{1}{R_F C_F} \qquad \left(f_H = \frac{1}{2\pi R_F C_F} \right)$$

and

$$\omega_L = \frac{1}{R_I C_I} \qquad \left(f_L = \frac{1}{2\pi R_I C_I} \right)$$

With these definitions, we can then write the gain as

$$|G(\omega)| = \frac{R_F/R_I}{\sqrt{1+(\omega/\omega_H)^2}\sqrt{1+(\omega/\omega_L)^{-2}}}$$

For a band-pass filter, the high-pass break frequency is greater than the low-pass break frequency. Thus,

$$\omega_H > \omega_L$$

For $R_I = 1\ k\Omega$, $C_F = 1\ \mu F$, $R_F = 10\ k\Omega$, and $C_F = 0.01\ \mu F$

$$f_H = \frac{1}{2\pi R_F C_F} = \frac{1}{2\pi (10^4)(10^{-8})} = \frac{10^4}{2\pi} = 1,592\ \text{Hz}$$

and

$$f_L = \frac{1}{2\pi R_I C_I} = \frac{1}{2\pi (10^3)(10^{-6})} = \frac{10^3}{2\pi} = 159\ \text{Hz}$$

Inserting these component values into the previous gain equation produces

$$|G(f)| = \frac{10}{\sqrt{1 + (6.28 \times 10^{-4} f)^2}\sqrt{1 + (6.28 \times 10^{-3} f)^{-2}}}$$

Clearly, the first denominator term decreases $|G|$ as $f \to \infty$, and the second denominator term decreases $|G|$ as $f \to 0$. Thus, it forms a band-pass filter that passes frequency components in the range $f_L \le f \le f_H$.

If a single filter section does not provide sufficient differentiation between the pass and stop bands, additional sections can be *cascaded* or connected in series. Op-amp filters have desirable electrical properties that allow such cascading to occur without sections interfering with each other, thus producing the anticipated results. If two sections are not sufficient, N sections can be cascaded to produce

$$|G_N(\omega)| = \left| \frac{V_{\text{OUT}}(\omega)}{V_{\text{IN}}(\omega)} \right|^N = \left(\frac{|Z_F(\omega)|}{|Z_I(\omega)|} \right)^N \tag{4.22}$$

In practice, when filter sections are cascaded, they are not identical, but typically have different break frequencies. Design rules for different filters types are called *Butterworth* and *Chebyshev* and have performance trade-offs. However, we leave these filter designs for advanced courses.

Cascaded band-pass filters EXAMPLE 4.14

Consider the op-amp circuit shown in Figure 4.19. When two identical filters are cascaded in series, the combined filter gain is given by

$$|G_2(\omega)| = \left(\frac{|Z_F(\omega)|}{|Z_I(\omega)|} \right)^2$$

$$|G_2(\omega)| = \left(\frac{\dfrac{R_F}{R_I}}{\sqrt{1 + (\omega R_F C_F)^2}\sqrt{1 + (\omega R_I C_I)^{-2}}} \right)^2$$

Let $R_I = 1\ k\Omega$, $C_F = 1\ \mu F$, $R_F = 10\ k\Omega$, and $C_F = 0.01\ \mu F$. Inserting these component values into the previous gain equation produces

$$|G_2(f)| = \left(\frac{10}{\sqrt{1 + (6.28 \times 10^{-4} f)^2}\sqrt{1 + (6.28 \times 10^{-3} f)^{-2}}} \right)^2$$

$$= \frac{100}{(1 + (6.28 \times 10^{-4} f)^2)(1 + (6.28 \times 10^{-3} f)^{-2})}$$

$|G_2(f)|$ decreases more quickly than a single $|G_1(f)|$ component when $f \to \infty$, because the $|G_2(f)|$ decreases as $1/f^2$ in the first denominator term, while $|G_1(f)|$ decreases as $1/f$. A similar behavior occurs as $f \to 0$ because of the $1/f^{-2}$ term in $|G_2(f)|$.

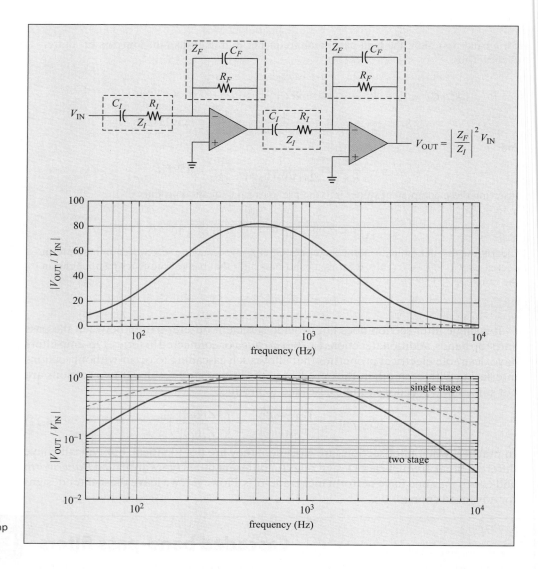

Figure 4.19

Cascade connection of two op-amp band-pass filters.

 4.3 FIELD-EFFECT TRANSISTOR (FET)

The field-effect transistor is a common device that is used in both analog and digital circuits. Figure 4.20 shows an *n-channel field-effect transistor*, which is also called an n-channel FET, or more concisely *FET* has three terminals: drain D, source S, and gate G. Current (positive charge) flows from D to S, controlled by the voltage on G relative to the source. The arrow on G pointing *in*ward indicates the FET is an "iN" or *negatively-charged electron-rich* (n-channel) device. The separation in the lines between G and the D-S channel indicates an insulating layer that does not permit direct current flow between G and the channel. This last feature has the important consequence that connecting an FET gate to a circuit has almost no effect on the circuit. This allows many FETs to be connected to a single wire connection, a feature that is crucial for the large digital memories in your computer, smartphone, and solid-state drives.

Figure 4.21 illustrates FET behavior. The FET contacts are shown as heavy lines that connect with a *semiconductor* channel (dashed box), whose electrical properties are determined by the gate voltage. V_{GS} is the voltage at the gate terminal relative to the source terminal. When $V_{GS} = 0$ there are no free electrons in the semiconductor near

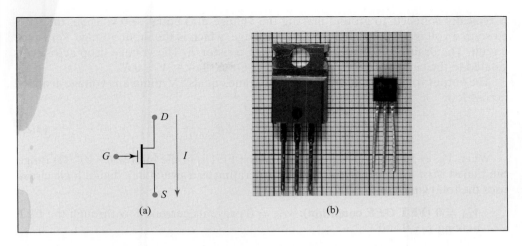

Figure 4.20

Field-effect transistor (n-channel). (a) D - drain, G - gate, and S - source. Current I flows from D to S. (b) Image shows a high-power FET (left) and a low-power FET. The grid in the image has a 1 mm spacing.

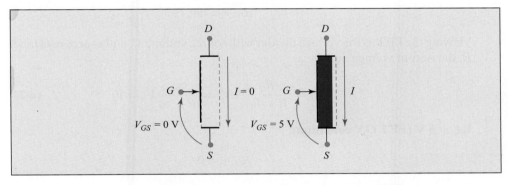

Figure 4.21

Gate to source voltage (V_{GS}) controls current flow from D to S. When $V_{GS} = 0$, no current can flow. $V_{GS} = 5$ V forms a channel in the semiconductor that allows current flow.

the gate and no current can flow. Applying $V_{GS} = 5$ V forms a dense semiconductor channel near G (blue region) containing free electrons that allows current to flow from D to S.

FET as a Digital Switch

The On/Off properties of FETs designed for switching applications make them ideal components for digital logic circuits. This section describes the basic operation that implements a digital logic gate. Digital circuits operate using only two voltage levels, 0 V and 5 V (although lower voltages, such as 3.3 V, are becoming more common).

Figure 4.22 shows a simple circuit implemented with an FET and resistor R. The FET operation is similar to an ideal switch, which is shown for comparison. The source

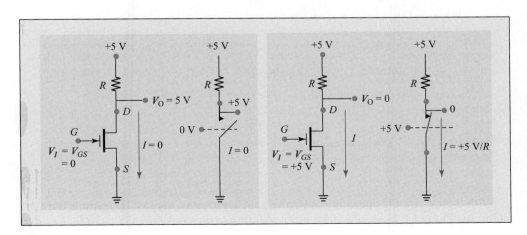

Figure 4.22

Simple digital FET circuit with the source connected to ground (0 V), input voltage V_{IN}, and output voltage V_{OUT}. When $V_{IN} = 0$, the current $I = 0$ and $V_{OUT} = 5$ V. When $V_{IN} = 5$ V, the current I is large, causing voltage drop across resistor R, making $V_{OUT} \approx 0$.

S typically connects to ground making the voltage at S equal to 0 V. Then, the gate-to-source voltage V_{GS} is simply the gate voltage, which is the input voltage V_{IN} to the circuit. The drain D connects to 5 V through resistor R. The voltage drop across R is related to the current I flowing through it by Ohm's law as $V = IR$.

The output voltage V_{OUT} is the voltage at D and equals 5 V minus any voltage dropped across R or

$$V_{OUT} = 5 - IR \tag{4.23}$$

When $V_{IN} = 5$ V, the "ON" resistance of the FET is in the milli-Ω ($= 10^{-3}\Omega$) range. Simplified (*first-order*) analysis of a FET operating as a switch in a digital logic circuit uses the following values.

$V_{IN} = 0$ **(FET *OFF* condition):** $V_{GS} = 0$ prevents current flow through the FET, making $I = 0$ and

$$R_{D-S} = \infty$$

Viewing the FET as the voltage divider with source voltage V_S and source resistance R, the output voltage equals

$$V_{OUT} = \frac{R_{D-S}}{R + R_{D-S}} V_S = \frac{\infty}{R + \infty} V_S = V_S \tag{4.24}$$

$V_{IN} = 5$ **V (FET *ON* condition):**

$$R_{D-S} = 0$$

Viewing the FET as the voltage divider, the output voltage equals

$$V_{OUT} = \frac{R_{D-S}}{R + R_{D-S}} V_S = \frac{0}{R + 0} V_S = 0 \tag{4.25}$$

A switching FET is designed so that there is a particular gate voltage, say $V_G = 1.5$ V at which there is a fast transition from no current flow to full current flow. This is similar to the fast transition in the output voltage in an op-amp threshold circuit.

EXAMPLE 4.15 **FET digital switch**

Let the supply voltage $V_S = 5$ V and $R = 1\,k\Omega$.

When $V_{IN} = 0$, $I = 0$, and we have

$$V_{OUT} = 5 - \overbrace{I}^{=0} R = 5\,\text{V}$$

When $V_{IN} = 5$ V, $R_{D-S} = 0$, and we have

$$I = \frac{V_s}{R + \underbrace{R_{D-S}}_{=0}} = \frac{5\,V}{1\,k\Omega} = 5\,\text{mA}$$

With digital circuits operating at two voltage levels, and the fast transitions between them, switching FETs are ideal for implementing logic gates.

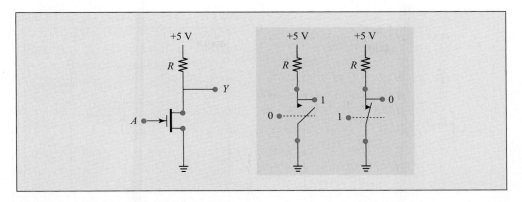

Figure 4.23

NOT gate implemented with a FET acting as a switch.

NOT gate   EXAMPLE 4.16

The simplest logic gate is the NOT gate, which produces a logic 1 at its output only if a 0 is present at the input, thus forming the logic complement. Figure 4.23 shows how a simple FET circuit implements a NOT gate. Let 0 V correspond to a logic 0 and 5 V to a logic 1. The shaded box indicates the state of the FET acting as a switch with gate voltage at either 0 V (logic 0) or 5 V (logic 1). When the input $A = 0$ V (logic 0) the output $Y = 5$ V (logic 1). When the gate input $A = 5$ V (logic 1), the output $Y = 0$ V (logic 0). Hence, the output is the logical complement of the input, thus forming a NOR gate.

AND gate   EXAMPLE 4.17

The AND logic element (or simply the AND gate) produces a logic 1 at its output only when logic 1's are applied to all of its inputs. Figure 4.24 show how to implement an AND gate using FETs. Two FETs are connected in series with the source of the top FET with input A applied to its gate, connected to the drain on the bottom one, and input B connected to its gate. These two FETs produce the intermediate logic variable X in the figure. Note that both FETs must be turned ON with logic 1's (5 V) at their gate inputs for current to flow, in which case $X = 0$. If either FET is OFF with a logic 0 (0 V) at its gate, current will not flow, making $X = 1$.

The logic variable X is applied to the gate of the third FET, which forms a NOT gate, producing the logic output Y. Hence $Y = 1$ only when $X = 0$, which occurs only when $A = 1$ and $B = 1$.

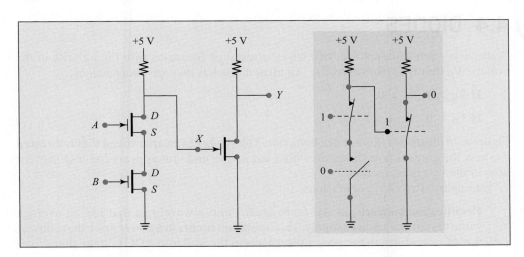

Figure 4.24

AND gate implemented with FETs. Shaded box shows operation with FETs acting as switches for $A = 1$ and $B = 0$.

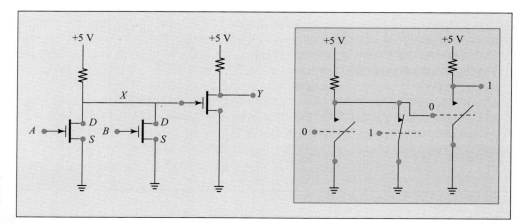

Figure 4.25

OR gate implemented with FETs. Shaded box shows operation with FETs acting as switches for $A = 0$ and $B = 1$.

The shaded box illustrates the operation when the FETs are viewed as switches with a logic 1 input producing an closed switch and a logic 0 producing an open switch. When either input (A or B) is a logic 0 at the gate, the channel of the corresponding FET does not conduct current (it acts as an open switch). Then the logic variable X is a 1 (5 V), thus causing output $Y = 0$. Note that the output FET implements a NOT gate. The shaded box in Figure 4.24 shows operation with FETs acting as switches for $A = 1$ and $B = 0$.

EXAMPLE 4.18

OR gate

An OR gate produces an output logic 1 when any of the inputs is a logic 1. Figure 4.25 indicates how an OR gate can be implemented using FETs. Two FETs are connected in parallel with the drains of the two FETs connected together. These two FETs produce the logic variable X in the figure, which in turn is applied to the gate of the FET (forming a NOT gate) producing the logic output Y. The shaded box indicates the operation when the FETs are viewed as switches with a logic 1 input producing a closed switch and a logic 0 producing an open switch. When either input (A or B) is a logic 1, the channel of the corresponding FET conducts current. Then the logic variable X is a 0, causing the output to be $Y = 1$. Note that the output FET implements a NOT gate. The shaded box shows operation with FETs acting as switches for $A = 0$ and $B = 1$.

4.4 DIODES

A diode is a semiconductor device whose resistance R_D varies with the polarity of the voltage V_D that is applied across it. An ideal diode has two resistance values:

If $V_D \geq 0$, $R_D = 0$.

If $V_D < 0$, $R_D = \infty$.

Figure 4.26 illustrates this strange behavior. The symbol is an arrowhead that is directed to show the current flow direction with a bar at one end that appears to block current flow in the reverse direction.

We explore two diode operations.

Rectification converts an AC (zero-mean) into a waveform that has an average value that forms a power supply. This operation occurs in a *power brick* that convert the AC waveform in the power receptacle in the wall into a DC voltage that drives your smart device.

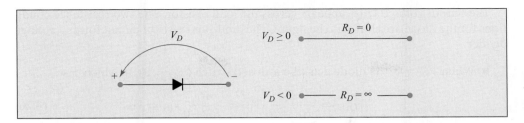

Figure 4.26

Ideal diode. $R_D = 0$ when $V_D \geq 0$ and $R_D = \infty$ when $V_D < 0$.

Current-to-light conversion generates the light that is produced by an LED. An LED is a semiconductor diode that is manufactured to produce light.

Rectifier

The diode is employed in circuits that rectify bipolar waveforms (that is, retain positive voltages and eliminate the negative). The main application for this process is called *rectification*, and converts alternating-current (AC) voltages into direct-current (DC) voltages that are used in every power supply that you plug into a wall socket to charge your phone.

Figure 4.27 illustrates rectification using a diode in series with a load resistor R_L. The input voltage V_{IN} is an AC waveform, which is one that has a zero average value and swings through positive and negative voltages.

Power brick operation   EXAMPLE 4.19

The waveform in a U.S. wall socket is a sinusoid that has a frequency $f = 60\,\text{Hz}$ and amplitude $A = 155\,\text{V}$. The RMS value is the square root of the average of the square of the sinusoidal value and is also called the *effective* voltage level because it produces the same power as a DC voltage of that amplitude. The RMS value of the AC signal at the outlet is $110\,V_{RMS}$.

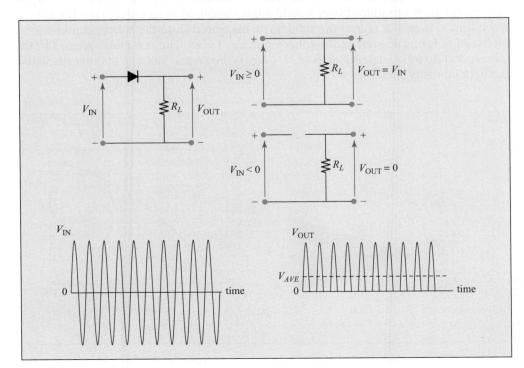

Figure 4.27

Diodes rectify bipolar waveforms to convert AC to DC. Dashed line indicates the average (DC) voltage value V_{AVE}.

The output voltage V_{OUT} appears across the load resistor. The two resistance conditions for the diode are shown in the figure. The diode and resistor circuit forms a voltage divider.

- When $V_{IN} \geq 0$, the diode acts like a closed switch ($R_D = 0$), in which case

$$V_{OUT} = \frac{R_L}{\underbrace{R_D}_{=0} + R_L} V_{IN} = \frac{R_L}{R_L + 0} V_{IN} = V_{IN} \qquad (4.26)$$

- When $V_{IN} < 0$, the diode acts like an open switch ($R_D = \infty$), in which case

$$V_{OUT} = \frac{R_L}{\underbrace{R_D}_{=\infty} + R_L} V_{IN} = \frac{R_L}{\infty} V_{IN} = 0 \qquad (4.27)$$

The average of V_{OUT} is shown in the dashed line, indicating $V_{AVE} > 0$. Hence, this circuit produces a DC voltage from an AC signal.

Practical rectifiers are more complicated circuits involving four diodes to exploit the negative-going $V_{IN} < 0$, but this simple single-diode circuit illustrates the rectification operation.

Light-Emitting Diode

Light-emitting diodes (LEDs) are diodes that produce light when current flows through them. A small arrow from the symbol indicates that light is emitted when current flows through an LED. Because the voltage across an LED is constant and independent of current, the light intensity is proportional to the current amplitude

$$I L = \kappa I_{LED} \qquad (4.28)$$

where κ is a proportionality constant that includes an efficiency factor.

Figure 4.28 shows a typical LED circuit, with R_L being a current limiting resistor, and $V_{LED} \approx 1.5 \, V$ is the constant voltage drop across the LED. The light (LI) produced by an LED is called *Radiant Intensity* and is measured in typical units of mW/sr, or light power (mW) produced over a solid angle measured in steradian (sr). If a sphere has radius r, then 1 sr equals the solid angle measured from the sphere center whose intersection forms an area on the sphere surface $A = r^2$. The complete sphere surface corresponds to solid angle equal to 4π sr. $LI(r)$ is shown as a function of range r because the light intensity decreases with r.

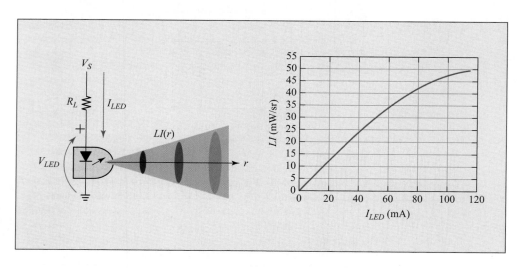

Figure 4.28

An LED converts current (mA) into light intensity (mW/sr), which is proportional to current.

LED light sources | EXAMPLE 4.20

LEDs serve various needs in smart devices:

Indicators: Red, green, yellow, and blue LEDs indicate the operational states of many devices. These LEDs have $I_{LED} \approx 10$ mA. Red, green, and blue LEDs allow you to set any color you choose in LED billboards. The small physical size of LEDs makes them ideal for smart device displays.

Communication: Wireless optical communication uses infrared (IR) light that has larger wavelengths than visible light. An infrared LED current $I_{LED} = 20$ mA produces a light intensity measured at a range of one meter, denoted as $LI(1 \text{ m})$, and equal to $LI(1 \text{ m}) = 10$ milli Watt per steradian (mW/sr) that is typical for your TV remote control.

Illumination: White LED provide illumination for taking photos. The higher intensity produced by these LEDs requires larger currents $I_{LED} \approx 100$ mA or more. Ultra-bright LEDs are sufficiently bright to serve as car headlights when $I_{LED} \approx 1,000$ mA.

LED electrical and light characteristics | EXAMPLE 4.21

With $V_{LED} = 1.5$ V, if $V_S = +5$ V and the desired $I_{LED} = 10$ mA produces sufficient light intensity, then

$$R_L = \frac{V_S - V_{LED}}{I_{LED}} = \frac{5 \text{ V} - 1.5 \text{ V}}{10 \times 10^{-3} \text{ A}} = 350 \, \Omega$$

Factoid: LEDs provide light more efficiently than incandescent and fluorescent lamps, making them the light source of the future.

Photodetectors

A photodetector acts as a partner to the LED in an optical sensor or communication system by converting light energy back into electrical current. Photodetectors come in various varieties, from photo-transistors to photo-diodes that convert light intensity into an electric current. A close cousin is the photo-voltaic cell that converts the sun's energy into electrical energy. Photodetectors used in practical devices are purchased as building-block units that are designed to be insensitive to environmental variations. This topic is described in advanced electronics classes.

4.5 Research Challenges

4.5.1 Minimizing Power Dissipation

Systems are designed to reduce the power usage in order to prolong battery life, which is desirable for portable systems, and to reduce the heat that is generated from such use. Computer speed usually correlates with its power consumption with faster computers requiring and dissipating more power. With devices reaching their maximum temperature capabilities, this increased heat threatens to limit the trend predicted by Moore's law. Techniques to lower power, such as lower operating voltages, lower clock frequencies, and using multiple cores that can be turned off when not needed (a technique called *dark silicon*) are being investigated to maintain the growth predicted by Moore's law.

4.5.2 Flexible Electronics

Flexible electronics are replacing printed circuit boards with new form factors, such as plastics for flexible displays (e-paper) and fabric for wearable solar cells. Such new materials require new construction techniques that involve printing with conducting and organic semiconductor inks (similar to an ink-jet printer) rather than using standard wire connections. Conventional methods for making transistors involves baking impurities into silicon to achieve desirable electrical properties. New processing techniques and materials allow transistors to be printed using special conductive inks.

4.5.3 Ion-Sensitive FETs

Field-effect transistors employ a gate voltage to effect the drain-to-source channel properties. Recall that voltage can also result because of a charge imbalance, such as in a capacitor. Charged molecules or ions that stick to a specially constructed gate surface in an ion-sensitive FET (ISFET) can also change channel properties and, hence, act as the sensor for these ions. Such devices have wide applications in biomedicine and fluid analysis.

4.6 Summary

This chapter described the basics of analog electronics whose operations include amplification, filtering, rectification, and conversion between light and electrical signals. A flexible and popular analog circuit that performs amplification is the negative-feedback *operational amplifier* (op amp), whose gain is set by the ratio of feedback to input resistance values. Adding capacitors to an op-amp circuit produces a *filter* whose gain changes with frequency to reduce noise in signals, limits the frequencies in a signal for analog-to-digital conversion, and extracts signals that occupy a specified frequency band in a communication channel. Field-effect transistors (FET) are voltage-controlled switches that are fundamental components in digital logic circuits. Diodes are non-linear devices whose resistance depends on the voltage across the diode. Standard diodes are useful for rectifying AC voltages to provide DC power, while LEDs offer an efficient means to generate light for displays. Photodetectors convert light to electrical signals to implement optoelectronic systems.

4.7 Problems

4.1 **Threshold detection input range.** Let op-amp gain $A = 10^4$ with supply voltages $V_{CC} = 5$ V and $V_{EE} = 0$ and threshold $V_\tau = 2.5$ V. What V_{IN} ranges produce $V_{OUT} = 5$ V and $V_{OUT} = 0$?

4.2 **Required A for threshold detection.** Use $V_{CC} = 5$ V, $V_{EE} = 0$, and $V_\tau = 2.5$ V. What is the minimum required op-amp gain A that produces $V_{OUT} = 5$ V and $V_{OUT} = 0$ for a digitally ambiguous range in V_{IN} less than 1 μV?

4.3 **Feedback amplifier gain.** Design an op-amp negative feedback amplifier with a gain $G = -100$ using $R_I = 200$ Ω.

4.4 **Amplifier gain for finite A.** Using the results in Example 4.5, compute amplifier gain G when $R_I = 1\,k\Omega$ and $R_F = 100\,k\Omega$ if op-amp gain $A = 1,000$.

4.5 **Feedback amplifier with positive gain.** Design an amplifier with a gain $G = 900$ using $R_I = 200$ Ω and two inverting op-amp feedback amplifiers having equal gains connected in series.

4.6 **Feedback amplifier with positive gain.** Design an amplifier with a gain $G = 200$ using $R_I = 200$ Ω and one op-amp feedback amplifier.

4.7 **Adding additional bit to obtain 8-level DAC.** Extend Example 4.10 by adding a third bit D_2. Compute the staircase values for the eight possible binary values 000_2 to 111_2.

4.8 **Op amp to match waveform to ADC.** An analog signal lies between -0.5 V and 0.5 V. The op-amp filter connects to an analog-to-digital converter (ADC) that has an input range from 0 to 5 V. Design an amplifier using op amps with a gain to match the input signal to the ADC so the signal at the DAC input varies between 0 and 5 V. Use $R_I = 1,000$ Ω. Draw the circuit.

4.9 **FET implementation of a 3-input AND gate.** Sketch the FET circuit of a 3-input AND gate. Draw the switch circuit when all three logic inputs equal 1. Write the truth table for the 3-input AND gate, making sure that all possible input combinations are represented in the truth table input section.

4.10 **FET implementation of a 3-input OR gate.** Sketch the FET circuit of a 3-input OR gate. Draw the switch circuit when all three logic inputs equal 1. Write the truth table for the 3-input OR gate, making sure that all possible input combinations are represented in the truth table input section.

4.11 **Diode rectification of a bipolar square waveform.** In Figure 4.27, consider V_{IN} to be a square waveform with period $T_P = 1$ ms that varies between -5 V and $+5$ V. Sketch V_{OUT} and compute V_{AVE}.

4.12 **Current to light an LED.** Assume we use two AAA batteries to provide $V_S = 3$ V to light an LED that has constant voltage drop $V_{LED} = 1.5$ V. Determine R_L that causes 10 mA to flow through the LED.

4.13 **Designing a circuit to drive 12 LEDs.** A poster showing six luxury automobiles uses a set of 12 LEDs to form tail lights. The poster is powered by a 9 V (500 mA) supply (a power brick). Assume each LED has a constant 1.6 V voltage drop and draws 20 mA current when lit. Design a circuit that uses current-limiting resistors to light the 12 LEDs with the available power brick. What is the power dissipation in the LEDs and resistors. What is the total power provided by the supply?

4.8 Matlab Projects

4.1 **Comparator for restoring digital data signals.** Extend Example 16.19 to increase the noise SD until errors occur approximately 20% of the time (approximately 10 errors in 50 samples).

4.2 **Variation of $|G|$ with varying A.** Extend Example 16.20 to display the variation in $|G|$ as A varies for a design specification $|G| = 100$. What is the minimum value of A for which $|G| > 95$?

4.3 **Op amp low-pass filter.** Extend Examples 4.12 and 16.21 to design a low pass filter having $f_B = 10$ kHz. Plot the gain using a log-log scale and the phase using a log-linear scale.

4.4 **Simulating diode rectification.** Using Example 16.22 as a guide, simulate rectification of a sinusoidal signal by an ideal diode (rectifying voltages greater than zero) and determine the average voltage of the rectified signal for input signal $s(t) = 10 \sin(\omega t)$.

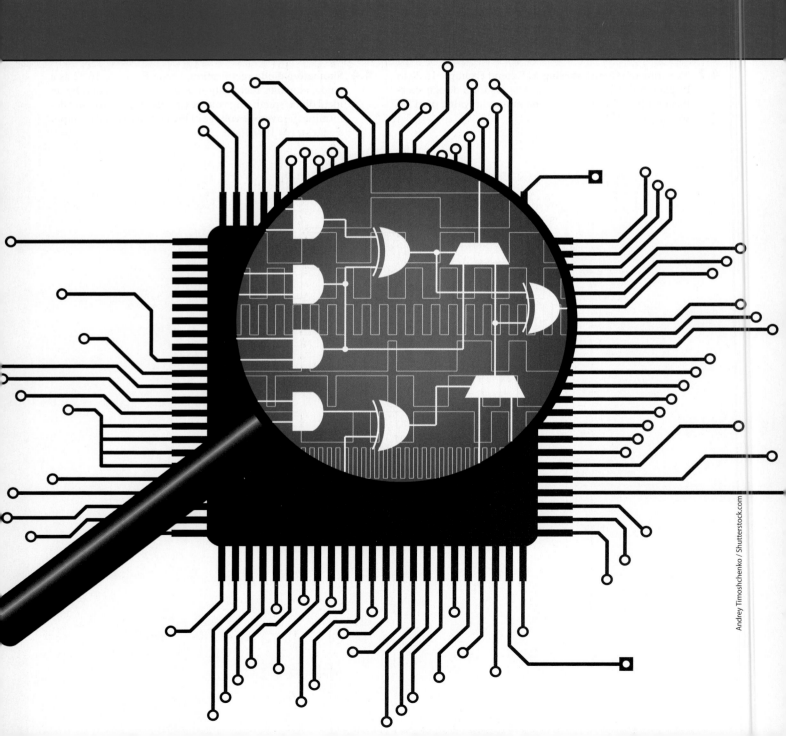

COMBINATIONAL LOGIC CIRCUITS

LEARNING OBJECTIVES

After completing this chapter, the reader should be able to:

- Relate voltage levels to logic variable values.
- Form truth tables, logic circuits, and logic equations.
- Understand the process of designing and analyzing a combinational logic circuit.
- Design logic circuit alternatives that use fewer logic gates.
- Implement digital logic circuits employed in smart devices, robots, and communication systems.

In this chapter, digital *signals* are considered to be *logic symbols* that are manipulated with logic circuits. Elementary logic gates are then interconnected to implement digital logic circuits that perform useful tasks. These include controlling a seven-segment display and performing arithmetic operations.

5.1 INTRODUCTION

The main topics in this chapter are covered in the following sequence:

Digital logic gates—The binary voltage levels in digital circuits lend themselves for analysis using AND, OR, and NOT logic operations that form the basic building blocks of digital logic circuits.

Digital logic circuits—Digital logic circuits implement a truth table that defines the desired output values when considering all possible input combinations of input logic variables.

Logic equations—Logic circuits are implemented on a digital computer using logic equations that define the desired logic operations.

Efficient digital logic circuits—Digital logic circuits can be implemented with a fewer number of gates by manipulating logic equations and by applying simplifying observations.

5.2 LOGIC VARIABLES AND LOGIC EQUATIONS

Digital signals represent binary data (0/1) by having only two discrete values, such as 0 and 5 V. Such digital signals were the earliest electrical signals, typically produced by switch closures in telegraphs. When a switch is *open*, no electrical current was transmitted. When the switched *closed*, current began to flow from the transmitter to the receiver.

A *logic variable* is an abstract descriptor that converts digital signals expressed in volts to logic levels that equal either 0 or 1 and represents mutually-exclusive alternatives, such as true/false, yes/no, or closed/open. We denote logic variables with capital letters, such as A, B, and Y. Logic variables have their own mathematical operations that are described using *Boolean algebra*.

Combinational (also called *combinatorial*) *logic* produces an output logic variable that is related to only the current values of input logic variables. In other words, the input logic variables *combine* to form the output logic variable value. The relationship between the input values and the output logic value that they produce is specified in three ways.

1. A *logic equation* uses Boolean algebra. While there are various ways of expressing logic operators, this book uses the operators and their symbols given in Figure 5.1. There are many ways to write the logic operators, and we chose

Figure 5.1

Logic operations performed by the elementary NOT, AND, and OR gates.

Logic Operation	Symbol	Example
NOT	–	$Y = \overline{A}$
AND	·	$Y = A \cdot B$
OR	+	$Y = A + B$

row #	A	B	C	Y
0	0	0	0	0
1	0	0	1	0
2	0	1	0	0
3	0	1	1	0
4	1	0	0	0
5	1	0	1	0
6	1	1	0	0
7	1	1	1	1

Figure 5.2

Truth table with input logic variables A, B, and C and output logic variable Y.

these for their simplicity and their relation to the arithmetic operations of addition (+) and multiplication (·). The problem with this notation is that the symbols, especially the (¯) cannot be typed in the instructions used in programming languages.

2. A *truth table* consists of two sections, as shown in Figure 5.2. The input section lists all of the possible logic conditions of the input variables denoted by A, B, and C. The output section indicates the output value Y that is produced for each set of inputs as required by the desired task.

3. A *logic circuit* is the interconnection of elemental gates. The elemental gates include the AND, OR, and NOT gates. All other combinational logic functions can be implemented using these three gates.

While the basic binary nature of data signals remain unchanged in their analysis, the devices that produce these signals have changed dramatically with improvements in technology. Early implementations of digital circuits used mechanical switches that controlled current flow. Big and manually operated switches were replaced with electrically operated switches called relays. These relatively slow mechanical relays were replaced by faster electrical switches, initially implemented with vacuum tubes. Large, hot, and energy inefficient, vacuum tubes were replaced by semiconductor transistors. Transistors have continued to shrink in size and energy usage, currently with billions embedded within integrated circuit chips. Digital electronics that implement these gates in modern computers consist of integrated circuits using energy-efficient field effect transistors (FETs).

5.3 ELEMENTARY LOGIC GATES

To illustrate elementary logic equations, truth tables, and logic circuits, this section describes the most basic combinational logic gates: the AND, OR, and NOT gates. All other combinational logic functions can be implemented with circuits using these three elementary gates.

5.3.1 NOT Gate

The simplest logic gate is the NOT gate, which produces a 1 at its output only if a 0 is present at the input, thus forming the logic complement. The NOT gate symbol, logic equation, and truth table are shown in Figure 5.3. By convention, inputs enter the circuit elements from the left, and the outputs are on the right. Because there are only two values for our binary logic variables, it is often said that one value is *not* the other—hence, the name NOT gate.

Figure 5.3

NOT gate symbol, logic equation, and truth table. The input is A and the output is $Y = \overline{A}$. The output of the NOT gate is the logical opposite or *complement* of the input.

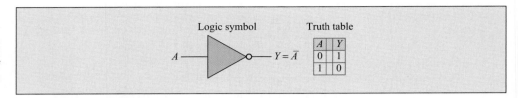

The logic equation *Y equals NOT A* is written as

$$Y = \overline{A} \tag{5.1}$$

with the overbar indicating the logical complement. The NOT operation is *negation* (or *logical complement*): $Y = 1$ when $A = 0$ and $Y = 0$ when $A = 1$, which is clearly shown in its simple truth table. The input section is labeled A, with its two possible values, 0 and 1. The output section, labeled Y, shows the value of the NOT gate output when the input is the value of A in the same row.

5.3.2 AND Gate

The AND logic element (AND gate) produces a logic 1 at its output only when logic 1's are applied to all of its inputs. Figure 5.4 shows a two-input AND gate logic symbol, logic equation, and truth table.

The logic equation for two input logic variables A and B, producing output logic variable Y, is written

$$Y = A \cdot B \tag{5.2}$$

In words, *Y equals A and B*.

The truth-table input section lists all of the possible values of inputs A and B using a binary counting order: 00 (=0), 01 (=1), 10 (=2), and 11 (=3). This counting order is the convention for structuring the input section of any truth table. This convention allows the input section to be written down first, before the values of the output logic variable are determined. We note that $Y = 1$ only when $A = 1$ *and* $B = 1$. If either $A = 0$ or $B = 0$, or both equal 0, then $A = 0$. The logical AND is identical to algebraic multiplication.

Figure 5.5 illustrates how two switches can be connected to form an AND circuit. A 1 designates a closed switch and a 0 designates an open switch. The output $Y = 1$, which

Figure 5.4

AND gate symbol, logic equation, and truth table. Logic variables A and B are inputs and $Y = A \cdot B$ is the output. Y equals 1 only when both A and B equal 1.

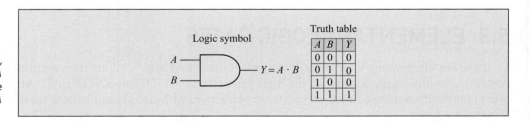

Figure 5.5

An AND gate formed with a pair of switches. (a) The output Y is 1, which is a closed switch only when both switches A and B are closed (= 1). (b) The output is 0, which is an open switch when either switch or both switches are open (= 0).

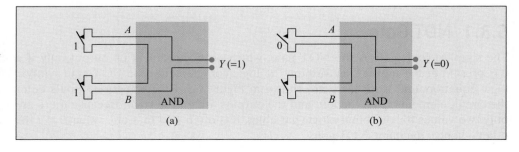

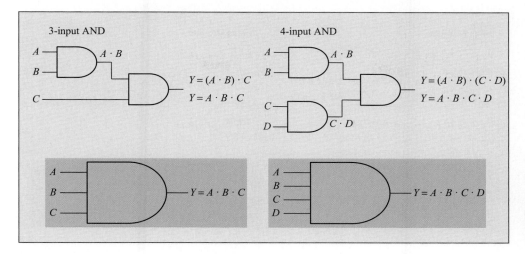

Figure 5.6

Standard 2-input AND gates combine to form 3-input and 4-input AND gates. Shaded boxes shows simplified symbols.

is a closed switch, occurs only when both switches A and B are closed, corresponding to $A = 1$ and $B = 1$. The output $Y = 0$ is an open switch that occurs when either switch or both switches are open ($=0$).

Many applications require more than two inputs to an AND gate. Figure 5.6 shows how to combine basic 2-input AND gates to include additional inputs. Three inputs A, B, and C applied to an AND gate would be expressed with the logic equation

$$Y = A \cdot B \cdot C \qquad (5.3)$$

Figure 5.6 shows parentheses added to illustrate how the outputs of the initial AND gates are combined at the output of the last AND gate. Although not necessary in this simple case, parentheses are used in more complex logic equations. The simplified logic symbols are shown below the interconnected 2-input AND gates. *The output of a multiple-input AND gate equals 1 only when all the inputs are 1's.*

5.3.3 OR Gate

An OR gate produces an output logic 1 when any of the inputs is a logic 1. An OR gate having two inputs is shown in Figure 5.7. $Y = 1$ when $A = 1$ OR $B = 1$. If both $A = 0$ and $B = 0$, then $Y = 0$. If both $A = 1$ and $B = 1$, then $Y = 1$—not 2 as would the case in normal addition, because logic variables take on only 0 and 1 values. The logic equation Y *equals A OR B* is written

$$Y = A + B \qquad (5.4)$$

The OR gate can be extended to accommodate additional logic variable inputs, as shown in Figure 5.8. As in the 2-input OR gate, the output of the multiple-input OR gate is a 1 when any of the inputs is a 1.

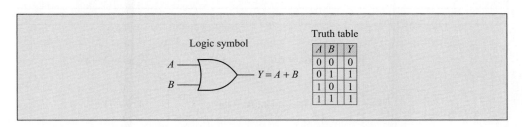

Figure 5.7

OR gate.

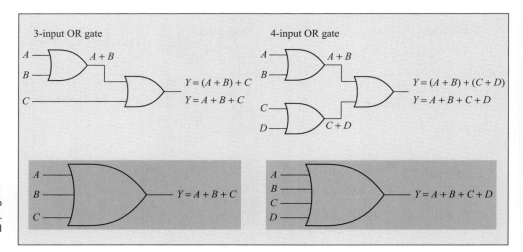

Figure 5.8

Two-input OR gates combine to form 3- and 4-input OR gates. Shaded boxes shows simplified symbols.

5.4 BUILDING-BLOCK GATES

The NOT, AND, and OR gates are the basic gates from which all other gates and combinational logic circuits are implemented. However, these are not the most flexible in terms of implementing a logic circuit using the smallest number of chips. This section describes the logic gates that are commonly used to actually construct logic circuits because either they are the easiest to implement with digital electronics or because they can result in the smallest chip count.

5.4.1 NAND Gate

The NAND (NOT-AND) gate is the most common gate in digital logic implementations because of its flexibility and simple structure. Figure 5.9 shows a NAND gate and a digital integrated circuit package that contains four NAND gates.

The logic equation for a NAND gate is

$$Y = \overline{A \cdot B}$$

The flexibility of the NAND gate is shown in Figure 5.10 in that it can form a NOT gate when the two inputs are connected together. The resulting NOT gate can be connected to the output of a NAND gate to form an AND gate.

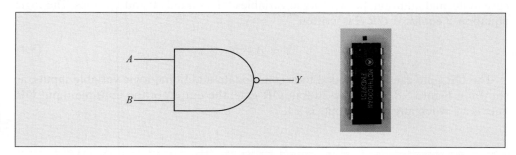

Figure 5.9

NAND gate and digital integrated circuit that contains four NAND gates.

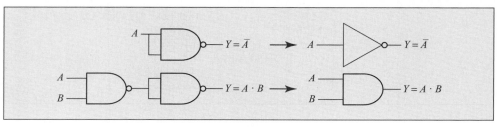

Figure 5.10

A NAND gate can form a NOT gate, and two NAND gates can form an AND gate.

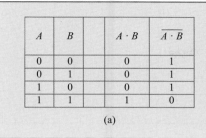

A	B	$A \cdot B$	$\overline{A \cdot B}$
0	0	0	1
0	1	0	1
1	0	0	1
1	1	1	0

(a)

A	B	$\overline{A}$	$\overline{B}$	$\overline{A} \cdot \overline{B}$
0	0	1	1	1
0	1	1	0	0
1	0	0	1	0
1	1	0	0	0

(b)

Figure 5.11

Truth tables for Example 5.1.

NAND gate   EXAMPLE 5.1

While it is tempting to think $\overline{A \cdot B} = \overline{A} \cdot \overline{B}$, it is wrong. The best way to remove doubt is to compare truth tables. Figure 5.11a shows the truth table of $A \cdot B$ and its NOT value $\overline{A \cdot B}$. In contrast, Figure 5.11b shows the truth table for $\overline{A} \cdot \overline{B}$ that gives values of inputs $\overline{A}$ and $\overline{B}$ for forming the output value. Comparing the values in the output columns for the same A and B values demonstrates that

$$\overline{A \cdot B} \neq \overline{A} \cdot \overline{B}$$

5.4.2 NOR Gate

Figure 5.12 shows a two-input NOR (NOT-OR) gate. The NOR gate is described by the logic equation

$$Y = \overline{A + B}$$

The truth table of $A + B$ and its NOT value $\overline{A + B}$ are given in Figure 5.13.

NOR gate using NOT and AND gates   EXAMPLE 5.2

The NOR truth table shows only a single 1 in the output section. Hence, we can write the logic equation of the NOR gate, $\overline{A + B}$, and equate it to the condition of the inputs that produce the logic 1 at the input, $\overline{A} \cdot \overline{B}$, which results in

$$\overline{A + B} = \overline{A} \cdot \overline{B}$$

This logic equation is one of *DeMorgan's laws*, which describes the equivalents of logical expressions. These laws are described in computer engineering courses to provide flexibility in implementing efficient logic circuits. If we consider the right side of the equation, we see the AND of two complemented logic variables. Figure 5.14 shows how the NOR gate is implemented from two NOT gates and one AND gate.

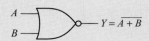

$$A \\ B \qquad Y = \overline{A+B}$$

Figure 5.12

Two-input NOR gate.

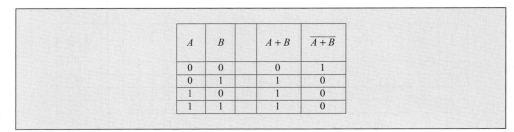

Figure 5.13

NOR truth table that starts with the OR truth table.

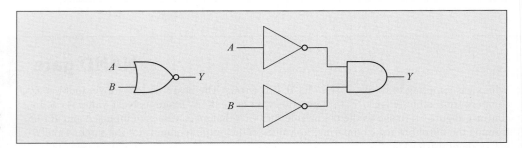

Figure 5.14

Equivalent logic circuits. NOR gate can also be implemented using two NOT gates and one AND gate.

5.4.3 ExOR Gate

The exclusive-OR gate (ExOR) is a useful gate for comparing two binary numbers, for encrypting data, and for arithmetic operations. The ExOR gate produces an output 1 only when one—but not both—of the inputs is 1. Let A and B be the inputs and Y be the output. The ExOR logic operation is denoted by the symbol $\oplus$ and is expressed as

$$Y = A \oplus B \tag{5.5}$$

Figure 5.15 shows the ExOR logic symbol and its truth table. The following examples show the ExOR gate used in comparison and in encryption tasks.

EXAMPLE 5.3

Comparing two binary numbers

Consider two 4-bit binary numbers given by $A_3 A_2 A_1 A_1$ and $B_3 B_2 B_1 B_0$. We want a logic circuit that compares these two numbers and produces a logic output NE with $NE = 1$ only when the two numbers are not equal (that is, there is a difference in at least one bit position in the two numbers).

Figure 5.16 shows an ExOR gate at each bit position A_i and B_i for $0 \leq i \leq 3$. If $A_i = B_i$, both are either 0 or 1, and the output of the ExOR gate is 0. If $A_i \neq B_i$, the output of their ExOR gate is 1. If the two 4-bit binary numbers are not equal, they must differ in at least one of the bit positions, causing that ExOR output to equal 1. The OR gate combines all of the ExOR gate outputs, and hence it will be a 1 if any of the ExOR outputs is a 1. Hence, $NE = 1$ indicates that the two numbers are not equal.

Figure 5.15

Exclusive OR (ExOR) gate produces an output that is a logical 1 when only one of its two inputs is a logical 1.

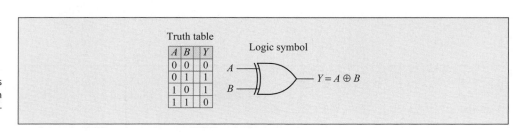

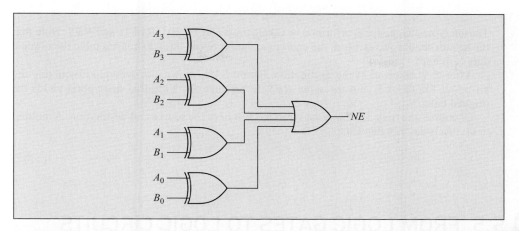

Figure 5.16

ExOR comparison circuit. $NE = 1$ when $A_3 A_2 A_1 A_1 \neq B_3 B_2 B_1 B_0$.

Performing encryption with ExOR gate

EXAMPLE 5.4

The easiest way to implement an encryption scheme is with ExOR gates. For encrypting, we apply the binary data and a random binary sequence—one bit position at a time—as inputs to an ExOR gate. The ExOR gate then produces the encrypted data sequence. Applying the encrypted data and the same random binary sequence—again one bit position at a time—to an ExOR gate yields the original data sequence.

To demonstrate this let D_i for $i = 0, 1, 2, \ldots, n_x - 1$ denote the original binary data produced by a source and RBS_i for $i = 0, 1, 2, \ldots, n_x - 1$, which is a random binary sequence having the same number of bits as in the original data. Then, using this bit-wise ExOR operation, the encrypted message, denoted by $\mathcal{E}_i$ for $i = 0, 1, 2, \ldots, n_x - 1$ is formed by

$$\mathcal{E}_i = D_i \oplus RBS_i \text{ for } i = 0, 1, 2, \ldots, n_x - 1$$

The encrypted message $\mathcal{E}_i$ is transmitted to the destination, where the receiving party either knows or can generate the same RBS_i used in the encryption. To retrieve the original data D_i, the ExOR operation is performed on $\mathcal{E}_i$ with RBS_i, or

$$D_i = \mathcal{E}_i \oplus RBS_i \text{ for } i = 0, 1, 2, \ldots, n_x - 1$$

thus reproducing the original data at the destination.

Consider encrypting hexadecimal numbers (0 through F that represent 4-bit sequences) generated by a source that produces the D_i for $i = 0, 1, 2, \ldots, n_x - 1$ ($n_x = 28$) shown in Figure 5.17a. The randomly-generated RBS_i contains the same number of bits as in D_i.

5	7	E	4	0	2	1	(Original data)
0101	0111	1110	0100	0000	0010	0001	(D_i)
$\oplus$	$\oplus$	$\oplus$	$\oplus$	$\oplus$	$\oplus$	$\oplus$	$\oplus$
0010	1010	1110	0100	1011	0010	1101	(RBS_i)
0111	1101	0000	0000	1011	0000	1100	$(\mathcal{E}_i)$
7	D	0	0	0	0	C	(Encrypted data)

(a)

7	D	0	0	0	0	C	(Encrypted data)
0111	1101	0000	0000	1011	0000	1100	$(\mathcal{E}_i)$
$\oplus$	$\oplus$	$\oplus$	$\oplus$	$\oplus$	$\oplus$	$\oplus$	$\oplus$
0010	1010	1110	0100	1011	0010	1101	(RBS_i)
0101	0111	1110	0100	0000	0010	0001	(D_i)
5	7	E	4	0	2	1	(Original data)

(b)

Figure 5.17

Tables showing (a) data encryption and (b) decryption in Example 5.4.

The encrypted sequence $\mathcal{E}_i$ is formed by taking the bit-by-bit ExOR of D_i and RBS_i. Note that the hexadecimal equivalents of the encrypted data (bottom line) do not resemble the original data (top line).

After $\mathcal{E}_i$ is received at the destination, Figure 5.17b shows it is decrypted by taking the bit-by-bit ExOR of $\mathcal{E}_i$ and the same RBS_i to produce D_i. Note that decryption yields the original data.

Of course, the trick is to have the same RBS_i at both the source and destination. A method to do this is described in Chapter 12.

5.5 FROM LOGIC GATES TO LOGIC CIRCUITS

Using the three basic gates (AND, OR, and NOT), we show that any combinational logic function can be implemented as a logic circuit. This construction of a logic circuit proceeds from the truth table, which explicitly lists all the possible sets of input logic variables and their associated output values. The truth table input section is formed using the conventional binary counting sequence, while the output section meets the requirements of the task. Several tasks are considered to illustrate the process, including digital arithmetic operations and controlling a seven-segment display.

The design of a logic circuit includes the following steps.

1. Construct a truth table of all the possible input logic combinations and their desired output logic values.

2. Identify the input logic patterns that produce logic 1 outputs.

3. Implement a logic circuit that recognizes these input patterns.

5.5.1 Constructing a Truth Table

If a logic circuit has n input logic variables, there are 2^n possible patterns of 0's and 1's. These patterns are listed in numerical order with the first row starting with zero expressed in binary form using one bit per input logic variable. In this way, the input patterns for all truth tables are the same and depend only on n, which is the number of input logic variables.

EXAMPLE 5.5

Truth table input section

The elementary AND and OR logic gates described above had two inputs A and B ($n = 2$), and the $2^n = 2^2 = 4$ possible binary input patterns are given by

$$00 \quad 01 \quad 10 \quad \text{and } 11$$

Each pattern represents one row in the truth table input section. Figure 5.18 shows truth tables with $n = 2$ and $n = 3$ input logic variables. A logic circuit having four inputs ($n = 4$), denoted A, B, C, and D, has $2^4 = 16$ possible input binary patterns, from 0000 (0_{10}) to 1111 (15_{10}). The corresponding truth table then has 16 rows.

Generalizing, n input logic variables produce 2^n rows in the truth table.

Inputs		Output
A	*B*	*Y*
0	0	
0	1	
1	0	
1	1	

Inputs			Output
A	*B*	*C*	*Y*
0	0	0	
0	0	1	
0	1	0	
0	1	1	
1	0	0	
1	0	1	
1	0	1	
1	1	0	
1	1	1	

Figure 5.18

Truth tables describing two and three input logic variables.

5.5.2 Binary Pattern Recognition

The output section of a truth table is specified to satisfy a particular task. We consider specifying the logic output outputs for various tasks later in this chapter. The next step in implementing a logic circuit involves recognizing each particular *binary pattern* of input variables that generates an output equal to 1. For example, a 4-bit pattern can be represented as

$$b_3 \; b_2 \; b_1 \; b_0$$

The standard method of recognizing a specific bit pattern is by using an AND gate. Because the output of an AND gate is a 1 only when all of its inputs are equal to 1, we need to transform the specific pattern we are trying to recognize into one having all 1's. We do this by converting each 0 in the pattern into a 1 by using a NOT gate. The logic variables that are equal to 1 in the desired pattern are applied directly to an AND gate, while those that equal 0 are first applied to a NOT gate and then to the AND gate. Then when the desired bit pattern occurs, the AND gate input will have all inputs equal to 1 and produce a logic 1 output. This procedure is illustrated in the following example.

Logic circuit to recognize the 4-bit pattern $b_3b_2b_1b_0 = 1100$

EXAMPLE 5.6

We want to design a logic circuit that produces the output P_{1100}, which equals 1 only when $b_3 = 1$ **and** $b_2 = 1$ **and** $b_1 = 0$ **and** $b_0 = 0$. The logic equation that corresponds to this pattern recognition is given by

$$P_{1100} = b_3 \cdot b_2 \cdot \overline{b_1} \cdot \overline{b_0}$$

To implement the logic circuit that has $b_3b_2b_1b_0$ as inputs and P_{1100} as the output, we connect logic variables b_3 and b_2 directly to an AND gate, because their values in the desired pattern equal 1. The logic variables b_1 and b_0 equal 0 in the desired pattern and hence are first applied to NOT gates to convert them to 1, as shown in Figure 5.19. The output of the AND gate then equals 1 only when the desired input pattern occurs.

Figure 5.19

Logic circuit to recognize the 4-bit pattern $b_3b_2b_1b_0 = 1100$. NOT gates convert the 0's in the desired pattern to 1's. $P_{1100} = 1$ only when $b_3b_2b_1b_0 = 1100$.

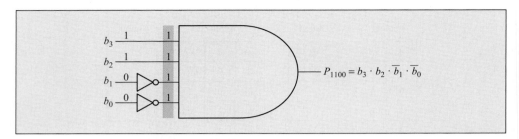

$$P_{1100} = b_3 \cdot b_2 \cdot \overline{b_1} \cdot \overline{b_0}$$

5.6 LOGIC CIRCUIT IMPLEMENTATION USING SUM OF PRODUCTS

Logic circuits typically need to recognize a set of input patterns with each pattern being an AND product of logic variables. When more than one pattern needs to be recognized, each pattern must be recognized separately by its own AND gate. The output of each AND gate is then connected to one of the inputs of an OR gate to produce their sum. Hence, the OR gate output then indicates when one of the desired patterns is present at an AND gate input. Such an implementation is called a *sum-of-products* logic circuit.

EXAMPLE 5.7

Recognizing binary pattern corresponding to 7_{10} or 11_{10}

Four bits $b_3b_2b_1b_0$ can represent decimal integers between 0 and 15. The two numbers to be recognized, 7_{10} and 11_{10}, have the 4-bit representations given by

$$7_{10} = 0111_2 \quad \text{and} \quad 11_{10} = 1011_2$$

A separate logic circuit is implemented to recognize each pattern. The outputs of these two logic circuits are then connected to a 2-input OR gate. The circuit that recognizes 7_{10} has output P_7. From the binary pattern representing 7, the logic equation for this circuit is

$$P_7 = \overline{b_3} \cdot b_2 \cdot b_1 \cdot b_0$$

The circuit that recognizes 11_{10} has output P_{11}. From the binary pattern representing 11, the logic equation for this circuit is

$$P_{11} = b_3 \cdot \overline{b_2} \cdot b_1 \cdot b_0$$

The output of the OR gate, indicated by $P_{7 \text{ or } 11}$, equals 1 only when one of the desired bit patterns occurs. The corresponding logic equation is written as

$$P_{7 \text{ or } 11} = P_7 + P_{11}$$

Figure 5.20 shows this logic circuit. As is evident from the figure, logic circuit diagrams can get complicated rather quickly with some signal lines crossing over others. To indicate the crossing lines that are actually electrically connected, a solid circle is drawn at their intersection. No solid circle indicates that there is no connection. Sometimes, a curved bridge in one of the crossing wires indicates that there is no electrical connection. In simple logic circuits it is often obvious when crossing wires electrically connect, so no additional notation is necessary.

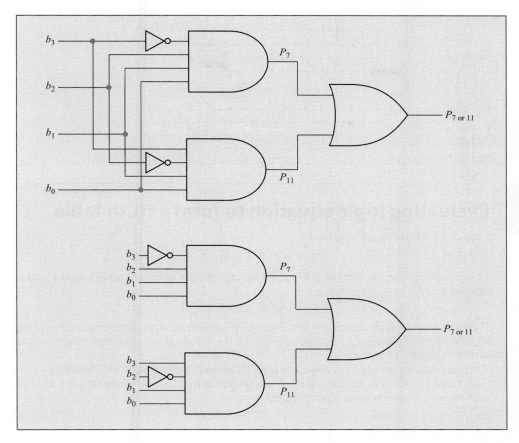

Figure 5.20

Equivalent drawings of a logic circuit to recognize binary patterns that represent either 7_{10} or 11_{10}.

 ## 5.7 TRUTH TABLE FROM LOGIC EQUATION AND LOGIC CIRCUIT

The previous topics described how to implement a logic circuit from a truth table. For analyzing a given logic equation or logic circuit, the inverse operation needs to be done. This section describes determining the truth-table output section either from a given logic equation or from a given logic circuit.

Forming a Truth Table from a Logic Equation

Logic equations can become very confusing. The evaluation of the logic operators in a logic equation follows the convention used for the operations in a complicated algebraic expression, which I recall as *My Dear Aunt Sally*, or first perform multiplications, then divisions, then additions, and finally the subtractions. Considering a complemented logic value as an individual logic variable, a logic equation is evaluated by first performing the AND (product) followed by the OR (addition).

For our purpose of implementing a truth table, parentheses can make logic equations simpler to understand by expressing them as the sum (OR) of product (AND) terms. Use parentheses to enclose AND-ed logic variables into a single logic expression. This will then leave a set of terms that are combined with a logic OR operation. Evaluating the products containing the fewest logic variables first, each product term in parentheses leads to one or more rows in the truth table being set = 1. The remaining rows that do not evaluate to 1 become 0. The following examples illustrate this approach.

A	B	C		Y	Reason
0	0	0		0	Remaining term $= 0$
0	0	1		1	$(\overline{A} \cdot \overline{B} \cdot C) = 1$
0	1	0		1	$(B \cdot \overline{C}) = 1$
0	1	1		0	Remaining term $= 0$
1	0	0		1	$A = 1$
1	0	1		1	$A = 1$
1	1	0		1	$A = 1$
1	1	1		1	$A = 1$

Figure 5.21

Evaluating logic equation to form a truth table.

EXAMPLE 5.8

Evaluating logic equation to form a truth table

Consider the logic equation given by

$$Y = A + B \cdot \overline{C} + \overline{A} \cdot \overline{B} \cdot C$$

Use parentheses to enclose the product terms to express the logic equation as the sum of product terms:

$$Y = (A) + (B \cdot \overline{C}) + (\overline{A} \cdot \overline{B} \cdot C)$$

Evaluating the products containing the fewest logic variables first, we set the truth table output rows to 1 by first evaluating (A), then $(B \cdot \overline{C})$, and finally $(\overline{A} \cdot \overline{B} \cdot C)$. The remaining rows that do not evaluate to 1 become 0.

Figure 5.21 shows the resulting truth table. Note that the row $ABC = 110$ satisfies both $A = 1$ and $(B \cdot \overline{C}) = 1$. Evaluation of $A = 1$ produced four 1's and eliminated the need for further evaluations for those input patterns.

Distributive Law in Logic Equations

To assist in expressing a logic equation as the sum of product terms, logic equations can be simplified by using a *distributive law* applied to the AND operation:

$$A \cdot (B + C) = A \cdot B + A \cdot C = (A \cdot B) + (A \cdot C) \qquad (5.6)$$

The distributive law also holds in standard arithmetic multiplication:

$$2 \times (3 + 4) = 2 \times 3 + 2 \times 4 \qquad (5.7)$$

Note that the distributive law does not hold for the OR operation. For example,

$$A + (B \cdot C) \neq A + B \cdot A + C \qquad (5.8)$$

To show this, consider why the distributive law does not hold for standard arithmetic addition:

$$2 + (3 \times 4) \neq (2 + 3) \times (2 + 4) \qquad (5.9)$$

EXAMPLE 5.9

Distributive law in logic equation to form a truth table

Consider the logic equation given by

$$Y = A \cdot (B + \overline{C})$$

Use the distributive law and then parentheses to enclose the product terms to express the logic equation as the sum of product terms:

$$Y = A \cdot B + A \cdot \overline{C} = (A \cdot B) + (A \cdot \overline{C})$$

Figure 5.22 shows the resulting truth table. Note that the row $ABC = 110$ satisfies both $(A \cdot B) = 1$ and $(A \cdot \overline{C}) = 1$.

A	B	C	Y	Reason
0	0	0	0	Remaining term = 0
0	0	1	0	Remaining term = 0
0	1	0	0	Remaining term = 0
0	1	1	0	Remaining term = 0
1	0	0	1	$(A \cdot \overline{C}) = 1$
1	0	1	0	Remaining term = 0
1	1	0	1	$(A \cdot B) = 1$
1	1	1	1	$(A \cdot B) = 1$

Figure 5.22

Distributive law applied to logic equation to form a truth table.

Forming a Truth Table from Logic Circuit

Any combinational logic circuit is constructed by interconnecting elemental logic gates. The simplest way of forming the truth table from a logic circuit is to translate the logic circuit into its equivalent logic equation and then using the method in the previous section. The following example illustrates this method.

Logic circuit to truth table   EXAMPLE **5.10**

Consider the logic circuit shown in Figure 5.23. The logic equation is given by

$$Y = \overbrace{(A + \overline{B})}^{\text{OR gate}} \cdot C$$

Use the distributive law and then parentheses to enclose the product terms to express the logic equation as the sum of product terms:

$$Y = A \cdot C + \overline{B} \cdot C = (A \cdot C) + (\overline{B} \cdot C)$$

Figure 5.23 shows the resulting truth table. Note that the row $ABC = 101$ satisfies both $(A \cdot C) = 1$ and $(\overline{B} \cdot C) = 1$. The table shows the reason to be $(A \cdot C) = 1$ because it was the first term considered, and once an output becomes 1 no other input terms need to be considered.

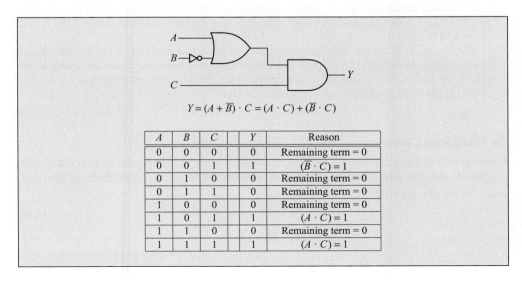

$$Y = (A + \overline{B}) \cdot C = (A \cdot C) + (\overline{B} \cdot C)$$

A	B	C	Y	Reason
0	0	0	0	Remaining term = 0
0	0	1	1	$(\overline{B} \cdot C) = 1$
0	1	0	0	Remaining term = 0
0	1	1	0	Remaining term = 0
1	0	0	0	Remaining term = 0
1	0	1	1	$(A \cdot C) = 1$
1	1	0	0	Remaining term = 0
1	1	1	1	$(A \cdot C) = 1$

Figure 5.23

Logic circuit converted to logic equation for determining the truth table.

5.8 DESIGNING EFFICIENT LOGIC CIRCUITS

When implementing a logic circuit, the number of gates should be kept as small as possible to make the logic circuit simple and inexpensive. This section considers two techniques that may reduce the number of gates in a logic circuit.

1. Compare original and complemented output logic variables to find the one that has the smaller number of 1's.

2. Appropriately assign "don't care" values ($\times$) to logic 0 or 1, that is, it doesn't matter if the logic value is 0 or 1 because that input pattern is not observed in practice.

To implement an efficient logic circuit that produces the output Y, the design procedure starts by comparing the number of 1's in the truth table output section with the number of 0's to find the smaller count. If the number of times Y equals 1 is smaller, we proceed with the implementation of the logic circuit using the procedure just described.

If the number of times Y equals 0 is smaller than the number of times it equals 1, we consider implementing the NOT of the output variable or $\overline{Y}$. We do this by first complementing the 0's in the output section to 1's and the 1's to 0's (the NOT operation) to form the values of $\overline{Y}$. Having done this, the number of 1's is smaller in this *complementary* output of the truth table. We then proceed with the logic circuit implementation of $\overline{Y}$ with the procedure previously described. After the design is finished, the output of this *complementary* logic circuit, $\overline{Y}$, is connected to a NOT gate that then produces the original Y values specified in the truth table.

The next example illustrates that implementing a complementary logic circuit can lead to a much simpler design.

EXAMPLE 5.11

Implementing a truth table

Consider the truth table shown in Figure 5.24. The truth table has two inputs A and B and the output Y. The complement of Y, or $\overline{Y}$, is also shown as the complementary output of the truth table.

We implement two logic circuits using the two methods previously described.

In the first design, we find the values of A and B that make $Y = 1$ and implement an AND gate to recognize each pattern. There are three such patterns, and the outputs of the corresponding AND gates are connected to the OR gate to implement the logic circuit.

In the second design, it was noted that there were fewer cases when $Y = 0$ in the truth table, so a logic circuit to implement $\overline{Y}$ was designed. Because $\overline{Y} = 1$ for only one set of A and B values, this design is very simple. To implement the original Y function, $\overline{Y}$ was applied to a NOT gate. Note that a simpler logic circuit results. This second logic circuit is the NAND (Not-AND) gate.

This example illustrates that an engineering solution to a problem is not unique (that is, several different systems can be built to perform the same function). Each particular implementation has its own advantages and disadvantages. Designing a system is really an *art* that often results in an elegant and clever implementation.

de Morgan's Laws

Considering alternative means of implementing a logic circuits leads to two interesting equations used in *propositional logic* and simplifying logic circuits called *de Morgan's laws*, which are given by

$$\overline{A \cdot B} = \overline{A} + \overline{B}$$ (5.10)
$$\overline{A + B} = \overline{A} \cdot \overline{B}$$

The following example illustrates the proof of the first law.

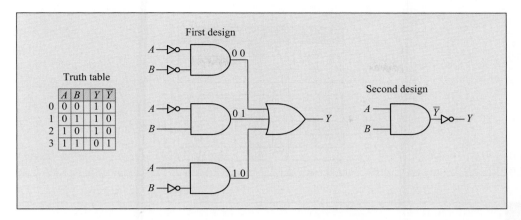

Truth table

	A	B		Y	$\overline{Y}$
0	0	0		1	0
1	0	1		1	0
2	1	0		1	0
3	1	1		0	1

Figure 5.24

Two logic circuits having different structures implemented from the truth table shown at left.

Proving one of de Morgan's laws EXAMPLE 5.12

This example proves de Morgan's first law

$$\overline{A \cdot B} = \overline{A} + \overline{B}$$

by using truth tables.

First, consider the logic equation given by

$$Y = \overline{A} + \overline{B}$$

In words, Y equals not A or not B. Hence, $A = 0$ or $B = 0$ makes $Y = 1$, otherwise $Y = 0$, generating the truth table shown in Figure 5.25. With $Y = \overline{A} + \overline{B}$ having three 1's and one 0, a more efficient logic circuit would result by implementing $\overline{Y}$, which has one 1 and three 0's, and then negating the output of that logic circuit with a NOT gate.

The logic equation for $\overline{Y}$ from the truth table gives

$$\overline{Y} = A \cdot B$$

Negating this logic equation to obtain Y gives

$$Y = (\overline{Y}) = \overline{A \cdot B}$$

Equating the initial logic equation with this last result, we arrive at de Morgan's first law:

$$\overline{A \cdot B} = \overline{A} + \overline{B}$$

Assigning Don't Care Values

Some truth tables have input sequences that will never occur in practice. In such cases, the output value for these *impossible* sequences is assigned an × symbol, which indicates a "don't care" condition. The output value can then be assigned either a 0 or 1 value—whichever is more convenient for implementing an efficient logic circuit.

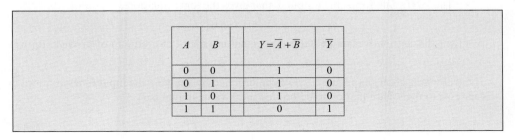

A	B		$Y = \overline{A} + \overline{B}$	$\overline{Y}$
0	0		1	0
0	1		1	0
1	0		1	0
1	1		0	1

Figure 5.25

Truth table for Example 5.12.

A	B	C	Y	$\overline{Y}$
0	0	0	1	0
0	0	1	1	0
0	1	0	1	0
0	1	1	1	0
1	0	0	1	0
1	0	1	1	0
1	0	1	1	0
1	1	0	X	X
1	1	1	0	1

Figure 5.26

Truth table with the complement of the output shown and containing don't-care ("X") values.

EXAMPLE 5.13 — Efficient logic circuit implementation

Consider the truth table shown in Figure 5.26. We consider two logic circuits that implement a truth table, the first with $\times = 0$ and the second with $\times = 1$.

Clearly, the number of 1's in $\overline{Y}$ is smaller than in Y—no matter what the value of $\times$. Hence we implement $\overline{Y}$ and then use a NOT gate to obtain Y.

For the $\times$ value, let us consider each value separately to determine the number of 1's.

For $\times = 1$, $\overline{Y}$ has a single input pattern that makes it 1,

$$\overline{Y} = A \cdot B \cdot C$$

and the logic equation for the truth table is

$$Y = \overline{A \cdot B \cdot C}$$

For $\times = 0$, $\overline{Y}$ has a two input patterns that makes it 1: $A = 1$ and $B = 1$, while C can equal either 0 or 1.

$$\overline{Y} = (A \cdot B \cdot C) + (A \cdot B \cdot \overline{C})$$

Hence, the value of C does not matter and the logic equation can be simplified to be

$$\overline{Y} = A \cdot B$$

Since $(\overline{\overline{Y}}) = Y$, the logic equation for the truth table is then

$$Y = \overline{A \cdot B}$$

Hence, making $\times = 0$ produces a more efficient design—one that uses fewer gates.

EXAMPLE 5.14 — Logic circuit analysis

Consider the logic circuit shown in Figure 5.27.

The logic equation for the logic circuit is found by proceeding from left-to-right forming the output of each gate.

- The NOT gate forms $\overline{B}$.
- The OR gate forms $A + \overline{B}$.
- The AND gate forms the output Y that gives the logic equation

$$Y = (A + \overline{B}) \cdot C$$

Applying the distributive law gives the logic equation with the desired sum-of-products form as

$$Y = A \cdot C + \overline{B} \cdot C$$

The truth table shown in Figure 5.27 was determined by applying the input binary pattern in each row to the logic equation to determine the Y value in that row.

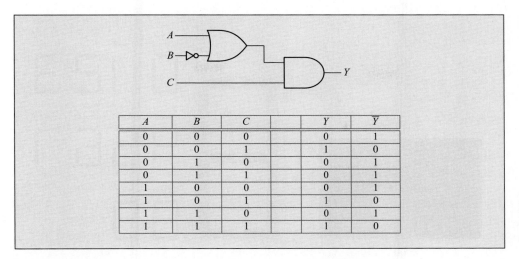

Figure 5.27

Logic circuit and truth table for Example 5.14.

A	B	C		Y	$\overline{Y}$
0	0	0		0	1
0	0	1		1	0
0	1	0		0	1
0	1	1		0	1
1	0	0		0	1
1	0	1		1	0
1	1	0		0	1
1	1	1		1	0

5.9 USEFUL LOGIC CIRCUITS

This section describes logic circuits that implement some interesting tasks that are often performed by smart devices.

5.9.1 Implementing an ExOR Gate

To implement the ExOR truth table in Figure 5.15, we use an AND gate to recognize each of the two patterns of A and B for which $Y = 1$. This occurs when $A \cdot \overline{B} = 1$ or when $\overline{A} \cdot B = 1$. This is expressed in logic equation form as

$$Y = (A \cdot \overline{B}) + (\overline{A} \cdot B)$$

To implement the ExOR gate the outputs of the two AND gates are applied to an OR gate as shown in Figure 5.28.

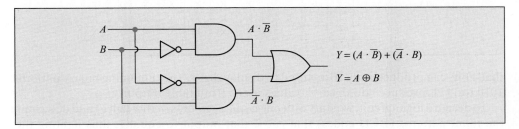

$$Y = (A \cdot \overline{B}) + (\overline{A} \cdot B)$$

$$Y = A \oplus B$$

Figure 5.28

Implementing an ExOR gate with AND, OR, and NOT gates.

5.9.2 Seven-Segment Display Driver

A common digital display used in clocks, scales, and smart sensors is the 7-segment display shown in Figure 5.29. The 7-segment display device has seven inputs, by convention labeled a, b, c, through g, with each connected to an LED segment. Making an input 1 lights the corresponding segment. Let us implement a logic circuit that displays digits.

It is not unusual to have a digital logic circuit that has more than one output. In this case, the initial design starts with a separate logic circuit design for each output. Then if it is noticed that some gates are common to both logic circuits, they can be used for both.

Decimal digits are typically represented in a 4-bit binary code, called *binary-coded-decimal (BCD)*. To form digits on the display, we need to implement a logic circuit that converts the 4-bit BCD pattern input into the appropriate seven logic circuit outputs with each one controlling its respective segment. The truth table relating the input BCD code to the seven outputs is shown in Figure 5.30. Note that, of the possible 16 patterns

Figure 5.29

(a) A 7-segment display. (b) Logic variables *a* through *g* identify the segments. When the logic variable of a particular segment equals 1, the segment is lit. (c) The logic variable values that produce the ten digits are shown.

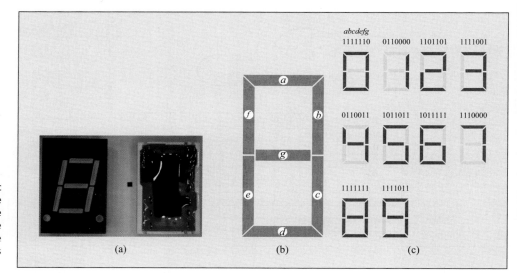

Figure 5.30

Truth table to convert binary-coded-decimal (BCD) codes to a 7-segment display.

		Inputs				Outputs						
row	digit	b_3	b_2	b_1	b_0	a	b	c	d	e	f	g
0	0	0	0	0	0	1	1	1	1	1	1	0
1	1	0	0	0	1	0	1	1	0	0	0	0
2	2	0	0	1	0	1	1	0	1	1	0	1
3	3	0	0	1	1	1	1	1	1	0	0	1
4	4	0	1	0	0	0	1	1	1	0	1	1
5	5	0	1	0	1	1	0	1	1	0	1	1
6	6	0	1	1	0	1	0	1	1	1	1	1
7	7	0	1	1	1	1	1	1	0	0	0	0
8	8	1	0	0	0	1	1	1	1	1	1	1
9	9	1	0	0	1	1	1	1	0	0	1	1
10	.	1	0	1	0	X	X	X	X	X	X	X
11	.	1	0	1	1	X	X	X	X	X	X	X
12	.	1	1	0	0	X	X	X	X	X	X	X
13	.	1	1	0	1	X	X	X	X	X	X	X
14	.	1	1	1	0	X	X	X	X	X	X	X
15	.	1	1	1	1	X	X	X	X	X	X	X

that 4 bits can produce, only 10 are used to display digits. The remaining binary patterns, 1010 to 1111, produce "don't care" values in the truth table output.

To design a logic circuit, we start with one of the segments (*a* through *g*) and determine the input patterns (BCD codes) that cause that output to equal 1, thus lighting that segment. Each such BCD code could be recognized using a 4-input AND gate pattern recognition circuit. The outputs of these AND gates then connect to an OR gate that becomes a logic 1 whenever any one of these BCD codes appears. The OR gate output connects to the segment LED. Each segment can be treated in this manner to implement the complete BCD to 7-segment decoder.

EXAMPLE 5.15

Lighting the *a* segment

From the truth table shown in Figure 5.30, we note that the a segment output contains eight 1's and two 0's. Hence, it is more efficient to implement a logic circuit for $\bar{a}$. Then, we can add a NOT gate to obtain the logic circuit for the a segment. By doing this, $\bar{a} = 1$ when the pattern 0001 (digit 1) or 0100 (digit 4) appears at the input. The logic circuit that implements this function is shown in Figure 5.31.

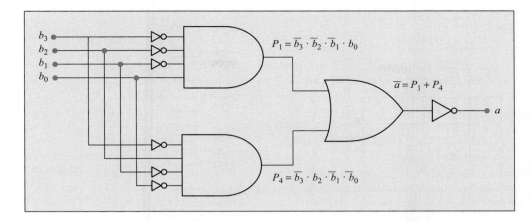

Figure 5.31

Logic circuit that lights segment a. Segment a is not lit only when either the digit 1 or 4 occurs. $P_1 = 1$ when a $1_{10} = b_3 b_2 b_1 b_0 = 0001_2$ occurs and $P_4 = 1$ when a $4_{10} = 0100_2$ occurs.

5.9.3 Binary Adder

One common task performed by a computer is to perform arithmetic operations, such as addition. We use the rules for doing addition on decimal numbers to determine how to perform the same operations on binary numbers. The references at the end of the chapter describe how the other arithmetic operations are performed.

When we add two groups of digits, we start with the least significant digits (the ones on the far right) and add these to compute the partial sum. If the sum S is less than the base of the number system (here $B = 10$), the sum is entered and the next significant digits are added. If the sum is greater than the base, the base is subtracted, the result is entered, and a carry digit C is added to the the sum of the next significant digits. The procedure is continued until the partial sum and the carry are both zero-valued.

The same procedure is applied to binary numbers, but the process is simpler because the numbers consist of only 0's and 1's. Let the first 3-bit binary sequence be denoted by $A_2 A_1 A_0$ and the second by $B_2 B_1 B_0$. The sum is denoted by $S_3 S_2 S_1 S_0$. Note that the sum can have one more bit than the two inputs because of the possibility of a carry. The carry bits are denoted by $C_2 C_1 C_0$.

Binary addition EXAMPLE 5.16

Consider adding $A_2 A_1 A_0 = 110_2$ to $B_2 B_1 B_0 = 111_2$.

Starting with the least significant bit (LSB) and expressing the result as the binary pattern $C\ S$ (because it directly expresses the value of a binary number), we get

$$A_0 + B_0 = 0 + 1 = 1_{10} = 1_2 \rightarrow C_0 = 0, S_0 = 1$$

The sum of the next significant bits plus the previous carry bit is

$$A_1 + B_1 + C_0 = 1 + 1 + 0 = 2_{10} = 10_2 \rightarrow C_1 = 1, S_1 = 0$$

The sum of the next significant bits plus the previous carry bit is

$$A_2 + B_2 + C_1 = 1 + 1 + 1 = 3_{10} = 11_2 \rightarrow C_2 = 1, S_2 = 1$$

C_2 is not zero, so we need to continue one more step. If the most significant bit (MSB) position has already been evaluated, the next carry value being $= 1$ indicates an *overflow* condition. The final sum equals the carry bit from the previous stage, as

$$S_3 = C_2$$

Hence, the sum is

$$S_3 S_2 S_1 S_0 = 1101_2$$

As a verification, adding $110_2 = 6_{10}$ and $111_2 = 7_{10}$ produces $13_{10} = 1101_2$.

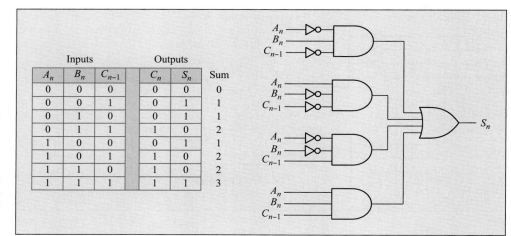

Figure 5.32

Truth table for a binary adder. Direct implementation of the sum bit S_n in a binary adder is shown. A similar logic circuit can be implemented to produce the carry bit C_n.

Inputs			Outputs		Sum
A_n	B_n	C_{n-1}	C_n	S_n	
0	0	0	0	0	0
0	0	1	0	1	1
0	1	0	0	1	1
0	1	1	1	0	2
1	0	0	0	1	1
1	0	1	1	0	2
1	1	0	1	0	2
1	1	1	1	1	3

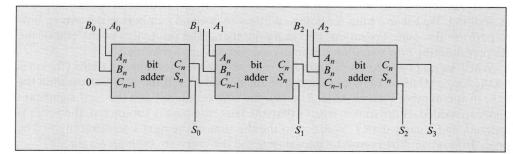

Figure 5.33

Repeating the same logic circuit, known as a *full adder*, can add two 3-bit numbers to form a 4-bit result.

Using this procedure, we can implement a logic circuit that adds two multi-bit binary numbers. In the description of binary addition just given, the main operation was the addition of the three bits formed by the two input bits, A_n and B_n, and the carry bit C_{n-1} from the previous sum to produce a sum bit S_n and carry bit C_n. The truth table that accomplishes this operation is shown in Figure 5.32. The output columns were formed with C_n followed by S_n to express binary numbers with the least-significant bit (LSB) being rightmost.

The logic circuit has three inputs and two outputs. Each output is implemented separately using the standard procedure. The logic circuit for the S_n output of the binary adder is shown in Figure 5.32.

A binary adder logic circuit is implemented by combining individual bit-adder sections. The circuit that adds two 3-bit numbers is shown in Figure 5.33. Note that the carry input to the first stage is a 0. The carry result from the last stage is the most significant bit in the sum.

5.9.4 Binary Multiplication

The direct implementation of digital addition can be extended easily to digital multiplication. The approach is the same as in the direct implementation of addition: For each combination of binary input patterns, we manually determine and specify the resulting output binary pattern.

Consider the multiplication of two 2-bit numbers $A_1 A_0$ and $B_1 B_0$ to produce product P:

$$P = A \times B \tag{5.11}$$

Inputs		Product		Inputs				Outputs			
A	B	P		A_1	A_0	B_1	B_0	P_3	P_2	P_1	P_0
0	0	0		0	0	0	0	0	0	0	0
0	1	0		0	0	0	1	0	0	0	0
0	2	0		0	0	1	0	0	0	0	0
0	3	0		0	0	1	1	0	0	0	0
1	0	0		0	1	0	0	0	0	0	0
1	1	1		0	1	0	1	0	0	0	1
1	2	2		0	1	1	0	0	0	1	0
1	3	3		0	1	1	1	0	0	1	1
2	0	0		1	0	0	0	0	0	0	0
2	1	2		1	0	0	1	0	0	1	0
2	2	4		1	0	1	0	0	1	0	0
2	3	6		1	0	1	1	0	1	1	0
3	0	0		1	1	0	0	0	0	0	0
3	1	3		1	1	0	1	0	0	1	1
3	2	6		1	1	1	0	0	1	1	0
3	3	9		1	1	1	1	1	0	0	1

Figure 5.34

Truth tables for direct implementation of a 2-bit multiplier. Decimal values are given in the left table and their binary equivalents in the right table.

Because A and B vary between 0 and 3, the product P varies between 0 and 9. Hence, 4 bits are required to represent the product $P = P_3 P_2 P_1 P_0$. Figure 5.34 shows the decimal table and binary truth table that describes the truth table.

5.9.5 Binary Division

The direct implementation of digital division requires the following two considerations.

1. The division implemented in a digital logic circuit is an *integer division* (that is, the result is an integer that is obtained by ignoring the remainder). For example, the integer division result of 3 divided by 2 is not 1.5, it is 1.

2. Division by 0 is not allowed and should generate an error—if and when it occurs. This is the purpose of a *divide-by-zero* error flag bit, which is a bit that is set to 1 when such a case occurs.

These considerations for integer division lead to the formulation of the truth table shown in Figure 5.35. Each combination of binary input patterns determines the resulting output binary pattern.

For example, consider the integer division of two 2-bit numbers $A_1 A_0$ and $B_1 B_0$ with the quotient being

$$Q = \frac{A}{B} \tag{5.12}$$

Because the A value of a 2-bit number varies between 0 and 3 and the non-zero denominator B varies between 1 and 3, the quotient Q varies between 0 (when $A = 0$) and 3 (when $A = 3$ and $B = 1$). Hence, 2 bits are required for the quotient $Q_1 Q_0$.

When $B = 0$, an error condition occurs. Use logic variable E to indicate an error with $E = 1$ indicating that a division by zero occurs. When $E = 1$, the values of $Q_1 Q_0$ do not matter and become a "don't care" condition, which can simplify a logic circuit design. Figure 5.35 shows the decimal and binary truth tables.

Inputs		Outputs		Inputs				Outputs		
A	B	Q	E	A_1	A_0	B_1	B_0	Q_1	Q_0	E
0	0	X	1	0	0	0	0	X	X	1
0	1	0	0	0	0	0	1	0	0	0
0	2	0	0	0	0	1	0	0	0	0
0	3	0	0	0	0	1	1	0	0	0
1	0	X	1	0	1	0	0	X	X	1
1	1	1	0	0	1	0	1	0	1	0
1	2	0	0	0	1	1	0	0	0	0
1	3	0	0	0	1	1	1	0	0	0
2	0	X	1	1	0	0	0	X	X	1
2	1	2	0	1	0	0	1	1	0	0
2	2	1	0	1	0	1	0	0	1	0
2	3	0	0	1	0	1	1	0	0	0
3	0	X	1	1	1	0	0	X	X	1
3	1	3	0	1	1	0	1	1	1	0
3	2	1	0	1	1	1	0	0	1	0
3	3	1	0	1	1	1	1	0	1	0

Figure 5.35

Truth tables for direct implementation of 2-bit division. Along with the quotient Q, and error condition $E = 1$ indicates division by zero. The "X" indicates a don't-care condition, and can take on either a 0 or 1 value.

5.10 Research Challenges

5.10.1 Molecular Logic Circuits

As an alternative to the logic circuits described in this chapter, researchers are investigating molecular logic gates that work using chemical signals rather than the electrical signals. Such devices could interact directly with biological systems, eliminating the need for chemical sensors and opening up new devices for biomedical engineering systems.

5.11 Summary

This chapter described combinational logic circuits that operate on binary-valued logic variables to accomplish a desired task. A combinational logic circuit produces an output logic variable that is related only to the current values of the input logic variables. Combinational logic includes the AND, OR, and NOT gates. Using these three gates, any logic function can be implemented as a logic circuit. The behavior of a combinational logic circuit was described with a truth table. The construction of a digital logic circuit from the truth table was illustrated with several applications.

5.12 Problems

5.1 Implementing an OR gate using AND and NOT. Implement a 2-input OR gate using only AND and NOT gates. (Hint: Start with the truth table that implements $\overline{Y}$.)

5.2 Using only NAND gates. This problem demonstrates that only a NAND gate is needed to implement the basic gates and, hence, any combinatorial logic circuit. Using only two-input NAND gates, implement a NOT, an OR, and an AND gate.

5.3 From logic equation to truth table. Generate the truth table that corresponds to the logic equation given by

$$Y = A \cdot \overline{B} + B \cdot \overline{C} + C \cdot \overline{A}$$

Implement a logic circuit that has a small number of gates.

5.4 Distributive law. Use the distributive law to generate the truth table that corresponds to the logic equation given by

$$Y = A \cdot (\overline{B} + C)$$

Implement one logic circuit directly from the logic equation. Implement a second logic circuit from the truth table using the sum-of-products approach. Advanced courses in logic design teach how to implement logic circuits having the minimum number of gates.

5.5 **Logic circuit analysis.** Generate the truth table that corresponds to the logic circuit shown in Figure 5.36. Draw a different logic circuit that implements this truth table.

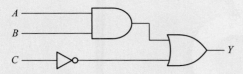

Figure 5.36

Logic circuit for Problem 5.5.

5.6 **Proving de Morgan's second law.** Starting with the truth table that implements $\overline{A + B}$, implement an alternative logic circuit that proves de Morgan's second law given in Eq. (5.10).

5.7 **7-Segment display: Lighting the b segment.** Implement the logic circuit that recognizes the BCD codes that light the b segment in a 7-segment display.

5.8 **Digital adder: Carry bit logic circuit.** Extending Figure 5.32, implement the logic circuit that produces the carry bit C_n from the inputs A_n, B_n, and C_{n-1}.

5.9 **Direct implementation of binary multiplication.** Using the truth table in Figure 5.34, implement the four logic circuits that produce each term in the product $P_3 P_2 P_1 P_0$.

5.10 **Direct implementation of binary division.** Using the truth table in Figure 5.35, design the logic circuit to produce the quotient $Q_1 Q_0$ and divide-by-zero error E.

5.13 Matlab Projects

5.1 **Implementing logic equations.** Using Example 16.26 as a guide, compose a Matlab script that forms the following logic equations, and generates and displays the truth table.

1. $Y = (A \cdot B) + (\overline{A} \cdot B)$

2. $Y = (A + \overline{B}) \cdot C$

3. $Y = A + B \cdot \overline{C} + \overline{A} \cdot \overline{B} \cdot C$

5.2 **Truth table for 7-segment display.** Using Example 16.26 as a guide, compose a Matlab script that forms the logic equation, and generates and displays the truth table that lights the segments in a 7-segment display having outputs a, b, c, d, e, f, g. Assume a logic variable equal to 1 lights the corresponding segment.

5.3 **Simulating a 7-segment display.** Compose a Matlab script that simulates a 7-segment display by plotting line segments with large *linewidth* values. Demonstrate your design by displaying a one-second-per-digit sequence of each digit from 0 to 9. (Hint: Write a function that draws an individual segment and call that function when a digit requires that segment.)

5.4 **Designing a digital display.** Compose a Matlab script that illustrates your creative design of digital display having more than 7 segments.

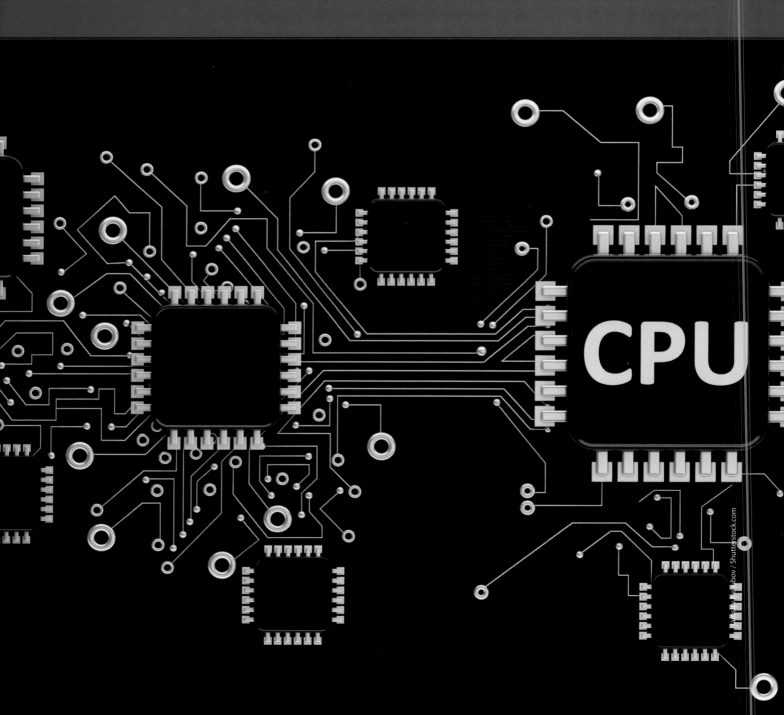

CHAPTER

6

CPU

SEQUENTIAL LOGIC CIRCUITS

LEARNING OBJECTIVES

After completing this chapter, the reader should be able to:

- Draw timing diagrams and construct state transition tables that illustrate the operation of flip-flops.

- Understand the process of designing and analyzing sequential logic circuits that implement digital memories and counters.

- Understand the functioning of a general-purpose computer.

- Compare computers of various sizes and computational abilities.

While a combinational logic circuit produces an output value that is a function of only the current input logic variables, the output of a sequential logic circuit also depends on its *state*, which is determined by the past input values. This chapter describes two elementary sequential circuits that are important components in smart devices: the *set-reset flip-flop* (SR-FF) is the basic memory element in many digital circuits and the *toggle flip-flop* (T-FF) is the basic counter element. The behavior of a flip-flop is described with a *timing diagram* and a state transition table that indicate the flip-flop output value in response to the changes in the inputs. The chapter then describes an alternative to logic circuit implementation by considering a general purpose computer that is *programmed* with a set of instructions to accomplish a specific task. The basic hardware and software elements in computers are presented.

6.1 INTRODUCTION

The main topics in this chapter are covered in the following sequence.

Set-Reset flip-flop—The SR-FF is the basic memory unit in digital logic circuits. Its implementation uses *positive feedback* that is illustrated with a pair of interconnected NOR gates.

Toggle flip-flop—The T-FF is the basic counting element in digital logic circuits.

Computers—Hardware implementation of logic circuits is replaced by computer instructions to simplify changes and upgrades. Computers are described in terms of their computational power and power consumption.

6.2 SEQUENTIAL CIRCUITS

A combinational logic produces an output that is a function of only the input logic variables. The sequential logic circuit output is a function of the inputs and also depends on its current *state*, which is a type of memory that causes the sequential circuit to produce different output values—even to the same input values.

A sequential circuit is similar to a human's *state of mind* that may elicit very different reactions to the same greeting. For example, if you studied all night, your response to a cheerful *"Good Morning"* may be quite different than if you had a full night's sleep.

This chapter describes an important class of sequential logic circuits called *flip-flops*. Flip-flops get their name from the behavior of their outputs flipping from one state to the other. While there are many different types of flip-flops used in practical logic circuits, this chapter describes the two shown in Figure 6.1 that are fundamental: the set-reset flip-flop (SR-FF), which is the basic memory unit in the computer, and the toggle flip-flop (T-FF), which is the computer's counting and pulse-generation element.

Figure 6.1

Flip-flops are an important class of sequential logic circuits. Set-reset flip-flops (RS-FFs) form memory circuits. Toggle flip-flops (T-FFs) form counting and pulse-generating circuits.

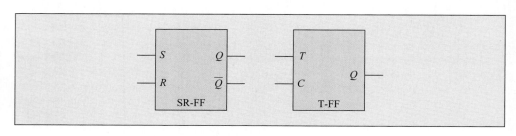

6.3 SET-RESET FLIP-FLOP

The set-reset flip-flop (SR-FF) is the basic *memory* element in digital devices. The SR-FF can be implemented using two OR and two NOT gates (or two NOR gates). What is novel and *non-combinational* is the presence of *feedback*, which connects an output back to the input. This feedback *reinforces* the asserted input values to implement the memory function.

Figure 6.2 shows the SR-FF has two inputs denoted S for *set* and R for *reset*, and two outputs conventionally denoted Q and its complement $\overline{Q}$. Having access to a logic variable and its complement often results in a simpler digital circuit because it may eliminate the need for a NOT gate. The SR-FF provides both for free.

While the truth table plays a major role in the design of combinational logic circuits, the *timing diagram* illustrates how the sequential circuit output changes in response to changes in the input. Figure 6.2 shows the timing diagram for the SR-FF. It is typical to use dashed arrows to associate the transition in an input, in this case $0 \rightarrow 1$, that cause the output change.

A *state transition table* is a alternate (but equivalent) description of a sequential logic circuit. It has a standardized form that contains all of the possible values of the inputs (along with the current output values) to determine the next output value. In this way, its input structure is similar to that of a truth table in combinational logic circuits, except the input section also includes the current output value. Figure 6.2 shows the state transition table for the SR-FF.

To describe the SR-FF behavior, let initial inputs $S = 0$ and $R = 0$ and output $Q = 0$, as shown in the first row of the state transition table. Let us examine both the timing diagram and each row of the state transition table:

(S, R, Q(cur) = 000): The next value of the output is $Q(next) = 0$, which is the same as the current state. Its complement is $\overline{Q} = 1$ (as expected). Since the output values remain at these values, this is a *stable state* of the sequential circuit.

(S, R, Q(cur) = 001): $Q(next) = 1$. Since it the same as the current state, this is another *stable state* as long as both $S = 0$ and $R = 0$.

(S, R, Q(cur) = 010): The $0 \rightarrow 1$ transition on the R input keeps $Q(next) = 0$.

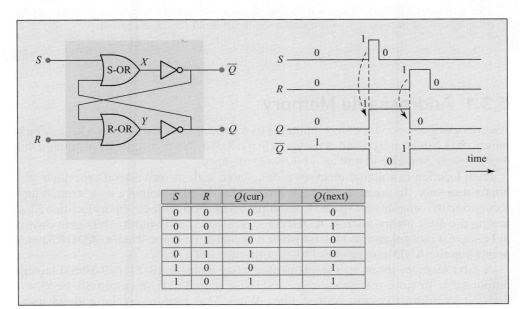

S	R	Q(cur)		Q(next)
0	0	0		0
0	0	1		1
0	1	0		0
0	1	1		0
1	0	0		1
1	0	1		1

Figure 6.2

The SR-FF implemented with OR and NOT gates, its timing diagram, and its state transition table. It is not allowed to have $S = 1$ and $R = 1$ at the same time.

$(S, R, Q(cur)) = 011)$: The $0 \to 1$ transition on the R input makes $Q(next) = 0$.

$(S, R, Q(cur)) = 100)$: The $0 \to 1$ transition on the S input makes $Q(next) = 1$.

$(S, R, Q(cur)) = 101)$: The $0 \to 1$ transition on the S input keeps $Q(next) = 1$.

The transition table does not have the full 8 $(= 2^3)$ rows, because it is not allowed to have $S = 1$ and $R = 1$ at the same time.

This SR-FF behavior can be explained by examining the logic circuit. The output Q also connects (or *feeds back*) to the S-OR gate input. Thus, with $Q = 0$ and $S = 0$, the output of the S-OR is $X = 0$. As X passes through the NOT gate, $\overline{Q} = \overline{X} = 1$. The $\overline{Q}$ output also feeds back to an input of R-OR having output Y. With $\overline{Q} = 1$, $Y = 1$, which passes through its NOT gate to produce $Q = \overline{Y} = 0$. This completes the feedback loop that results in a stable set of input and output values. Thus, the SR-FF is in a stable condition when $S = 0$ and $R = 0$.

Lets examine what happens when a transition occurs on the R or S inputs:

R $0 \to 1$: When the R input on the R-OR gate changes from $0 \to 1$, it makes $Y = 1$, which in turn causes $Q = 0$. With $S = 0$ and $Q = 0$, $X = 0$, making $\overline{Q} = 1$, which feeds back to the other input of the R-OR, reinforcing the $R = 1$ value. This is the reset condition that drives $Q \to 0$.

S $0 \to 1$: When the S input on the S-OR gate changes from $0 \to 1$, it makes $X = 1$, which in turn causes $\overline{Q} = 0$. With $R = 0$ and $\overline{Q} = 0$, $Y = 0$, making $Q = 1$, which feeds back to the other input of the S-OR, reinforcing the $S = 1$ value. This is the set condition that drives $Q \to 1$.

In summary, the value of Q *remembers* which input, S or R, was 1 last. This is the *memory feature* of the SR-FF. The following example shows the SR-FF used in a car alarm system, converting a momentary pulse produced by a the car movement sensor into a continuous blaring sound.

EXAMPLE 6.1 Car burglar alarm

When a parked car experiences a sudden bump, a sensitive motion sensor (an accelerometer) produces a short-duration $0 \to 1 \to 0$ electronic pulse. This momentary pulse connects to the S input the SR-FF, as shown in Figure 6.3 and causes its Q output to become and remain at 1. This activates an alarm and keeps it blaring (at least for a while). The alarm is turned off by using a key that provides a $0 \to 1$ transition at the R input that resets $Q = 0$.

6.3.1 Addressable Memory

An important component of a computer is its *random access memory* (RAM) in which binary data can be stored and retrieved when needed. We first consider a simple 1-bit memory and then extend it to an 8-bit (byte) memory.

Each location in a digital memory is designated with its own address specified as a binary sequence. To access a particular location in our 1-bit memory, its address is first recognized by using an AND gate. For example, let us consider the memory location 53_{10}, having the 6-bit binary address ADDR53 $= 110101$. The six-input AND gate shown in Figure 6.4 recognizes this address when it occurs at its input. That is, ADDR53 $= 1$ only when the AND gate input is 110101, otherwise ADDR53 $= 0$.

A 1-bit memory location is implemented using a single SR-FF (SR-FF53) having output Q53. To store the binary value INDATA (equal to 0 or 1) into SR-FF53, we first put its address on the address lines. When 53_{10} appears on the address lines,

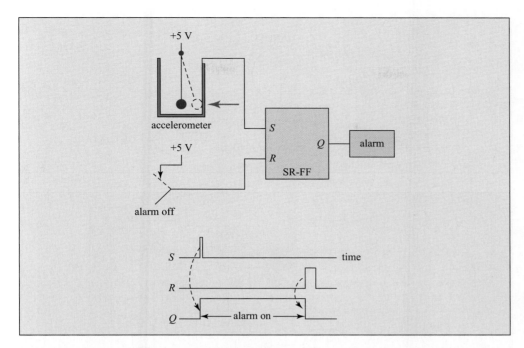

Figure 6.3

SR-FF used in a car alarm.

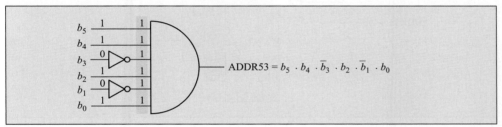

Figure 6.4

Address recognition logic circuit. The 6-input AND gate shown recognizes the 6-bit address 110101 (location 53_{10}).

ADDR53 $= 1$, and this output connects to two AND gates, with the first having output X that connects to the S input of SR-FF53 and other output Y that connects to R, as shown in Figure 6.5. The data INDATA is applied to *all* the SR-FFs in the memory, but only one (SR-FF53) has its address AND gate output equal to one. If INDATA $= 1$, X becomes 1 and SR-FF53 is set, making Q53 $= 1$. If INDATA $= 0$, it passes through a NOT gate making $Y = 1$ and SR-FF53 is reset, making Q53 $= 0$. When 53_{10} is removed from the address line ADDR53 $= 0$, then $X = 0$ and $Y = 0$, thus preventing any changes to occur in SR-FF53. Thus, the INDATA value is stored in SR-FF53.

To retrieve the data value Q53 stored in SR-FF53, we apply ADDR53 $= 110101$ to all of the address recognition circuits, making only ADDR53 $= 1$. ADDR53 and Q53 are applied to the *data output AND gate* AND53 with output MEM53, as shown in Figure 6.6. If Q53 $= 1$ and ADDR53 $= 1$, the output MEM53 $= 1$. If Q53 $= 0$ and ADDR53 $= 1$, MEM53 $= 0$. All of the other data output AND gates at address $XX_{10} \neq 53$ produce 0 outputs because their ADDRXX $= 0$. The outputs of all data output AND gates are applied to the input of one large OR gate (at least conceptually)

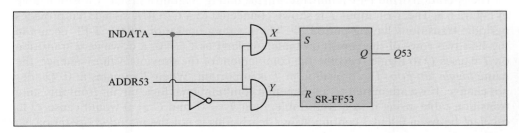

Figure 6.5

Writing the data bit value INDATA into SR-FF53 at memory location ADDR53.

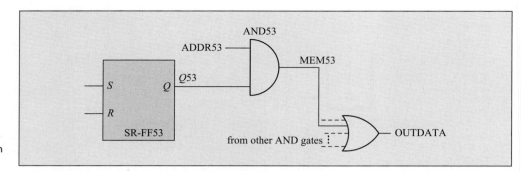

Figure 6.6

Reading the data bit value stored in memory location 53.

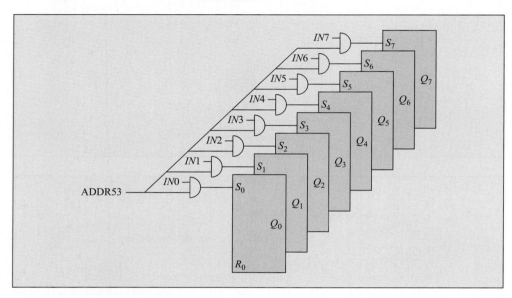

Figure 6.7

A byte in computer memory is organized as a stack of eight SR-FFs.

to produce the output data OUTDATA. Note that OUTDATA = 1 only if Q53 = 1. That is, the retrieved value OUTDATA has the same binary value as that stored in SR-FF53.

Binary data are typically stored as 8-bit units (bytes). As shown in Figure 6.7, a byte in an 8-bit memory can be considered to be a stack of eight SR-FFs sharing a common address.

 ## 6.4 TOGGLE FLIP-FLOP

The toggle flip-flop (T-FF) is the basic counting element of the computer and is shown in Figure 6.8. The T-FF has two inputs, T for toggle and C for clear, and one output Q. The value of Q switches from 0 to 1 or from 1 to 0 (*toggles*) whenever there is a downward ($1 \rightarrow 0$) transition at its T input. When a 1 is applied to the C input, $Q = 0$ immediately and maintained at 0 as long as $C = 1$.

The operation of the T-FF is illustrated in the timing diagram and state transition table in Figure 6.8. The T-FF input T is shown connected to a *CLOCK* signal that provides periodic transitions having period T_C. The toggling operation of the T-FF occurs in the first two rows of the transition table: As long as $C = 0$, a downward transition on T causes Q to toggle (become the complement of the previous value)—hence, the name *toggle flip-flop*. The transition in T is important: When T remains at 0, Q does not change. If we attempted to implement a combinational logic circuit from this state transition table, using it as a truth table, then $T = 0$, and $C = 0$ would cause Q to oscillate between 0 and 1 continuously. Clearly, this is not the intended operation, so

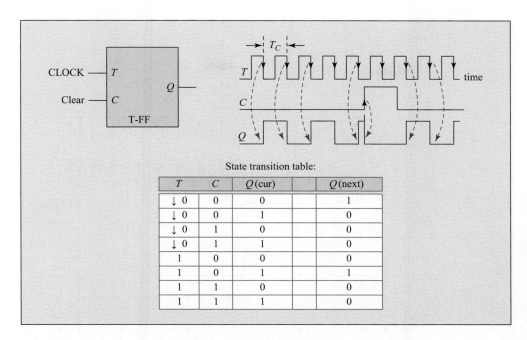

Figure 6.8

Toggle flip-flop (T-FF) symbol, timing diagram, and transition table. The Q output switches its value whenever there is a downward transition ($1 \rightarrow 0$) on the T input.

the digital electronic circuit that implements a T-FF incorporates a transition sensing device in the T input, which is beyond the level of this course.

The timing diagram and transition table show that whenever $C = 1$, $Q = 0$—no matter what occurs on the T input.

 # 6.5 COUNTING WITH TOGGLE FLIP-FLOPS

Toggle flip-flops are connected in a chain for counting the downward transitions in pulses. In a chain, the Q output of one T-FF connects to the T input of the next T-FF. A $1 \rightarrow 0$ transition at the T input of any T-FF switches its output Q to the complementary value. The T-FF chain counts the $1 \rightarrow 0$ transitions that occur at the T input of the first T-FF. The T-FF outputs are indexed starting with 0, making Q_0 correspond to the coefficient of 2^0, Q_1 corresponds to the coefficient of 2^1, and so on. The count of downward ($1 \downarrow 0$) CLOCK transitions is encoded in the T-FFs outputs in the binary form

$$CNT = Q_3 \ Q_2 \ Q_1 \ Q_0$$

For a chain containing M T-FFs, numbered 0 to $M - 1$, the CNT value is determined by

$$CNT = Q_0 + 2Q_1 + 4Q_2 + 8Q_3 + \cdots + 2^{M-1}Q_{M-1} \tag{6.1}$$

Thus, a chain containing M T-FFs produces unique CNT values in the range

$$0 \leq CNT \leq 2^M - 1 \tag{6.2}$$

This is called a *modulo-M counter*. At the $2^{M^{th}}$ input transition, all Q values reset to 0, making $CNT = 0$.

Circuits typically determine the count value CNT at the upward ($0 \uparrow 1$) transition of the clock, which occurs during the stable period between downward transitions. The clock period T_C is chosen so that $T_C/2$ is sufficiently long for all the T-FFs in the chain to settle to their final stable value. This settling time is one consideration that limits the minimum clock period $T_{c,\min}$, or maximum clock frequency of

$$f_{c,\max} = \frac{1}{T_{c,\min}} \tag{6.3}$$

for a digital device.

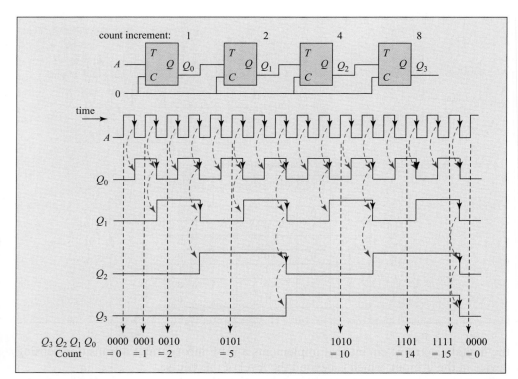

Figure 6.9

Counting with a chain of four T-FFs. T-FFs change on downward transition in A and count value is observed on upward transition.

EXAMPLE 6.2

Modulo-16 binary counter

Figure 6.9 shows 4 T-FFs connected in a chain. All C inputs are equal to 0. The input to the first T-FF is denoted A, which contains the downward transitions we want to count.

The T-FFs are initialized by resetting all Q values to 0. This is typically done by a short pulse $(0 \rightarrow 1 \rightarrow 0)$ applied simultaneously to all C inputs (not shown in the figure). The initial count is shown as $Q_3 Q_2 Q_1 Q_0 = 0000$ on the lower left.

The $1 \rightarrow 0$ transitions are the ones that cause changes, and these are indicated with downward arrows. The first $1 \rightarrow 0$ transition on A toggles Q_0 from $0 \rightarrow 1$. Q_0 also connects to T on the next T-FF, but its upward transition has no effect because a T-FF changes only on a $1 \rightarrow 0$ transition. After the first $1 \rightarrow 0$ transition on A, $Q_3 Q_2 Q_1 Q_0 = 0001$, corresponding to $CNT = 1$.

The second $1 \rightarrow 0$ transition on A toggles Q_0 from $1 \rightarrow 0$, and this transition toggles Q_1 from $0 \rightarrow 1$. After the second $1 \rightarrow 0$ transition on A, $Q_3 Q_2 Q_1 Q_0 = 0010$, making $CNT = 2$.

This procedure continues until $Q_3 Q_2 Q_1 Q_0 = 1111$, or $CNT = 15$. The next (16th) $1 \rightarrow 0$ transition on A toggles Q_0 from $1 \rightarrow 0$, and this transition toggles Q_1 $1 \rightarrow 0$, and so on, producing the final stable values $Q_3 Q_2 Q_1 Q_0 = 0000$, for $CNT = 0$. This is the initial state of the T-FF chain, so CNT takes on values from 0 to 15, thus producing a modulo-16 counter.

6.5.1 Modulo-K Counters

This section shows how to implement a modulo-K counter when K is not a power of 2 by including a combinational logic circuit in the counter. The previous section shows that a chain of M T-FFs counts from 0 to $2^M - 1$. If we want a modulo-K with $K < 2^M - 1$, we need to reset the counter when it reaches K.

To form a modulo-K, we need M T-FFs where

$$M \geq \log_2 K = \frac{\log_{10} K}{\log_{10} 2} \tag{6.4}$$

Number of T-FFs in a modulo-100 counter EXAMPLE 6.3

A modulo-100 counter counts from 0 to 99. The number of T-FFs needed in the chain (M) can be determined by either *direct calculation* or using *intuition*.

Direct calculation: To form a modulo-K counter with $K = 100$, we compute

$$M = \text{ceiling}(\log_2 100) = \text{ceiling}\left(\overbrace{\frac{\log_{10} 100}{\log_{10} 2}}^{=2/0.3}\right) = \text{ceiling}(6.7) = 7$$

Thus, we need seven T-FFs in the chain.

Intuition: To count modulo-100, we need a number of T-FFs to count at least that number. Because M T-FFs is a modulo-2^M counter, we need the smallest M such that $2^M \geq 100$. A little experience with powers of 2 provides the answer: because $2^6 = 64(< 100)$, $M > 6$. Adding one more T-FF to the chain produces a modulo-2^7 counter. Because $2^7 = 128$ (> 100), $M = 7$ is both necessary and sufficient.

A modulo-K counter starts the count at 0 and resets it to 0 at the Kth count. This is accomplished using the following three steps.

1. Determine the binary pattern that represents K.
2. Recognize this binary pattern with an AND gate circuit whose inputs connect to $Q_0, Q_1, Q_2, \ldots, Q_{M-1}$. The Q values that equal 1 in the pattern for K connect directly from Q to the AND gate. For the Q values that equal 0, the connection is through a NOT gate to make them 1.
3. Connect the AND output to the C inputs of all M T-FFs.

When the count equals K, the AND gate output becomes 1, which immediately resets all the T-FFs to zero. The all-zero pattern of the T-FFs is then no longer the pattern that the AND recognizes, making the AND gate output 0. Making $C = 0$ allows the T-FF chain to restart the count at the next downward transition. The pulse at the AND output is very short—so short that the Q values momentarily representing K can be neglected. The next example illustrates this operation.

Modulo-6 counter EXAMPLE 6.4

A modulo-6 counter produces counts 0, 1, 2, 3, 4, 5, 0, 1, 2, ... To count six downward transitions, we need

$$M = \text{ceiling}(\log_2 6) = \text{ceiling}\left(\overbrace{\frac{\log_{10} 6}{\log_{10} 2}}^{=0.78/0.3}\right) = \text{ceiling}(2.59) = 3$$

or three T-FF's. By intuition, $2^2 = 4$ (too small) and $2^3 = 8$ (necessary and sufficient).

Figure 6.10 shows the three T-FF chain. The modulo-6 counter is implemented using the following three steps.

1. The binary pattern for $6_{10} = 110_2$ ($= Q_2 Q_1 Q_0$).
2. The outputs of the T-FFs are connected to an AND gate that recognizes this binary pattern. When this pattern occurs, the AND gate output $Y = 1$.
3. The AND gate output connects to the C inputs of all the T-FFs, resetting them all to zero when $Q_2 Q_1 Q_0 = 110$, making $Y = 1$. This resets the counter to zero, which makes $Y = 0$ because $Q_2 Q_1 Q_0 = 000$.

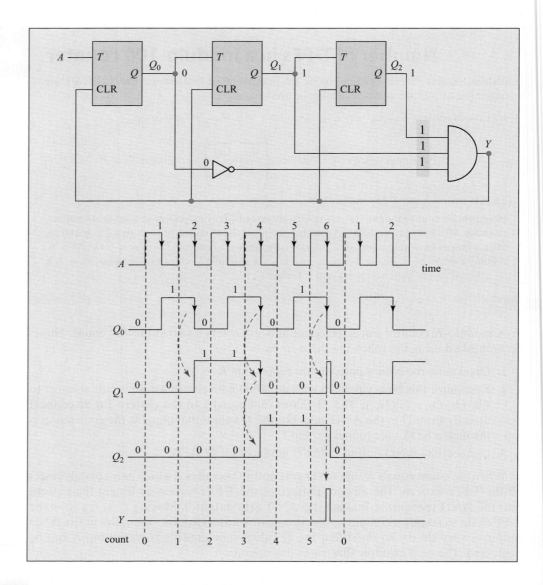

Figure 6.10

A modulo-6 counter circuit.

EXAMPLE **6.5**

Modulo-44 counter

To design a modulo-44 counter we need six T-FFs, as shown in Figure 6.11, because $2^6 = 64$ and $2^5 = 32$ makes five T-FFs insufficient.

The input A contains the $1 \downarrow 0$ transitions that are counted and connects to the first (Q_0) T-FF T input. The Q outputs connect to the T inputs of the next T-FF in the chain. The AND gate recognizes the pattern

$$44_{10} = 32 + 8 + 4 = 101100_2$$

with the logic equation as

$$C = Q_5 \cdot \overline{Q_4} \cdot Q_3 \cdot Q_2 \cdot \overline{Q_1} \cdot \overline{Q_0}$$

The AND gate output C connects to the CLR inputs of all of the T-FFs.

Figure 6.11 shows the T-FF output values at the count of 44. The chain is cleared immediately after this count appears.

The next section describes a method that implements a counter that uses a smaller gate count.

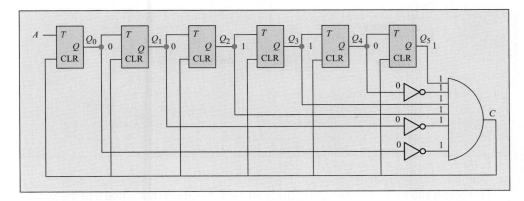

Figure 6.11

Modulo-44 counter in Example 6.5.

6.5.2 Efficient Modulo-K Counter

The previous example of a modulo-6 counter used a 3-input AND gate to recognize the binary pattern for 6 (= 110). While this works, a more efficient counting circuit is implemented by observing the counting sequence and noting that the T-FF Q values that equal 0 do not need to be connected to the AND gate. This is the case because the pattern of 1's in the binary representation of K in a modulo-K counter is sufficient to reset the counting chain at the correct time. The other binary sequences containing the same pattern of 1's occurs at count values greater than K, and hence never occur in the modulo-K counter.

This observation eliminates one or more NOT gates and uses a pattern recognition AND gate with fewer inputs in the efficient counter implementation. The following example illustrates the approach.

Efficient modulo-6 counter implementation   EXAMPLE 6.6

Consider the values of $Q_2 Q_1 Q_0$ in the counting sequence shown in Figure 6.12. Note that $Q_2 Q_1 = 11$ in the binary pattern for six (=110) and for the pattern for seven (=111).

Because the modulo-6 counter resets at the count of six, **the count never reaches seven.** Thus, the value of Q_0 does not matter. Thus, its NOT gate can be eliminated and an AND gate with fewer inputs is used. Figure 6.13 shows the efficient implementation of the modulo-6 counter.

Count	Q_2	Q_1	Q_0
0	0	0	0
1	0	0	1
2	0	1	0
3	0	1	1
4	1	0	0
5	1	0	1
6	1	1	0
7	1	1	1

Figure 6.12

Values of $Q_2 Q_1 Q_0$ in the counting sequence.

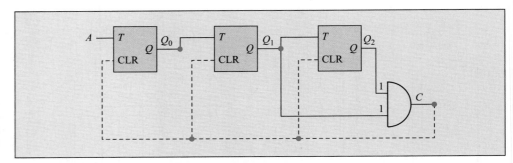

Figure 6.13

Efficient implementation of modulo-6 counter that recognizes only the 1 values in 110 to reset the counting chain.

This result can be generalized to include all T-FF Q values that equal 0 in the pattern for K in a Modulo-K counter; They do not need to be included in the AND gate inputs. For example, a modulo-85 counter has binary pattern equal to

$$85_{10} = 1010101$$

If we replace any 0 in this pattern with a 1, the resulting number is greater than 85. For example, replacing the middle 0 produces

$$101\ 1\ 101 = 85 + 8 = 93_{10}$$

However, this pattern will not be observed in the modulo-85 counter, because the count resets back to 0 at count 85.

 EXAMPLE 6.7 **Efficient modulo-44 counter**

Figure 6.14 shows an efficient implementation of the modulo-44 counter of Example 6.5. The AND gate recognizes the pattern

$$44_{10} = 32 + 8 + 4 = 101100_2$$

by applying only the bit positions that equal 1 to an AND gate to form the logic equation as

$$C = Q_5 \cdot Q_3 \cdot Q_2$$

The pattern recognition AND gate used previously in Example 6.5 had six inputs. The efficient implementation uses a 3-input AND gate and eliminates three NOT gates.

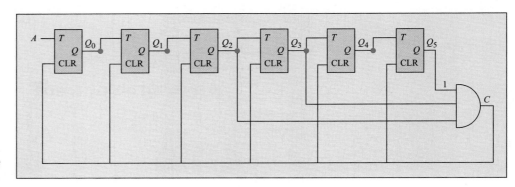

Figure 6.14

Efficient implementation of the modulo-44 counter in Example 6.7.

6.6 FROM LOGIC CIRCUITS TO COMPUTERS

The combinational and sequential logic circuits described previously implemented logic equations using hardware logic gates. Inexpensive microcontrollers and microcomputers offer the alternative to implement logic equations in software (as computer instructions). The remainder of this chapter describes the operation of a general computer.

6.6.1 Computer Components

Although computer designs cover a wide spectrum—from those in smartphones to the massive collection of servers in a computer cloud cluster (discussed in Chapter 15)—four components remain constant from the earliest computers to those in the cloud.

CPU: Each computer has a *central processing unit (CPU)* at its core. CPUs come in different architectures that form trade-offs depending on your computing needs:

- Some are fast but require a lot of power.

- Some are power efficient (to extend battery life) at the expense of speed.

- Some are designed to display video while connecting to a network to obtain the content.

Memory: Computer memory is often divided into three categories.

- *Random-access memory (RAM)* can store and retrieve data. RAM implemented with T-FFs is volatile in that its data is lost if power is removed. Current RAM capacity of a computer is measured in billions of bytes or *gigabytes* (GB). It is common to find PCs with at least 2 GB of RAM with some models accommodating up to 16 GB or more. Additional memory can be added to these computers by plugging in additional memory modules.

- *Read-only memory (ROM)* can only retrieve permanently-stored data. ROM is like a book, in that previously stored data are read from it but cannot be written into it. Because it only retrieves stored data, the ROM has a simpler structure than RAM. Thus, the ROM has a greater capacity for a given physical size. Popular forms of ROM are the optical compact disk (CD-ROM) that stores 640 MB and the digital versatile disk read-only memory (DVD-ROM) that can store up to 4.7 GB.

- *Electrically-erasable and programmable read-only memory (EEPROM)* can be written into and retained—even after power is removed. The EEPROM memory is what allows your smartphone to remember the phone numbers of your contacts—even if your battery discharges completely. While EEPROM can address individual bytes in memory, *flash* memory is an EEPROM that is erased and written in blocks for simplifying the connections used for addressing the memory. Thumb drives are implemented using flash, because a block containing 512 bytes is manipulated as a single unit.

> *Factoid:* A well-worn joke among EE's is that an engineer fresh out of school is assigned to design a Write-Only Memory (WOM).

Operating System: The *operating system (OS)* controls computer operation by structuring the various tasks it needs to perform, such as processing the scenes in a video display, while simultaneously receiving Internet packets. Current operating systems use icons and touch screens to receive user commands.

Apps: An *application program (app)* is software that is written to perform a particular task, such as word processing, computer-aided graphic design, or even assisting your golf game on a smartphone. Prior to the availability of such *shrink-wrap* (or now more typically, *down-loadable*) apps, computer users needed to be *computer programmers* who were familiar with the OS, and at least one computer language, such as C or Java. Programming knowledge is seldom necessary, because it is possible to purchase apps that make your computational activity more efficient, easier to use, and more flexible. So you can be on your way to being a *computer user* without having any idea of how the computer works or what language was used to write the app. Software programmers by the thousands try to make their fortunes by creating *killer apps* that users will find appealing. Thus, newer versions of software appear on a regular basis with each version offering more *bells and whistles*.

NIC: A *network interface card (NIC)* connects the CPU to a network, either through a cable or wireless connection. The NIC has permitted the personal computer to change from a *personal computing device* to the following main directions:

- An *archive* that provides access to almost any imaginable information.

- An *entertainment device* that displays video and interactive games.

- A *social communication means* for the general public through the smartphone.

- A *portal* to a larger computer system having any desired computing power and memory capacity.

Simple CPU Architecture

Figure 6.15 shows the essential components of a CPU that are typically contained on a single integrated circuit chip.

- The *memory* contains instructions and data. The input to the memory is the address of a memory location and the output is the contents of that location. Data include results of computation and the pixel values for your display.

- The *program counter* maintains the address of the current instruction being executed. Upon completion of the instruction, the program counter contains the address of the next instruction.

- The *control unit* interprets instructions by moving data to appropriate registers and performing the desired function.

- The *arithmetic logic unit (ALU)* receives numerical data, processes it according to the control unit instruction, and produces a result.

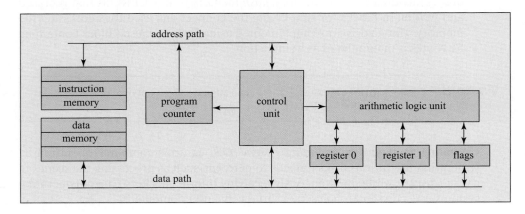

Figure 6.15

Components in a simple computer.

- The *registers* are a set of fast byte-size memories that temporarily store the data that are processed by the ALU. Two registers (R0 and R1) are shown, but there may be eight or more.

- The *flags* are bit-sized memory locations that indicate the outcome of the just-completed arithmetic operation. For example, the ALU operation producing a zero result would set the Z (for zero) flag to 1. If the result was negative, the N (for negative) flag would be set to 1. If the result was too large to be stored in the register, the O (for overflow) flag would be set to 1. These flags are used by instructions that make comparisons and for indicating error conditions.

6.6.2 Computer Computational Power

Computers come in a wide variety of types, depending on the application specifications regarding cost, size, power consumption, memory, and speed. The list from simplest to most complex includes:

Microcontroller (MCU): The MCU is the smallest stand-alone computer, typically contained on a single IC chip. An MCU has the basic computer components and operates by following a sequence of instructions. An MCU is typically designed for a particular application and is simple enough not to require an operating system. The low cost of the low-end devices (some less than one dollar) make them common components in intelligent systems. For example, an automobile may contain as many as 30 MCUs for controlling individual components such as anti-lock braking systems (ABS) or ignition timing. High-end MCUs control smart devices, such as your smartphone, that perform a number of specialized tasks, such as communication, encryption, and displaying video. MCUs typically contain EEPROM memories that are updated as features change.

Microprocessor (MPU): The MPU, when compared to the MCU, has a faster clock, accesses a larger memory that resides in external chips, and is controlled by an operating system. On the down side, the MPU has a greater power consumption, which in some cases is 100 times more than an MCU.

Personal computer (PC): The typical PC (either a desktop, laptop, or tablet) typically contains two separate specialized computers with each implemented on its own integrated circuit chip.

> **CPU:** The CPU is the general-purpose processor that executes a wide variety of instructions. The instructions are written by application programmers and the CPU executes them serially—one after another. Current-generation CPUs are typically implemented using several MPU *cores*. These cores operate in parallel: one executing OS instructions and others executing apps. When the chip temperature exceeds a threshold, execution is switched from hot operating cores to cool inactive cores, maintaining the chip temperature below a damaging level.

> **GPU:** The GPU (Graphic Processor Unit) is designed for one purpose: To generate an image on the display quickly (every 5 ms) no matter how large the display size. For example, a 22 inch monitor having 1680×1050 resolution contains 1.8 million pixels, each pixel having red, green, and blue LEDs, and each LED brightness is specified with 8 bits. The GPU computes these five million values for each frame that is displayed every 5 ms. The GPU achieves this performance through an architecture that exploits the parallel nature of the processing. One may think of a GPU that consists of 1.8 million elementary CPUs—one per pixel. In practice, elementary CPUs are basically simple arithmetic logic units with as many as 2,500 per GPU. These elementary CPUs are in turn controlled by MCUs, typically numbering from 8 to 64 per GPU.

Servers: Servers are specialized CPUs that are used in internet nodes for transmitting packets arriving their input, verifying their integrity, buffering them in memory to accommodate traffic congestion, and sending them off to their next destination node. While a standard CPU can perform these functions, servers are designed with architectures and instruction sets for efficient communication and storage to handle the high volume of data that occur in today's networks. While high-speed communications are important, power consumption and reliability are major concerns because servers are typically packed together in thousands in one service center.

Supercomputer: Supercomputers are the computer giants and are necessary for high-performance computing (HPC) applications. These involve modeling complex chemical and biological processes, simulating the response of airplane wings made of novel materials to random wind gusts, and determining how to arrange production parameters so potato chips can be made to stack into their cans during packaging. Supercomputers are measured and compared in terms of the number of floating-point operations (flops) that they can perform in one second. The record in 2013 is 35 Teraflops/sec. Supercomputer design involves more than simply interconnecting of a large number of fast CPUs. It also requires a co-design of both hardware and software along with the architecture of the interconnections. Different arrangements perform better for certain types of computational tasks. Some tasks for example are CPU intensive, such as optimizations and sorting, while others are memory intensive, such as big database searches. Thus, a specific architecture optimized for each type is needed.

6.6.3 Computer Power Dissipation

To examine the power consumed by a computer, we examine the two main sources of significant power dissipation (conversion to heat): Switching logic levels and leakage current.

Switching Logic Levels

Switching a logic gate value from 1 to 0 and from 0 to 1 requires charging and discharging the capacitance in the transistors and metal interconnections between supply voltage V_S and ground. The energy stored in a capacitor is

$$\mathcal{E}_S = \frac{1}{2} C V_S^2 \tag{6.5}$$

where C is the capacitance in the transistors and wires and V_S is the supply voltage. Discharging this capacitor through a resistor dissipates this stored energy as heat, thus making the resistor hot. This discharge occurs every clock cycle (at the clock frequency f_{clk}). Thus, the power dissipated in a computer (that is, converted to heat) by switching is given by

$$P_S = \mathcal{E}_S f_{clk} = \frac{1}{2} C V_S^2 f_{clk} \tag{6.6}$$

Let us examine how each of these terms affect the power dissipation.

C: P_S can be reduced by making the capacitance smaller. Recall that $C = \epsilon A/d$. Making computer chips smaller reduces the A of each device, and hence reduces the power dissipated.

f_{clk}: P_S can be reduced by using a slower clock frequency. In recent designs the f_{clk} value has increased to 4 GHz to improve computer performance, but power concerns have recently decreased f_{clk} to 3 GHz and less. Smartphone computers have f_{clk} between 1 GHz and 2 GHz to conserve battery power.

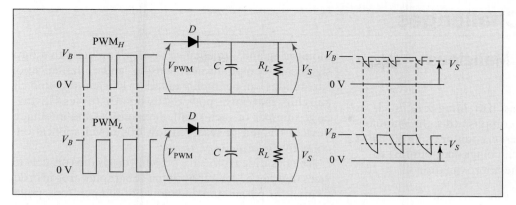

Figure 6.16

PWM switching lowers the average computer supply voltage V_S from the battery voltage V_B.

V_S: P_S is reduced by reducing the supply voltage. This reduction is typically accomplished by using pulse-width modulation (PWM) to charge a capacitor to the desired voltage, as shown in Figure 6.16. PWM_H has a duty cycle that produces a higher V_S, while PWM_L has a lower duty cycle that produces a lower V_S. The capacitor charges up to V_B through diode D quickly and discharges through load resistance R_L.

The square of the voltage makes it a more important parameter than C or f_{clk} for reducing P_S. Typical computer supply voltages vary between 5 V and 3.3 V.

> **Factoid:** My research friends tell me that (in the future) fast computers will work at $V_S = 1.2$ V, while slower computers will require only $V_S = 0.8$ V.

Leakage Current

While making computer chips smaller lowers the switching power P_S, the non-ideal characteristics of smaller transistors to accommodate smaller computer chips causes a *leakage current* that results in a second form of power dissipation. This leakage current I_L occurs as long as a supply voltage is connected to the circuit, even when the computer is doing *nothing*. The power due to leakage current is

$$P_L = V_S I_L \tag{6.7}$$

The leakage occurs in each transistor and hence is proportional to transistor count, which follows Moore's law. Thus, the newer computers experience greater leakage current, and the associated power dissipation P_L is becoming comparable to the switching power dissipation P_S. To reduce leakage current, unused cores in a multi-core computer are switched off by removing supply voltage, which is a technique called *dark silicon*. This core switching also allows the old core to cool down while the new core heats up. This, however, requires complex techniques to determine how and when to perform the switching in addition to the additional circuits needed to implement the switching.

The total power dissipated by a computer chip is the sum of these two power components

$$P_T = P_S + P_L \tag{6.8}$$

Computer engineers are challenged to minimize P_T to reduce cooling needs in large computers and to increase battery life in portable devices and robots.

> **Factoid:** It is during the switching times of digital circuits that generates most of the heat in a computer. Even though the switching time is short, switching occurs very often, especially when the computer is operating a high rate (GHz). Modern CPU chips embed a thermometer circuit that reduces the operating frequency when the chip becomes too hot.

6.7 Research Challenges

6.7.1 Discovering Malicious Hardware

Assume a manufacturer has bad intentions, such as planned obsolescence. To accomplish this, the manufacturer includes a separate (secret) counter using 40 T-FFs that is not included in the circuit diagram. Counting to 2^{40} with a 1 MHz clock (using the approximation $2^{10} = 10^3$) takes

$$\overbrace{2^{40}}^{=10^{12}} \times 10^{-6}\,s = 10^6\,s = 11 \text{ days of operation} \qquad (6.9)$$

When the counter highest T-FF makes a $1 \rightarrow 0$ transition, a switch connects the power supply directly to ground, destroying the chip and necessitating the purchase of a new device. Research into discovering the presence of such malicious circuits is a research challenge.

6.7.2 Smart Dust

At the other end of the spectrum from a supercomputer is a tiny computer, about the size of its battery that performs monitoring and surveillance tasks. Computers that have significant computational, sensing, and communication capabilities can be implemented on single integrated circuit chips that can be powered by small batteries. Having a large number of such small, inexpensive, and intelligent devices has led to research towards novel applications, collectively termed *smart dust*.

The basic idea is to have possibly thousands of such devices that can be scattered over a geographic area in order to monitor parameters indicative of temperature, pollution, human activity, or other interesting phenomena. Because each device is designed to operate efficiently, the battery life may be measured in terms of days to years in some applications. During this operational time period, the devices localize themselves in space and interconnect as a network to communicate the sensed data back to base stations, which are implemented as a few larger (and more expensive) digital devices.

Active research investigates organizational, computational, and communication tasks and relies on sophisticated theories in mathematics and computer science.

6.8 Summary

Sequential logic circuits produce an output value that is related not only to the current values of the inputs but also to their past values. The behavior of a sequential circuit is described with a timing diagram or a state transition table. A common sequential circuit, the flip-flop, was presented in its two basic forms: The set-reset flip-flop (SR-FF) is the basic memory unit in the computer, and the toggle flip-flop (T-FF) is the computer's elemental counting element. Combinational and sequential logic were combined to implement efficient counters.

Rather than implementing a specific logic circuit, a computer is *programmed* with a set of instructions to accomplish a specific task. This chapter described the basic hardware and software elements in computers.

6.9 Problems

6.1 SR-FF using two NOR gates. It is not allowed to have $S = 1$ and $R = 1$ at the same time for a SR-FF that is implemented with NOR gates as shown in Figure 6.2. What are the values of Q and $\overline{Q}$ when $S = 1$ and $R = 1$ if by mistake or during power-up?

6.2 Implementing an SR-FF using two NAND gates. Implement an SR-FF using two NAND gates and describe its behavior using a timing diagram.

6.3 Random-access digital memory input. Design a random-access digital memory that *stores* single bits stored in two addressable SR-FFs having 3-bit addresses 5 and 7. The memory has the address input lines $A2$, $A1$, $A0$ and data input line DIN.

6.4 Random-access digital memory output. Design a random-access digital memory that *reads* single bits stored in two addressable SR-FFs having 3-bit addresses 5 and 7. The memory has address input lines $A2$, $A1$, $A0$, and data output line DOUT.

6.5 Random-access digital memory input/output. Design a random-access digital memory that can write or read single bits stored in two addressable SR-FFs having 3-bit addresses 5 and 7. The memory has the following inputs and outputs

- Address input lines $A2$, $A1$, $A0$
- Data input line DIN
- Data output line DOUT

■ Read/Write input line RW (RW = 0 stores DIN into addressed memory. RW = 1 puts addressed memory data on DOUT).

6.6 Counting chain containing two T-FFs. Sketch the counting chain containing two T-FFs and its timing diagram, starting with $Q_1 Q_0 = 00$. Show the transitions and the count values over a complete count cycle.

6.7 Modulo-365 counter. Implement an efficient modulo-365 counter using the following steps.

1. How many T-FFs are needed?
2. What is the binary pattern representing 365?
3. Show that only the 1's in the pattern representing 365 need to be inputs to the recognition AND gate by replacing one of the 0's in the pattern with a 1 and computing the resulting decimal value.
4. Draw the simplified modulo-365 counter circuit.

6.10 Matlab Projects

6.1 (Implement a Mod-8 counter). Modify Example 16.29 to implement a mod-8 counter. Verify by applying a CLK input sequence that contains ten $1 \rightarrow 0$ transitions. Start the count by setting zero initial conditions.

6.2 (Implement a Mod-13 counter). Modify Example 16.30 to implement a mod-13 counter. Verify by applying a CLK input sequence that contains 15 $1 \rightarrow 0$ transitions. Start the count by setting zero initial conditions.

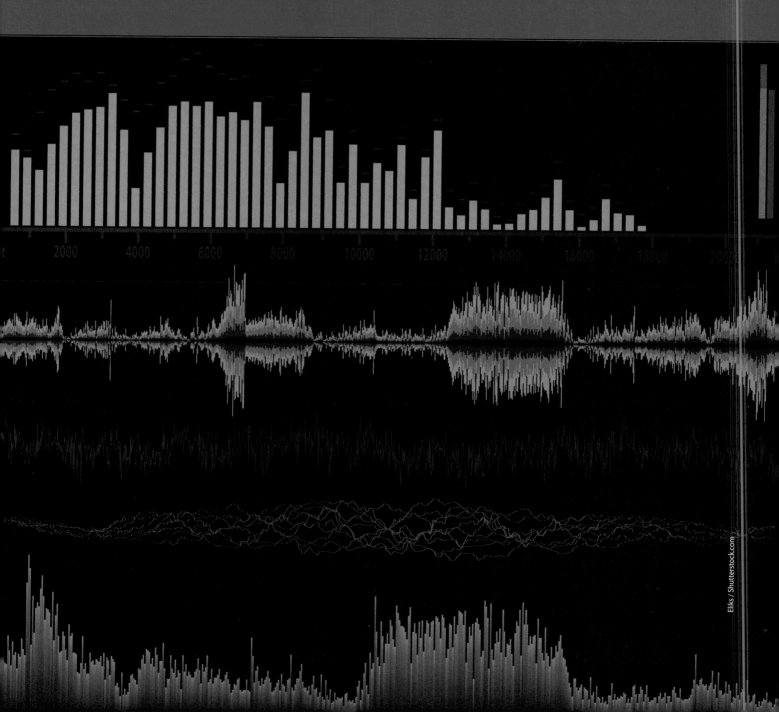

CHAPTER 7

CONVERTING BETWEEN ANALOG AND DIGITAL SIGNALS

LEARNING OBJECTIVES

After completing this chapter, the reader should be able to:

■ Understand how to sample analog waveforms without losing information.

■ Represent samples using finite-precision numbers.

■ Reconstruct analog waveforms from digital samples.

Chapter 2 described the role sensors play in producing digital information. This chapter describes a second common source of digital data: sampling analog signals. Such sampling processes occur in smart phones for converting audio waveforms into digital data for transmission and storage in memory and in digital cameras for capturing images.

7.1 INTRODUCTION

The main topics in this chapter are covered in the following sequence.

Waveform sampling—Sampling an analog waveform without losing information must consider how quickly the waveform changes in time. Time and image waveforms are characterized by the highest frequency component that is present. All of the information is retained by sampling the waveform at a rate that is higher than twice this highest frequency component. Sampling at a higher rate helps to visualize the waveform from the samples but produces unnecessary data.

Quantization—For digital transmission, storage, or processing, waveform samples must be represented in a binary form used by computers. Quantization is the process of representing the infinite-precision sample values with a finite number of bits. The number of bits employed is a trade-off between the errors introduced and the size of the data that is generated.

Digital-to-analog conversion—Digital data produced by the analog-to-digital conversion process must typically be converted back to analog waveforms. Computer animation generates digital sounds and images using formulas and models, and displaying them requires the same digital-to-analog conversion. Two common techniques are described for producing analog values from digital data.

Practical quantizer designs—This chapter concludes by describing considerations for producing digital data that achieve a desired performance, for example, making a digital system indistinguishable from an analog system.

7.2 ANALOG-TO-DIGITAL CONVERSION

Consider the digital audio system shown in Figure 7.1 that is a part of your smartphone. Acoustic energy in a microphone produces an analog signal that is a continuous-time waveform. An analog-to-digital converter (ADC) samples the waveforms to produce a sequence of samples that are then transformed by a quantizer within the ADC to produce binary data. These data are either stored into a digital memory for later playback (as in recorded music) or transmitted over a digital communication channel (as during a telephone conversation). At the receiving end, a digital-to-analog converter (DAC)

Figure 7.1

In a digital audio system, signals are present in both analog and digital forms.

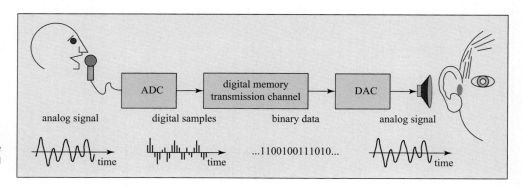

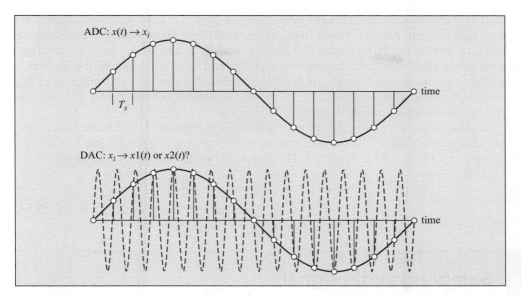

Figure 7.2

Reconstructing an analog waveform from samples presents a problem. Sampling an analog waveform with sampling period T_s produces a unique set of numbers. Converting this set of numbers back into an analog waveform could produce waveform $x1(t)$ (solid line) or $x2(t)$ (dashed line). Which waveform is correct?

reproduces the analog signal that then drives a speaker to produce acoustic energy. The goal of such a system is to have the reproduction perceptually indistinguishable from the original energy. This chapter describes the various steps that occur in such systems.

7.2.1 Sampling Analog Waveforms

In the sampling process, an analog signal is converted into a sequence of numbers by recording the values of the analog signal at a set of time-sampling points. We find that, although an analog waveform is converted to a unique set of numbers in the sampling process, returning from the samples to an analog waveform is problematic because there are an infinite number of analog signals that could have produced the same set of samples. Figure 7.2 illustrates this problem. So, which analog waveform should we choose? This section considers one-dimensional waveforms that are functions of time, but the same principle holds for sampling images with the function of position.

Figure 7.3 shows three examples of the sampling process with the original analog waveform shown in solid line and the samples as dots. Typically, the sampling times are regularly spaced, occurring in time every *sampling period T_s*. Intuition tells us that T_s should be small enough so that all of the important features of the analog signal waveform are represented in the sample values. For example, when T_s is sufficiently small, it should be possible to *visualize* the analog signal from the sample values. The figure shows values for T_s that are too small, appropriate for the signal, and too large.

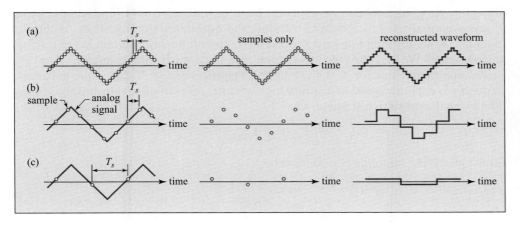

Figure 7.3

Sampling and reconstructing an analog signal. When the sampling points accurately represent the analog signal waveform, the sampling period T_s is appropriate for the signal. (a) T_s is too small, producing more samples than necessary. (b) T_s is sufficiently small. (c) T_s is too large.

If T_s is chosen too small, the sampling is being done densely, and more samples are extracted from the analog signal than necessary. This leads to wasteful use of digital memory. For example, an audio CD that is capable of storing 10 musical selections would then store only 5 if the sampling is performed using sampling period $T_s/2$. If T_s is chosen too large, the sampling is done too coarsely, and we lose information. In this case, 20 musical selections could be stored on an audio CD rather than 10, but these would not sound as good as the original recording.

In sampling applications, the term *sampling rate* is denoted f_s with units *samples per second* (sample/s), or commonly referred to as a *frequency* (Hz) and is used to indicate how often a signal is to be sampled. The sampling rate is equal to the reciprocal of the sampling period or

$$f_s = \frac{1}{T_s} \tag{7.1}$$

EXAMPLE 7.1 Sampling audio signals

Consider two common cases of sampling audio signals.

Digital telephone. In a digital telephone system, the speech signal is sampled at 8,000 times per second, making $f_s = 8,000$ Hz. The sampling period is then

$$T_s = \frac{1 \text{ second}}{8,000 \text{ samples}} = 1.25 \times 10^{-4} \text{ s/sample} = 125 \ \mu\text{s/sample}$$

Audio CD. In an audio CD system, music is sampled at 44,000 times per second, $f_s = 44,000$ Hz. The sampling period is then

$$T_s = \frac{1 \text{ second}}{44,000 \text{ samples}} = 0.0000227 \text{ s/sample} = 22.7 \ \mu\text{s/sample}$$

In a stereo system, each channel is sampled at this rate.

> **Factoid:** Our rule of thumb for adequately sampling $x(t)$ to produce x_i is that we should be able to visualize (by eye) the original waveform $x(t)$ from x_i.

7.2.2 Nyquist Sampling Criterion

Since the sampling process can generate huge quantities of data (one sample every T_s seconds) and too much data is wasteful, the question naturally arises: Is there a minimum value for sampling rate f_s that still retains all the information in the original analog signal? We first state the answer and then illustrate it with examples.

The minimum value for f_s is given by the *Nyquist criterion*: If $f_{\max}$ is the maximum frequency component present in an analog waveform, the sampling rate must be greater than twice the maximum frequency or

$$f_s > 2 f_{\max} \tag{7.2}$$

Equivalently, the sampling period must be

$$T_s < \frac{1}{2 f_{\max}} \tag{7.3}$$

Telephone-quality speech

EXAMPLE **7.2**

Telephone-quality speech is processed by the telephone company to remove the frequencies above the telephone bandwidth. The highest frequency in telephone-quality speech is

$$f_{max} = 3,300 \text{ Hz}$$

The sampling period used in the telephone system is chosen so

$$f_s > 2f_{max} \quad \text{making} \quad f_s > 6,600 \text{ Hz}$$

The telephone company commonly uses

$$f_s = 8,000 \text{ Hz}$$

for digital telephone applications.

Audio CD system

EXAMPLE **7.3**

Let us consider an audio CD system. Most people cannot hear frequencies higher than 20,000 Hz (20 kHz). Thus, storing any higher frequency components is futile, as these components will not be heard or appreciated anyway. Hence, 20 kHz is often taken as an upper bound on high fidelity music, so

$$f_{max} = 20 \text{ kHz}$$

In this case, the Nyquist criterion states that the sampling rate is

$$f_s > 40 \text{ kHz}$$

In practice, audio CD systems employ the sampling rate

$$f_s = 44 \text{ kHz}$$

for producing digitally recorded music. All standard CD players use this rate to play back music.

7.2.3 Aliasing

The Nyquist criterion determines the minimum sampling rate f_s that prevents loss of information. We now describe the problems encountered if the analog signal is not sampled sufficiently often. When $f_s \leq 2f_{max}$ (that is, if the Nyquist criterion is not obeyed) a new strange frequency will be perceived from the samples called the *alias frequency* f_a.

Let us consider sampling the sinusoidal waveform $\cos(2\pi f_o t)$. When the sample values are viewed, the phenomenon of aliasing causes the eye to see a sinusoid in the sample values having a frequency that is different than that present in the analog signal. Because $\cos(2\pi f_o t)$ contains only one frequency component, the maximum frequency is equal to that frequency or $f_{max} = f_o$. Let this sinusoid be sampled with a sampling rate f_s that is too small (that is, $f_s \leq 2f_o$). Then the Nyquist criterion is not obeyed, and a sinusoid having an alias frequency f_a is observed, where

$$f_a = |f_s - f_o| \tag{7.4}$$

EXAMPLE 7.4   Demonstration of aliasing

Consider the analog waveform $x(t) = \cos(2\pi f_o t)$ that has $f_{max} = f_o$. With sampling period $f_s > 2 f_o$, the sampling period equals

$$T_s = \frac{1}{f_s}$$

The resulting sample values equal

$$x_i = x(t = iT_s) = \cos(2\pi f_o\, iT_s)$$

To demonstrate aliasing, increase the analog waveform frequency

$$f_o \to f_o' = f_o + f_s$$

while keeping the same sampling frequency. Clearly, this violates the Nyquist criterion, $f_s < 2 f_o'$. The resulting sample values equal

$$x_i' = x'(t = iT_s) = \cos\left(2\pi f_o'\, iT_s\right) = \cos(2\pi(f_o + f_s)iT_s)$$

Simplifying, we get

$$x_i' = \cos\left(2\pi f_o\, iT_s + \overbrace{2\pi f_s iT_s}^{=2\pi i}\right)$$

Because adding an integer number of 2π to the argument of a cosine function does not change its value, the last equation simplifies to

$$x_i' = \cos(2\pi f_o\, iT_s) = x_i$$

These last values correspond to those that are observed for the original waveform. Hence, frequency f_o is the aliased frequency for f_o'.

EXAMPLE 7.5   Graphical demonstration of aliasing

Let us consider the sinusoid with frequency $f_o = 2$ Hz,

$$s(t) = \sin(2\pi f_o t) = \sin(4\pi t)$$

This waveform is shown in solid line in Figure 7.4. Because this waveform has only one frequency component ($f_o = 2$ Hz), its maximum frequency is $f_{max} = 2$ Hz. To retain all the information the sampling rate must be

$$f_s > 2 f_{max} = 4 \text{ Hz} \quad \to \quad T_s = \frac{1}{f_s} = \frac{1}{4} = 0.25 \text{ s}$$

The sampled time sequence with index i equals the values of the analog waveform at the sampling times:

$$s_i = s(t = iT_s) = \sin(4\pi\, i\, T_s)$$

Consider the value $f_s = 16$ Hz ($T_s = 1/16$ s), shown in Figure 7.4a. These samples form a good perceptual representation of the analog waveform, in that the waveform can be visualized from the samples connected by a dashed line. (The human eye performs a linear interpolation between sample points.)

Figure 7.4b shows $f_s = 8$ Hz. No aliasing occurs because $f_s > 2 f_o$ ($f_s = 8$ Hz > 4 Hz). The dashed curve *looks* similar to solid curve but the deviations between the original waveform and linear interpolation are noticeable.

Figure 7.4c shows $f_s = 5$ Hz. No aliasing occurs because $f_s > 2 f_o$ ($f_s = 5$ Hz > 4 Hz). The dashed curve differs significantly from solid curve but techniques exist to recover solid curve from samples. Although sampling the analog signal at slightly more than twice the highest

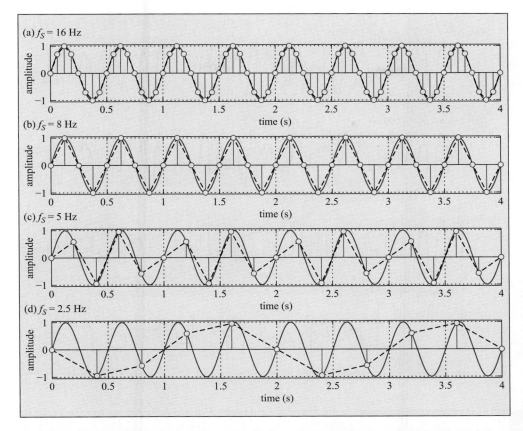

Figure 7.4

Demonstration of aliasing. The analog waveform $s(t) = \sin(2\pi f_o t) = \sin(4\pi t)$, ($f_o = 2$ Hz) for $0 \leq t \leq 4$ s (solid curve), is sampled using different sampling rates f_s. The samples shown as circles are connected by dashed lines to form a linear interpolation.

frequency component in the signal retains the information, it is difficult to visualize the waveform from the samples.

Hence, for a reasonably accurate linear interpolation, the sampling rate must be much higher, at least eight times the highest frequency, or 4 times the Nyquist rate.

When $f_s < 2 f_{max}$, aliasing will occur. The alias frequency in the linear interpolation of the sample values is a frequency $f_a < f_s/2$. Consider $f_s = 2.5$ Hz shown in Figure 7.4d. Then, for $f_o = 2$ Hz, we have

$$f_a = |f_s - f_o| = |2.5 - 2| = 0.5 \text{ Hz}$$

The waveform having this aliased frequency is shown as a dashed line.

To quantify the alias effect, compare the values at the sample points for both the original and the alias frequencies for $f_s = 2.5$ Hz ($T_s = 2/5$ s). Starting at $t = 0$, the sample times occur at the instants

$$t = 0 \quad T_s \quad 2T_s \quad 3T_s \quad 4T_s \quad 5T_s \cdots$$
$$= 0 \quad \tfrac{2}{5}\text{ s} \quad \tfrac{4}{5}\text{ s} \quad \tfrac{6}{5}\text{ s} \quad \tfrac{8}{5}\text{ s} \quad 2\text{ s} \cdots$$

The values of the original analog cosine waveform at these time instants equal

$$\sin(2\pi f_o \times 0) \quad \sin(2\pi f_o \times T_s) \quad \sin(2\pi f_o \times 2T_s) \quad \sin(2\pi f_o \times 3T_s) \quad \cdots$$
$$\sin(0) \quad \sin(\tfrac{8}{5}\pi) \quad \sin(\tfrac{16}{5}\pi) \quad \sin(\tfrac{24}{5}\pi) \quad \cdots$$
$$0 \quad -0.951 \quad -0.588 \quad 0.588 \quad \cdots$$

The values of the alias signal at these time instants equal

$$\sin(2\pi f_a \times 0) \quad \sin(2\pi f_a \times T_s) \quad \sin(2\pi f_a \times 2T_s) \quad \sin(2\pi f_a \times 3T_s) \quad \cdots$$
$$\sin(0) \quad \sin(\tfrac{2}{5}\pi) \quad \sin(\tfrac{4}{5}\pi) \quad \sin(\tfrac{6}{5}\pi) \quad \cdots$$
$$0 \quad -0.951 \quad -0.588 \quad 0.588 \quad \cdots$$

Note that the sample values of both the original and the alias signal are equal.

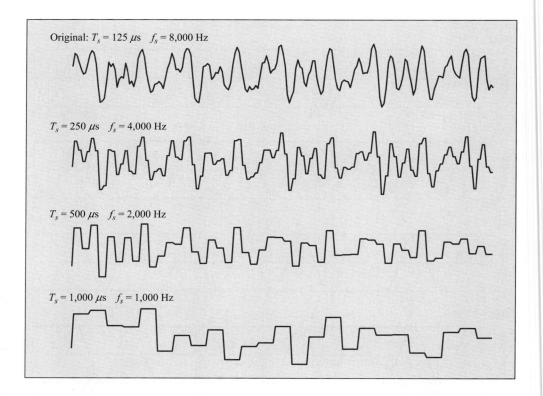

Figure 7.5

Demonstration of aliasing in microphone speech.

EXAMPLE   7.6

Aliased microphone speech waveform

Figure 7.5 shows a speech waveform that was acquired using the microphone on a laptop. The signal is sampled at $f_s = 8{,}000$ Hz. The sampling period T_s was doubled in each successive waveform to demonstrate aliasing. As T_s increases, more details in the speech waveform are lost. When $T_s = 1{,}000\,\mu$s, the speech is unintelligible.

Anti-Aliasing Filters

To prevent unintended aliasing, a waveform is low-pass filtered to reduce frequency components that are greater than half of the sampling frequency f_s using an *anti-aliasing filter*. Ideally, this filter should pass all frequency components $f < f_s/2$ and eliminate all components $f \geq f_s/2$. Such higher frequency components are typically present in the noise contained in the waveform.

▶ 7.3 SPATIAL FREQUENCIES

Images contain spatial frequencies. Consider a Cartesian (x, y) coordinate system that contains a gray-scale image with x index $0 \leq i \leq n_x - 1$ and y index $0 \leq j \leq n_y - 1$. Figure 7.6 shows images with $n_x = 256$ and $n_y = 256$. The pixel in the upper left-hand corner has an index of $(0,0)$. A gray-scale image contains pixel values $v_{i,j}$ in the range $0 \leq v_{i,j} \leq A_{\max}$, where 0 corresponds to black and $A_{\max}$ to white.

Horizontal spatial frequencies vary in the x direction, and vertical frequencies vary in the y direction. The sinusoidal period is related to the size of the image in that direction. For example, horizontal frequencies have a fundamental period containing n_x points.

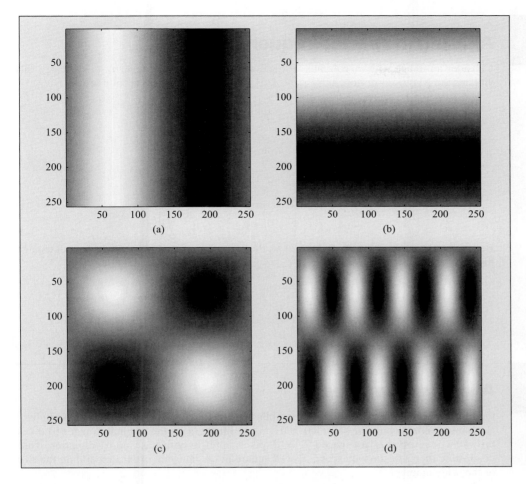

Figure 7.6

Spatial frequencies. (a) $f_x = 1$, $f_y = 0$. (b) $f_x = 0$, $f_y = 1$. (c) $f_x = 1$, $f_y = 1$. (d) $f_x = 4$, $f_y = 1$.

A horizontal frequency $0 \leq f_x \leq f_{x,\max}$ is given by the x vector as

$$v_i = \frac{A_{\max}}{2} \left[1 + \sin \left(\frac{2\pi f_x i}{n_x} \right) \right] \quad \text{for } 0 \leq i \leq n_x - 1 \quad (7.5)$$

adding 1 to the sine values makes them vary between 0 and 2, making $0 \leq v_i \leq A_{\max}$, required to display the gray-scale image.

The image is sampled at every pixel. Hence, following the Nyquist criterion, the maximum spacial frequency produces at least two samples per period. Therefore, with n_x pixels, the maximum frequency equals

$$f_{x,\max} < \frac{n_x}{2} \quad (7.6)$$

Similarly, a vertical frequency $0 \leq f_y \leq f_{y,\max}$ is given by the y vector as

$$v_j = \frac{A_{\max}}{2} \left[1 + \sin \left(\frac{2\pi f_y j}{n_y} \right) \right] \quad \text{for } 0 \leq j \leq n_y - 1 \quad (7.7)$$

The maximum vertical spacial frequency is

$$f_{y,\max} < \frac{n_y}{2} \quad (7.8)$$

EXAMPLE   7.7   **Aliasing in spacial frequencies**

Consider an image with n_x columns. To demonstrate aliasing, let $f_x = n_x + 1$. Clearly, this violates the Nyquist criterion, as $f_{x,max} < n_x/2$. The values in the x direction are

$$v_i = \frac{A_{max}}{2}\left[1 + \sin\left(\frac{2\pi f_x i}{n_x}\right)\right]$$

$$= \frac{A_{max}}{2}\left[1 + \sin\left(\frac{2\pi(n_x + 1)i}{n_x}\right)\right]$$

$$= \frac{A_{max}}{2}\left[1 + \sin\left(\overbrace{\frac{2\pi n_x i}{n_x}}^{=2\pi i} + \frac{2\pi i}{n_x}\right)\right]$$

Because adding an integer number of 2π to the argument of a sine function does not change its value, the last equation simplifies to

$$v_i = \frac{A_{max}}{2}\left[1 + \sin\left(\frac{2\pi i}{n_x}\right)\right]$$

These last values correspond to those observed for $f_x = 1$, which is the alias frequency.

EXAMPLE   7.8   **Aliasing in images**

Figure 7.7 shows the effects of spatially sampling an image using increasingly coarser sampling. Each image contains $(n_x, n_y) = (500, 500)$ points. Image (a) is the original. Image (b) takes the value of pixels that are spaced $T_s = 10$ pixels apart in both the x and y direction. That pixel value then sets the values of a $T_s \times T_s$ square in the image in order to maintain the size of the original image.

As T_s increases, the detail of finer (*higher spatial frequency*) objects, such as the whiskers, begin to degrade. When the viewer *squints*, it forms a spatial low-pass filter that removes the higher frequencies in all of the images, and they begin to look the same, although without fine details.

 ## 7.4 QUANTIZATION

Analog-to-digital conversion in practice involves two steps:

Sample-and-hold: Maintaining the sample value at a constant level for a sufficient time interval to allow the conversion process to be completed.

Quantization: Representing the sampled analog value with a finite number of bits.

7.4.1 Sample-and-Hold Operation

Because the quantization process takes time to accomplish, the analog sample value must be *held* constant during the conversion process. Figure 7.8 shows that the sample value is held by storing its voltage value at the sampling time in a capacitor. Prior to the sample-and-hold time $t < t_{SaH}$, the sampling switch is closed, making the capacitor voltage $V_a(t)$ equal to the varying input voltage $v_{IN}(t)$

$$V_a(t) = V_{IN}(t) \text{ for } t < t_{SaH} \tag{7.9}$$

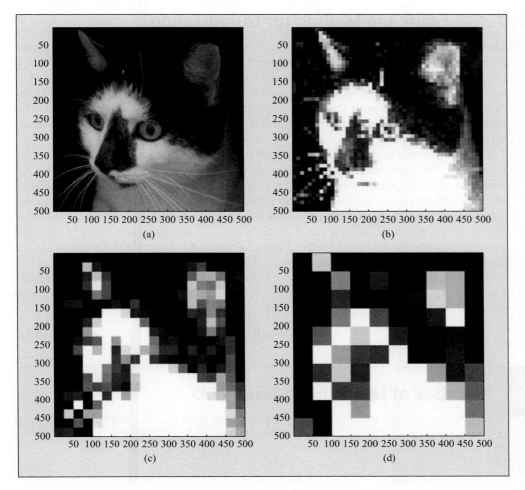

Figure 7.7

Sampling an image. (a) original 500×500 pixels. (b) $T_s = 10$ pixels. (c) $T_s = 25$ pixels. (d) $T_s = 50$ pixels.

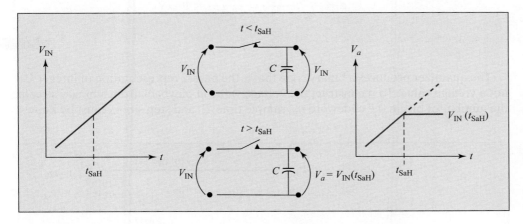

Figure 7.8

Sampling and holding $V_{IN}(t)$ to produce the infinite-precision value V_a to be quantized.

At the sample time $t = t_{SaH}$, the switch opens and the capacitor voltage V_a remains at the same voltage value because there is no place for the capacitor to discharge. Hence,

$$V_a = V_{IN}(t_{SaH}), \text{ for } t \geq t_{SaH} \tag{7.10}$$

Having a constant, infinite-precision, voltage value V_a, we can proceed to convert it into a finite-precision approximation V_q, which is described next.

7.4.2 Staircase Method for Quantization

Quantization is the process of transforming a sample value having infinite precision, such as $s_i = 3.1415926\ldots$, into a finite-precision value, such as $sq_i = 3.14$, that is expressed with a finite number of bits, such as the 7-bit binary code word 0110010. The number of bits must be sufficiently large to provide a reasonable approximation to the original sample value with the error representing an added noise to the signal. This section describes how quantization is performed in your smartphone and for sensor waveform processing in robotics.

One simple method of explaining how a sample value is quantized is shown in Figure 7.9. The figure illustrates the conversion of two sample values $V_a(1)$ and $V_a(2)$ obtained from an analog signal from two consecutive sampling periods. At the start of every sample period T_s, the quantizer generates a staircase pattern of values having step-size Δ. This staircase continues until its value V_q exceeds the analog signal value V_a. When this condition occurs, the staircase stops, and the quantized value is set equal to the final value of V_q. By the end of sample period duration T_s, the conversion has been completed with V_q being the quantized approximation of V_a.

A quantizer having b bits produces 2^b output levels. The kth quantizer level for integer $k = 0, 1, 2, \ldots, 2^b - 1$ corresponds to approximating the analog input voltage V_a with the quantized value

$$V_q = k\Delta \text{ for } k = 0, 1, 2, \ldots, 2^b - 1 \tag{7.11}$$

EXAMPLE 7.9   Number of levels in an audio CD

In an audio CD, each quantized sample is represented by 16 bits. The number of steps in the staircase is then

$$n_{\text{steps}} = 2^{16} = 65,536$$

If the voltage range is from $V_{\min} = 0$ V to $V_{\max} = +2$ V, the step size is

$$\Delta = \frac{V_{\max} - V_{\min}}{n_{\text{steps}} - 1} = \frac{2 \text{ V}}{65,535} \approx 10 \ \mu\text{V}$$

The quantizer produces a binary code that is the base-2 representation of integer k—not a voltage value. To reconstruct the voltage value at playback from a binary stream, the number of bits in the codeword n_b, sample time T_s, and step size Δ must be known.

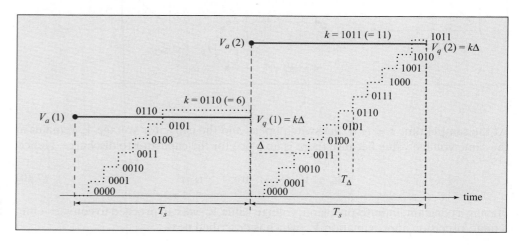

Figure 7.9

Quantizing infinite-precision values $V_a(1)$ and $V_a(2)$ using the staircase method.

To generate a staircase with 2^b requires a fast clock. If T_s is the sample period, a new step must be produced every T_Δ seconds, where

$$T_\Delta = \frac{T_s}{2^b} \qquad\qquad (7.12)$$

Staircase quantizer   EXAMPLE 7.10

Microcontrollers (MCUs) typically include 10-bit ADCs as part of the chip design. Each ADC then produces $2^{10} = 1{,}024$ levels. An MCU typically uses a voltage supply $V_s = 5$ V, which determines the voltage range of the ADC. The corresponding step size is then

$$\Delta = \frac{V_s}{2^b - 1} = \frac{5\,\text{V}}{1023} = 0.0049\,\text{V}$$

An easier approximation to remember is to use $2^{10} \approx 1{,}000$, to produce

$$\Delta \approx 5\,V/1{,}000 = 5\,\text{mV}$$

The ADC produces sample value k with $0 \le k \le 1023$. If $k = 500$, the corresponding quantized value to V_a equals

$$V_q = k\Delta = 500 \times 0.0049\,\text{V} = 2.45\,\text{V}$$

The approximation would produce the value $V_q = 500 \times 5\,\text{mV} = 2.5\,\text{V}$.

If a staircase conversion must occur within $T_s = 1$ ms, the time to generate each step in a 10-bit quantizer is

$$T_\Delta = \frac{T_s}{2^b} = \frac{10^{-3}\,\text{s}}{1024} = 9.77 \times 10^{-7}\,\text{s} \approx 1\,\mu\text{s}$$

This is too short for most practical ADCs. The successive approximation method, described next, can perform the same conversion 100 times faster.

7.4.3 Successive Approximation Method

Most practical ADCs use a more complex digital circuit that performs a faster conversion than the staircase method by using *successive approximations* in b comparisons (rather that 2^b steps used in the staircase quantizer). The process compares V_a with a threshold voltage V_τ, using one comparison per each bit in the binary representation and starting with the most significant bit (MSB). The V_τ value is variable and depends on the results of the previous comparisons: hence, the term *successive approximation*. The circuit required to accomplish this conversion is more complex, but successive approximation has become the preferred method because of its speed.

Quantization by successive approximation   EXAMPLE 7.11

To illustrate the successive approximation method, let a 4-bit quantizer span the voltage range from 0 V to Q V, with $Q = 5$ V for a typical MCU. Figure 7.10 shows the four steps that occur for a 4-bit quantizer, with the result encoded in $b_3\, b_2\, b_1\, b_0$.

STEP 1. Determining the value of MSB b_3, V_a is compared to $V_\tau = Q/2$, which is the midpoint of the quantizer range. If $V_a \ge Q/2$, the MSB $b_3 = 1$, otherwise $b_3 = 0$. The figure shows the comparison produces $b_3 = 0$.

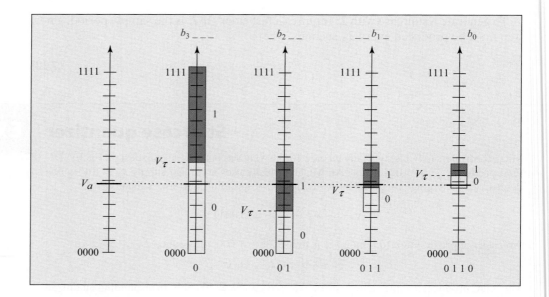

Figure 7.10

Successive approximation for quantizing V_a.

STEP 2. Determining the value of b_2 depends on the previous b_3 result.

- If $b_3 = 0$, $V_\tau = Q/4$.
- If $b_3 = 1$, $V_\tau = Q/2 + Q/4 = 3Q/4$.
- Comparison: If $V_a \geq V_\tau$ then $b_2 = 1$, otherwise $b_2 = 0$. Figure 7.10 shows the comparison produces $b_2 = 1$.

STEP 3. Determining the value of b_1 depends on $b_3 b_2$.

- If $b_3 b_2 = 00$, $V_\tau = Q/8$.
- If $b_3 b_2 = 01$, $V_\tau = Q/4 + Q/8 = 3Q/8$.
- If $b_3 b_2 = 10$, $V_\tau = Q/2 + Q/8 = 5Q/8$.
- If $b_3 b_2 = 11$, $V_\tau = Q/2 + Q/4 + Q/8 = 7Q/8$.
- Comparison: If $V_a \geq V_\tau$ then $b_1 = 1$, otherwise $b_1 = 0$. The figure shows the comparison produces $b_1 = 1$.

STEP 4. Determining the value of b_0 depends on $b_3 b_2 b_1$:

- If $b_3 b_2 b_1 = 000$, $V_\tau = Q/16$.
- If $b_3 b_2 b_1 = 001$, $V_\tau = Q/8 + Q/16 = 3Q/16$.
- If $b_3 b_2 b_1 = 010$, $V_\tau = Q/4 + Q/16 = 5Q/16$.
- If $b_3 b_2 b_1 = 011$, $V_\tau = Q/4 + Q/8 + Q/16 = 7Q/16$.
- If $b_3 b_2 b_1 = 100$, $V_\tau = Q/2 + Q/16 = 9Q/16$.
- If $b_3 b_2 b_1 = 101$, $V_\tau = Q/2 + Q/8 + Q/16 = 11Q/16$.
- If $b_3 b_2 b_1 = 110$, $V_\tau = Q/2 + Q/4 + Q/16 = 13Q/16$.
- If $b_3 b_2 b_1 = 111$, $V_\tau = Q/2 + Q/4 + Q/8 + Q/16 = 15Q/8$.
- Comparison: If $V_a \geq V_\tau$ then $b_0 = 1$, otherwise $b_0 = 0$. Figure 7.10 shows the comparison produces $b_0 = 0$.

These comparisons are performed using a threshold detection circuit, and the V_τ values are generated by adding the appropriate value using an adder circuit. Both are implemented with op amps.

7.4.4 Setting Number of Quantizer Bits

A discrete voltage value produced by a quantizer, x_q, is encoded as the binary count of the number of steps used in the conversion. If a small number of bits is used, a staircase pattern having a small number of large-sized steps is needed to cover the desired voltage range. The large step size can produce large errors between the analog value x_a and its approximation x_q. Using the staircase quantizer described previously, the quantized value x_q is always larger than the corresponding analog value x_a. Then the quantizer error is determined by the step size Δ by

$$x_q - x_a < \Delta \qquad (7.13)$$

Quantizing a sinusoid using 3 bits is shown in Figure 7.11a. This figure shows a graph (x_a, x_q) that indicates the transformation produced by the staircase quantizer in

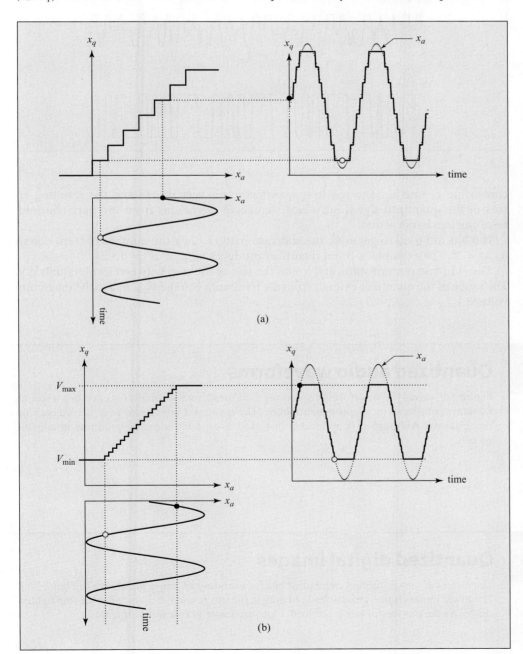

(a)

(b)

Figure 7.11

Quantizing a sinusoid. (a) Using 3 bits (8 levels). (b) Using 4 bits (16 levels). Note that the insufficient quantizer range causes severe clipping.

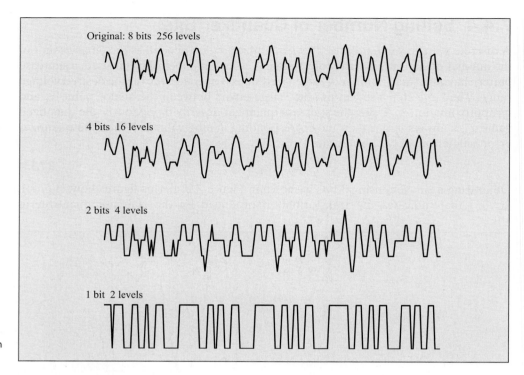

Original: 8 bits 256 levels

4 bits 16 levels

2 bits 4 levels

1 bit 2 levels

Figure 7.12

Varying the number of bits in an audio waveform.

converting x_a into x_q. The resulting quantization is very coarse and the resulting errors in the quantized signal introduce significant deviations from the pure sinusoid, resulting in a harsh sound.

If b bits are used to generate the staircase pattern, then the number of steps equals $n_{steps} = 2^b$. For example, a 10-bit quantizer produces $n_{steps} = 2^{10} = 1,024$.

The staircase pattern starts at 0 V and the size of the step between levels equals ΔV. The range of the quantizer extends from the minimum voltage $V_{min} = 0$ to the maximum voltage $V_{max} = (n_{steps} - 1)\Delta$.

EXAMPLE 7.12

Quantized audio waveforms

Figure 7.12 shows the waveforms when using a different number of bits for encoding a speech waveform produced by a laptop microphone. The quantized waveforms were played back on the speakers. Although it is noticeably distorted, even 1-bit encoding produces intelligible speech.

EXAMPLE 7.13

Quantized digital images

The effects of using different number of bits for encoding an image are shown in Figure 7.13. The lower limit sets the darkest shade of gray in the image and the upper limit sets the lightest shade. As the number of bits is reduced, there are fewer gray levels in the image.

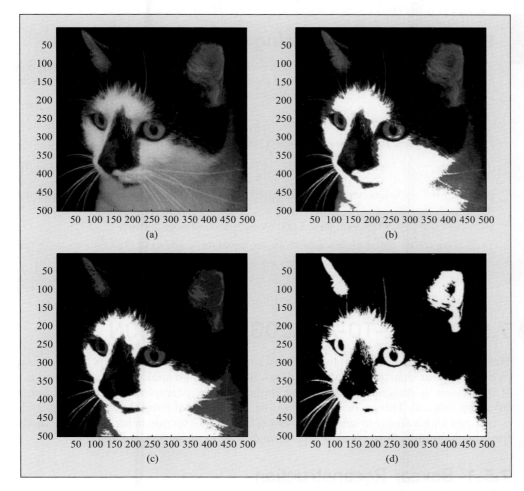

Figure 7.13

Varying the number of bits in displaying an image. (a) Image gray scale quantized to 8 bits (256 levels). (b) 3 bits (8 levels). (c) 2 bits (4 levels). (d) 1 bit (2 levels).

Digital memory on a greeting card | EXAMPLE 7.14

A musical greeting card contains a 1.5 V battery and plays a song that lasts 5 seconds. The sampling rate was $f_s = 5,000$ Hz and the quantizer has 64 levels.

The size of the digital memory (in bits) on the card is found by first computing the number of bits per sample as

$$\log_2 64 = 6 \text{ bits/sample}$$

This gives a bit rate equal to

$$6 \text{ bits/sample} \times 5,000 \text{ samples/second} = 30,000 \text{ bits/second (bps)}$$

Multiplying this bit rate by the song duration gives the memory size as

$$30,000 \text{ bps} \times 5 \text{ s} = 150,000 \text{ bits}$$

The quantizer step size is computed from the battery voltage as

$$\Delta = \frac{1.5 \text{ V}}{63} = 0.024 \text{ V} = 24 \text{ mV}$$

EXAMPLE 7.15

Better-sounding greeting card

The quality of the sound produced by a musical greeting card is improved by reducing the quantizer Δ. If the card in the previous example is improved by reducing the step size to $\Delta = 10$ mV, the number of quantizer steps increases to

$$n_{steps} = \frac{1.5 \text{ V}}{0.01 \text{ V}} = 150$$

The number of bits b to provide 150 levels is found from

$$\log_2 150 = \frac{\log_{10} 150}{\log_{10} 2} = 7.22$$

Because 7 bits are insufficient, $b = 8$ is required for this quantizer. The resulting bit rate using the same f_s equals

$$8 \text{ bits/sample} \times 5{,}000 \text{ samples/second} = 40{,}000 \text{ bps}$$

The 5-second song then requires a memory size equal to

$$40{,}000 \text{ bps} \times 5 \text{ s} = 200{,}000 \text{ bits}$$

7.5 DIGITAL-TO-ANALOG CONVERSION

Having stored an audio signal in quantized samples as a collection of bits in a memory, like that in your smartphone, we would now like to hear an analog replica of the audio. If the system is designed properly, the replica will be perceptually indistinguishable from the original. This reconstruction of the analog signal from the digital samples is performed with a *digital-to-analog converter* (DAC). This section describes two methods for implementing a DAC: *boxcar reconstruction* and *pulse-width modulation*.

7.5.1 Boxcar Reconstruction

To reconstruct the analog signal from the samples, the *boxcar DAC* produces a rectangular pulse of duration T_s having an amplitude corresponding to the quantized binary value stored in memory. There is a separate pulse for each binary code word in memory, and each pulse lasts for the duration of the original sampling period T_s. When plotted over time, these pulses produce a step-like constant or *boxcar* approximation to the original analog waveform. Figure 7.14 shows a reconstruction of a sinusoidal waveform

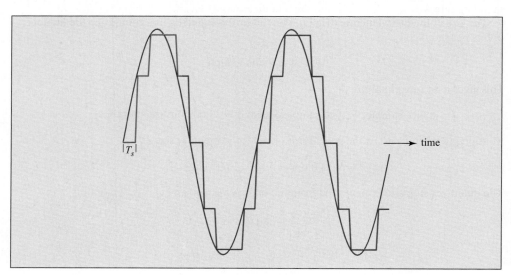

Figure 7.14

A boxcar reconstruction of a sinusoidal waveform. Original waveform is shown in blue, boxcar DAC reconstruction in red.

from the digital values stored in memory, along with knowledge of sampling time T_s and step size Δ. Note that these values could have been obtained by sampling a sinusoidal waveform or *synthesized* by computing the sinusoidal values from the trigonometric function.

ADC–DAC EXAMPLE 7.16

Consider the analog waveform reconstructed from digital samples shown in Figure 7.15. The same sampling rate and quantizer were used in both sampling and reconstruction.

The sampling rate f_s is determined from the duration of the constant-valued segments, which are observed to last two small divisions on the oscilloscope trace. With each large division corresponding to 0.01 ms, the sample period is

$$T_s = \frac{2}{5} \times 0.01 \text{ ms} = 0.04 \text{ ms}$$

The sample rate is then the reciprocal of the sample period

$$f_s = \frac{1}{T_s} = \frac{1}{4 \times 10^{-5}} = 2.5 \times 10^4 \text{ Hz}$$

The quantizer step size Δ is determined from the smallest vertical increment that is displayed, which is observed to be one small division. With each large division being 0.5 V and containing five small divisions, we have

$$\Delta = \frac{1}{5} \times 0.5 \text{ V} = 0.1 \text{ V}$$

If the boxcar DAC uses the same Δ and a 6-bit quantizer, consider the reconstruction of the boxcar approximation from the binary values

001001111001

To do the decoding, the sequence is first grouped into 6-bit codes to yield

$$\underbrace{001001}_{=9} \ \underbrace{111001}_{=57}$$

Then these values are scaled by the step size to produce

$$9\Delta = 0.9 \text{ V} \quad \text{and} \quad 57\Delta = 5.7 \text{ V}$$

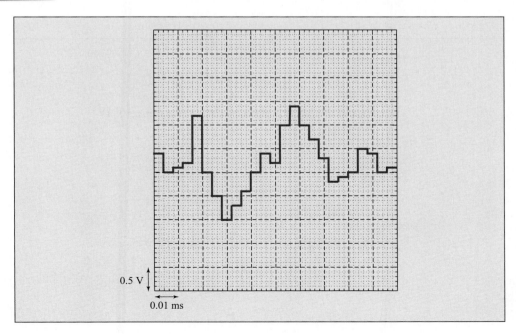

0.5 V

0.01 ms

Figure 7.15

Sample waveform produced by DAC as displayed by an oscilloscope.

7.5.2 Pulse-Width Modulation

Pulse-width modulation (PWM) is an alternate, less expensive technique that varies the ON/OFF time of a binary (digital) waveform to produce a desired average (analog) value. This technique is commonly used to dim LED displays and turn motors at variable speeds.

Figure 7.16 shows pulse trains with variable duty cycles that produce variable average voltages. The period of the PWM waveform T_{PWM} is constant and is usually much shorter than sampling time T_s to allow an average voltage to be determined. For example, if $T_s = 1$ ms, a reasonable value would be $T_{PWM} = 0.1$ ms.

A PWM system that operates with b bits divides the T_{PWM} period into sub-intervals with each having duration δT, so

$$\delta T = \frac{T_{PWM}}{2^b - 1} \tag{7.14}$$

For example, for $n_b = 4$, $\delta T = T_{PWM}/15$. For $T_{PWM} = 0.25$ ms, $\delta T = 0.017$ ms (17 μs).

A desired average voltage is obtained by maintaining the ON state for duration T_{ON} by specifying the number of intervals k, as

$$T_{ON} = k\delta T \tag{7.15}$$

where $0 \leq k \leq 2^b - 1$. The average voltage produced by the PWM waveform for a specified k value equals

$$V_q(k) = \frac{T_{ON}}{T_{PWM}} V_{max} = \frac{k\delta T}{T_{PWM}} V_{max} \tag{7.16}$$

The effective PWM step size equals the smallest increment in V_q, as

$$\Delta_{PWM} = \frac{\delta T}{T_{PWM}} V_{max} = \frac{V_{max}}{2^b - 1} \tag{7.17}$$

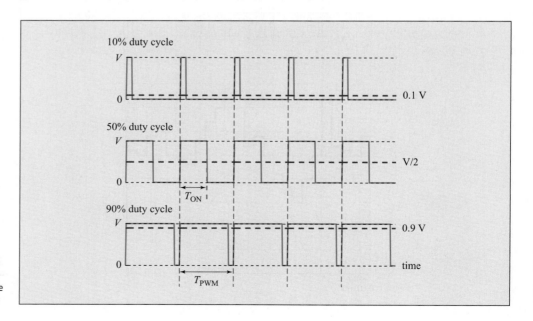

Figure 7.16

PWM waveform duty cycles produce a variety of average voltages.

The average voltage for a specified k value is equal to

$$V_q(k) = k\Delta_{\text{PWM}} \tag{7.18}$$

which varies from 0 V ($k = 0$) to V_{max} ($k = 2^b - 1$).

The *duty cycle* of a binary waveform is the percentage of the period that the waveform spends in the ON state. The duty cycle of the PWM waveform is given by

$$\text{Duty cycle} = \frac{T_{\text{ON}}}{T_{\text{PWM}}} \times 100\% \tag{7.19}$$

PWM voltage   EXAMPLE 7.17

A 5-bit PWM system operates with $T_{\text{PWM}} = 2$ ms and $V_{\text{max}} = 5$ V. The sub-interval duration equals

$$\delta T = \frac{T_{\text{PWM}}}{2^5 - 1} = \frac{2\,\text{ms}}{31} = 0.065\,\text{ms}\ (65\,\mu\text{s})$$

The effective step size equals

$$\Delta_{\text{PWM}} = \frac{\delta T}{T_{\text{PWM}}} V_{\text{max}} = \frac{0.065\,\text{ms}}{2\,\text{ms}}\,5\,\text{V} = 0.16\,\text{V}$$

The value of k can be specified between 0 and 31 to yield an average voltage, as

$$V_q(k) = k\Delta_{\text{PWM}} = 0.16\,k\,\text{V}$$

When $k = 10$, $V_q = 1.6$ V.

To produce $V_q = 2.4$ V the value of k equals

$$k = \frac{V_q}{\Delta_{\text{PWM}}} = \frac{2.4\,\text{V}}{0.16\,\text{V}} = 15$$

The duty cycle value is then

$$\text{Duty cycle} = \frac{k\delta_T}{T_{\text{PWM}}} \times 100\% = \frac{15 \times 0.065\,\text{ms}}{2\,\text{ms}} \times 100\% = 49\%$$

By controlling the duty cycle and having a sufficiently high frequency of repetition, the PWM waveform can produce V_q values equivalent to the boxcar DAC. Figure 7.17 shows the averages over several cycles produce samples of a sinusoidal waveform.

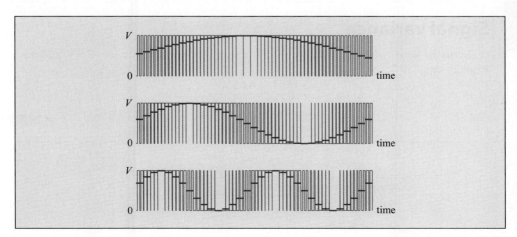

Figure 7.17

Pulse trains with variable duty cycles produce good approximations to sinusoidal waveforms. Solid lines are averages given over two T_{PWM} periods.

7.6 QUANTIZER DESIGN CONSIDERATIONS

Figure 7.18 shows the number of bits and sampling rates for commonly available ADCs. The main concerns when designing a quantizer include the following.

1. How is the quality of a quantized signal measured?

2. How many bits should be used to encode the quantized samples?

3. How are the quantizer limits adjusted to the analog signal to optimum performance?

The design of a quantizer involves a trade-off in how the staircase pattern having a fixed number of steps is realized: If each step is small, the staircase does not go very high, and if each step is large, the staircase may be much larger than needed. Each concern is now considered separately.

7.6.1 Quantized Signal Quality

The quality of ADC systems is usually measured by comparing the signal level to the level of noise or error that corrupts the signal. Because noise is a random quantity, this comparison uses a parameter called the *signal-to-noise ratio* (SNR) that is equal to the signal variance divided by the noise variance. Noise can be introduced by sources other than a quantizer, so the SNR is a general quality factor that is encountered in different contexts in the book. This section considers the SNR due to only the noise (approximation error) contributed by the quantizer.

Quantizer Signal-to-Noise Ratio

An AC voltmeter measures sinusoidal voltages in terms of its *root-mean-squared (RMS)* voltage value that equals the *square root of the mean (average) of the squared signal value*. The signals considered in this book have a zero average value (ignoring the offset to match the signal waveform to the quantizer). A sinusoidal signal with amplitude A has an *RMS* value denoted σ_s and equal to

$$\sigma_s = \frac{A}{\sqrt{2}} \tag{7.20}$$

The signal variance is simply the square of the RMS value, so

$$\sigma_s^2 = \frac{A^2}{2} \tag{7.21}$$

EXAMPLE 7.18   **Signal variance**

A sinusoidal signal with amplitude $A = 2.5$ V varies between -2.5 V and $+2.5$ V, and has a variance equal to

$$\sigma_s^2 = \frac{(2.5\,\text{V})^2}{2} = \frac{6.25\,\text{V}^2}{2} = 3.125\,\text{V}^2$$

Figure 7.18

Commonly available ADCs.

ADC Type	Number of Bits	f_s (kHz)
Cell phone & laptop microphones	8	8
MCU	10	10
CD	16	44

The quantizer converts an infinite-precision analog value (s_i) into a finite-precision quantized value (sq_i). This approximation results in a rounding error ϵ_i, as

$$\epsilon_i = sq_i - s_i \tag{7.22}$$

The perceptual effect of this error is analogous to adding noise ϵ_i to s_i to produce the (perceived) quantized value sq_i, as

$$sq_i = s_i + \epsilon_i \tag{7.23}$$

A quantizer with step size Δ is considered to produce a *quantization noise variance* (σ_ϵ^2) equal to

$$\sigma_\epsilon^2 = \frac{\Delta^2}{12} \tag{7.24}$$

Quantization noise variance EXAMPLE 7.19

A 10-bit quantizer extends from 0 V to 5 V and has step size

$$\Delta = \frac{V_{max}}{2^{10} - 1} = \frac{5\,V}{1023} = 0.0049\,V$$

The variance of the quantization noise it produces equals

$$\sigma_\epsilon^2 = \frac{\Delta^2}{12} = \frac{(0.0049\,V)^2}{12} = 2 \times 10^{-6}\,V^2$$

The signal-to-noise ratio (SNR) is the ratio given by

$$SNR = \frac{\sigma_s^2}{\sigma_\epsilon^2} \tag{7.25}$$

The SNR values for audio systems can vary from a few tens (for a poor telephone connection) to several million (for high-quality audio systems). For comparing numbers that exhibit such a large range, the logarithmic scale is often used. The most common base for expressing quantities in logarithmic units is 10. When measuring power levels at the output of audio equipment, the decibel (dB) scale is used. Given the signal power level σ_s^2 and the noise power level σ_ϵ^2, the signal-to-noise ratio in dB equals

$$SNR_{dB} = 10 \log_{10} \frac{\sigma_s^2}{\sigma_\epsilon^2} \tag{7.26}$$

SNR of quantized signals EXAMPLE 7.20

The previous example shows that a 10-bit quantizer extending from 0 V to 5 V has a variance equal to $\sigma_\epsilon^2 = 2 \times 10^{-6}\,V^2$. A sinusoidal signal with amplitude $A = 2\,V$ has variance

$$\sigma_s^2 = \frac{(2\,V)^2}{2} = 2\,V^2$$

The signal is offset by 2.5 V, so that it varies from 0.5 V to 4.5 V within the quantizer range. The SNR of the quantized samples is computed as

$$SNR = \frac{\sigma_s^2}{\sigma_\epsilon^2} = \frac{2\,V^2}{2 \times 10^{-6}\,V^2} = 10^6$$

The SNR value in dB equals

$$SNR_{dB} = 10 \log_{10} \frac{\sigma_s^2}{\sigma_\epsilon^2} = 10 \log_{10} 10^6 = 60\,dB$$

7.6.2 Setting Quantizer Range in Practice

Most interesting signals, such as speech and music, do not have well-defined maximum and minimum values. The strength of such signals is measured in terms of its average RMS voltage denoted X_{RMS}. The signal variance equals

$$\sigma_S^2 = X_{RMS}^2 \tag{7.27}$$

The audio strength can be amplified by adjusting its volume. Consider setting the signal level for a music waveform whose amplitude varies widely over time, from very small (silence) to very large (loud). The trade-off in matching this signal to a quantizer is between the quantization errors and the clipping errors. The quiet signals occupy a small number of quantizer levels result in large quantization errors and the signals that exceed the quantizer limits too often cause significant clipping errors. Increasing X_{RMS} reduces quantization errors, but also increases the clipping errors. Decreasing X_{RMS} reduces the clipping errors, but increases quantization errors.

Experience has shown that an acceptable perceptual trade-off between quantization errors and clipping is achieved by adjusting the X_{RMS} value to quantizer range (0 V to V_{max}) such that

$$X_{RMS} = \frac{V_{max}}{8} \tag{7.28}$$

With this setting, a typical audio signal rarely exceeds the range of the quantizer (about once in every 10,000 samples). This frequency of occurrence is negligible and usually acceptable for commercial music applications. A b-bit quantizer spans the range from $-4X_{RMS}$ to $4X_{RMS}$ with the step size equal to

$$\Delta = \frac{8X_{RMS}}{2^b - 1} \tag{7.29}$$

EXAMPLE 7.21

Quantizer setting in a digital audio recorder

A digital audio recorder has a 16-bit quantizer that ranges from 0 V to 8 V. The number of steps is

$$n_{steps} = 2^{16} = 65,536$$

The step size is

$$\Delta = \frac{8\,V}{65,536 - 1} = 1.22 \times 10^{-4}\,V$$

Note that when $2^b \gg 1$, the effect of subtracting one is negligible.
The variance of the quantization noise it produces equals

$$\sigma_\epsilon^2 = \frac{\Delta^2}{12} = \frac{1.49 \times 10^{-8}\,V^2}{12} = 1.24 \times 10^{-9}\,V^2$$

To avoid significant clipping, the music to be recorded is adjusted to have a long-term average level equal to $2\,V_{RMS}$. The SNR of the quantized samples is computed as

$$SNR = \frac{X_{RMS}^2}{\sigma_\epsilon^2} = \frac{4\,V_{RMS}^2}{1.24 \times 10^{-9}\,V^2} = 3.22 \times 10^9$$

The SNR value in dB units equals

$$SNR_{dB} = 10\log_{10}\left(3.22 \times 10^9\right) = 90.5\,dB$$

7.7 Research Challenges

7.7.1 Sparse Sampling

The analog-to-digital conversion process described in this chapter samples analog waveforms using a periodic sampling rate—even when the waveform produces zero valued samples, such as during silent periods. Clearly, this is wasteful, because it produces many data values that are not informative. Sparse data sampling converts analog waveforms only when something interesting (an *event*) occurs. Events have characteristic features, such as large amplitudes, that facilitate their recognition. Sparse sampling and its reconstruction to retrieve the original waveform are being investigated for efficient data storage.

7.8 Summary

This chapter described and analyzed the procedure that transforms an analog signal, such as speech or music, into a sequence of numbers using an analog-to-digital converter (ADC). Analog-to-digital conversion was found to occur in two steps: first the analog waveform is sampled and then the samples are quantized by being represented by a finite number of bits. The Nyquist criterion states that the sampling rate must be greater than twice the highest frequency present in the signal in order to prevent information loss. A sampling rate lower that the Nyquist rate produces alias signals.

Quantization converts the infinite-precision analog values into a finite-precision digital values. The quantization error increases with step size Δ. A staircase model illustrated the quantization process. Successive approximation performs quantization in b comparisons rather than 2^b steps in the staircase method.

Two methods of digital-to-analog conversion were described. The boxcar digital-to-analog converter (DAC) reconstructs the analog waveform using piece-wise constant values. Pulse-width modulation (PWM) is an efficient DAC that uses a digital waveform with a variable duty cycle to form an analog voltage.

Quantization error is perceived as noise added to the original analog signal. A trade-off exists between quantization and clipping errors.

7.9 Problems

7.1 Sampling audio waveforms. A talking book generates a digital audio waveform that was generated by sampling the original audio waveform 2,000 times a second. What is the maximum allowed frequency for proper operation?

7.2 Cell phone audio waveforms. The audio signals on your cell phone occupy the frequency range from 300 Hz to 3,400 Hz. What sampling frequency would you specify for digitizing the voice waveform?

7.3 Cell phone HD Voice waveforms. The HD Voice systems for cell phone systems occupy the frequency range from 50 Hz to 7,000 Hz and permit high-quality speech to be transmitted. What sampling frequency would you specify for digitizing the voice waveform?

7.4 Aliasing. Let the sampling frequency $f_s = 10,000$ Hz with the result that the alias frequency $f_a = 1,000$ Hz is observed. What two values of the original frequency f_o could have produced this alias frequency?

7.5 Maximum spacial frequencies. Your old cell phone display has a 180 rows and 320 columns. What are the maximum horizontal and vertical spacial frequencies present in an icon image that would produce an un-aliased image on that display?

7.6 2014_{10} in binary. What is the binary representation of 2014_{10}?

7.7 01011010_2 in decimal. What is the decimal representation of 010101010_2?

7.8 Magic cards. *Magic cards* allow you to guess a number between 0 and 15 that a person was thinking of by answering whether the number appeared on a set of four cards. These cards appear in Figure 7.19 with one number

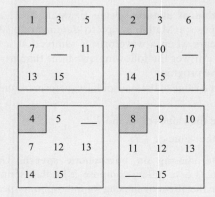

Figure 7.19

Magic cards used in Problem 7.8.

replaced by dashes in each card. What are the missing numbers?

7.9 Video game audio. Your computer game uses digitized audio stored as 10 bits per sample. How many voltage levels does this represent? If the audio is reproduced over a range of 0 V to 5 V, what is step size Δ? If the audio samples were generated 44,000 times per second in each of two stereo channels, how many bits per second does the system produce to give your game sound?

7.10 Binary addresses on the Internet. Current binary addresses of Websites use 4 bytes. What is the number of possible unique addresses?

7.11 CD audio duration. An audio CD stores 650 megabytes (650 MB) of data, where $1\ MB = 2^{20}$ bytes $= 1,048,576$ bytes, and 1 byte is an 8-bit data unit. The sampling rate $f_s = 44$ kHz is used with 16-bit quantization. What duration of stereo music (two separate waveforms) can be stored on a CD? Give answer in minutes.

7.12 ADC with staircase quantizer. An 8-bit ADC performs a conversion every sampling period $T_s = 0.1$ ms. What is the period of the staircase T_Δ?

7.13 Quantizing with successive approximation. An 8-bit ADC performs a conversion using successive approximation with sampling period $T_s = 0.1$ ms. What is the conversion time per bit T_Δ?

7.14 Boxcar DAC output waveform. A digital memory produces the following bit sequence that goes to a 8-bit boxcar DAC that ranges between 0 V and 5 V, and pro-

duces a reconstruction every $T_s = 0.1$ ms. Sketch the reconstructed analog waveform produced by the following bits, providing amplitude and time values in the sketch.

$$1010000000110011100000111$$

7.15 Musical greeting card with boxcar DAC. A musical greeting card plays a 10-second segment of a song when it is opened. If the original audio was processed to remove all frequencies above 3 kHz and 5 bits quantize the samples, what is the number of bits stored on the card digital memory?

7.16 PWM duty cycle. A PWM waveform extends from 0 V to 5 V and has a period $T_{PWM} = 10$ ms. What is the duty cycle of the PWM waveform that produces a 1 V average?

7.17 PWM DAC on-time. Let $V_{max} = 5$ V and $T_{PWM} = 2$ ms. What value of T_{ON} produces $V_{ave} = 3.4$ V?

7.18 PWM DAC value. Let $V_{max} = 5$ V and $T_{PWM} = 2$ ms in a PWM DAC that uses 8 bits. What V_q is closest to $V_{ave} = 3.4$ V?

7.19 SNR of quantized signals. The signal variance $\sigma_s^2 = 4V_{rms}^2$ and the quantizer has step size $\Delta = 0.01$ V. What is the SNR expressed as a ratio of powers and in dB units?

7.20 Maximum SNR of a sinusoidal signal. A sinusoidal signal with amplitude A is offset and applied to a 10-bit quantizer with range from 0 to 5 V. What value of A maximizes the SNR of the samples? What is the SNR?

7.10 Matlab Projects

7.1 Sampling sinusoids. Using Example 16.31 as a guide, compose a Matlab script to plot 4 cycles of a sinusoidal waveform that are sampled 16 times per period.

7.2 Demonstration of aliasing. Using Example 16.32 as a guide, compose a Matlab script to demonstrate aliasing that occurs when a waveform is sampled at a rate $f_s < 2f_{max}$. Answer the following questions that are displayed by the program:

   a. What is the value of the sinusoidal waveform frequency f_o?

   b. What are the values of f_s and T_s?

   c. What is the value of f_a?

7.3 Demonstrate aliasing on microphone speech. Using Example 16.33 as a guide, compose a Matlab script to demonstrate the effect of aliasing for microphone speech. Answer the following questions that are displayed by the program.

   a. For what under-sampling factor value does the speech become unintelligible?

   b. What is the equivalent sampling rate for this factor?

7.4 Synthesizing a sinusoidal image. Following Example 16.36, compose a Matlab script to generate a gray scale image that has n_x periods in the x direction and n_y periods in the y direction. Enter the values of n_x and n_y.

7.5 Under-sampling a jpg image. Following Example 16.37, compose a Matlab script to sample a jpg image and determine the sampling period that begins to deteriorate important features in the image. Answer the following questions that are illustrated by the program.

   a. For what sampling period value does the image become unrecognizable?

   b. What is the equivalent sampling rate for this sampling period?

7.6 **Quantizing waveforms.** Using Example 16.34 as a guide, compose a Matlab script to quantize the waveform

$$s_i = \sin(2\pi i/64) + \sin(2\pi i/48) \text{ V for } 0 \leq i \leq 512$$

a. What is the maximum amplitude of s_i?

b. What is the number of bits that results in a quantization error magnitude $|\epsilon_i| < 0.01$ V?

7.7 **Demonstrate quantization of microphone speech.** Using Example 16.35 as a guide, compose a Matlab script to demonstrate the effect of quantization of microphone speech. For what quantization does the speech become unintelligible?

7.8 **PWM DAC.** Using Example 16.38 as a guide, compose a Matlab script to form a 7-bit PWM DAC waveform with $T_{\text{PWM}} = 10$ ms and $V_{\text{max}} = 5$ V. Generate a plot that shows an upward ramp from 0 V to 5 V in 1 V steps.

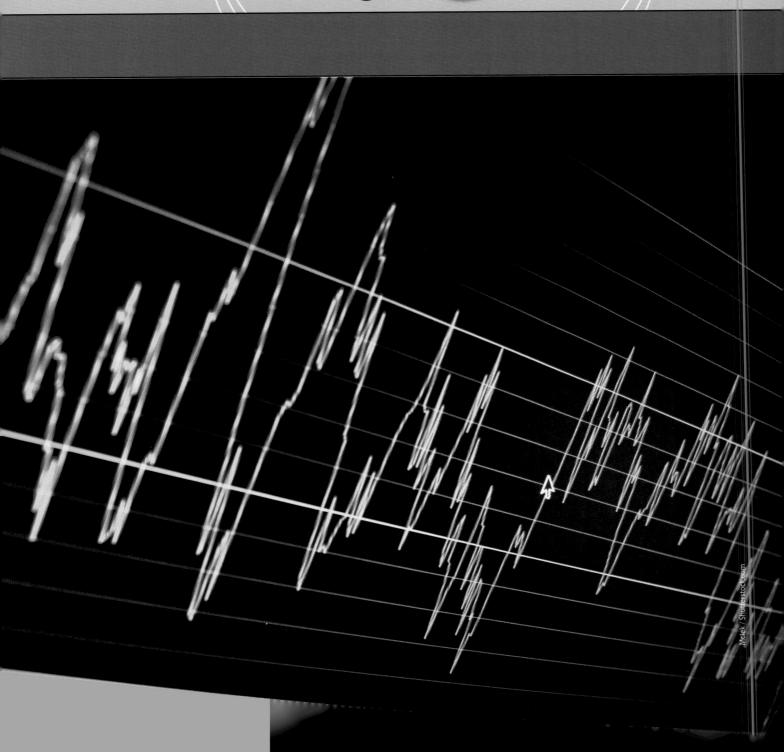

DIGITAL SIGNAL PROCESSING

LEARNING OBJECTIVES

After completing this chapter, the reader should be able to:

- Understand the fundamentals of digital filtering techniques.

- Construct a digital filter from its difference equation.

- Implement moving-average (MA), autoregressive (AR), and ARMA digital filters.

Digital signals are data sequences typically generated through analog-to-discrete-time conversion that contain information. A digital *filter* is a computational process that converts an input sequence x_i into an output sequence y_i that has more desirable features, such as having only relevant components or containing less noise. This chapter describes the basics of digital filter processing in the time domain.

8.1 INTRODUCTION

This chapter provides a general introduction to digital filter operation in the time domain by describing the following topics.

Digital processing—Processing signals with computers using *digital filters* has replaced analog techniques using *analog filters*.

Digital filter anatomy—Digital filters are numerical algorithms whose operation is described using three simple components: delays, multipliers, and summing junctions.

Difference equations—Difference equations are the main mathematical tool for analyzing and synthesizing digital filters.

Digital filter types—Moving-average (MA), auto-regressive (AR), and ARMA digital filters are described. The MA filter motivates the matched processor used for detecting data signals in digital data communication systems.

Figure 8.1 shows a digital filter that processes a sinusoidal sequence that contains random components to produce an output that retains the sinusoidal form and reduces noise. In most cases, a digital filter is implemented by writing a computer program containing instructions that perform numerical calculations on data sequences. With the decreasing cost of computers and the flexibility of changing instructions, digital filters are commonly replacing analog filters that are implemented with physical components and electronics. Digital filters offer flexibility in that changes can be made with software updates that previously could only be done by replacing analog filter hardware, which is an expensive proposition. Figure 8.2 shows a digital filter transforming signal $x(t)$ into $y(t)$ by using an analog-to-discrete time converter (ADC) to convert analog waveform $x(t)$ into digital sequence x_i, the digital filter that processes x_i to form output sequence y_i, and a digital-to-analog converter (DAC) to transform sequence y_i back into analog waveform $y(t)$. To a user, it is impossible to tell whether the transformation of analog waveforms $x(t)$ to $y(t)$ was performed by an analog filter or by a digital filter.

This chapter describes three standard processing approaches that digital filters employ: moving average (MA), autoregressive (AR), and autoregressive-moving average

Figure 8.1

A digital filter is a computational process that converts input sequence x_i into a more desirable output sequence y_i.

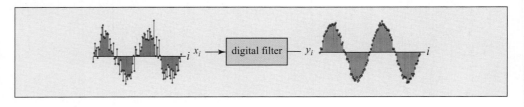

Figure 8.2

A digital replacement of an analog filter transforms $x(t)$ into $y(t)$ by using an ADC, a digital filter to process x_i to produce y_i, and a DAC to convert y_i into $y(t)$.

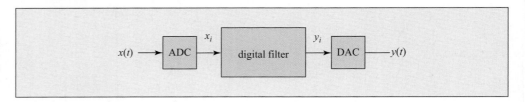

(ARMA). These digital filters are formulated and analyzed using adders, multipliers, and delays in block diagrams that illustrate their operation, although their practical implementation involves writing a computer program that performs these functions. Digital filters are analyzed using difference equations that also allow the direct implementation of the digital filter block diagram. A digital filter is characterized in the time domain by its response to a unit-sample sequence called the *unit-sample response*, which is a time sequence that describes the performance in the time domain. In the next chapter, this unit-sample response is examined in the frequency domain to compute the behavior of digital filters in the frequency domain.

8.2 DIGITAL FILTER ANATOMY

A digital-filter block diagram is implemented by interconnecting the three simple elements shown in Figure 8.3.

Summing junctions: A summing junction has two or more inputs and one output that is equal to the sum of the inputs. Figure 8.3 shows the simplest summing junction having two inputs. A summing junction having m inputs $x_1, x_2, \ldots, x_m$ forms the sum

$$y = \sum_{i=1}^{m} x_i \tag{8.1}$$

The summing junction is readily implemented using a simple computer program instruction, such as the addition of two term performed as

```
y = x1 + x2;
```

Multipliers: A multiplier has a single input, a single output, and a coefficient value that multiplies the input value to produce the output. If x is the input and a is the coefficient, the output y equals

$$y = ax \tag{8.2}$$

The multiplier is implemented using a simple computer program instruction, such as

```
y = a * x;
```

Delays: The delay has a single input and single output. Indicated with a small block labeled δ, the delay is a digital memory that stores the value at the input for one index value, making the past value available for the computations. At index value i, if the input to the memory is x_i or the ith sample of the x_i sequence, the output is the past value x_{i-1}.

These delay elements are typically connected in a cascade, forming a *shift register* in which the output of one delay connects to the input of the next delay. A shift register of size m is implemented in computer instructions defining a buffer array, such as $buf(1), buf(2), \ldots, buf(m)$, with index j. At each increment in time index i, the values in the buffer are shifted down by one element using

$$buf(j) = buf(j-1) \text{ for } 2 \le j \le m \tag{8.3}$$

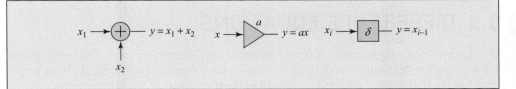

Figure 8.3

Digital filter block diagrams are implemented using summing junctions, multipliers, and delays.

Figure 8.4

Block diagram of a digital filter that computes the running average of the current and past value of the input. Dashed box indicates a shift-register memory that provides the values needed by the multipliers.

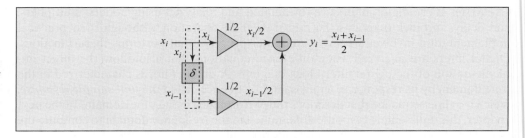

and then inserting the current value (say x_i) into the first buffer position as

$$buf(1) = x_i \qquad (8.4)$$

The design of a digital filter involves determining the number of elements that are needed, their interconnection, and the coefficient values. These three elements are interconnected to form the digital filter structure using the connection rule:

While values can be directed to the inputs of one or more other elements, the only valid method of combining values is through a summing junction.

For example, the output of a delay can connect to the input of another delay, the input of a multiplier, and to a summing junction, but two or more outputs cannot go directly into the input of a multiplier or input of a delay. If such a connection is necessary, a summing junction is introduced to combine these values to form a single output value.

EXAMPLE 8.1

Digital filter that computes the two-sample average

Figure 8.4 shows a digital filter that computes the running average of the current and past value of the input.

- i is current time index value.
- x_i and y_i are the current values of the input and output, respectively.
- x_{i-1} is the value of the previous input.

The current input x_i connects to two elements: a multiplier with gain of $1/2$ and a delay. The multiplier produces $x_i/2$. The current output value of the delay is the input at the previous time x_{i-1}. The delay output connects to a multiplier with gain $1/2$ that produces $x_{i-1}/2$. The outputs of the two multipliers connect to a summing junction, which produces the current value of the digital filter output y_i:

$$y_i = \frac{1}{2} \overbrace{x_i}^{\text{current value}} + \frac{1}{2} \overbrace{x_{i-1}}^{\text{past value}}$$

Thus, the current output is the average of the current and past inputs.

 ## 8.3 DIFFERENCE EQUATIONS

The operation of a digital filter is described using a *difference equation*. These equations are very useful. Not only can they be written by inspection from the digital filter structure, but the digital filter structure—as well as the numerical procedure that implements the digital filter—can be implemented directly from the difference equation.

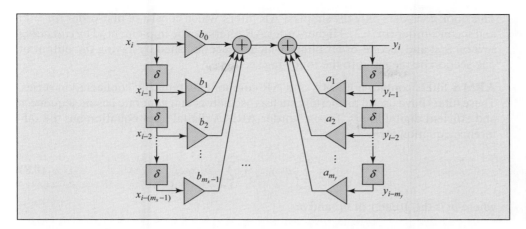

Figure 8.5

Block diagram of the general digital filter.

The general form of a difference equation is given by

$$\overbrace{y_i}^{\text{current output value}} = \overbrace{\sum_{j=1}^{m_y} a_j\, y_{i-j}}^{\text{past output values}} + \overbrace{\sum_{j=0}^{m_x-1} b_j\, x_{i-j}}^{\text{current and past input values}} \tag{8.5}$$

That is, the current output y_i equals the sum of past m_y outputs that are scaled by *feedback* coefficients a_j plus the sum of current and past inputs (m_x values) that are scaled by *feedforward* coefficients b_j. Figure 8.5 shows a block diagram that implements this general difference equation. While there are alternate structures that are more memory efficient, this text describes simplified digital filters based on this general structure.

In analyzing digital filter operation, it is typical to assume *zero initial conditions*, where the values of past outputs or those that occur before an input is applied are equal to zero. *Zero initial condition means that the outputs of delays in the digital filter must be initialized to zero.*

> **Factoid:** The term *order* is often used to indicate the complexity of the filter and is equal to the maximum number of values, either of the input (m_x) or output (m_y), that occur in its difference equation calculation.

General Digital Filter Types

The general difference equation can implement three common digital filter types.

MA (moving-average) filter has $a_j = 0$ for all j (that is, no feedback but only feedforward coefficients b_j). The mth-order MA digital filter has the difference equation given by

$$y_i = \sum_{j=0}^{m-1} b_j\, x_{i-j} \tag{8.6}$$

The set of b_j coefficients are specified to perform a particular task that can range from computing the running average, detecting the presence of a known sequence, or to computing a frequency transform.

AR (*auto-regressive* or *auto-recursive*) filters have at least one non-zero a_j coefficient along with the single non-zero b_0 coefficient. AR filters are useful for designing efficient digital filter replacements for analog filters. The mth-order AR digital filter equation has the difference equation form

$$y_i = \sum_{j=1}^{m} a_j\, y_{i-j} + b_0\, x_i \tag{8.7}$$

This book considers only the simplest AR filters, which consist of first-order ($m = 1$) and second-order ($m = 2$). Higher-order AR filters can be implemented by *cascading* several first and second-order filters, which is accomplished by having the output of one section be the input to the following section.

ARMA filters contain both MA and AR components typically connected in series. These filters have useful properties, such as oscillators that generate cosine sequences and efficient digital filters. The mth-order ARMA digital filter equation has the difference equation form

$$y_i = \sum_{j=1}^{m_y} a_j\, y_{i-j} + \sum_{j=0}^{m_x-1} b_j\, x_{i-j} \tag{8.8}$$

where m is the greater of m_y and m_x.

 ## 8.4 DIGITAL FILTER INPUTS

Digital filters are characterized by their time and frequency properties. This section describes the probing of a digital filter in the time domain by applying an input sequence and observing the response.

Figure 8.6 shows three signal sequences that start at $i = 0$ to probe digital filters.

Unit-sample sequence d_i: This most elementary sequence has a single non-zero element

$$d_i = 1 \text{ for } i = 0 \tag{8.9}$$

$$= 0 \text{ otherwise}$$

Figure 8.7 shows the significance of d_i when it is applied to a digital filter (that is, when $x_i = d_i$). The output y_i is called the *unit-sample response* and is denoted h_i. Thus, when $x_i = d_i$

$$y_i = h_i \text{ for } i = 0, 1, 2, \ldots, n_h - 1 \tag{8.10}$$

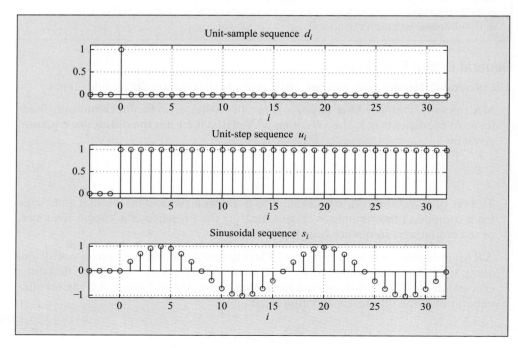

Figure 8.6

Three signal sequences that start at $i = 0$ to probe digital filters.

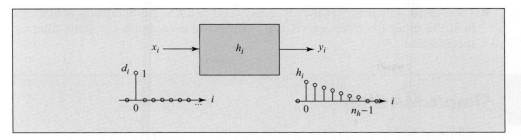

Figure 8.7

A digital filter is characterized by its response to a unit-sample sequence. When $x_i = d_i$, then $y_i = h_i$.

The unit-sample response is used to determine the characteristics of a digital filter. In this chapter, the time-domain response is discussed. Chapter 9 employs h_i to compute the frequency transfer function that describes the filter frequency-domain behavior.

Unit-step sequence u_i: This sequence makes a single transition from 0 to a 1, typically occurring at $i = 0$, and remains constant thereafter. Thus,

$$u_i = 1 \text{ for } i \geq 0 \qquad (8.11)$$
$$= 0 \text{ otherwise}$$

This step sequence is useful for observing the performance of digital filters that model the operation of mechanical systems (such as motors) through their *unit-step response*. For example, the impulse response of a motor is difficult to observe, while its step response displays all of the important dynamic behavior of the motor (such as the time that it takes to reach its final speed).

Sinusoidal sequences: The sinusoidal sequences include the sine sequence

$$s_i = \sin(2\pi f i) \text{ for } i \geq 0 \qquad (8.12)$$
$$= 0 \text{ otherwise}$$

and cosine sequence

$$c_i = \cos(2\pi f i) \text{ for } i \geq 0 \qquad (8.13)$$
$$= 0 \text{ otherwise}$$

where f is the frequency. If the sinusoidal sequence has a period equal to n_P samples,

$$f = \frac{1}{n_P} \qquad (8.14)$$

 # 8.5 MOVING-AVERAGE (MA) FILTERS

The MA digital filter uses only the current and past input values and does not use the past output values. The difference equation of an MA digital filter is given by

$$y_i = \sum_{j=0}^{m_x-1} b_j x_{i-j} \qquad (8.15)$$

MA filters have the following desirable properties.

- MA filters are useful in practice, forming the motivation for data processors that determine transmitted data values from detected signals, as discussed in Chapters 10 and 11.

- MA filters always produce stable operation in that output values will not tend to ∞ if the input values are bounded.

- Their disadvantage is that they typically require more multiplication operations than the other filter types (AR and ARMA) to accomplish the same filtering specification.

EXAMPLE   **8.2**

Simple MA filter

Example 8.1 showed a digital filter described by the difference equation given by

$$y_i = \frac{x_i}{2} + \frac{x_{i-1}}{2}$$

Comparing this formula to the general MA difference equation, $m_x = 2$ with $b_0 = 1/2$, $b_1 = 1/2$, and $b_j = 0$ for other values of j.

The MA digital filter structure can be implemented directly from the difference equation using the following steps.

1. Draw one point on the left side of the page corresponding to input x_i and another on the right side of the page corresponding to output y_i.

2. From the x_i point, draw two lines:

   - A downward vertical line that connects to the first of a series of $m_x - 1$ delays. The output of the first delay is x_{i-1}, and this delay feeds into the second delay (if $m_x > 2$) having output x_{i-2}, and so on, until the $m_x - 1$ delay output produces the past value $x_{i-(m_x-1)}$.

   - Right-going horizontal lines connect the input and each delay output to its own multiplier, that is, x_i connects to multiplier having coefficient b_0, x_{i-1} connects to multiplier having b_1, and so on with $x_{i-(m_x-1)}$ connecting to multiplier having b_{m_x-1}.

3. Draw lines connecting the output of each multiplier to the summing junction.

4. Draw a line from the summing junction to the output point y_i.

EXAMPLE   **8.3**

Implementing an MA filter from the difference equation

Consider the difference equation given by

$$y_i = \frac{x_i}{2} + \frac{x_{i-1}}{2}$$

Figure 8.8 shows the resulting MA filter following the application of the previous steps.

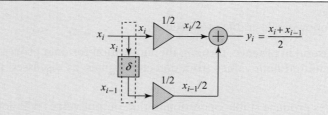

Figure 8.8

Implementing an MA filter from the difference equation.

Unit-Sample Response of MA Filter

For MA filters, the unit-sample response is particularly simple. As the input unit-sample sequence passes through the MA filter, it successively selects each coefficient value and places it at the output. Hence,

$$h_i = b_i \text{ for } i = 0, 1, 2, \ldots, m_x - 1 \tag{8.16}$$

Figure 8.9 illustrates the relationship of b_j to h_j by specifying the feedforward coefficient b_j as the corresponding h_j value. An MA digital filter can be implemented directly from its unit-sample response by setting $b_j = h_j$ for $0 \le j \le n_h - 1$. *Thus, an MA filter can be easily implemented to have any desired unit-sample response.*

Consider an MA filter having $m_x = 5$ coefficients equal to

$$b_0 = 1, b_1 = 2, b_2 = 3, b_3 = 2, b_4 = 1$$

Figure 8.10 shows that the unit-sample response produces values equal to the coefficient sequence. Any signal can be thought of as the unit-sample response of some digital

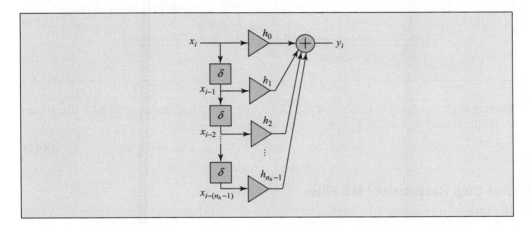

Figure 8.9

The unit-sample response of an MA filter h_i is simply equal to the coefficient set.

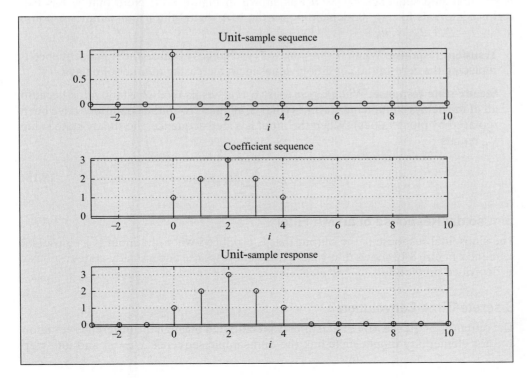

Figure 8.10

MA filter unit-sample response when $x_i = d_i$.

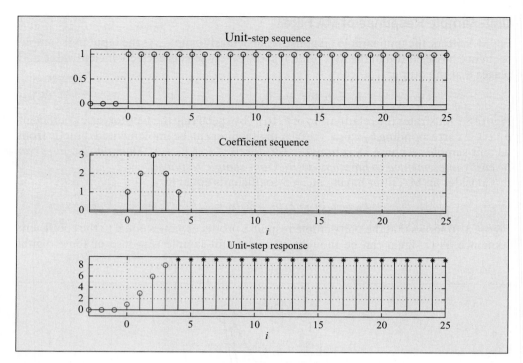

Figure 8.11

MA filter unit-step response when $x_i = u_i$. Transient response is shown in red, and steady-state response in black asterisks.

filter. A signal sequence s_i for $i = 0, 1, 2, \ldots, n_x - 1$ can formed by an MA filter with a coefficient set b_j for $0 \le j \le m_x - 1$ by setting $m_x = n_x$, and

$$b_i = s_i \text{ for } i = 0, 1, 2, \ldots, n_x - 1 \tag{8.17}$$

Unit-Step Response of MA Filter

The unit-step response of a digital filter is the output that is produced when the input is the unit-step sequence ($x_i = u_i$) as shown in Figure 8.11. Note that y_i has two distinct intervals for $i \ge 0$, depending on the MA digital filter unit-sample response h_i for $i = 0, 1, 2, \ldots, n_h - 1$:

Transient response: When $0 \le i \le n_h - 2$, y_i changes because the input sequence is replacing the zero-initial condition values in an increasing number of delays.

Steady-state response: When $i \ge n_h - 1$, y_i reaches its final constant value because all of the delays contain input values (that is, all the zero-initial conditions have been replaced by input values). When the input is a step-sequence, the steady-state value y_∞ equals

$$y_\infty = \sum_{j=0}^{n_h-1} h_j \tag{8.18}$$

Sinusoidal Response of an MA Filter

The sinusoidal response is the output that is produced when the input is a sinusoidal sequence. Figure 8.12 shows that y has a transient response and a steady-state response, as described previously.

Discrete-Time Convolution

The output y_i of the the MA digital filter was computed through the difference equation and for elementary inputs, including the unit-sample sequence $x_i = d_i$ and unit-step

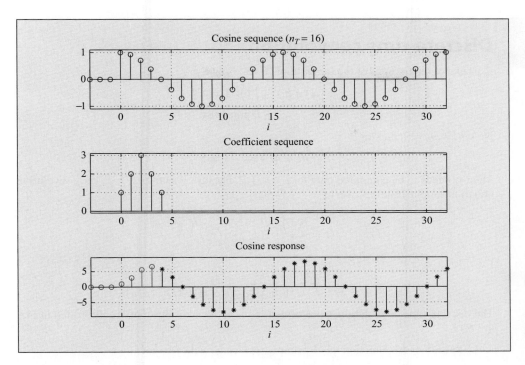

Figure 8.12

MA filter sinusoidal response when x is a cosine sequence. Transient response is shown in red and steady-state response in black asterisks.

sequence $x_i = u_i$. An important alternate view of the difference equation is computing the output y_i from a general input sequence x_i and the unit-sample response h_i through discrete-time *convolution*, as

$$y_i = h_i * x_i = \sum_{j=0}^{n_h-1} h_j x_{i-j} \text{ for } i = 0, 1, 2, \ldots, n_y - 1 \tag{8.19}$$

The duration of y_i is related to those of x_i and h_i by

$$n_y = n_x + n_h - 1 \tag{8.20}$$

Comparing the convolution equation to the MA difference equation, we have

$$y_i = b_o x_i + b_1 x_{i-1} + b_2 x_{i-2} + \cdots + b_{m_x-1} x_{i-m_x-1} \tag{8.21}$$

$$= \sum_{j=0}^{m_x-1} b_j x_{i-j}$$

Note that

$$h_i = b_i \text{ for } 0 \leq i \leq m_x - 1 \tag{8.22}$$

Hence, the discrete-time convolution is equivalent to evaluating the output sequence of an MA filter through the difference equation. The computation of the discrete-time convolution is illustrated in the next example.

Factoid: It is good practice to use simple examples for remembering specific operations and testing your numerical procedures.

EXAMPLE **8.4**

Discrete-time convolution

Consider x_i and h_i to have the same simple form:

$$x_i(= h_i) = 1 \text{ for } i = 0, 1$$

$$= 0 \text{ otherwise}$$

Note $n_x = n_h = 2$, making

$$n_y = n_x + n_h - 1 = 3$$

Then y_i needs to be computed only for $i = 0, 1, 2$, because $y_i = 0$ for $i > 2$. The convolution equation simplifies in this case to

$$y_i = \sum_{j=0}^{1} h_j \, x_{i-j} \text{ for } i = 0, 1, 2$$

or

$$y_i = \overbrace{h_0 \, x_i}^{j=0} + \overbrace{h_1 \, x_{i-1}}^{j=1} \text{ for } i = 0, 1, 2$$

The discrete-time convolution is performed in the following steps that are illustrated in Figure 8.13.

1. Specify h_i as a function of j by simply replacing i with j to get

$$h_j = 1 \text{ for } j = 0, 1$$

$$= 0 \text{ otherwise}$$

   Figure 8.13 shows h_j as the top curve.

2. To perform the term-by-term multiplication of x_{i-j} with h_j, we need to specify x_{i-j} as a function of j. As a function of $-j$, x_{-j} is the time-reversed version of x_j (that is, it starts at $j = 0$ and proceeds in the $-j$ direction). Figure 8.13 shows the form of x_{-j} in the second curve.

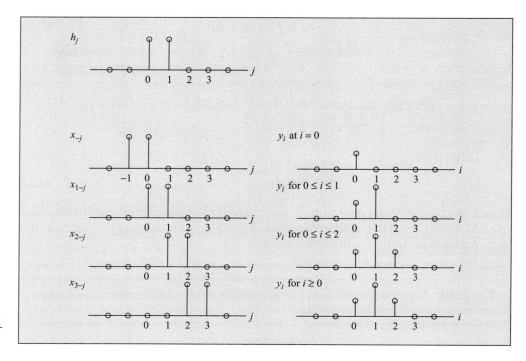

Figure 8.13

Illustrating discrete-time convolution $y_i = h_i * x_i$.

3. The value of x_{i-j} is a delayed version of x_j with i setting the *end time*. That is, when $i = 0$, we have x_{-j}, which ends at $j = 0$. When $i = 1$, x_{1-j} ends at $j = 1$. As i increases, x_{i-j} shifts to more positive values of j. Figure 8.13 shows the progression of x_{i-j} as i increases from 0 to 3.

4. Having formed x_{i-j}, we can proceed to computing the value of y_i from the overlapping h_j and x_{i-j} values. The results are shown in the right-side curves in Figure 8.13. Let us consider the computations for each value of i:

   $i = 0$: Note that h_j and x_{-j} have only one non-zero term in common: $h_0 = 1$ and $x_0 = 1$. Hence,

   $$y_0 = h_0 x_0 = 1 \times 1 = 1$$

   $i = 1$: h_j and x_{1-j} have two non-zero terms in common, producing

   $$y_1 = h_0 x_1 + h_1 x_0 = 1 \times 1 + 1 \times 1 = 2$$

   $i = 2$: h_j and x_{2-j} have only one non-zero term in common: $h_1 = 1$ and $x_1 = 1$. Thus,

   $$y_2 = h_1 x_1 = 1 \times 1 = 1$$

   $i = 3$: This term shows why it evaluates to zero. Note that h_j and x_{i-j} have no non-zero terms in common for $i \geq 3$, producing $y_3 = 0$. Hence, $y_i = 0$ for $i \geq 3$.

Note that the convolution of two square sequences results in a triangular sequence. This is a useful simple result for testing your program.

8.6 AUTOREGRESSIVE (AR) FILTERS

The AR digital filter uses only the current input value and the past m_y output values. The difference equation of an MA digital filter is given by

$$y_i = \sum_{j=1}^{m_y} a_j \, y_{i-j} + b_0 x_i \qquad (8.23)$$

Figure 8.14 shows the block diagram of a general AR digital filter.

AR filters have the following desirable properties:

■ AR filters describe physical systems whose unit-sample responses decay with time. First-order AR systems model simple exponential decays, while second-order systems model the decay of systems that exhibit a resonant frequency.

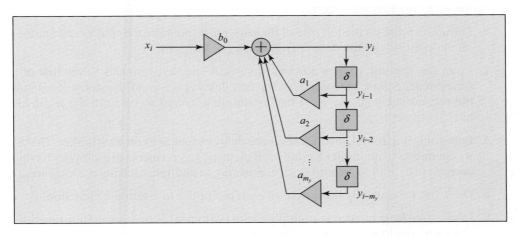

Figure 8.14

Block diagram of a general autoregressive (AR) digital filter.

Figure 8.15

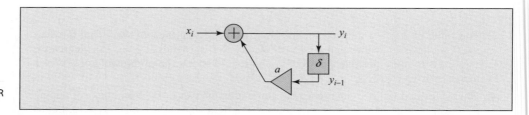

Block diagram of a first-order AR filter.

- AR filters have simple structures, but if not carefully implemented, they may produce unstable operation in that output values will tend to ∞—even if the input values are bounded.

- Most practical digital filter implementations combine first-order and second-order AR digital filters.

8.6.1 First-Order AR Filter

The simplest *first-order AR filter* shown in Figure 8.15 has the difference equation form

$$y_i = a\,y_{i-1} + x_i \tag{8.24}$$

where a_1 is simplified to a and $b_0 = 1$. Practical digital filters are those with $|a| \leq 1$. When $a = 1$, the AR filter implements an *ideal integrator*, while $|a| > 1$ results in the AR filter output becoming infinite, making the filter *unstable*.

EXAMPLE 8.5   First-order AR filter

The first-order AR filter with $0 < a < 1$ is the digital filter that corresponds to an analog RC low-pass filter. An example of difference equation for a practical digital filter is given by

$$y_i = 0.9y_{i-1} + x_i$$

Comparing with the general difference equation, we find $m_y = 1$, $a_1 = 0.9$, and $a_j = 0$ for other values of j, and $m_x = 1$ with $b_0 = 1$, and $b_j = 0$ for other values of j.

The AR digital filter structure can be implemented directly from the difference equation using the following steps.

1. Draw one point on the left side of the page corresponding to input x_i and another on the right side of the page corresponding to output y_i.

2. From the y_i point, draw a downward vertical line that connects to the first of a series of m_y delays. The output of the first delay is y_{i-1}, and this delay feeds into the second delay (if $m_y > 1$) having output y_{i-2} and so on until the m_y delay having output y_{i-m_y}.

3. Left-going horizontal lines connect each delay output to its own multiplier. That is, y_{i-1} connects to multiplier having coefficient a_1, y_{i-2} connects to multiplier having coefficient a_2, and so on with y_{i-m_y} connecting to multiplier having coefficient a_{m_y}.

4. Draw lines connecting the output of each multiplier to a summing junction.

5. Draw a line from input x_i to multiplier with coefficient b_0 to the summing junction. The summing junction output connects to point y_i.

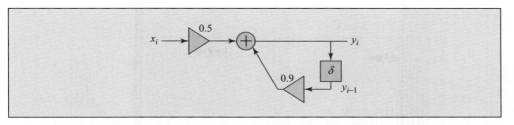

Figure 8.16

Example of implementing an AR filter.

Implementing an AR filter from the difference equation

EXAMPLE 8.6

Consider the difference equation given by

$$y_i = 0.9y_{i-1} + 0.5x_i$$

Figure 8.16 shows the implementation of an AR filter using the previous steps.

Unit-Sample Response of First-Order AR Filter

We determine h_i for the first-order AR filter as

$$y_i = a\,y_{i-1} + x_i$$

by assuming zero initial conditions ($y_i = 0$ for $i < 0$), applying a unit-sample sequence $x_i = d_i$, and computing the output $y_i = h_i$ as a function of i. Consider the following values of i.

For $i < 0$: With the zero initial condition, $y_{i-1} = 0$ and $d_i = 0$. Hence,

$$h_i = a \times 0 + 0 = 0 \text{ for } i < 0$$

For $i = 0$: With the zero initial condition, $y_{-1} = 0$ and $d_0 = 1$. Hence,

$$h_0 = a \times 0 + 1 = 1$$

For $i = 1$: $y_{i-1} = y_0 = 1$ and $d_1 = 0$. Hence,

$$h_1 = a \times 1 + 0 = a$$

For $i = 2$: $y_{i-1} = y_1 = a$ and $d_2 = 0$. Hence,

$$h_2 = a \times a + 0 = a^2$$

For $i = 3$: $y_{i-1} = y_2 = a^2$ and $d_3 = 0$. Hence,

$$h_3 = a \times a^2 + 0 = a^3$$

By induction, we have

$$h_i = a^i \text{ for } i \geq 0 \qquad (8.25)$$

and $h_i = 0$ otherwise.

Note:

- $h_i \to 0$ as i increases when $|a| < 1$. Practical (stable) first-order AR filters will always have $|a| < 1$.

- $h_i \to \infty$ as i increases if $|a| > 1$. This results in an *unstable system*, which is undesirable for practical filters. Computed sequences always have finite durations, so h_i never reaches ∞ but can exceed the number system of your computer software.

- Theoretically, h_i has an infinite duration because $|h_i| > 0$ for all $i \geq 0$. For large i and $|a| < 1$, $h_i \to 0$, so that eventually the values will fall below the precision that can be represented in a computer. So *effectively*, even AR filters have finite-duration unit-sample responses.

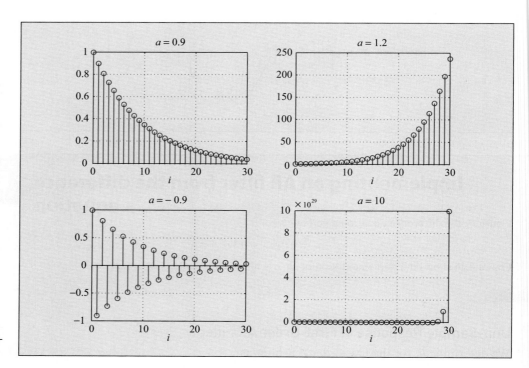

Figure 8.17

First-order AR filter unit-sample responses for different values of a.

EXAMPLE 8.7

First-order AR digital filter unit-sample responses

Figure 8.17 shows how the unit-sample response of a first-order AR filter varies with a. So

$$h_i = a^i \text{ for } i \geq 0$$

Note that $h_i \to 0$ for $|a| < 1$ and $h_i \to \infty$ for $a > 1$.

The $a = 10$ curve shows the cryptic tell-tale sign that a digital filter is unstable. Typically, only the last few values appear on the graph with the scale (in this case 10^{29}) being very large.

EXAMPLE 8.8

Replacing an RC low-pass analog filter with a first-order AR digital filter

Replace an analog RC low-pass filter having $h(t) = e^{-t/(\text{RC})}$ for $t \geq 0$ with a digital filter using sampling period T_s.

Sampling $h(t)$ every T_s seconds, we obtain the sequence values

$$h_i = h(t)|_{t=iT_s} = e^{-i\,T_s/(\text{RC})} \text{ for } i \geq 0$$

Equating with the general form of the first-order AR unit-sample response $h_i = a^i$ for $i \geq 0$, we get

$$a^i = e^{-i\,T_s/(\text{RC})} = \left(e^{-T_s/(\text{RC})}\right)^i \quad \rightarrow \quad a = e^{-T_s/(\text{RC})}$$

Figure 8.18 compares the analog filter impulse response $h(t)$ with h_i, which is the unit-sample response produced by the digital filter replacement. Inexpensive micro-controllers make such analog filters with digital filter replacements very common in practice.

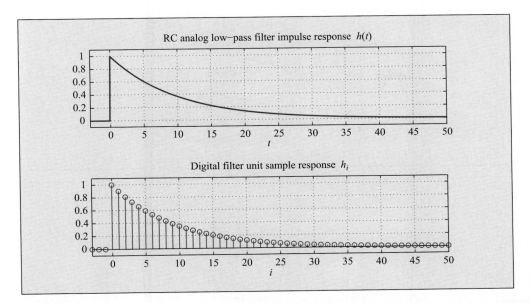

Figure 8.18

Replacing an RC analog low-pass filter with a first-order AR digital filter.

Modeling low-pass transmission channels

EXAMPLE **8.9**

Transmission channels are often modeled with AR filters to indicate their limitations for transmitting binary data signals. The high frequency content of binary data signals occurs in their transitions from one voltage limit to the other.

Figure 8.19 shows a channel that is a moderate low-pass filter ($a = 0.8$) that allows a binary signal to reach its steady-state value during the data transmission interval ($n = 32$) and a channel that is a severe low-pass filter ($a = 0.95$) that does not allow a binary signal to reach its steady-state value during the data transmission interval.

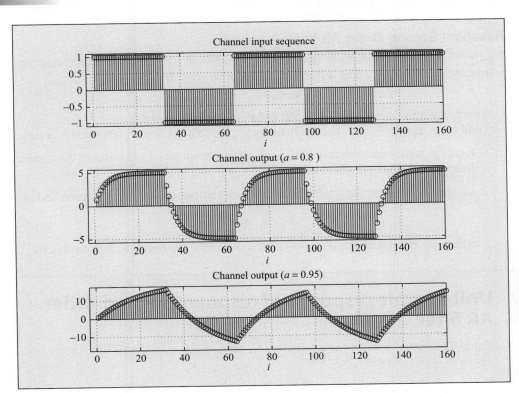

Figure 8.19

Modeling a band-limited transmission channel with first-order AR filter ($a = 0.8$) and a severely band-limited transmission channel ($a = 0.95$). The output amplitudes were those produced by the AR filter having the corresponding coefficient value, but are inconsequential otherwise.

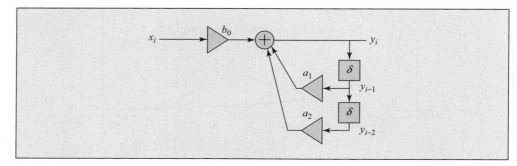

Figure 8.20

Block diagram of a second-order AR filter.

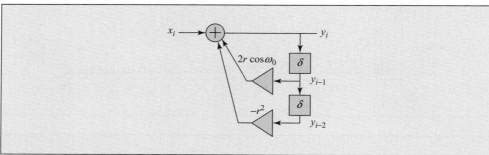

Figure 8.21

Block diagram of a second-order AR filter having resonant frequency ω_o.

8.6.2 Second-Order AR Filters

The *second-order AR filter* is the other important AR filter and exhibits a *resonance* frequency. Figure 8.20 shows the second-order AR filter has the difference equation form

$$y_i = a_1\, y_{i-1} + a_2 y_{i-2} + b_0 x_i \tag{8.26}$$

The second-order AR digital filter is a building-block component that together with the first-order AR filter can form an digital filter of any order merely by cascading individual first-order and second-order sections.

Resonant Second-Order AR filter

Figure 8.21 shows an AR filter that has a resonance at frequency ω_o with a damping term r, where $0 < r < 1$. The difference equation is given by

$$y_i = 2r \cos(\omega_o)\, y_{i-1} - r^2 y_{i-2} + x_i$$

Comparing with the general difference equation, we find $m_y = 2$, $a_1 = 2r \cos \omega_o$, $a_2 = -r^2$ and $m_x = 1$, with $b_0 = 1$. The coefficients are specified in terms of its r and ω_o values:

> r specifies the rate of decay in its unit-sample response. In this sense, it is similar to a in first-order AR filters.

> ω_o specifies the frequency of the oscillatory component. Its value is specified as
> $$\omega_o = \frac{2\pi}{n_P} \tag{8.27}$$
> where n_P is the number of points in one period of the oscillatory component.

EXAMPLE 8.10

Unit-sample response of resonant second-order AR filter

Figure 8.22 shows the unit-sample response of a resonant second-order AR filter with $r = 0.9$ and $\omega_o = 2\pi/16$ ($n_p = 16$). Note that

$$|h_{35}|_{\max} = 0.9^{35} < 0.01$$

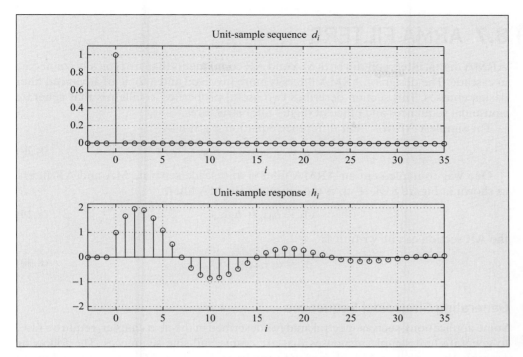

Figure 8.22

Resonant second-order AR filter unit-sample sequence.

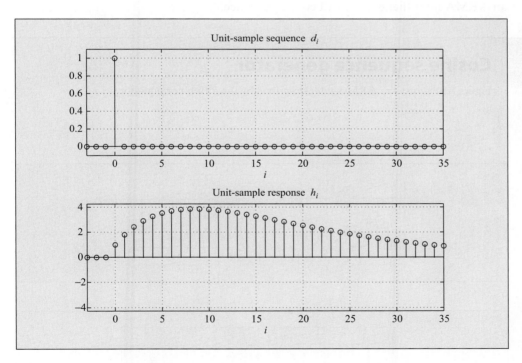

Figure 8.23

Non-resonant second-order AR filter unit-sample sequence.

That is, h_i values for $i \geq 35$ are negligibly small (≈ 0). Hence, even an AR filter has an effective finite duration.

Figure 8.23 shows the unit-sample response of a non-resonant second-order AR filter with $r = 0.9$ and $\omega_o = 0$.

8.7 ARMA FILTERS

ARMA digital filters contain both MA and AR components that are typically connected in cascade. The design of ARMA filters is more involved and a topic of advanced filter design courses. This section describes two useful properties: oscillators that generate sinusoidal sequences and efficient digital filter structures.

The simplest ARMA filter difference equation is given by

$$y_i = a_1 y_{i-1} + b_0 x_i + b_1 x_{i-1} \tag{8.28}$$

One way to implement an ARMA filter is to cascade separate MA and AR filters, as shown in Figure 8.24. If $y1_i$ is the output of the MA filter,

$$y1_i = b_0 x_i + b_1 x_{i-1} \tag{8.29}$$

the AR section can be written as

$$y_i = a_1 y_{i-1} + y1_i \tag{8.30}$$

Generating Sinusoidal Sequences

Some applications, such as spectral analysis described in the next chapter, require a filter to generate unit-sample responses that are cosine and sine sequences. The following example shows that an MA filter section can be added to an AR system to implement an ARMA filter that generates a cosine sequence.

EXAMPLE 8.11

Cosine sequence generator

Figure 8.25 shows the ARMA filter described by the difference equation

$$y_i = 2\cos(\omega_o)y_{i-1} - y_{i-2} + x_i - \cos(\omega_o)x_{i-1}$$

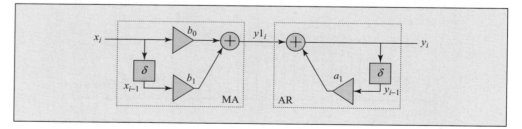

Figure 8.24

ARMA filter implemented as a cascade of MA and AR filters.

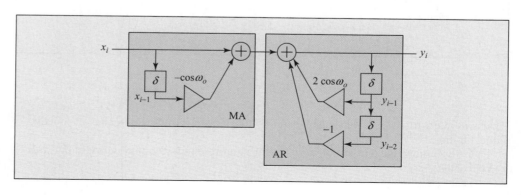

Figure 8.25

ARMA filter that generates a cosine sequence.

has a cosine sequence as its unit-sample response. For input $x_i = d_i$ and assuming zero initial conditions, the output is the cosine sequence

$$y_i = \cos(\omega_o i) \text{ for } i \geq 0$$

Note the damping term $r = 1$ forms a unit-sample response that does not decay with time. The second-order MA section provides the modification that produces the cosine sequence.

Efficient ARMA Filter Structures

A memory-efficient ARMA filter can be implemented by switching the order of the MA and AR sections to implement the ARMA filter with an AR section followed by the MA section, as shown in Figure 8.26.

If $y1_i$ is the output of the AR filter, the AR section can be written as

$$y1_i = a_1 y1_{i-1} + x_i \tag{8.31}$$

and the MA section produces the output

$$y_i = b_0 y1_i + b_1 y1_{i-1} \tag{8.32}$$

Note the memories connected to $y1_i$ remember the same values. Hence, both can be replaced by a single memory, resulting in a simpler, more memory efficient, filter structure.

Efficient ARMA cosine generator EXAMPLE 8.12

This example describes an efficient ARMA filter structure that generates a cosine sequence. Starting with the difference equation in Example 8.11:

$$y_i = 2\cos(\omega_o) y_{i-1} - y_{i-2} + x_i - \cos(\omega_o) x_{i-1}$$

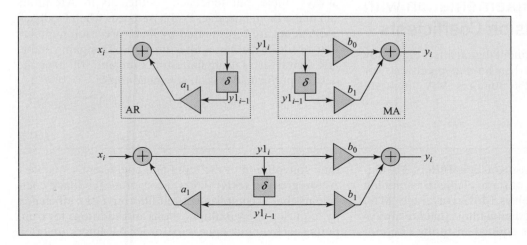

Figure 8.26

Efficient ARMA filter shares memory of AR and MA sections.

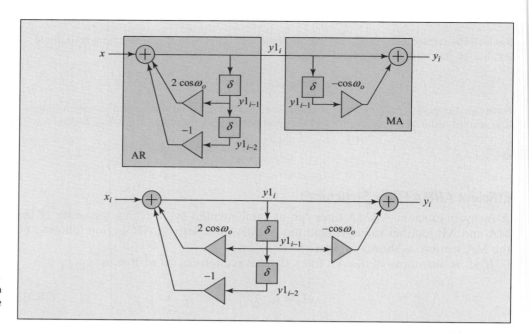

Figure 8.27

ARMA filter implemented as a oscillator that generates a cosine sequence.

We reverse the order of filters by placing the AR section first, having a difference equation of

$$y1_i = 2\cos(\omega_o)y1_{i-1} - y1_{i-2} + x_i$$

This output then becomes the input to the MA section, having a difference equation of

$$y_i = y1_i - \cos(\omega_o)y1_{i-1}$$

Combining common shift-register memory values produces the ARMA filter shown in Figure 8.27.

8.8 Research Challenges

8.8.1 AR Filter Implementation with Finite-Precision Coefficients

Micro-controllers that implement digital filters efficiently use integer arithmetic that has limited precision. For example, 8 bits represent 256 numbers. MA filters are always stable, but feedback coefficients in AR filters can make them unstable. Problems occur when feedback coefficients are represented by finite-precision numbers that result in making a filter unstable. Designing stable AR digital filters using finite-precision and still have the desired performance is an advanced topic.

8.9 Summary

This chapter presented the basics of digital filters. The digital filter anatomy consisted of three elements: summing junctions, multipliers, and delays. Moving-average (MA) filters form a flexible class of digital filters that are always stable. The MA operation is equivalent to the discrete-time convolution of the input and coefficient sequences. Autoregressive (AR) filters incorporate feedback coefficients that accomplish certain filtering tasks efficiently by requiring a few multiplications and additions for computing each output value. First-order AR filters are often used to replace RC analog low-pass filters. First-order AR filters often model the band-limited characteristic of data communication channels. Second-order AR digital filters exhibit a resonance that forms the basis of band-pass fil-

ters. Autoregressive moving-average (ARMA) digital filters are implemented by cascading MA and AR filters to form efficient filters and oscillators having sinusoidal unit-sample responses.

8.10 Problems

8.1 MA digital filter from description. Write the difference equation and draw the block diagram of the digital filter that computes the average of the current and past two samples of input sequence x_i.

8.2 Difference equation from MA digital filter structure. Write the difference equation of the digital filter shown in Figure 8.28.

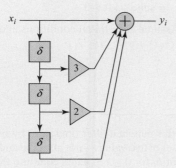

Figure 8.28

MA filter for Problem 8.2.

8.3 AR digital filter from description. Write the difference equation and draw the block diagram of the digital filter whose current output is the sum of twice the current input and half the previous output.

8.4 First-order AR digital filter with positive feedback coefficient. Draw the digital filter structure and compute the unit-sample response h_i for $0 \leq i \leq 3$ of the digital filter defined by the difference equation

$$y_i = 0.9 y_{i-1} + x_i$$

What is the smallest value of i that produces $h_i < 0.01$?

8.5 Difference equation from AR filter structure. Write the difference equation of the digital filter shown in Figure 8.29.

8.6 Second-order AR system. Assuming zero initial conditions, compute the unit-sample response h_i for $0 \leq i \leq 2$ for the difference equation

$$y_i = 2r \cos(\omega_o) y_{i-1} - r^2 y_{i-2} + x_i$$

where $r = 0.9$ and $\omega_o = \pi/10$.

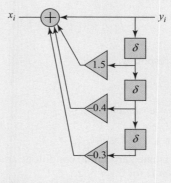

Figure 8.29

AR filter for Problem 8.5.

8.7 ARMA filter from difference equation. Draw two ARMA digital filters that correspond to the difference equation

$$y_i = 2r \cos(\omega_o) y_{i-1} - r^2 y_{i-2} + x_i - r \cos(\omega_o) x_{i-1}$$

for $r = 0.9$ and $\omega_o = \pi/10$. The first structure is configured as an initial MA filter that connects to an AR filter. The second structure is configured as an initial AR filter that connects to an MA filter.

8.8 ARMA filter from difference equation. Draw two ARMA digital filters that correspond to the difference equation

$$y_i = 2 \cos(\omega_o) y_{i-1} - y_{i-2} + \sin(\omega_o) x_{i-1}$$

for $r = 0.9$ and $\omega_o = \pi/10$. The first structure is configured as an initial MA filter that connects to an AR filter. The second structure is configured as an initial AR filter that connects to an MA filter.

8.9 Difference equation from ARMA filter block diagram. Write the difference equation of the digital filter shown in Figure 8.30.

8.10 Difference equation from efficient ARMA filter block diagram. Write the difference equation of the digital filter shown in Figure 8.31.

8.11 Verification of cosine sequence generator. Verify Example 8.11 that the difference equation

$$y_i = 2 \cos(\omega_o) y_{i-1} - y_{i-2} + x_i - \cos(\omega_o) x_{i-1}$$

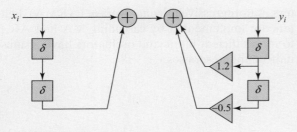

Figure 8.30

ARMA filter for Problem 8.9.

has unit-sample response

$$h_i = \cos(\omega_o i) \text{ for } 0 \leq i \leq 2$$

for $\omega_o = \pi/8$. Assume zero initial conditions and compare h_i with $\cos(\omega_o i)$ for $0 \leq i \leq 2$.

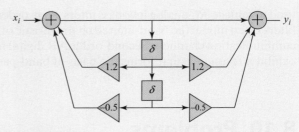

Figure 8.31

ARMA filter for Problem 8.10.

8.12 Sine sequence generator. Verify that the difference equation

$$y_i = 2\cos(\omega_o)y_{i-1} - y_{i-2} + \sin(\omega_o)x_{i-1}$$

has unit-sample response

$$h_i = \sin(\omega_o i) \text{ for } 0 \leq i \leq 2$$

for $\omega_o = \pi/8$. Assume zero initial conditions and compare h_i with $\sin(\omega_o i)$, for $0 \leq i \leq 2$.

8.11 Matlab Projects

8.1 MA digital filter. Using Example 16.39 as a guide, compose a Matlab script to implement a general MA digital filter. Verify your program by computing 10 samples of the unit-sample response (recall $x_i = d_i$) of a five-sample averager defined by

$$y_i = \sum_{j=0}^{4} h_j x_{i-j} \text{ for } 0 \leq i \leq 9$$

8.2 MA filter implementing a linear interpolation. The unit-sample response of a three-sample MA digital filter that acts as linear interpolator is given by

$$h_0 = 0.5, h_1 = 1, h_2 = 0.5, h_i = 0, \text{ otherwise}$$

Compose a Matlab script extending Example 16.39 to show that this MA filter linearly interpolates the values between specified sequence values. First, generate $s_i = \sin(\omega_o i)$ for $\omega_o = 2\pi/32$ and $0 \leq i \leq 64$. The system input $x_i = s_i$ for even i, and $x_i = 0$ for odd i. Using subplot(3,1,n),stem(.), plot x_i, y_i and the error $\epsilon_i = s_i - y_i$.

8.3 MA matched filter. Using Example 16.39 as a guide, compose a Matlab script to compute the output of a MA filter whose filter coefficient sequence h_i equals the input sequence x_i in *time-reversed order*. For example, if $x_0 = 1, x_1 = 2$, and $x_2 = 3$, then $h_0 = 3, h_1 = 2$, and $h_2 = 1$. Such a filter is called a *matched filter*. The output sequence of a matched filter is symmetric about the time point at which the entire signal sequence is in the MA filter (about $i = 2$).

Demonstrate this matched filter operation by specifying your favorite set of five values—not all equal and some less than zero—for h_i for $0 \leq i \leq 4$ and compute the output. The matched filter is the basis for the *matched processor* for detecting data signals in the presence of noise, described in Chapter 10.

8.4 Cascading MA filters. Compose a Matlab script that cascades two MA filters with the output of the first forming the input to the second. Verify your program by computing the unit-sample response of the cascaded connection of two identical three-sample linear interpolators.

8.5 AR digital filter. Using Example 16.40 as a guide, compose a Matlab script to implement a general AR digital filter. Verify your program by computing h_i for $0 \leq i \leq 50$ for the first-order AR filters following coefficients.

1. $a = -0.9$
2. $a = 1$
3. $a = 1.1$
4. $a = 10$

8.6 Cascading first-order AR digital filters. Compose a Matlab script that cascades two AR filters with the output of the first forming the input to the second. to the unit-sample response of the cascaded connection of two identical first-order AR systems with $a = 0.95$.

8.7 **Second-order AR digital filter.** Compose a Matlab script to compute h_i for $0 \leq i \leq 50$ for the second-order AR filters having the following coefficients.

1. $r = 0.9, \omega = \pi/32$
2. $r = 1.1, \omega = \pi/32$
3. $r = 0.9, \omega = 0$

8.8 **Second-order ARMA filter.** Using Example 16.41 as a guide, compose Matlab code to implement an ARMA system that has

$$h_i = \sin(\omega_o i) \text{ for } 0 \leq i \leq 64$$

for $\omega_o = \pi/8$. Hint: the difference equation is

$$y_i = 2\cos(\omega_o)y_{i-1} - y_{i-2} + \sin(\omega_o)x_{i-1}$$

Using *subplot(3,1,n),stem(.)*, plot h_i computed directly from the sine function, y_i, and the error $h_i - y_i$.

Equalizer collection

SPECTRAL ANALYSIS

LEARNING OBJECTIVES

After completing this chapter, the reader should be able to:

- Understand and perform spectral analysis on discrete-time data.
- Interpret results produced by the fast Fourier transform.
- Analyze systems in the frequency domain.

Electrical engineers describe signals in terms of their frequency components and analyze systems in terms of their effect in passing different frequencies. A signal is described by its *spectrum* and a system by its *frequency transfer function*. These frequency-domain descriptions are facilitated by the time-to-frequency numerical transformation called the *discrete-time Fourier transform* (DFT). Spectral analysis is routinely performed using the efficient DFT algorithm called the *fast Fourier transform* (FFT). This chapter describes these frequency-domain techniques and their application in practice.

9.1 INTRODUCTION

This chapter provides a general introduction to spectral analysis by describing the following sequence of topics.

Frequency domain analysis—Analog signals and the behavior of systems are often described in terms of their frequency properties.

Discrete Fourier transform—The frequency-domain representation of discrete-time sequences and the unit-sample responses of digital filters are computed with the numerical procedure called the discrete Fourier transform. Its popularity and wide application is due to an efficient implementation of this algorithm called the fast Fourier transform (FFT), which performs the same calculations up to 100 times faster.

Signal spectrum and transfer function—The spectrum describes a signal in terms of its frequency components, while the transfer function describes the effect of a system on these components.

Frequency domain processing—Engineers familiar with the frequency properties of signals and systems often choose to process signals in the frequency domain, in large part due to the efficiency of the FFT. Careful attention must be paid to the sequences applied to the FFT to obtain the correct results.

9.2 FREQUENCY DOMAIN ANALYSIS

Waveforms that have complex shapes in the time domain are often simpler to express in terms of their frequency components. Figure 9.1 shows time waveform $x(t)$ that may be the acoustic signal produced by a plucked musical instrument, such as a piano or guitar, and sensed by a microphone. The waveform is initially large and decays with time. Spectrum analysis displays the energy of the frequency components present in the waveform. The signal spectrum clearly shows that the waveform contains three

Figure 9.1

The time waveform $x(t)$ composed of three harmonically related frequencies has a simple frequency-domain representation in its spectrum $S_x(f)$.

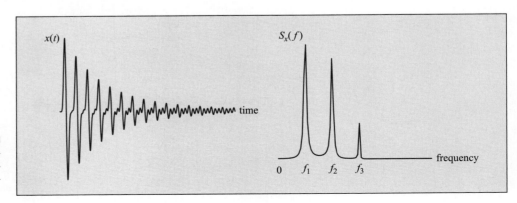

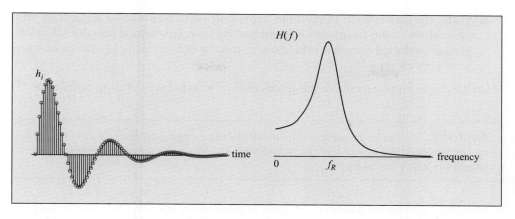

Figure 9.2

The unit-sample response h_i of a second-order AR digital filter and its frequency-domain representation in the transfer function $H(f)$.

frequencies: a fundamental component f_1, a component at second harmonic $f_2 = 2f_1$, and a third harmonic $f_3 = 3f_1$.

Systems are also often simpler to express in terms of their effects on frequency components. Figure 9.2 shows the unit-sample response of a second-order AR digital filter h_i. Such a digital filter may simulate the time-domain behavior of a pendulum or other resonant system. Frequency analysis computes the transfer function that displays how frequency components are modified as they pass through the digital filter. The transfer function $H(f)$ shows that frequencies around the digital filter resonant frequency f_R are passed, while frequencies less than and greater than f_R are attenuated. In the case of the pendulum, any pushing force time profile causes a pendulum to swing at its resonant frequency.

The analysis of signals and systems in the frequency domain is often simpler because the convolutions that govern their time-domain behaviors convert to simpler multiplication operations. When discussing system behavior, it is easier to describe the behavior in the frequency domain than in the time domain. The frequency-domain behavior is determined by transforming time sequences, such as x_i, into their frequency representation, X_k, through the discrete Fourier transform (DFT). But this simplification comes at a cost: The analysis of time-domain signals in the frequency domain requires three additional considerations, shown in Figure 9.3.

1. Transformations are required to convert a time-domain sequence x_i into its frequency-domain counterpart X_k, which is called the *discrete Fourier transform (DFT)*, and then to convert the frequency-domain result Y_k back into the time-domain sequence y_i, which is called an *inverse discrete Fourier transform (IDFT)*.

2. Multiplication of two sequences, such as $Y_k = H_k X_k$, requires that H_k and X_k be defined over the same range of k (frequency) values. There is no similar constraint in performing the convolution operation in the time domain, which uses the duration of an input sequence n_x and that of the digital filter unit-sample response n_h. Given those two values, the digital filter produces y_i for $0 \leq i \leq n_y - 1$ with $n_y = n_x + n_h - 1$. For the equivalent operation in the frequency domain, H_k and X_k must be defined for $0 \leq k \leq n_x + n_h - 1$.

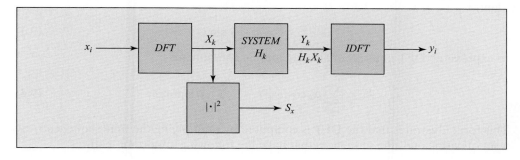

Figure 9.3

Block diagram that summarizes operations in the frequency domain.

3. While the convolution in the time domain involves real-valued sequences, multiplications in the frequency domain involve complex-valued variables. Luckily, Matlab performs complex-valued operations by default. (The Matlab developers were very clever.)

Many electrical engineers make these concessions to avoid convolution operations.

> **Factoid:** The trade-off preferred by many electrical engineers: *convolution* involving two real-valued time waveforms can be accomplished by a *multiplication* of two complex-valued frequency functions.

Figure 9.3 also shows a second common application of the DFT for determining the frequency content of time sequences in the form of the signal spectrum S_x, which is computed from the magnitude of the X_k. Spectral analysis is routinely performed using the efficient algorithm that computes the DFT, which is called the fast Fourier transform (FFT).

9.3 DISCRETE FOURIER TRANSFORM

The *discrete Fourier transform (DFT)* is a numerical algorithm that converts a time-domain sequence into a frequency-domain sequence comprised of discrete-time frequency components. The DFT is popular for two main reasons:

1. Electrical engineers often analyze signals and systems in the frequency domain.

2. The *fast Fourier transform* computes the DFT quickly using symmetry properties of complex-valued calculations.

The first time you encounter the DFT, it can seem a bit complicated. Using the radian frequency ω simplifies the notation. Given the time sequence x_i for $i = 0, 1, 2, \ldots, n_x - 1$, where n_x is an even-valued integer to simplify the analysis, its DFT is a complex-valued sequence X_k computed with

$$X_k = \sum_{i=0}^{n_x-1} x_i e^{-j\omega_k i} \quad \text{for } k = 0, 1, 2, \ldots, n_x - 1 \tag{9.1}$$

where the index k corresponding to radian frequency

$$\omega_k = \frac{2\pi k}{n_x} \tag{9.2}$$

and the imaginary number $j = \sqrt{-1}$. It is conventional to denote the DFT of a sequence by using the upper-case letter version of the lower-case sequence designation in the time domain. Note that an n_x-point time sequence produces an n_x-point frequency sequence.

The computation of the DFT becomes a bit more familiar when we apply the Euler identity

$$e^{-j\omega} = \cos \omega - j \sin \omega \tag{9.3}$$

to express the DFT in terms of sine and cosine components

$$X_k = \sum_{i=0}^{n_x-1} x_i \cos(\omega_k i) - j \sum_{i=0}^{n_x-1} x_i \sin(\omega_k i) \tag{9.4}$$

This form illustrates that the DFT is computed by multiplying the time sequence x_i by a set of cosine and sine sequences that have the following important features.

Complete Sinusoidal Periods The frequencies in the DFT all have integral number of periods in the n_x interval. The frequency ω_k is specified by integer k, for $0 \leq k \leq n_x - 1$.

The $k = 1$ term is the fundamental frequency ω_1. The argument of the sinusoidal sequences is then

$$\omega_1 i = \frac{2\pi i}{n_x} \tag{9.5}$$

and varies as i increases from 0 to $n_x - 1$ as

$$\overset{i=0}{\overbrace{0}} \leq \frac{2\pi i}{n_x} \leq 2\pi \overset{i=n_x-1}{\overbrace{\frac{n_x - 1}{n_x}}} < 2\pi \quad \text{radians} \tag{9.6}$$

That is, as i goes from 0 to $n_x - 1$, the argument of the sinusoids goes from 0 to almost 2π radians. If $i = n_x$, the argument becomes 2π, which produces the same values of the sinusoids as $i = 0$. Hence, the range of n_x argument values defines one complete period of the fundamental frequency.

Harmonic Frequencies The value of k defines the kth harmonic

$$\omega_k = k\omega_1 = \frac{2\pi}{n_x}k \tag{9.7}$$

As previously, as i goes from 0 to $n_x - 1$, $\omega_k i$, which is the argument of the kth harmonic, increases from 0 to almost $(2\pi)k$ radians.

The frequency index k varies from 0 to $n_x - 1$. Hence, this range of k values defines *a necessary and sufficient set of harmonic frequencies*. That is, a k value outside this range produce the same values of the sinusoids as some value within the range $0 \leq k \leq n_x - 1$.

A k value outside $0 \leq k \leq n_x - 1$ EXAMPLE 9.1

Consider the frequency component corresponding to the value of $k = n_x + 1$, as

$$\omega_{n_x+1} = (n_x + 1) \overset{=2\pi/n_x}{\overbrace{\omega_1}} = \frac{2\pi n_x}{n_x} + \frac{2\pi}{n_x} \tag{9.8}$$

Increasing the argument of a sinusoidal function by a multiple of 2π does not change its value. For integer n,

$$\sin(\theta \pm 2n\pi) = \sin(\theta)$$

so for $k = n_x + 1$,

$$\sin(\omega_{n_x+1}i) = \sin[(n_x + 1)\omega_1 i] = \sin(\overset{=2\pi i}{\overbrace{n_x\omega_1 i}} + \omega_1 i) = \sin(\omega_1 i)$$

That is, the value $k = n_x + 1$ produces the same values as $k = 1$.

Orthogonal Sequences Sequences a_i and b_i for $i = 0, 1, 2, \ldots, n_x - 1$ are orthogonal if the sum of their term-by-term products equals 0, or

$$\sum_{i=0}^{n_x-1} a_i b_i = 0 \tag{9.9}$$

Figure 9.4

Three orthogonal conditions. The bottom sequence is the product of the top two sequences. Note that equal positive and negative values sum to zero. From left to right: $\sin(\omega_n i) \perp \cos(\omega_m i)$, $\sin(\omega_n i) \perp \sin(\omega_m i)$, and $\cos(\omega_n i) \perp \cos(\omega_m i)$.

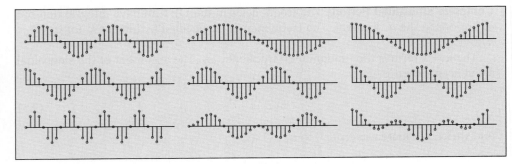

Applied to frequency analysis, harmonically related sinusoidal sequences that contain complete periods are orthogonal, as shown in Figure 9.4, with $\perp$ indicating orthogonality. Consider two harmonic frequencies ω_m and ω_n with integers m and n, where $0 \leq m, n \leq n_x - 1$, as

$$\omega_m = m\omega_1 = \frac{2\pi m}{n_x} \tag{9.10}$$

$$\omega_n = n\omega_1 = \frac{2\pi n}{n_x}$$

The following sequences are orthogonal.

1. Cosine and sine sequences defined over complete periods are orthogonal because

$$\sum_{i=0}^{n_x-1} \cos(\omega_m i) \sin(\omega_n i) = 0 \tag{9.11}$$

2. Harmonically-related sine sequences defined over complete periods are orthogonal because

$$\sum_{i=0}^{n_x-1} \sin(\omega_m i) \sin(\omega_n i) = 0 \tag{9.12}$$

3. Harmonically-related cosine sequences defined over complete periods are orthogonal because

$$\sum_{i=0}^{n_x-1} \cos(\omega_m i) \cos(\omega_n i) = 0 \tag{9.13}$$

The $k = 0$ and $k = n_x/2$ Terms The $k = 0$ term of the DFT is as follows.

■ For $k = 0$, the argument of the sinusoidal sequences is then

$$\omega_k i = \frac{2\pi k i}{n_x} = 0 \tag{9.14}$$

■ For $k = 0$, $\cos(0) = 1$ and $\sin(0) = 0$ for $i = 0, 1, 2, \ldots, n_x - 1$. In this case,

$$X_0 = \sum_{i=0}^{n_x-1} x_i \text{ for } i = 0, 1, 2, \ldots, n_x - 1 \tag{9.15}$$

or the sum of the sequence values. Note that X_0 is real when x_i is a real-valued sequence.

X_0 for the 2-sample sequence EXAMPLE 9.2

The simple sequence that I always use to refresh my understanding of the DFT is

$$x_0 = 1, \ x_1 = 1, \ \text{and} \ x_i = 0, \ \text{otherwise with} \ n_x = 2^K \ \text{for integer} \ K$$

No matter how large n_x is, only the two non-zero values contribute to the sum in the DFT. Thus,

$$X_k = \sum_{i=0}^{n_x-1} x_i e^{-j\omega_k i} = x_0 + x_1 e^{-j\omega_k} = 1 + e^{-j\omega_k}$$

For $k = 0$

$$X_0 = 1 + e^0 = 2$$

Hence, the DFT equals 2 at $k = 0$.

The $k = n_x/2$ term of the DFT is as follows.

- For $k = n_x/2$, where n_x is an even-valued integer, the argument of the sinusoidal sequences is then

$$\omega_k i = \frac{2\pi k i}{n_x} = \frac{2\pi n_x i}{2 n_x} = \pi i \tag{9.16}$$

- For $k = n_x/2$, $\cos(\pi i) = (-1)^i$ and $\sin(\pi i) = 0$ for $i = 0, 1, 2, \ldots, n_x - 1$. In this case,

$$X_{n_x/2} = \sum_{i=0}^{n_x-1} (-1)^i x_i \ \text{for} \ i = 0, 1, 2, \ldots, n_x - 1 \tag{9.17}$$

$$= x_0 - x_1 + x_2 - x_3 + \cdots - x_{n_x-1}$$

Note that $X_{n_x/2}$ is real when x_i is a real-valued sequence.

$X_{n_x/2}$ for a simple sequence EXAMPLE 9.3

Let

$$x_0 = 1, \ x_1 = 1, \ \text{and} \ x_i = 0, \ \text{otherwise with} \ n_x = 2^K \ \text{for integer} \ K$$

No matter how large n_x is, only the two non-zero values contribute to the sum in the DFT. Thus,

$$X_k = \sum_{i=0}^{n_x-1} x_i e^{-j\omega_k i} = x_0 + x_1 e^{-j\omega_k} = 1 + e^{-j2\pi k/n_x}$$

For $k = n_x/2$, where n_x is an even-valued integer, we have

$$X_{n_x/2} = 1 + e^{-j\frac{2\pi n_x}{2n_x}} = 1 + \overbrace{e^{-j\pi}}^{=-1} = 0$$

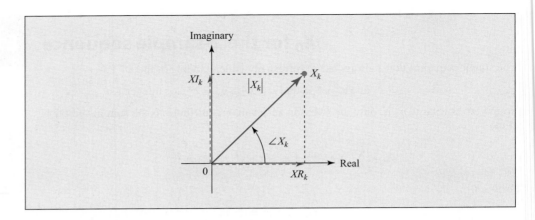

Figure 9.5

X_k in the complex plane.

9.3.1 Magnitude and Phase

The DFT X_k is a complex value and can be expressed as a magnitude and phase. Figure 9.5 shows X_k as a point in the complex plane that can be expressed in terms of its real and imaginary parts

$$X_k = XR_k + j\,XI_k \qquad (9.18)$$

and as a vector having a magnitude and angle with respect to the real axis as

$$X_k = |X_k|\,e^{j\angle X_k} \qquad (9.19)$$

The two forms are related with the magnitude equal to

$$|X_k| = \sqrt{XR_k^2 + XI_k^2} \ \ \text{for} \ \ k = 0, 1, 2, \ldots, n_x - 1 \qquad (9.20)$$

and the phase equal to

$$\angle X_k = \arctan\left(\frac{XI_k}{XR_k}\right) \ \ \text{for} \ \ k = 0, 1, 2, \ldots, n_x - 1 \qquad (9.21)$$

The following examples evaluates the DFT for simple sequences to illustrate the interpretation of the DFT.

EXAMPLE 9.4 DFT of unit-sample sequence

Consider the simplest time sequence

$$x_i = d_i \ \ \text{for} \ \ i = 0, 1, 2, \ldots, n_x - 1 \ (n_x = 20)$$

Recall $x_0 = 1$, and $x_i = 0$ otherwise. Figure 9.6 shows the results.
The real component is

$$XR_k = \sum_{i=0}^{n_x-1} x_i \cos(2\pi i k / n_x) \ \ \text{for} \ \ k = 0, 1, 2, \ldots, n_x - 1$$

with only the $i = 0$ term for each k:

$$XR_k = \overbrace{(1)}^{=x_0} \ \overbrace{\cos(2\pi(0)k/n_x)}^{=1} = 1 \ \text{for} \ \ k = 0, 1, 2, \ldots, n_x - 1$$

The imaginary component is

$$XI_k = -\sum_{i=0}^{n_x-1} x_i \sin(2\pi i k / n_x) \ \ \text{for} \ \ k = 0, 1, 2, \ldots, n_x - 1$$

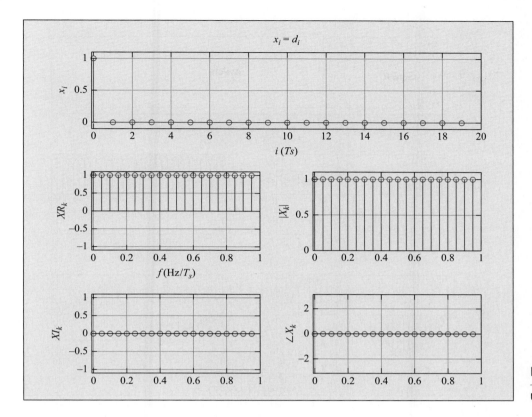

Figure 9.6

Unit-sample sequence and its DFT.

with only the $i = 0$ term for each k:

$$XI_k = - \overbrace{(1)}^{=x_0} \overbrace{\sin(2\pi(0)k/n_x)}^{=0} = 0 \text{ for } k = 0, 1, 2, \ldots, n_x - 1$$

The magnitude response is

$$|X_k| = \sqrt{\underbrace{XR_k^2}_{=1} + \underbrace{XI_k^2}_{=0}} = 1 \text{ for } k = 0, 1, 2, \ldots, n_x - 1$$

The phase response is

$$\angle X_k = \arctan\left(\frac{XI_k}{XR_k}\right) = \arctan(0) = 0 \text{ for } k = 0, 1, 2, \ldots, n_x - 1$$

This result demonstrates that unit-sample sequence has magnitude response that is constant over frequency and a phase response that is identically equal to zero.

DFT of delayed unit-sample sequence   EXAMPLE 9.5

Consider the delayed unit-sample time sequence

$$x_i = d_{i-1} \text{ for } i = 0, 1, 2, \ldots, n_x - 1 \ (n_x = 32)$$

Figure 9.7 shows $x_1 = 1$ and $x_i = 0$ otherwise.

The real component has only the $i = 1$ term for each k:

$$XR_k = \overbrace{(1)}^{=x_1} \cos(2\pi(1)k/n_x) = \cos(2\pi k/n_x) \text{ for } k = 0, 1, 2, \ldots, n_x - 1$$

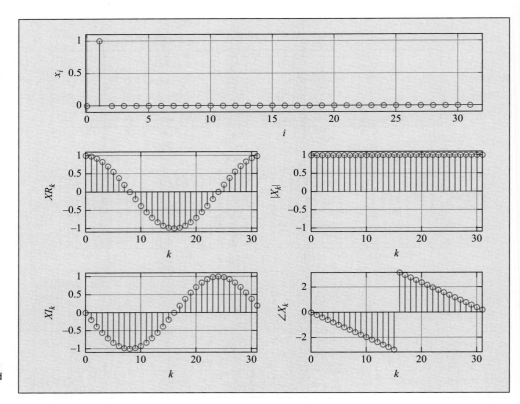

Figure 9.7

Delayed unit-sample sequence and its DFT.

The imaginary component has only the $i = 1$ term for each k:

$$XI_k = - \overbrace{(1)}^{=x_1} \sin(2\pi(1)k/n_x) = -\sin(2\pi k/n_x) \quad \text{for} \quad k = 0, 1, 2, \ldots, n_x - 1$$

The magnitude is

$$|X_k| = \sqrt{XR_k^2 + XI_k^2}$$

$$= \sqrt{\cos^2(2\pi k/n_x) + \sin^2(2\pi k/n_x)} = 1 \quad \text{for} \quad k = 0, 1, 2, \ldots, n_x - 1$$

The phase is

$$\angle X_k = \arctan\left(\frac{XI_k}{XR_k}\right)$$

$$= \arctan\left(\frac{-\sin(2\pi k/n_x)}{\cos(2\pi k/n_x)}\right) = -2\pi k/n_x$$

for $0 \le k \le n_x - 1$. Delaying the unit-sample sequence does not affect the magnitude spectrum. The phase response exhibits a negative slope due to the delay of one sample. The 2π jump from $-\pi$ to $+\pi$ is due to the arctangent function, which produces values in the range $[-\pi, \pi)$.

Factoid: Delaying a time sequence x_i by i_d samples does not affect the magnitude $|X_k|$, but it does affect the *phase* by introducing a negative linear slope $= -(2\pi i_d k/n_x)$. The 2π phase jumps are due to the principle value that is computed by the arctangent function, which lies in $[-\pi, \pi)$.

DFT of 2-sample sequence EXAMPLE 9.6

Consider the 32-point DFT of the 2-sample time sequence

$$x_0 = 1, x_1 = 1, x_i = 0 \text{ for } i = 0, 1, 2, \ldots, n_x - 1 \ (n_x = 32)$$

shown in Figure 9.8.

The DFT then equals

$$X_k = \sum_{i=0}^{n_x-1} x_i e^{-j\omega_k i} = 1 + e^{-j\omega_k}$$

The standard approach to analyzing a time sequence that exhibits a point of symmetry is to express the DFT about the point of symmetry. In this case, this sequence is symmetric about $i = 1/2$. Factoring $e^{-j\omega_k/2}$ produces the result

$$X_k = e^{-j\omega_k/2} \left(e^{j\omega_k/2} + e^{-j\omega_k/2} \right)$$

Applying the Euler identity

$$\cos(\theta) = \frac{e^{j\theta} + e^{-j\theta}}{2}$$

we get

$$X_k = 2e^{-j\omega_k/2} \cos(\omega_k/2)$$

The magnitude is

$$|X_k| = 2|\cos(\omega_k/2)| = 2 \left| \cos\left(\frac{2\pi k}{n_x} \right) \right| \text{ for } k = 0, 1, 2, \ldots, n_x - 1$$

Figure 9.8 shows the rectified cosine sequence that makes the magnitude ≥ 0.

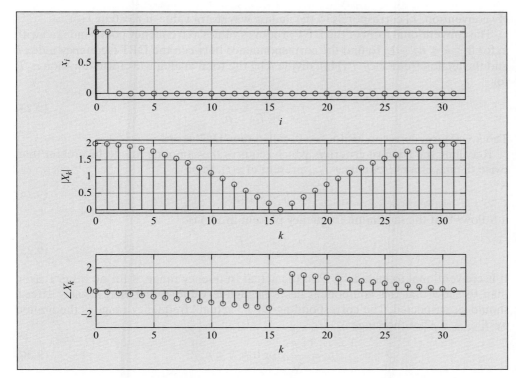

Figure 9.8

Two-sample sequence and its DFT.

The phase equals

$$\angle X_k = -\frac{2\pi k}{n_x} \quad \text{for } k = 0, 1, 2, \ldots, n_x - 1 \text{ when } \cos(\omega_k/2) > 0$$

$$= -\frac{2\pi k}{n_x} + \pi \quad \text{for } k = 0, 1, 2, \ldots, n_x - 1 \text{ when } \cos(\omega_k/2) < 0$$

and is complicated by two additional features:

> **The absolute values of negative quantities:** When the absolute value introduces a negative sign to obtain a positive value, as in the $|\cos|$, the phase compensates by adding a $\pm\pi$ to the computed phase result. This occurs in Figure 9.8 for $17 \leq k \leq 31$. Hence, the phase exhibits a jump of $\pm\pi$ when $|X(k)|$ passes through zero.

> **Zero absolute values:** When $|X(k)| = 0$, the phase does not matter. The arctangent function returns a zero in this case, as Figure 9.8 shows for $k = 16$.

> **Factoid:** The previous examples illustrate that, while $|X_k|$ has a clear interpretation, $\angle X_k$ exhibits jumps of $\pm\pi$ when $|X_k|$ passes through zeros and jumps of $\pm 2\pi$ to accommodate the result produced by the arctangent function. Hence, the complexity of $\angle X_k$ usually causes it to be ignored in a first-order frequency analysis.

"Actual" Frequencies

The DFT is often employed to compute the spectrum of a signal sequence x_i for $i = 0, 1, 2, \ldots, n_x - 1$ that is acquired by sampling waveform $x(t)$ with sampling rate $f_s = 1/T_s$. To find the correspondence between the sequence index i and the analog time t (s), we multiply i by T_s for

$$x_i = x(t)|_{t=iT_s} = x(i T_s) \tag{9.22}$$

By convention, x_0 corresponds to the analog waveform value at $t = 0$ or $x(0)$.

The conventional form of the DFT produces values at frequency points indexed with k, for $0 \leq k \leq n_x - 1$. To find the correspondence between the DFT frequency index k and the analog frequency f (Hz), divide k by the total analog waveform duration $n_x T_s$ for

$$f = \frac{k}{n_x T_s} = \frac{k f_s}{n_x} \tag{9.23}$$

The $k = 0$ term corresponds to $f = 0$, called the "DC" term.

Recalling the Nyquist criterion, which requires the sampling rate to be greater than twice the maximum frequency $f_{\max}$ present in $x(t)$, to be

$$f_s > 2 f_{\max} \tag{9.24}$$

it follows that the maximum frequency present in $x(t)$ is

$$f_{\max} < \frac{f_s}{2} \tag{9.25}$$

It is conventional to include $f_s/2$ in the actual frequency range, if for no other reason than to show that this component has a negligible value. Otherwise, aliasing effects should be suspected. The corresponding DFT "actual" frequency range is the limited by $f_{\max} \leq f_s/2$ or the range

$$f_k = \frac{k f_s}{n_x} \quad \text{for } 0 \leq k \leq n_x/2 \tag{9.26}$$

So, only half of the frequency components produced by the DFT correspond to "actual" frequencies present in $x(t)$. The other frequencies or

$$f_k = \frac{k f_s}{n_x} \quad \text{for } n_x/2 + 1 \le k \le n_x - 1 \tag{9.27}$$

are not "actual" frequencies but are artifacts due to sampling the time waveform $x(t)$ to form the time sequence x_i. The underlying mathematics of this artifact is described in advanced signal processing courses. Engineers who do not realize this upper frequency range is an artifact are very confused by the existence of these *higher* frequencies.

Actual frequencies   **EXAMPLE 9.7**

Consider performing a spectral analysis by sampling the analog signal shown in Figure 9.9 given by

$$x(t) = \sin(2\pi f_o t)$$

Let $f_o = 1$ kHz and sampling rate $f_s = 8 f_o = 8$ kHz ($T_s = 0.125$ ms), which satisfies the Nyquist criterion. This sampling rate produces eight samples per period, so the corresponding discrete sequence frequency

$$f_d = 1/8$$

With

$$\omega_d = 2\pi f_d = \frac{2\pi}{8}$$

the time sequence is given by

$$x_i = \sin(\omega_d i) \quad \text{for } i = 0, 1, 2, \ldots, n_x - 1 \ (n_x = 16)$$

The conventional DFT is computed at n_x frequencies

$$\omega_k = \frac{2\pi k}{n_x} = \frac{2\pi k}{16} \quad \text{for } 0 \le k \le 15$$

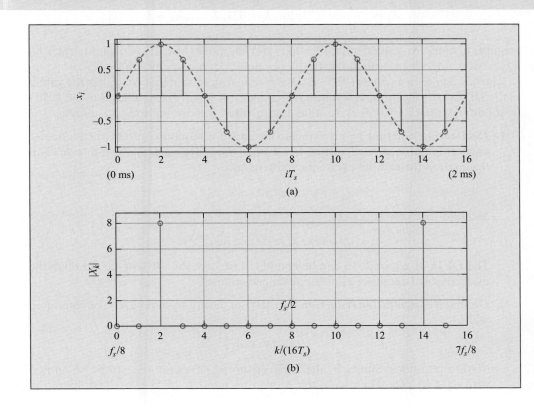

(a)

(b)

Figure 9.9

16-point DFT of sinusoidal samples. (a) $x(t) = \sin(2\pi f_o t)$ shown in red, x_i are samples acquired with sampling rate $f_s = 8 f_o$. (b) $|X_k|$.

In terms of the DFT coefficient k, the sequence x_i contains only one "actual" frequency component, that for $k = 2$ is

$$\omega_d = \frac{2\pi}{8} = \frac{2\pi(2)}{16} = \omega_2$$

The 16-point DFT X_k for $0 \leq k \leq 15$ is computed, and $|X_k|$ is displayed in Figure 9.9. The magnitude spectrum shows only two non-zero components:

$k = 2$:   This is reasonable because $x(t)$ is a sinusoid at frequency $\omega_d = \omega_2$.

$k = 14$:   This *non-actual* frequency is

$$f_{na} = \frac{7 f_s}{8} = 7 f_o = 7 \, \text{kHz}$$

Clearly, this frequency is an artifact caused by the sampling process, because x_i contains only a single frequency.

To find the correspondence between the sequence index i and analog time t (seconds), we multiply i by $T_s = 0.125$ ms for

$$x_i = x(t)|_{t=iT_s} = x(i T_s) = x(0.125 i \ \text{ms})$$

To find the correspondence between the DFT frequency index k and the analog frequency f_k (Hz), we divide k by the total analog waveform duration $n_x T_s$ for

$$f_k = \frac{k}{n_x T_s} = \frac{k f_s}{n_x} = \frac{k \, (8 \, \text{kHz})}{16} = 0.50 k \ \text{kHz}$$

9.3.2 Fast Fourier Transform

The direct calculation of the n_x-point DFT using the formula

$$X_k = \sum_{i=0}^{n_x-1} x_i e^{-j\omega_k i} \quad \text{for} \quad k = 0, 1, 2, \ldots, n_x - 1$$

requires n_x complex-valued multiplications and additions (MADs) to evaluate each X_k, a total of n_x^2 MADs to evaluate n_x frequency points, or a n_x-point DFT.

The *fast Fourier transform* (FFT) is the numerical algorithm that computes the values of the DFT very efficiently. It was discovered by Cooley and Tukey at IBM in the 1950's by modifying the original discrete-time data x_i when performing spectral analysis.

1. **Decimating** x_i in time by breaking up the entire x_i sequence into two sequences— each of length $n_x/2$—according to whether the index value is even or odd. This decimation formed the even-indexed sequence

$$xe_i : x_0, x_2, x_4, \ldots, x_{n_x-2}$$

and odd-indexed sequence

$$xo_i : x_1, x_3, x_5, \ldots, x_{n_x-1}$$

The DFTs of xo_n and xe_n can be combined to form the DFT of x_n, although the mathematical details are left for advanced courses.

2. **Observed computation reduction.** The DFT of the $n_x/2$-point xo_n can be computed with

$$(n_x/2)^2 = n_x^2/4$$

MAD operations. Similarly, the DFT of the $(n_x/2)$-point xe_n can be computed with $n_x^2/4$ MADs. Then from these two, the number of MADs to compute the

DFT of x_i can be computed with

$$n_x^2/4 + n_x^2/4 = n_x^2/2$$

That is, *in half the time*!

Original versions of the FFT required n_x to be a power of 2 or

$$n_x = 2^K \text{ for integer } K \tag{9.28}$$

Then the decimation process can be repeated $\log_2 n_x = K$ times. At each decimation step, n_x MADs need to be performed for a total operation count equal to

$$Kn_x = n_x \log_2 n_x \tag{9.29}$$

MADs to compute an n_x-point DFT. The following example illustrates that the reduction is significant for large n_x.

Significance of MAD count reduction for FFT · EXAMPLE 9.8

Consider computing the DFT of x_i for $i = 0, 1, 2, \ldots, n_x - 1$ with $n_x = 1024 = 2^{10}$. The FFT would complete this task using a MAD count of

$$n_x \log_2 n_x = 1,024 \log_2 1,024 = 1,024 \times 10 \approx 10^4$$

The direct calculation of the DFT would require a MAD count equal to

$$n_x^2 = (1,024)^2 \approx 10^6$$

The FFT is 100 times faster and produces the same result!

> *Factoid:* I once met Jim Cooley and asked how he got the idea of decimating the data into even-numbered and odd-numbered sequences, since this is not an obvious thing to do. He told me that when the calculations were performed manually with mechanical calculators, it was standard practice to divide data this way and give xe_i to one group of people for processing and xo_i to another group. The results were then compared to detect any errors, which would should up as significant differences. However, he noticed that the results came back in half the time, as compared to when only one group was given the entire data set, thus motivating the FFT!

Current FFT algorithms do not require n_x to be a power of 2, simplifying the interpretation of the spectral values, as the next example illustrates.

FFT for $n_x \neq 2^K$ · EXAMPLE 9.9

Consider computing the 20-point DFT of the two-sample sequence shown in Figure 9.10 using the FFT. The *actual* frequencies occur for $0 \le k \le n_x/2 = 10$.

For the correspondence between the DFT frequency index k and the analog frequency f (Hz), we divide kf_s by n_x for

$$f(k) = \frac{kf_s}{n_x}$$

More convenient values for frequencies occur when dividing by $n_x = 20$ than by a power of 2 such as $n_x = 16$ or 32.

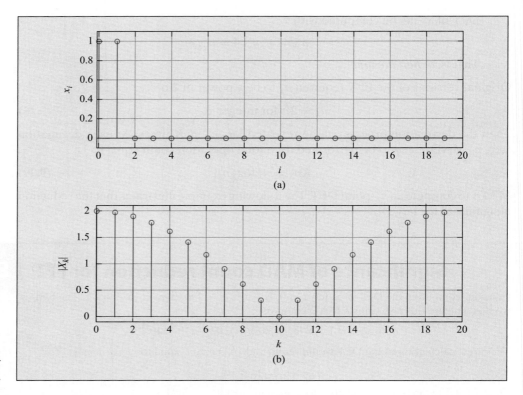

Figure 9.10

20-point DFT computed with FFT illustrates result with $n_x = 20 \neq 2^K$.

9.4 SIGNAL SPECTRUM

When the time sequence is a signal x_i for $i = 0, 1, 2, \ldots, n_x - 1$, its DFT is called the signal *spectrum*. Typically, only the magnitude of the spectrum is computed. We consider two real-valued spectra: the magnitude spectrum and the power spectrum. The magnitude spectrum is given by

$$|X_k| = \sqrt{XR_k^2 + XI_k^2} = \sqrt{X_k X_k^*} \quad \text{for} \quad k = 0, 1, 2, \ldots, n_x - 1 \qquad (9.30)$$

where X_k^* is the complex conjugate of X_k. The power spectrum describes the distribution of the energy over the frequency components and is equal to

$$Sx_k = |X_k|^2 = XR_k^2 + XI_k^2 \qquad (9.31)$$

The *log-spectrum* is the power spectrum expressed in decibel units (dB) is given by

$$Sx_{k,dB} = 10 \log_{10} Sx_k \qquad (9.32)$$

It is conventional to display the dB scale with the maximum value normalized to 0 dB. This is accomplished by dividing by the maximum value of Sx_k denoted max(Sx), as

$$Sx_{k,dB} = 10 \log_{10} \left(\frac{Sx_k}{\max(Sx)} \right) \qquad (9.33)$$

The range of displayed dB values is varied to show the desired details.

High-power details: The range from −10 to 0 dB shows details around the spectral peaks.

Low-power details: The range from −60 to 0 dB shows low-power details.

A problem occurs with the log-spectrum for values of Sx_k that approach 0 because their dB values approach −∞. If the spectrum equals zero, it typically does so between sample points, so it is rare (but possible) in practice to observe $Sx_k = 0$ at one of the k values.

Padding with zeros EXAMPLE **9.10**

Consider the rectangular sequence

$$x_i = 1 \ \text{ for } \ i = 0, 1, 2, \ldots, n_x - 1 \ (n_x = 7)$$

Figure 9.11a shows the results. Note that these results are difficult to interpret. We can increase the frequency resolution from $n_x = 7$ to $n_x = 100$ by adding 93 zeros to x_i to form the padded sequence xp_i:

$$xp_i = x_i \ \text{ for } \ 0 \leq i \leq 6$$

$$= 0 \ \text{ for } \ 7 \leq i \leq 99$$

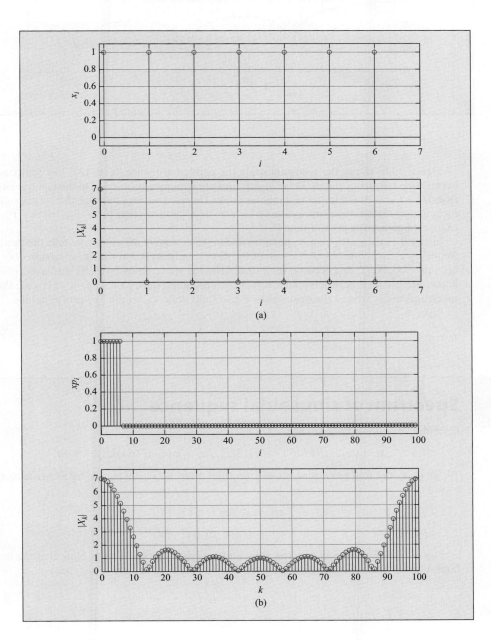

Figure 9.11

Spectrum of rectangular pulse. (a) Seven-point DFT is not very useful for interpreting the DFT. (b) A 100-point DFT of rectangular pulse having $n_x = 7$ non-zero values padded with 93 zeros.

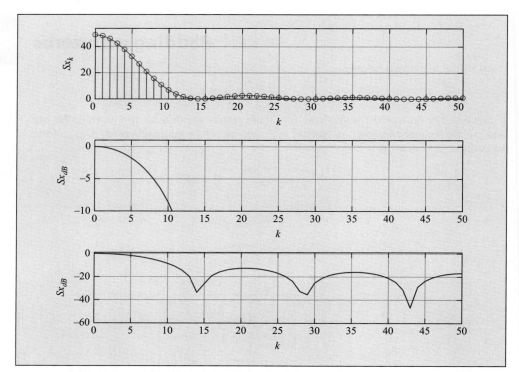

Figure 9.12

Power spectrum and log-spectrum of 7-sample rectangular pulse sequence displayed over the "actual" frequency range $0 \leq k \leq n_x/2$.

Figure 9.11b shows the spectrum using the padded sequence. Note $|X_k|$ has components over a wide frequency range. These high frequency components are due to the fast transitions from $0 \rightarrow 1$ and $1 \rightarrow 0$ in the rectangular pulse. These transitions indicate the corresponding rectangular analog waveform pulse $x(t)$ is not band-limited (that is, its $f_{max} = \infty$). Hence, $|X_{n_x/2}|$ is significantly greater than 0.

Figure 9.12 shows the power spectrum and log-spectrum in dB units over only the "actual" frequency range. Note $Sx_k \neq 0$ at one of the sample values, so that the log-spectrum does not become $-\infty$, but it can become very small. The log-spectrum is typically displayed using a linear interpolation between points, rather than using the stem plot. The -10 dB scale shows spectrum details around the peak, while the -60 dB scale shows the low-power details.

EXAMPLE 9.11

Spectrum of sinusoidal sequence

Consider the sinusoidal sequence

$$x_i = \sin(2\pi f i) \text{ for } i = 0, 1, 2, \ldots, n_x - 1 \ (n_x = 16 \text{ and } f = 1/8)$$

We increase the frequency resolution from $n_x = 16$ to $n_x = 100$ by adding 84 zeros to x_i to form xp_i:

$$xp_i = x_i \text{ for } 0 \leq i \leq 15$$

$$= 0 \text{ for } 16 \leq i \leq 99$$

Figure 9.13 shows the power spectrum and log-spectrum in dB units. Note that the linear interpolation between the points is misleading when the log-spectrum $\rightarrow -\infty$.

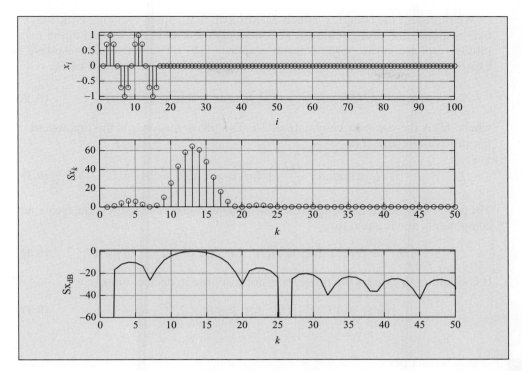

Figure 9.13

Power spectrum of 16-point sinusoidal pulse using 100-point FFT.

 # 9.5 FREQUENCY TRANSFER FUNCTION

When the time sequence is a unit-sample response h_i for $i = 0, 1, 2, \ldots, n_h - 1$ of a digital system, then its DFT is called the system *frequency transfer function* or more concisely its *transfer function*, which is denoted H_k for $k = 0, 1, 2, \ldots, n_f - 1$. Digital systems (or digital filters) fall into two categories that describe the duration of the filter unit-sample response.

MA Filters have Finite Unit-Sample Responses The h_i for $i = 0, 1, 2, \ldots, n_x - 1$ with n_x a finite integer is conventionally termed a *finite impulse response (FIR)*. As such, there is no difficulty in computing the MA transfer function with the DFT. With h_i for $i = 0, 1, 2, \ldots, n_h - 1$, we have

$$H_k = \sum_{i=0}^{n_h-1} h_i e^{-j\omega_k i} \text{ for } k = 0, 1, 2, \ldots, n_x - 1 \tag{9.34}$$

A higher-resolution H_k for $0 \leq k \leq n_f - 1$ can be produced (if needed) by padding h_i with $n_f - n_h$ zeros. Thus,

$$hp_i = h_i \text{ for } i = 0, 1, 2, \ldots, n_h - 1 \tag{9.35}$$

$$= 0 \text{ for } n_h \leq i \leq n_f - 1$$

The higher-resolution H_k is then computed as

$$H_k = \sum_{i=0}^{n_f-1} hp_i e^{-j\omega_k i} \text{ for } k = 0, 1, 2, \ldots, n_f - 1 \tag{9.36}$$

Note that the zero values in hp_i do not contribute to H_k. Hence, the sum can have upper limit equal $n_h - 1$. We set the upper limit to $n_f - 1$ because that is what the FFT algorithm computes. The frequency values are then

$$f_k = \frac{k}{n_f} \text{ for } k = 0, 1, 2, \ldots, n_f - 1 \tag{9.37}$$

with resolution $f_1 = 1/n_f$.

We represent the complex-valued H_k with two real-valued sequences: the *magnitude* response and *phase* response. A commonly used sequence related to the magnitude response is the power-transfer response $|H_k|^2$. The magnitude response is expressed in terms of the real (HR_k) and imaginary (HI_k) components of H_k.

$$|H_k| = \sqrt{HR_k^2 + HI_k^2} = \sqrt{H_k H_k^*} \quad \text{for} \quad k = 0, 1, 2, \ldots, n_f - 1 \qquad (9.38)$$

where H_k^* is the complex conjugate of H_k. The phase response is the arctangent of the ratio of HI_k and HR_k

$$\angle H_k = \arctan\left(\frac{HI_k}{HR_k}\right) \quad \text{for} \quad k = 0, 1, 2, \ldots, n_f - 1 \qquad (9.39)$$

The power-transfer function describes the transmission of the energy in the frequency components and is equal to

$$|H_k|^2 = HR_k^2 + HI_k^2 = H_k H_k^* \quad \text{for} \quad k = 0, 1, 2, \ldots, n_f - 1 \qquad (9.40)$$

It is common to express the power-transfer response in decibel units

$$|H_k|_{dB}^2 = 10 \log_{10} |H_k|^2 \quad \text{for} \quad k = 0, 1, 2, \ldots, n_f - 1 \qquad (9.41)$$

EXAMPLE 9.12 DFT of unit-sample response two-sample averager

Let h_i be the unit-sample response of the two-sample running average MA digital filter given by

$$h_i = 1 \text{ for } i = 0, 1$$

$$= 0 \text{ otherwise}$$

To increase the resolution in the plot of H_k, the value of n_f was set equal to 50 by padding with 48 zeros.

$$hp_i = h_i \text{ for } i = 0, 1$$

$$= 0 \text{ for } 2 \leq i \leq 49$$

The 48 zero-valued hp_i samples do not contribute to the DFT H_k, but this large n_f sets the frequency resolution

$$f_1 = \frac{1}{n_f} = \frac{1}{50}$$

Figure 9.14 shows $|H_k|$ plotted over the *actual frequencies* for $0 \leq k \leq n_f/2$. Recall that $0 \leq f_k \leq n_f/2$ represent the "actual" frequency values. Note that the two-sample averager implements a low-pass filter because the transfer function magnitude decreases as frequency increases.

For sampling period $T_s = 1$ ms, the corresponding analog frequency f_k equals

$$f_k = \frac{k}{n_f T_s} = \frac{k}{50 \times 10^{-3} \, s} = 20(k) \, \text{Hz} \quad \text{for} \quad k = 0, 1, 2, \ldots, (n_f - 1)/2$$

AR and ARMA Filters Have Infinite Unit-Sample Responses The DFT is then problematic to evaluate numerically, because the FFT requires the sequence size to be finite. The transfer function can still be approximated, as the following examples illustrate.

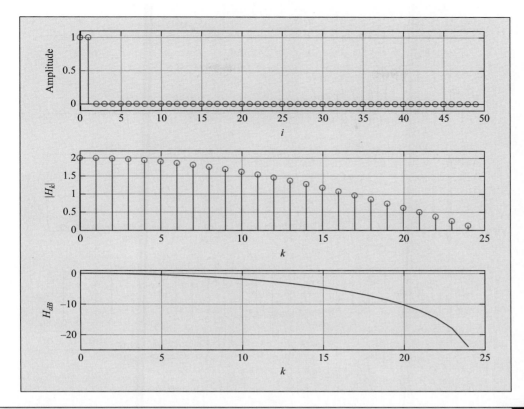

Figure 9.14

Transfer function of two-sample averager padded with 48 zeros to $n_f = 50$.

Transfer function of first-order AR system   EXAMPLE 9.13

Consider the first-order AR system with difference equation

$$y_i = a y_{i-1} + x_i$$

with the unit-sample response given by

$$h_i = a^i \text{ for } i \geq 0$$

We truncate the infinite-duration sequence to obtain the finite-duration sequence hf_i with n_h non-zero values for

$$hf_i = h_i \text{ for } i = 0, 1, 2, \ldots, n_h - 1$$
$$= 0 \text{ for } i \geq n_h$$

We choose n_h sufficiently large so that the truncation value is negligibly small, such as

$$h_{n_h} = a^{n_h} < \max(h_i)/100$$

For $a = 0.9$, $0.9^{50} = 0.005$, and $\max(h_i) = 1$. So, $n_h = 50$ is sufficiently large.
We then compute the DFT

$$H_k = \sum_{i=0}^{n_h-1} hf_i e^{-j\omega_k i} \text{ for } i = 0, 1, 2, \ldots, n_h - 1 \ (n_h = 50)$$

Figure 9.15 shows the transfer function magnitude $|H_k|$ for $0 \leq k \leq 25$. The transfer-function magnitude $|H_k|$ for $0 \leq k \leq 25$ clearly shows the low-pass filter character of the first-order AR filter.

The transfer function can be computed analytically for the infinite-duration sequence

$$H_k = \sum_{i=0}^{\infty} h_i e^{-j\omega_k i} = \sum_{i=0}^{\infty} a^i e^{-j\omega_k i}$$
$$= \sum_{i=0}^{\infty} \left(a e^{-j\omega_k} \right)^i$$

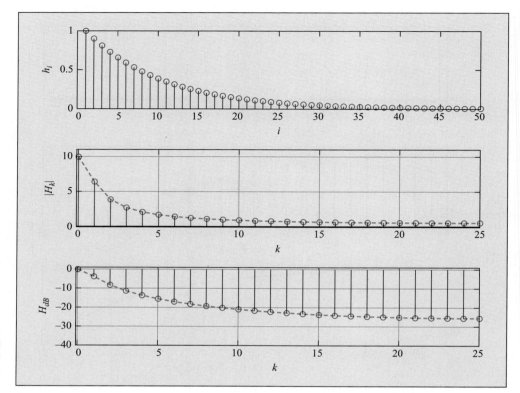

Figure 9.15

Analytically computed $|H_k|^2$ of AR digital filter having unit-sample response $h_i = a^i$ for $i = 0, 1, 2, \ldots, n_h - 1$ ($a = 0.9$ and $n_h = 50$).

with

$$\omega_k = \frac{2\pi k}{n_h}$$

We can apply the infinite geometric sum formula

$$\sum_{i=0}^{\infty} c^i = \frac{1}{1-c} \quad \text{for } |c| < 1$$

to get the result

$$H_k = \frac{1}{1 - ae^{-j\omega_k}} \quad \text{for } k = 0, 1, 2, \ldots, n_h/2$$

The power transfer function is then

$$|H_k|^2 = H_k H_k^*$$

$$= \left(\frac{1}{1 - ae^{-j\omega_k}} \right) \left(\frac{1}{1 - ae^{j\omega_k}} \right)$$

$$= \frac{1}{1 + a^2 - a(e^{j\omega_k} + e^{-j\omega_k})}$$

$$= \frac{1}{1 + a^2 - 2a\cos(\omega_k)} \quad \text{for } k = 0, 1, 2, \ldots, n_h/2$$

This analytic function is shown with a dashed line in Figure 9.15. Note that if $h_{n_f} < h_{\max}/100$, $|H_k|$ computed with the FFT is indistinguishable from the analytic form using the infinite unit-sample response.

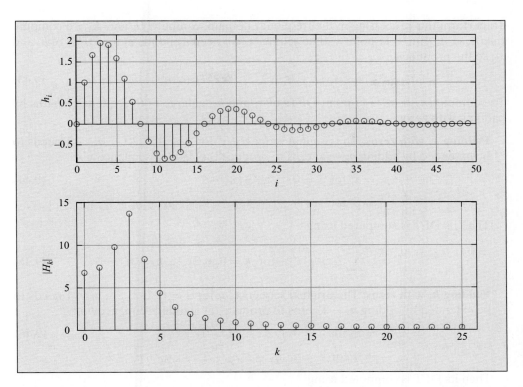

Figure 9.16

DFT of h_i for a second-order AR digital filter ($r = 0.9$, $\omega_o = 2\pi/16$).

Second-order AR system   EXAMPLE 9.14

Consider the unit-sample response h_i of the second-order AR system with difference equation given by

$$y_i = 2r \cos(\omega_o) \, y_{i-1} - r^2 y_{i-2} + x_i$$

with $r = 0.9$ and $\omega_o = 2\pi/16$.

Figure 9.16 shows the 50-point unit-sample response h_i for $0 \le i \le 49$. The transfer-function magnitude $|H_k|$ for $0 \le k \le 25$ clearly shows the band-pass filter character of the second-order resonant AR filter.

9.6 PROCESSING SIGNALS IN THE FREQUENCY DOMAIN

This section describes how to process data in the frequency domain to get the same result that is obtained by time-domain processing. While many applications perform digital filtering operations in the time domain, there may be cases, such as when one of the sequences is already expressed in the frequency domain, that processing in the frequency domain is more appropriate.

Chapter 8 described time-domain processing by a digital filter. To summarize, we have an input time sequence x_i for $i = 0, 1, 2, \ldots, n_x - 1$, and the filter unit-sample response h_i for $i = 0, 1, 2, \ldots, n_h - 1$. We use the convolution equation to compute

$$y_i = \sum_{j=0}^{n_h-1} h_j x_{i-j} \text{ for } i = 0, 1, 2, \ldots, n_y - 1 = n_x + n_h - 2 \qquad (9.42)$$

To perform this processing in the frequency domain, compute H_k and X_k. To compute the $(n_y = n_x + n_h - 1)$-point y_i time sequence, we need n_y points in the DFTs of h_i and x_i, so the product is

$$Y_k = H_k X_k \quad \text{for } k = 0, 1, 2, \ldots, n_y - 1 \ (= n_x + n_h - 2) \tag{9.43}$$

To compute $(n_x + n_h - 1)$-point DFTs of x_i and h_i, enlarge both sequence lengths by padding with zeros.

Padding x_i with zeros: The original sequence x_i for $i = 0, 1, 2, \ldots, n_x - 1$ needs to be extended by adding $n_h - 1$ zeros to produce the padded sequence xp_i:

$$xp_i = x_i \text{ for } i = 0, 1, 2, \ldots, n_x - 1 \tag{9.44}$$

$$= 0 \text{ for } i = n_x, n_x + 1, n_x + 2, \ldots, n_x + n_h - 2$$

Then its DFT is computed using

$$X_k = \sum_{i=0}^{n_x+n_h-2} xp_i e^{-j\omega_k i} \text{ for } k = 0, 1, 2, \ldots, n_x + n_h - 2 \tag{9.45}$$

Padding h_i with zeros: The original sequence h_i for $i = 0, 1, 2, \ldots, n_h - 1$ needs to be extended by adding $n_x - 1$ zeros to produce the padded sequence hp_i:

$$hp_i = h_i \text{ for } i = 0, 1, 2, \ldots, n_h - 1 \tag{9.46}$$

$$= 0 \text{ for } i = n_x, n_x + 1, n_x + 2, \ldots, n_x + n_h - 2$$

Then its DFT is computed using

$$H_k = \sum_{i=0}^{n_x+n_h-2} hp_i e^{-j\omega_k i} \text{ for } k = 0, 1, 2, \ldots, n_x + n_h - 2 \tag{9.47}$$

We can now compute the product $Y_k = H_k X_k$ for $k = 0, 1, 2, \ldots, n_x + n_h - 2$.

To determine y_i, compute the *inverse discrete Fourier Transform (IDFT)*, which has a similar analytic form to the DFT. If Y_k is the DFT of y_i, the IDFT is given by

$$y_i = \frac{1}{n_x + n_h - 1} \sum_{i=0}^{n_x+n_h-2} Y_k e^{j\omega_k i} \text{ for } i = 0, 1, 2, \ldots, n_x + n_h - 2 \tag{9.48}$$

The only differences are the division by the number of points and the $+j$ in the complex exponential term.

Advanced digital-signal processing courses teach how to process continuous streams of data with this method. The issue is that n_x input points produce $n_x + n_h - 1$ output points. A technique called *overlap-and-save* stores the last n_h output points of the current y_i time sequence and adds them to the first n_h point in the next y_i sequence. This additional complication results in many applications doing digital-filter processing in the time domain.

EXAMPLE 9.15

Processing time sequences in the frequency domain

Figure 9.17 shows processing in the frequency domain by using DFT and IDFT transformations. We use two simple (rectangular) sequences so that the result should be a triangular sequence and we can concentrate on the steps in the process.

Define two simple and identical rectangular sequences

$$x_i = 1 \text{ for } 0 \le i \le 9 \ (n_x = 10)$$

and

$$h_i = 1 \text{ for } 0 \le i \le 9 \ (n_h = 10)$$

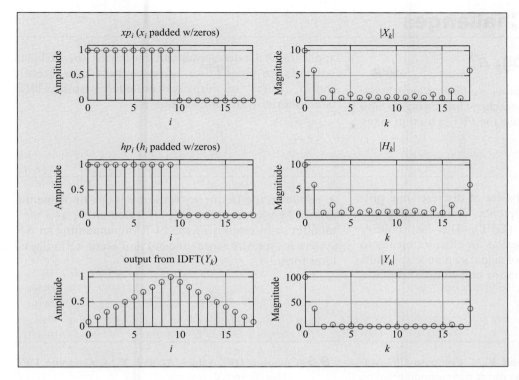

Figure 9.17

Processing time-domain sequences in the frequency domain.

The duration of the output y_i for $0 \leq n_y \leq n_y - 1$ is

$$n_y = n_x + n_h - 1 = 19$$

The padded sequence xp_i is generated from x_i by

$$xp_i = x_i \text{ for } 0 \leq i \leq 9$$

$$= 0 \text{ for } 10 \leq i \leq 18 = n_x + n_h - 2$$

The sequence X_k for $0 \leq k \leq 18$ is the DFT of xp_i, and its magnitude $|X_k|$ for $0 \leq k \leq 18$ is plotted in Figure 9.17.

The padded sequence hp_i is generated from h_i by

$$hp_i = h_i \text{ for } 0 \leq i \leq 9$$

$$= 0 \text{ for } 10 \leq i \leq 18$$

The sequence H_k for $0 \leq k \leq 18$ is the DFT of hp_i and its magnitude $|H_k|$ for $0 \leq k \leq 18$ is plotted.

The DFT of the output is then computed as

$$Y_k = H_k X_k \text{ for } 0 \leq k \leq 18$$

and its magnitude $|Y_k|$ for $0 \leq k \leq 18$ is plotted.

The output sequence y_i for $0 \leq i \leq 18$ is computed from the IDFT

$$y_i = \frac{1}{19} \sum_{k=0}^{18} Y_k e^{j\omega_k i} \text{ for } 0 \leq i \leq 18$$

where

$$\omega_k = \frac{2\pi k}{19} \text{ for } 0 \leq k \leq 18$$

The form of y_i is the expected triangular sequence is displayed in Figure 9.17, assuring us that the processing was performed correctly.

9.7 Research Challenges

9.7.1 Why only $n \log_2 n$?

The research question is: *Can we do better than $n \log_2 n$ in computing the* FFT? Researchers are trying to find more efficient methods because the FFT is used in many applications. Under certain circumstances, the FFT can be made ten times faster, as researchers have recently shown. Refer to http://web.mit.edu/newsoffice/2012/faster-fourier-transforms-0118.html

9.8 Summary

This chapter presented the basics of discrete-time processing of data in the frequency domain using the discrete Fourier transform (DFT). The fast Fourier transform (FFT) was discussed as an efficient means to compute the DFT. The DFT of signal sequence x_i results in the spectrum Sx_k, while that of unit-sample response h_i results in the frequency response $|H_k|^2$. Implementation of MA systems can be performed either as a digital filter or by employing the FFT. Implementing an AR system is typically more efficient and accurate in digital filter form.

9.9 Problems

9.1 **Verify DFT values X_0 and $X_{n_x/2}$.** Compute DFT values X_0 and $X_{n_x/2}$ using the direct computation given by either Eq. (9.1) or Eq. (9.4) for

$$x_0 = 1, x_1 = -1, \text{ and } x_i = 0 \text{ for } 2 \leq i \leq 63$$

9.2 **DFT of delayed unit-sample sequence.** Compute $|X_k|$ for $0 \leq k \leq 32$ for the unit-sample sequence delayed by two samples as

$$x_i = d_{i-2} \text{ for } 0 \geq i \geq 63$$

9.3 **DFT of 3-sample interpolator.** Compute $|X_k|$ for $0 \leq k \leq 32$ using

$$x_0 = 0.5, x_1 = 1, x_2 = 0.5, \text{ and } x_i = 0 \text{ for } 3 \geq i \geq 63$$

Hint: Simplify by expressing the DFT about the point of symmetry in the time domain.

9.4 **Compute DFT values X_0 and $X_{n_x/2}$.** Compute DFT values X_0 and $X_{n_x/2}$ using

$$x_i = 0.9^i \text{ for } 0 \geq i \geq 63$$

Hint: Use the finite geometric sum formula.

9.5 **Compute DFT values X_0 and $X_{n_x/2}$.** Compute DFT values X_0 and $X_{n_x/2}$ using

$$x_i = (-0.9)^i \text{ for } 0 \geq i \geq 63$$

9.6 **Compute the power transfer function for first-order AR system.** Extend Example 9.13 to compute $|H_k|^2$ for $k = 0, 1, 2, \ldots, n_x - 1$ for an AR system with the unit-sample response as

$$h_i = (-0.9)^i \text{ for } i \geq 0$$

Hint: Use the infinite geometric sum formula.

9.7 **Padding with zeros.** The convolution of two rectangular sequences, each defined by

$$x_i = 1 \text{ for } 0 \leq i \leq 15$$

is to be performed in the frequency domain with an n-point DFT. What is the minimum value of n that produces the expected triangular result.

9.10 Matlab Projects

9.1 **Direct computation of DFT.** Consider Example 16.42 to verify Examples 9.2 and 9.3 by composing a Matlab script to compute the DFT and plot the magnitude $|X_k|$. Evaluate the DFT using the direct computation given by either Eq. (9.1) or Eq (9.4) for

$$x_0 = 1, x_1 = -1, \text{ and } x_i = 0 \text{ for } 2 \leq i \leq 63$$

and plot $|X_k|$ for $0 \leq k \leq 63$.

9.2 **Non-conventional DFT.** Conventional DFT algorithms transform n_x time points into n_x frequency points of which only the points $0 \leq k \leq n_x/2$ are "actual." To show that the direct computation of the DFT given by either Eq. (9.1) or Eq. (9.4) can be done for only the "actual" frequencies, compute $|X_k|$ for $0 \leq k \leq n_x/2$ for the time sequence given by

$$x_0 = 1, x_1 = 1, \text{ and } x_i = 0 \text{ for } 2 \leq i \leq 63$$

and plot $|X_k|$ for $0 \leq k \leq 32$.

9.3 **Verifying Matlab's FFT algorithm.** Verify Matlab's FFT function for

$$x_0 = 1, x_1 = 1, \text{ and } x_i = 0 \text{ for } 2 \le i \le 63$$

by computing and plotting $|X_k|$ for $0 \le k \le 63$ and comparing to the results obtained in Project 9.1.

9.4 **Non-power of 2 n_x-point FFT.** Verify that n_x need not be a power of 2 by computing the 100-point FFT for

$$x_0 = 0.5, x_1 = 1, x_2 = 0.5, \text{ and } x_i = 0 \text{ for } 3 \le i \le 99$$

and extract the *actual* frequencies by plotting $|X_k|$ for $0 \le k \le 50$,

9.5 **Magnitude and phase of the 3-sample averager.** Let

$$y_i = \frac{1}{3}(x_i + x_{i-1} + x_{i-2})$$

and h_i be its unit-sample response.

$$hp_i = h_i \text{ for } 0 \le i \le n_h - 1$$

$$= 0 \text{ for } n_h \le i \le n_f - 1 \ (n_f = 80)$$

Use subplot(3,1,n),stem(.) to plot hp_i, $|Hp_k|$, and $\angle Hp_k$ for $0 \le k \le n_f/2$.

9.6 **Magnitude response of the 3-sample interpolator.** The zero-padded unit-sample response of a 3-sample MA digital filter that acts as linear interpolator is given by

$$h_0 = 0.5, h_1 = 1, h_2 = 0.5, \text{ and } h_i = 0 \text{ for } 3 \le i \le 63$$

Compute and plot the magnitude response $|H_k|$ for $0 \le k \le 63$.

9.7 **Padding with zeros.** Compute the convolution of two rectangular sequences, each defined by

$$x_i = 1 \text{ for } 0 \le i \le 15$$

in the frequency domain with n-point DFTs using the minimum value of n that produces the expected results.

9.8 **Signal processing in the frequency domain.** Extend Example 16.46 to compose Matlab code that performs digital filter operations in the frequency domain of a MA digital filter that computes the running average of three input samples, using the following steps:

a. Formulate the filter transfer function H_k for $k = 0, 1, 2, \ldots, n_x - 1 (n_x = 20)$.

b. Using the input command, enter x_i array containing fewer than n_x values. (See Example 16.47).

c. Pad x_i with zeros to form xp_i for $i = 0, 1, 2, \ldots, n_x - 1$.

d. Compute X_k for $k = 0, 1, 2, \ldots, n_x - 1$.

e. Compute $Y_k = H_k X_k$ for $k = 0, 1, 2, \ldots, n_x - 1$.

f. Compute and plot y_i for $i = 0, 1, 2, \ldots, n_x - 1$ from the inverse DFT of Y_k.

CHAPTER 10

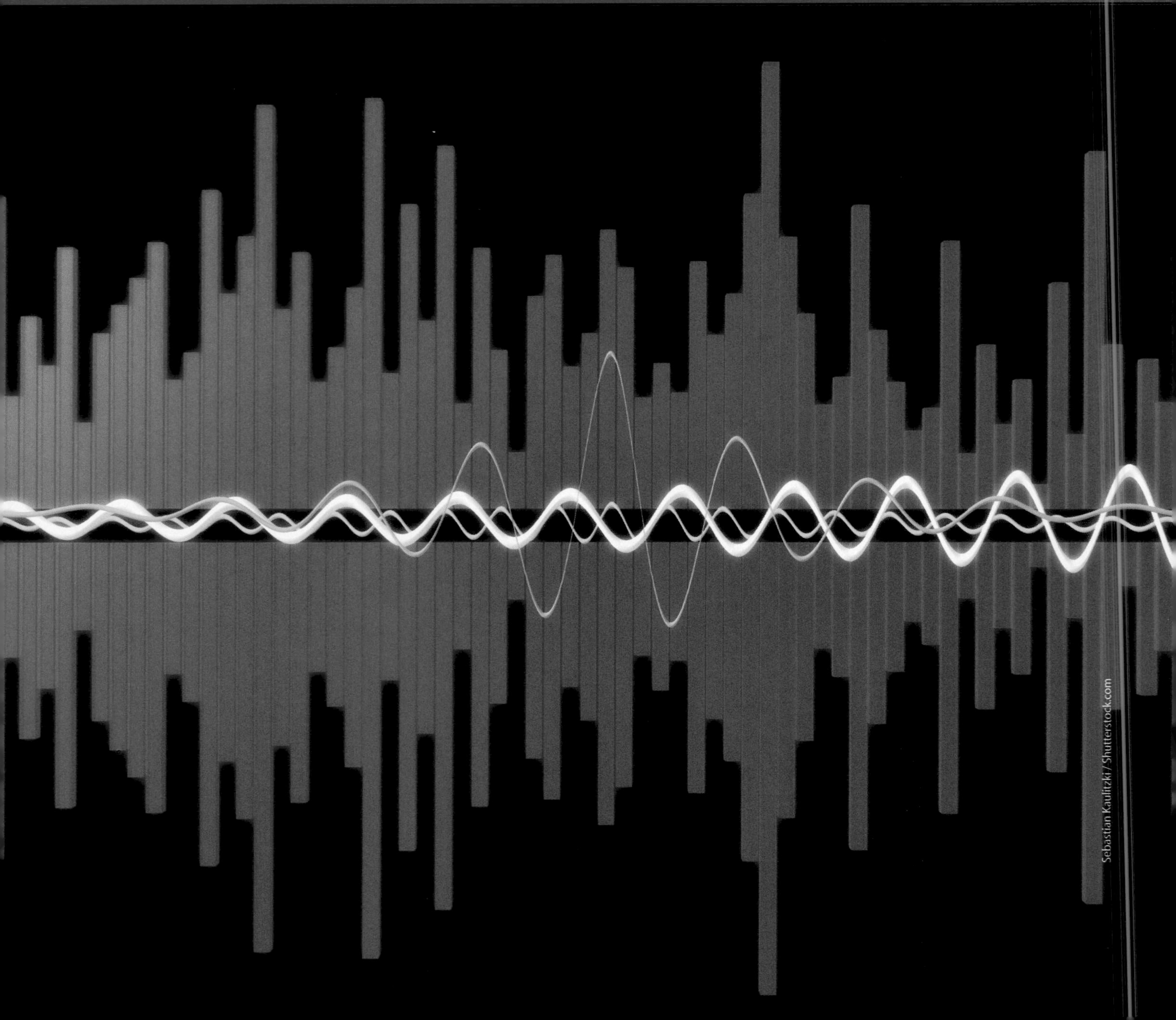

Sebastian Kaulitzki / Shutterstock.com

DETECTING DATA SIGNALS IN NOISE

LEARNING OBJECTIVES

After completing this chapter, the reader should be able to:

- Apply data processing techniques to extract data from signals.

- Understand the purpose of complementary signals for robust data transmission.

- Use a pseudo-random number generator to generate binary data and to simulate noise.

- Appreciate why the most important parameter in data transmission is the signal-to-noise ratio equal to the signal energy divided by the noise variance.

- Investigate the performance of data transmission signals in noise through simulations.

This chapter describes how data are transmitted reliably in the presence of noise—even from great distances, such as from Mars. Data signal transmission is a special case in which the signal values are known at both the transmitter and receiver (that is, a specified set of values are employed for transmitting a 0 and another set for transmitting a 1). This chapter describes how a processor matched to this set of values determines the bit value that was transmitted. Two signals having complementary values produce a maximal difference in the matched processor output. Reliable transmission is typically achieved at the cost of longer transmission times. To reduce the probability of error for reliable data transmission, the signal energy is increased by appropriate signal design.

10.1 INTRODUCTION

The main topics in this chapter are covered in the following sequence.

Data transmission model—A source produces binary data that is encoded using a signal that is transmitted over a channel. During transmission, the channel introduces random noise, and a processor in the receiver tries to extract the signal that was transmitted. Random number generators are employed to simulate random binary data produced by the source and the noise introduced by the channel.

Signal processing—The conventional processor that operates on received signals forms the sum of weighted signal values. Knowing the signal values that were transmitted leads to a processor whose weights are matched to these signal values.

Signal design—When a matched processor operates on received signals, we design the two signals that encode data values 0 and 1 to form a complementary pair (that is, the signal values of one are the negative values of the other). This choice maximizes the matched processor output differentiation of these two signals for signals having a particular energy.

Processing signals in noise—When a matched processor operates on complementary signals that are corrupted with noise, the reliability of detecting the correct (transmitted) signal value is determined solely by the ratio of signal energy to noise variance, which is known as the *signal-to-noise ratio (SNR)*.

Probability of error—Because noise is random, data transmission system performance is measured in terms of the *probability of error*. No data transmission system operates perfectly, causing this probability to be greater than zero. A simulation procedure is described to estimate the probability of error by performing multiple transmissions in the presence of noise and counting the number of errors.

10.2 DATA TRANSMISSION MODEL

Figure 10.1 shows the block diagram description of binary data transmission that includes the following elements.

Source produces a binary data value D_t, that is equal to either a 0 or 1, and is to be transmitted to a destination. This chapter considers the case of transmitting a

Figure 10.1

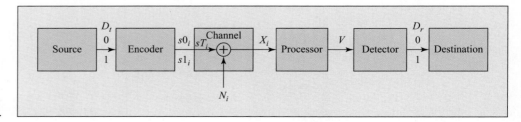

Model of a data transmission system.

single bit that is randomly generated. The upper-case letter indicates D_t is a random quantity (that is, its value is not predictable) that obeys the laws of probability.

Encoder encodes the data bit values using transmission signal waveform $s(t)$. A *signal sequence* is an indexed set of sample values denoted

$$s_i \text{ for } i = 0, 1, 2, \ldots, n_x - 1 \tag{10.1}$$

where i is the index and n_x is the signal duration. This sequence is obtained by converting the signal waveform $s(t)$ by sampling with period T_s, or

$$s_i = s(t)|_{t=iT_s} \text{ for } i = 0, 1, 2, \ldots, n_x - 1 \tag{10.2}$$

where s_0 corresponds to $s(t)$ at time $t = 0$ and $s(t)$ extends over $0 \leq t < n_x T_s$.

The encoder assigns signals according to the D_t bit value with

$$D_t = 1 \rightarrow s1_i \text{ for } i = 0, 1, 2, \ldots, n_x - 1 \tag{10.3}$$

$$D_t = 0 \rightarrow s0_i \text{ for } i = 0, 1, 2, \ldots, n_x - 1$$

Data signals are special in that both the encoder at the transmitter end of the channel and processor at the receiving end know their values. One constraint on these signals is that both $s0_i$ and $s1_i$ must have the same energy $\mathcal{E}_s$, which is computed as the sum of their squared values:

$$\mathcal{E}_s = \sum_{i=0}^{n_x-1} s0_i^2 = \sum_{i=0}^{n_x-1} s1_i^2 \tag{10.4}$$

Channel transports the transmitted signal sT_i for $i = 0, 1, 2, \ldots, n_x - 1$ from the transmitter to receiver, where

$$sT_i = s1_i, \quad \text{if } D_t = 1 \tag{10.5}$$

$$= s0_i, \quad \text{if } D_t = 0$$

A channel can be a length of wire or a medium that supports wireless transmission. In the transmission process, the channel adds random noise N_i to sT_i to form the signal at the channel output X_i for $i = 0, 1, 2, \ldots, n_x - 1$ where

$$X_i = s1_i + N_i \quad \text{if } D_t = 1 \tag{10.6}$$

$$X_i = s0_i + N_i \quad \text{if } D_t = 0$$

X_i is random and upper-case, because N_i being random makes the sum random as well.

Processor uses the observed sequence X_i to compute an output value V. Because noise makes X_i random, V is also random.

Detector applies V to a threshold detector to determine the received bit value, producing the receiver value $D_r = 0$ or 1.

Destination accepts data value D_r for accomplishing the desired task.

Data Processing Task

The task at hand is to process X_i to determine the data value D_t that it encodes. Because X_i contains random noise, there is a possibility of making a mistake. There are two types of errors that can occur

$$\text{If } D_t = 1 \text{ and } D_r = 0 \quad \rightarrow \quad \text{false negative} \tag{10.7}$$

$$\text{If } D_t = 0 \text{ and } D_r = 1 \quad \rightarrow \quad \text{false positive}$$

With random noise, we need to compute the probability of making an error. Figure 10.2 shows a simple model that describes data communication over a noisy channel.

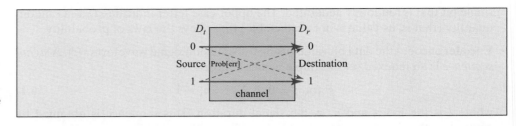

Figure 10.2

Error model when binary data are transmitted over a noisy channel.

Prob[error] indicates the probability that an error occurs. In a practical system, probability of error should very small, such as Prob[error] $< 10^{-3}$, or less than one error in 1,000 transmissions.

This chapter describes why it is not theoretically possible to design a system that is perfect (that is, one that has Prob[error] $= 0$). The engineering problem is to design a system that has an acceptably small Prob[error] by designing the transmission signals and processing them appropriately.

10.3 PROCESSING DATA SIGNALS

Our description of signal processing procedures progresses in the following steps:

1. We first describe the basic processing operation employed by most digital receivers. The general approach is termed a *linear processor* that stores samples of the received signal X_i for $i = 0, 1, 2, \ldots, n_x - 1$ and scales them by a set of coefficients to produce a single output value V.

2. The linear processor that achieves the best performance is called the *matched processor* where the coefficients are tuned to the data signals values. This processing is also known as *template matching*, *correlation detection*, and *matched filtering*. This is the most reliable method of extracting known data values from a detected signal corrupted with additive Gaussian noise.

3. Motivated by the matched processor, we design a pair of data signal sequences, $s1_i$ and $s0_i$, that produce maximally different output values V to differentiate a signal that encodes a 1 from that encoding a 0.

This section considers the ideal noiseless case to illustrate the processing that occurs. The next section extends the processing to signals that contain noise.

10.3.1 Linear Processor

Detected signal X_i for $i = 0, 1, 2, \ldots, n_x - 1$ that contains noise must be processed to enhance the information-carrying components and reduce the noise. The method used in practice is called a *linear processor*, which includes a sample memory that stores n_x values X_i for $i = 0, 1, 2, \ldots, n_x - 1$ (that is, all of the data samples that are in $s1_i$ or $s0_i$). These stored sample values are multiplied by delay-specific coefficients c_i for $i = 0, 1, 2, \ldots, n_x - 1$, and the products are summed to form an output value V according to

$$V = \sum_{i=0}^{n_x - 1} c_i X_i \qquad (10.8)$$

Figure 10.3 shows the structure of a general linear processor that operates on the detected input sequence X_i to produce output V. The blocks indicated with δ represent memory units that store an input value for one sample period. For example, at time i, the input to a memory unit is X_i, while the memory unit output equals the past value X_{i-1}. The value of X_i is multiplied by coefficient c_i with the multipliers indicated by triangles. The products at the multiplier outputs are added together at a summing junction to produce V.

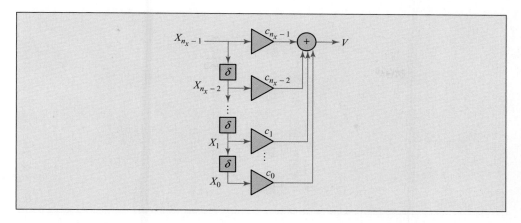

Figure 10.3

Block diagram of the general linear processor at the time that it observes all transmitted signal values X_i for $i = 0, 1, 2, \ldots, n_x - 1$.

This processor structure is a natural choice for data signal processing for the following reasons.

- The transmitted signal has a finite duration, n_x.

- Some signal values are more important than others, and larger magnitude coefficients are assigned to the more important signal sample values.

- The processor uses all n_x values to produce a single output value V that the detector uses to decide the received data value.

Linear processor EXAMPLE 10.1

The observed signal sequence X_i for $i = 0, 1, 2, \ldots, n_x - 1$ with $n_x = 3$ equals

$$X_0 = 2, \; X_1 = -1, \; \text{and} \; X_2 = 1$$

A linear processor operates on $n_x = 3$ samples with coefficients

$$c_0 = 1, \; c_1 = 2, \; \text{and} \; c_2 = 3$$

Then the linear processor produces output

$$V = \sum_{i=0}^{2} c_i \, X_i = c_0 \, X_0 + c_1 \, X_1 + c_2 \, X_2 = 1(2) + 2(-1) + 3(1) = 3$$

The coefficient magnitude $|c_i|$ measures the *importance* of a signal value. The largest magnitude coefficient c_2 indicates that X_2 has more weight in determining the V value than the other X_i values.

10.3.2 Matched Processor

The general linear processor applies weights to the observed input values through the coefficients. *In the presence of a constant random noise level, it is intuitive that X_i values that have larger magnitudes are detected more reliably than those that have smaller magnitudes.* Hence, larger magnitude signals should be given more weight by assigning larger coefficients to them. This is the motivation for the *matched processor*.

To implement a matched processor, the general processor undergoes the following modification: The filter coefficients c_i are *tuned* to the values of the signal that it is designed to detect. We start by designing a signal sequence s_i for $i = 0, 1, 2, \ldots, n_x - 1$ having known values to transmit data. We can then design the processor matched to s_i

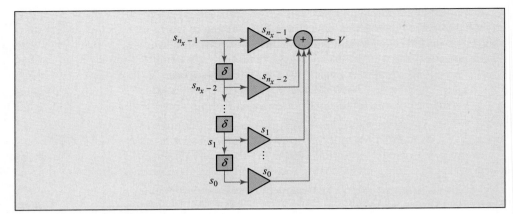

Figure 10.4

Block diagram of the matched processor when $X_i = s_i$ for $i = 0, 1, 2, \ldots, n_x - 1$ at the time that it observes all n_x signal values.

by specifying $c_i = s_i$. The processor output V then is given by

$$V = \sum_{i=0}^{n_x - 1} s_i \, X_i \qquad (10.9)$$

Figure 10.4 shows the structure of a matched processor when $X_i = s_i$ for $i = 0, 1, 2, \ldots, n_x - 1$ at the time that it observes all n_x signal values to produce output V.

When the signal to which the matched processor is *tuned* occurs (that is, when $X_i = s_i$ for $i = 0, 1, 2, \ldots, n_x - 1$), then

$$V = \sum_{i=0}^{n_x - 1} s_i \, \overbrace{X_i}^{=s_i} = \sum_{i=0}^{n_x - 1} s_i^2 = \mathcal{E}_s \qquad (10.10)$$

So, when the processor matched to a signal actually detects this signal, the output is large—equal to the sum of the squares of the signal values. This large positive value equals the signal energy $\mathcal{E}_s$. It can be shown that the matched processor is the best we can do to detect signals in the presence of noise.

EXAMPLE 10.2   **Matched processor in the ideal noiseless case**

Let the signal sequence with $n_x = 3$ be

$$s_0 = 2, s_1 = -1, \text{ and } s_2 = 1$$

The three coefficients matched to this signal are then

$$c_0 = 2, c_1 = -1, \text{ and } c_2 = 1$$

When the observed signal is $X_i = s_i$ for $i = 0, 1, 2, \ldots, n_x - 1$, the matched processor produces output value

$$V = \sum_{i=0}^{2} c_i \, X_i = \sum_{i=0}^{2} s_i \, X_i = s_0 \, s_0 + s_1 \, s_1 + s_2 \, s_2 = (2)^2 + (-1)^2 + (1)^2 = 6$$

The energy in the signal s_i is $\mathcal{E}_s = 6$.

Let us apply a different signal having the same energy to this matched processor. Consider the time-reversed signal

$$s_0' = 1, s_1' = -1, \text{ and } s_2' = 2$$

Because s_i' has the same three values as s_i (but in a different order) their energies are equal, as

$$\mathcal{E}_s = \mathcal{E}_{s'} = 6$$

When $X_i = s_i'$ is processed by the matched processor tuned to s_i, the output V' is

$$V' = \sum_{i=0}^{2} s_i \overbrace{X_i}^{=s_i'} = s_0 s_0' + s_1 s_1' + s_2 s_2' = (2)(1) + (-1)(-1) + (1)(1) = 4$$

Note that $V' < \mathcal{E}_s$, demonstrating that a matched processor tuned to a signal generates the largest possible value when that signal is observed, for signals having the same energy.

10.3.3 Complementary Signals

The previous section shows that a matched processor tuned to a signal generates the largest possible signal when that signal is observed. We want to apply this observation to signals that encode the transmitted binary value D_t with $s1_i$ encoding a transmitted 1 ($D_t = 1$) and $s0_i$ encoding a 0 ($D_t = 0$). We would like to specify these signals so that the matched processor output produces very different values, so that a detector can easily differentiate the occurrence of these two signals. It turns out that the largest difference in matched processor output occurs when $s1_i$ and $s0_i$ are *complementary*, that is

$$s0_i = -s1_i \text{ for } i = 0, 1, 2, \ldots, n_x - 1 \tag{10.11}$$

We form a complementary pair from specified signal s_i for $i = 0, 1, 2, \ldots, n_x - 1$ by defining

$$s1_i = s_i \text{ for } i = 0, 1, 2, \ldots, n_x - 1 \quad \text{when } D_t = 1 \tag{10.12}$$
$$s0_i = -s_i \text{ for } i = 0, 1, 2, \ldots, n_x - 1 \quad \text{when } D_t = 0$$

Doing this provides two signals having the same energy, as $\mathcal{E}_{s1} = \mathcal{E}_{s0} = \mathcal{E}_s$.

Signal energy of sinusoidal complementary signals

EXAMPLE 10.3

Figure 10.5 shows one period of a sinusoidal sequence

$$s_i = A \sin(2\pi i/32) \text{ for } i = 0, 1, 2, \ldots, n_x - 1 (n_x = 32)$$

that forms the pair of complementary signals

$$s1_i = s_i$$
$$s0_i = -s_i$$

The energy in a sinusoidal sequence having amplitude A and an integral number of periods equals

$$\mathcal{E}_s = \sum_{i=0}^{n_x-1} s_i^2 = \sum_{i=0}^{31} A^2 \sin^2(2\pi i/32) = A^2 \sum_{i=0}^{31} \frac{1 + \cos(4\pi i/32)}{2} = 16A^2 \quad (= n_x A^2/2)$$

This is true because the sum of sinusoidal values over an integral number of periods equals zero. In our case,

$$\sum_{i=0}^{31} \cos(4\pi i/32) = 0$$

This leaves n_x values of $A^2/2$ to produce the sum $\mathcal{E}_s = n_x A^2/2$.

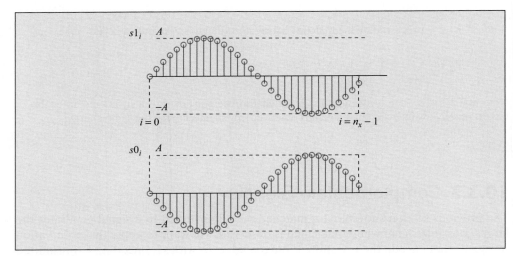

Figure 10.5

Complementary data signals for $s_i = A \sin(2\pi i / 32)$ for $i = 0, 1, 2, \ldots, n_x - 1$ with $n_x = 32$.

To show the utility of the matched processor, consider the ideal (noiseless) cases when $X_i = s1_i$ and $X_i = s0_i$.

$X_i = s1_i$: The output V of matched processor when $X_i = s1_i = s_i$ is denoted by $V_{|X=s1}$ (in words, "V given that $X = s1$") and equals

$$V_{|X=s1} = \sum_{i=0}^{n_x-1} s1_i \overbrace{X_i}^{=s1_i} = \sum_{i=0}^{n_x-1} s1_i^2 = \mathcal{E}_s \tag{10.13}$$

$X_i = s0_i$: The output V when $X_i = s0_i = -s_i$ is denoted by $V_{|X=s0}$ and equals

$$V_{|X=s0} = \sum_{i=0}^{n_x-1} s0_i \overbrace{X_i}^{s0_i=-s_i} = -\sum_{i=0}^{n_x-1} s_i^2 = -\mathcal{E}_s \tag{10.14}$$

which is a negative value with large magnitude.

Note that when $\mathcal{E}_s$ is large, the matched processor output $V_{|X=s1} = \mathcal{E}_s$ can be easily differentiated from $V_{|X=s0} = -\mathcal{E}_s$.

EXAMPLE 10.4   Processing complementary signals

Let $s_i = 1$ for $0 \le i \le 3$. We then form complementary signals

$$s1_i = s_i = 1 \text{ for } 0 \le i \le 3$$

$$s0_i = -s_i = -1 \text{ for } 0 \le i \le 3$$

The energy of s_i equals

$$\mathcal{E}_s = \sum_{i=0}^{3} s_i^2 = \sum_{i=0}^{3} (1)^2 = 4$$

The matched processor produces the outputs:

$$V_{|X=s1} = \sum_{i=0}^{n_x-1} s_i \overbrace{X_i}^{=s1_i} = \sum_{i=0}^{n_x-1} \overbrace{s_i}^{=1} \overbrace{s1_i}^{=1} = \sum_{i=0}^{3} 1 = 4$$

and

$$V_{|X=s0} = \sum_{i=0}^{n_x-1} s_i \overbrace{X_i}^{=s0_i} = \sum_{i=0}^{n_x-1} \overbrace{s_i}^{=1} \overbrace{s0_i}^{=-1} = \sum_{i=0}^{3} (-1) = -4$$

> *Factoid:* A data system does not need to use complementary signals, but one that does will work better (that is, produce fewer errors) when noise is present.

10.4 PROCESSING SIGNALS IN NOISE

Noise is typically considered to be a *Gaussian random process* (that is, it obeys the famous *bell-shaped curve*, shown in Figure 10.6). We model the noise sample N_i as a Gaussian random number.

- N_i is *random* because we cannot predict its value. The noise values obey the Gaussian distribution with most values falling around zero, and values having larger magnitudes being less probable.

- N_i is *Gaussian* because each noise sample can be considered to be the sum of many independent random factors.

The bell-shaped curve for noise has mean value $\mu_N = 0$ and standard deviation (SD) σ_N with the *variance* equal to σ_N^2. One important observation for Gaussian N_i is that observed noise values tend to cluster around zero with 95% of the values falling within the range

$$[-1.96\sigma_N, 1.96\sigma_N] \tag{10.15}$$

> *Factoid:* Note: The SD σ_N has the same units as the signal values s_i, while the variance σ_N^2 has the same units as signal energy $\mathcal{E}_s$.

10.4.1 Signal-to-Noise Ratio

A parameter that describes both the quality of the received signal and the probability of detecting the correct data value is given by the *signal-to-noise ratio (SNR)*, which is equal to the ratio of the signal energy $\mathcal{E}_s$ to the noise variance:

$$\text{SNR} = \frac{\mathcal{E}_s}{\sigma_N^2} \tag{10.16}$$

Electrical engineers express this value in dB units, as

$$\text{SNR}_{\text{dB}} = 10\log_{10}\left(\frac{\mathcal{E}_s}{\sigma_N^2}\right) \tag{10.17}$$

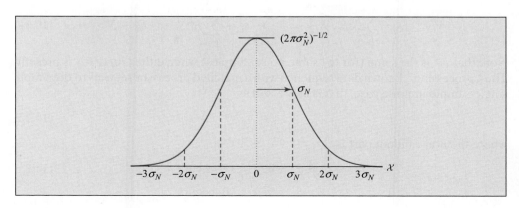

Figure 10.6

Gaussian distribution describing noise sample values.

EXAMPLE 10.5 Signal-to-noise ratio

Let the specified signal s_i for $i = 0, 1, 2, \ldots, n_x - 1$ have signal energy

$$\mathcal{E}_s = \sum_{i=0}^{n_x-1} s_i^2 = 100$$

If Gaussian noise with $\sigma_N = 2$ is added to the signal, 95% of the N_i values fall within $[-3.92, 3.92]$. The signal-to-noise power is

$$\text{SNR} = \frac{\mathcal{E}_s}{\sigma_N^2} = \frac{100}{4} = 25$$

and the value in decibels is

$$\text{SNR}_{\text{dB}} = 10 \log_{10} \left(\frac{\mathcal{E}_s}{\sigma_N^2} \right) = 10 \log_{10} 25 = 14 \text{ dB}$$

10.4.2 Matched Processor Output

We can now examine the output of the matched processor when noise is present in the two signals that can occur.

$X_i = s1_i + N_i$: In this case,

$$V_{|X=s1+N} = \sum_{i=0}^{n_x-1} s_i X_i = \sum_{i=0}^{n_x-1} s_i(s1_i + N_i) \tag{10.18}$$

Setting $s1_i = s_i$, we have

$$V_{|X=s1+N} = \sum_{i=0}^{n_x-1} s_i \overbrace{X_i}^{=s_i+N_i} = \sum_{i=0}^{n_x-1} s_i^2 + \sum_{i=0}^{n_x-1} s_i N_i = \mathcal{E}_s + N^* \tag{10.19}$$

where N^* is a random number that represents the random Gaussian noise component in the matched processor output.

$X_i = s0_i + N_i$: In this case,

$$V_{|X=s0+N} = \sum_{i=0}^{n_x-1} s_i X_i = \sum_{i=0}^{n_x-1} s_i(s0_i + N_i) \tag{10.20}$$

Setting $s0_i = -s_i$, we have

$$V_{|X=s0+N} = \sum_{i=0}^{n_x-1} s_i \overbrace{X_i}^{=-s_i+N_i} = -\sum_{i=0}^{n_x-1} s_i^2 + \sum_{i=0}^{n_x-1} s_i N_i = -\mathcal{E}_s + N^* \tag{10.21}$$

Note that N^* is the same (for the same noise samples) when either $s0_i$ or $s1_i$ is present. Thus, processing a known data sequence with a matched processor reduces to the simple single-sample in noise case, that is expressed as

$$V = s^* + N^* \tag{10.22}$$

where the non-random part is

$$s^* = \mathcal{E}_s \quad \text{when} \quad s1_i \text{ occurs} \tag{10.23}$$

$$s^* = -\mathcal{E}_s \quad \text{when} \quad s0_i \text{ occurs}$$

and the random part is the single value derived from processing the set of noise samples

$$N^* = \sum_{i=0}^{n_x-1} s_i N_i \tag{10.24}$$

When complementary signals are used, V is applied to a threshold detector to determine the received data value D_r:

$$\text{If } V \geq 0 \rightarrow D_r = 1 \tag{10.25}$$

$$\text{If } V < 0 \rightarrow D_r = 0$$

An analysis using probability theory shows three important features that describe the matched processor output V.

1. V is a random number that has a Gaussian distribution.

2. The mean of V depends on which signal was transmitted as

$$\mu_V = \mathcal{E}_s \text{ when } X_i = s1_i + N_i \tag{10.26}$$

$$\mu_V = -\mathcal{E}_s \text{ when } X_i = s0_i + N_i$$

3. The variance of V equals

$$\sigma_V^2 = \sigma_N^2 \sum_{i=0}^{n_x-1} s_i^2 = \sigma_N^2 \mathcal{E}_s \tag{10.27}$$

and is the same value independent of whether $s1_i$ or $s0_i$ was transmitted.

Figure 10.7 shows the Gaussian distribution of V when $\mathcal{E}_s \gg \sigma_V$ and when $\mathcal{E}_s \approx \sigma_V$. An error occurs when the V value falls on the wrong side of threshold $\tau = 0$. This happens when a signal that produces $V = +\mathcal{E}_s$ under ideal noiseless conditions is corrupted with a specific set of noise values that causes the matched processor to produce $V < 0$.

When $\mathcal{E}_s \gg \sigma_V$, we should expect almost error-free performance with V always falling on the correct side of the threshold τ. However, this is not always possible, because the tails of a Gaussian distribution theoretically extend in the negative direction to $x = -\infty$ and in the positive direction to $x = +\infty$. Even when $\mathcal{E}_s \gg \sigma_V$, it is theoretically possible for rare V values to fall on the wrong side of threshold τ, no matter how large we make $\mathcal{E}_s$.

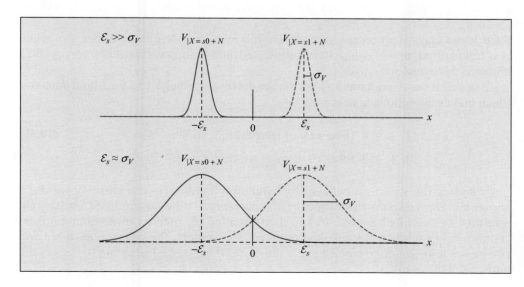

Figure 10.7

The Gaussian distribution of V when $\mathcal{E}_s \gg \sigma_V$ and when $\mathcal{E}_s \approx \sigma_V$.

10.4.3 Detecting Signals in Noise

We start with the simplest signal with $s0_i$ and $s1_i$ consisting of only one sample having magnitude s. Thus,

$$s1 = s \text{ if } D_t = 1 \quad \text{and} \quad s0 = -s \text{ if } D_t = 0 \tag{10.28}$$

The signal energy is then

$$\mathcal{E}_s = s^2 \tag{10.29}$$

The received signal is corrupted with the single noise sample N:

$$X = s + N \quad \text{if } s1 \text{ is transmitted} \tag{10.30}$$
$$X = -s + N \quad \text{if } s0 \text{ is transmitted}$$

where N is a zero-mean Gaussian random number having variance σ_N^2.

The signal-to-noise ratio is computed as a ratio of signal energy to noise variance

$$\text{SNR} = \frac{\mathcal{E}_s}{\sigma_N^2} = \frac{s^2}{\sigma_N^2} \tag{10.31}$$

The matched processor computes the output

$$V = \sum_{i=0}^{n_x - 1} s_i X_i = s X \tag{10.32}$$

In this simple case, the result depends on what signal was transmitted, as

$$V_{|X=s1+N} = (\overbrace{s1}^{=s} + N) = s^2 + s N \tag{10.33}$$

or

$$V_{|X=s0+N} = s(\overbrace{s0}^{=-s} + N) = -s^2 + s N \tag{10.34}$$

Then V is a Gaussian random number that has mean value equal to either $+s^2$, when $s1$ is transmitted, or $-s^2$, when $s0$ is transmitted, and it has a variance $\sigma_V^2 = s^2 \sigma_N^2$ and a standard deviation $\sigma_V = s \sigma_N$.

To decide if the X contains an $s1$ or $s0$, the detector applies $\mathcal{V}$ to a threshold detector, which makes the following decision.

$$\text{If } V \geq 0 \text{ then } s1 \text{ was transmitted, making } D_r = 1 \tag{10.35}$$
$$\text{If } V < 0 \text{ then } s0 \text{ was transmitted, making } D_r = 0$$

To display the statistical behavior of a random number (in this case V), we use a *histogram*, which displays the frequency of occurrence of random numbers. A histogram is formed by defining a set of bins and then counting the number of random numbers falling within each bin. Because the numbers are random, the count in each bin will also be random. The following example illustrates the use of a histogram to display the statistical behavior of V.

Simple signal with noise   **EXAMPLE 10.6**

To illustrate the problem, consider a source that transmits two hundred data signals: one hundred 0's (each using $s0 = -10$) followed by one hundred 1's (each using $s1 = 10$). Consider first the high SNR case, followed by a low SNR case.

Figure 10.8a shows the two hundred signals corrupted by Gaussian noise with $\sigma_N = 1$ (the high SNR case). The signal energy is

$$\mathcal{E}_s = s^2 = 100$$

and the signal-to-noise ratio

$$\text{SNR} = \frac{\mathcal{E}_s}{\sigma_N^2} = \frac{100}{1} = 100$$

and in decibel units

$$\text{SNR}_{\text{dB}} = 10\log_{10}100 = 20 \text{ dB}$$

These two hundred signals are each applied to the matched processor that computes V as

$$V = sX = 10X$$

When $X = s0 + N = -10 + N$, the signal $s0$ is constant and the random noise value N is different for each observed X value. In this case the matched-processor output for each observed X value equals

$$V_{|X=s0+N} = -100 + 10N = -\mathcal{E}_s + 10N$$

Because N has a Gaussian distribution, $V_{|X=s0+N}$ also has a Gaussian distribution with mean $-\mathcal{E}_s$ and SD $10\sigma_N$. These one hundred $V_{|X=s0+N}$ values generate the histogram component that is centered about $x = -\mathcal{E}_s$ in Figure 10.8a.

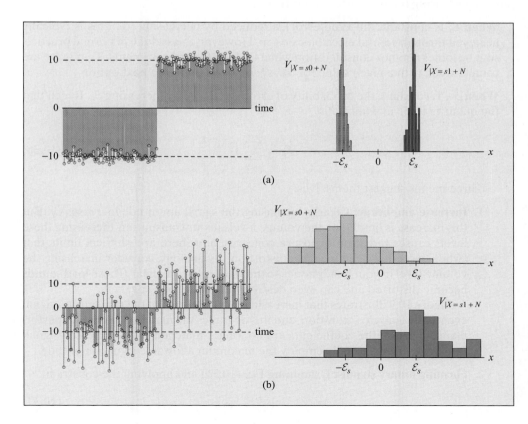

Figure 10.8

Two hundred data signals with noise and histogram of matched-processor output V. (a) High SNR (20 dB). (b) Low SNR (0 dB).

When $X = s1 + N = 10 + N$, the matched-processor output for each observed X value equals

$$V_{|X=s1+N} = 100 + 10N = \mathcal{E}_s + 10N$$

These one hundred $V_{|X=s1+N}$ values generate the histogram component that is centered about $x = \mathcal{E}_s$.

The histogram in Figure 10.8a indicates that a threshold detector using threshold value $\tau = 0$ would correctly separate the $s0$ and $s1$ signals almost all the time and, thus, have a Prob[error] close to zero.

Figure 10.8b shows the same data values and signal energy ($\mathcal{E}_s = 100$) as in Figure 10.8a but with more noise added ($\sigma_N = 10$). The signal-to-noise ratio in this higher-noise case equals

$$\text{SNR} = \frac{\mathcal{E}_s}{\sigma_N^2} = \frac{100}{100} = 1$$

and in decibel units

$$\text{SNR}_{\text{dB}} = 10 \log_{10} 1 = 0 \text{ dB}$$

The histograms of $V_{|X=s0+N}$ and $V_{|X=s1+N}$ in Figure 10.8b show values that overlap the zero value. A threshold detector using $\tau = 0$ would produce many errors in this case. Because the noise N is random, the errors also occur randomly and the system performance is measured by its Prob[error]. The system operating in the high noise case would have an unacceptably large Prob[error] (≈ 0.16).

Hence, the SNR is an important factor that determines the reliability of data transmission systems in terms of its Prob[error].

The previous example illustrates that we can reduce the number of errors by increasing the signal-to-noise ratio. The SNR can be improved either by increasing $\mathcal{E}_s$ or by reducing σ_N^2 using the following methods.

When $\mathcal{E}_s$ is constant, the Prob[error] is reduced by decreasing σ_N. This is typically achieved through careful electronic design. However, this is done in normal practice, so it becomes a minor consideration when looking for improvements. That is, we are usually stuck with a given value for σ_N, so we move on to the next option.

When σ_N is constant, the probability of error is reduced by increasing $\mathcal{E}_s$. Recall that the quantity to be maximized is

$$\mathcal{E}_s = \sum_{i=0}^{n_x-1} s_i^2 \tag{10.36}$$

Three means suggest themselves:

1. **Increase amplitude:** Clearly, increasing the signal amplitude increases $\mathcal{E}_s$. But this increase is limited by the voltage levels in our transmitter. Increasing these levels causes radiation exposure concerns, and there are physical limits that (when exceeded) lead to signal distortion. By analogy, consider increasing the volume level in your headphones so the whole class can hear. These load sounds become distorted and can even destroy the headphone speakers.

   Figure 10.9 illustrates that increasing the amplitude of the transmitted signal eventually leads to saturation, and this degrades the matched-processor performance because the coefficients are no longer matched to the detected signal. Practical systems usually employ the maximum allowable signal amplitude.

2. **Forming binary signals:** Examining Eq. (10.36) and applying the constraint

$$|s_i| \leq V_{\text{max}} \tag{10.37}$$

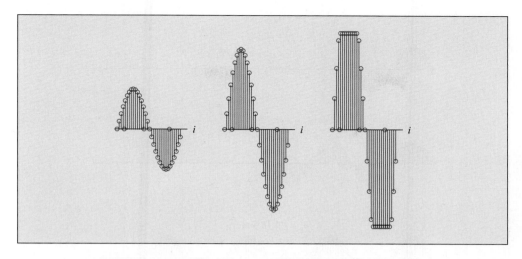

Figure 10.9

Increasing the amplitude of s_i eventually leads to saturation or other non-linear effects that make the received signal X_i no longer matched to the processor coefficients.

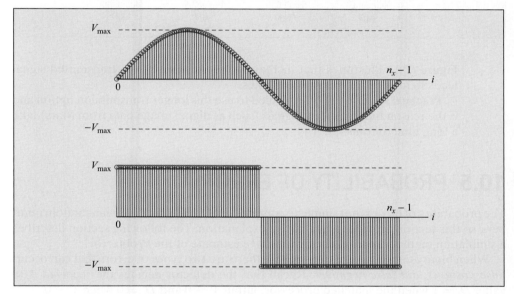

Figure 10.10

Increasing s_i to the amplitude limits imposed by channel maximizes $\mathcal{E}_s$. A sinusoidal s_i (top) has $\mathcal{E}_s = n_x V_{max}^2/2$. A binary signal is formed by maximizing amplitudes to $\pm V_{max}$ (bottom) and has $\mathcal{E}_s = n_x V_{max}^2$.

engineers developed *binary signals* in which s_i takes on only one of two values, either $+V_{max}$ or $-V_{max}$, as

$$s_i = \pm V_{max} \text{ for } i = 0, 1, 2, \ldots, n_x - 1 \qquad (10.38)$$

Conventional signals can be converted to binary signals by extending positive values to $+V_{max}$ and negative values to $-V_{max}$, as shown in Figure 10.10.

The energy of a binary signal is then

$$\mathcal{E}_s = \sum_{i=0}^{n_x-1} s_i^2 = \sum_{i=0}^{n_x-1} V_{max}^2 = n_x V_{max}^2 \qquad (10.39)$$

Since the signal is always at a maximum magnitude, the binary signal has the maximum possible energy that can be placed on a channel. Recall, a sinusoidal sequence having amplitude V_{max} has energy $\mathcal{E}_s = n_x V_{max}^2/2$, which is half that achievable with a binary signal.

3. **Increase n_x:** Finally, if a binary signal already uses the maximum-allowed magnitudes, we can still increase $\mathcal{E}_s$ by making the signal longer by increasing n_x, as indicated by Eq.(10.39). However, this means that it takes longer to transmit the signal s_i for $i = 0, 1, 2, \ldots, n_x - 1$, which encodes each binary data value.

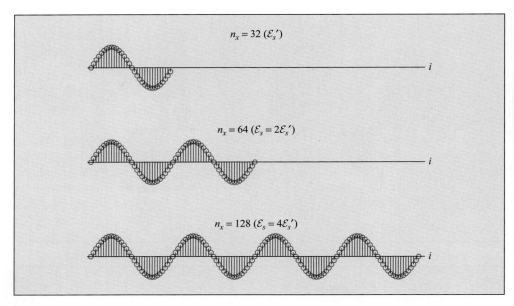

Figure 10.11

Increasing signal duration n_x lowers data transmission rate.

Figure 10.11 illustrates that increasing the duration of the transmitted signal leads to lower data transmission rates.

Practical systems are often forced to use this longer transmission option and is the reason that faint data signals (such as digital image data from Mars) take a long time to transmit.

10.5 PROBABILITY OF ERROR

The probability of error is not simple to compute analytically for Gaussian random numbers, so this section provides a qualitative explanation. The following section describes a simulation method to provide a quantitative estimate of the Prob[error].

When binary data (0/1) are transmitted, there are two types of errors that can occur: *false positives and false negatives*. Recall that the detector decides the received data value $D_r = 1$ when the matched processor output $V > 0$ and $D_r = 0$ when $V \le 0$.

False positive error. A false positive (FP) error occurs when the signal $s0_i$ is transmitted ($D_t = 0$) and the matched processor output V produces a value greater than zero ($D_r = 1$). This condition corresponds to the matched processor output being positive when the detected signal is $X_i = s0_i + N_i$, as

$$V_{|X=s0+N} > 0 \tag{10.40}$$

When $s0_i$ is transmitted, the matched processor output $V_{|X=s0+N}$ has a Gaussian distribution with mean $-\mathcal{E}_s$, as shown in Figure 10.12. The probability of a false positive is the area under this curve to the right of zero ($x > 0$). Computing this area exactly is an analytically difficult procedure.

False negative error. A false negative (FN) error occurs when the signal $s1_i$ is transmitted ($D_t = 1$) and the matched processor output V produces a value less than or equal to zero ($D_r = 0$). This condition corresponds to the matched processor output being negative when the detected signal is $X_i = s1_i + N_i$, as

$$V_{|X=s1+N} \le 0 \tag{10.41}$$

When $s1_i$ is transmitted, the matched processor output $V_{|X=s1+N}$ has a Gaussian distribution with mean $\mathcal{E}_s$, as shown in Figure 10.12. The probability of a false negative is the area under this curve to the left of zero ($x \le 0$).

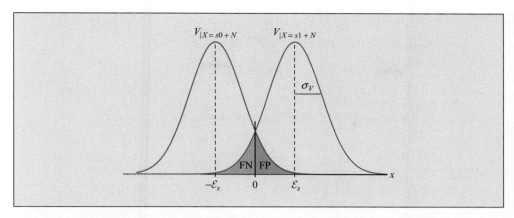

Figure 10.12

Probability of error equals the shaded areas FP (false positive) and FN (false negative) under the Gaussian curves.

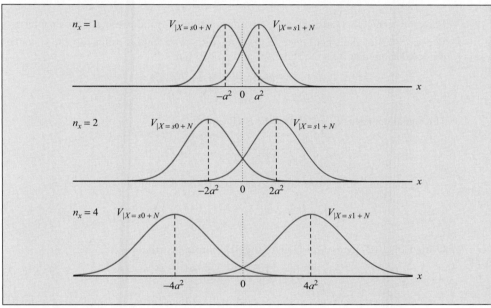

Figure 10.13

Probability of error reduction by increasing n_x.

Qualitatively, the Prob[error] equals the sum of the FP and FN areas and reducing these areas leads to a smaller Prob[error]. The careful reader would have noticed the FN and FP areas are equal when the threshold value $\tau = 0$. Shifting the τ value in the positive direction reduces FP while increasing FN.

The probability of error can be reduced either by reducing σ_V (making the bell-shaped curves narrower) or by increasing $\mathcal{E}_s$ (spreading the curves farther apart). The following example illustrates how increasing the signal duration n_x reduces the Prob[error].

Increasing n_x to reduce the probability of error   EXAMPLE 10.7

This example examines the Prob[error] qualitatively by determining the change in the Gaussian distributions as the signal duration n_x increases. The signal amplitude is set to be the largest possible value ($a = V_{\max}$) to maximize $\mathcal{E}_s$. The noise level σ_N is held constant, with σ_N also equal to a. The results are shown in Figure 10.13.

For $n_x = 1$: Let the signal encode the data with a single sample using signal $s = a$. The detected data X is the sum of the signal and noise sample N, as

$$X = a + N \quad \text{if} \quad D_t = 1$$

$$X = -a + N \quad \text{if} \quad D_t = 0$$

The signal energy is $\mathcal{E}_s = a^2$ and the SNR equals

$$\text{SNR} = \frac{\mathcal{E}_s}{\sigma_N^2} = \frac{a^2}{\sigma_N^2}$$

For $\sigma_N = a$, SNR $= 1$.

The matched processor output V has a Gaussian distribution described by

$$\mu_V = \mathcal{E}_s = a^2 \quad \text{if} \quad D_t = 1$$
$$= -\mathcal{E}_s = -a^2 \quad \text{if} \quad D_t = 0$$

The SDs of both Gaussian distributions are the same and equal to

$$\sigma_V = \sigma_N = a$$

For $n_x = 1$, the two Gaussian curves overlap significantly indicating a large Prob[error].

For $n_x = 2$: Let the signal encode the data with two samples using $s_0 = s_1 = a$. The detected sequence X_i is the sum of s_i and noise N_i, as

$$X_i = a + N_i \quad \text{for} \quad i = 0, 1 \quad \text{if} \quad D_t = 1$$
$$X_i = -a + N_i \quad \text{for} \quad i = 0, 1 \quad \text{if} \quad D_t = 0$$

The signal energy is $\mathcal{E}_s = 2a^2$ and the SNR equals

$$\text{SNR} = \frac{\mathcal{E}_s}{\sigma_N^2} = \frac{2a^2}{\sigma_N^2} = 2$$

The matched processor output V has a Gaussian distribution described by

$$\mu_V = \mathcal{E}_s = 2a^2 \quad \text{if} \quad D_t = 1$$
$$\mu_V = -\mathcal{E}_s = -2a^2 \quad \text{if} \quad D_t = 0$$

Using Eq. (10.27), $\sigma_V^2 = \mathcal{E}_s \sigma_N^2$, gives the SD of both distributions of V as

$$\sigma_V = \sqrt{\mathcal{E}_s}\, \sigma_N = \sqrt{2}\, a\, \sigma_N = \sqrt{2} a^2$$

The two Gaussian curves overlap less for $n_x = 2$ than for $n_x = 1$ indicating a smaller Prob[error].

For $n_x = 4$: Let the signal encode the data with four samples $s_0 = s_i = s_2 = s_3 = a$. The detected sequence X_i is the sum of s_i and noise N_i, as

$$X_i = a + N_i \quad \text{for} \quad 0 \leq i \leq 3 \quad \text{if} \quad D_t = 1$$
$$X_i = -a + N_i \quad \text{for} \quad 0 \leq i \leq 3 \quad \text{if} \quad D_t = 0$$

The signal energy is $\mathcal{E}_s = 4a^2$ and the SNR equals

$$\text{SNR} = \frac{\mathcal{E}_s}{\sigma_N^2} = \frac{4a^2}{\sigma_N^2} = 4$$

The matched processor output V has a Gaussian distribution described by

$$\mu_V = \mathcal{E}_s = -4a^2 \quad \text{if} \quad D_t = 0$$
$$\mu_V = \mathcal{E}_s = 4a^2 \quad \text{if} \quad D_t = 1$$

Using Eq. (10.27), $\sigma_V^2 = \mathcal{E}_s \sigma_N^2$ gives the SD of V as

$$\sigma_V = \sqrt{\mathcal{E}_s}\, \sigma_N = 2\, a\, \sigma_N = 2\, a^2$$

The two Gaussian curves overlap even less for $n_x = 4$ indicating a further reduction in the Prob[error].

Figure 10.13 shows that increasing n_x increases $\mathcal{E}_s$ (separating the two Gaussian curves) but also increases σ_V (broadening each Gaussian curve). Generalizing this result, as n_x increases, the Gaussian curves separate linearly with n_x and broaden with $\sqrt{n_x}$. Thus, increasing n_x provides a net reduction in the Prob[error]. Hence, any desired Prob[error] (>0) can be achieved by specifying a sufficiently large n_x. This explains the ability to transmit data from Mars—which requires very large n_x. The trade-off is that increasing n_x lowers the data rate— which explains why it takes so long to acquire a digital image from Mars.

The next section describes a simulation procedure that estimates the Prob[error] by varying n_x or the amplitude of s_i, or the noise variance σ_N^2.

10.6 ESTIMATING PROBABILITY OF ERROR WITH SIMULATIONS

Because noise is random, error performance is measured in terms of the *probability of error*, denoted Prob[error]. Computing Prob[error] analytically is difficult, because Gaussian functions produce area functions (integrals) that are difficult to evaluate. This section describes a simulation procedure to estimate the Prob[error] by performing multiple transmissions in the presence of noise and counting the number of errors.

While SNR is an important measure, what we really want to know is "*How many transmission errors can I expect to occur?*" Although we want no errors, such a system in practice would be too slow, because to accomplish this, the signal duration would need to be too large. An optimal system is one in which errors are rare, and these rare errors can be corrected using error-correction techniques.

Because errors are caused by random disturbances, we need to observe the system performance *statistically* by counting errors when we transmit a large number of signals. To illustrate this approach, let n_t denote the total number of transmissions and N_e (a random number) denote the number of errors that are observed. The *error probability* is estimated as

$$\text{Prob[error]} \approx \frac{N_e}{n_t} \qquad (10.42)$$

We explore this problem by describing the following components.

- A source that generates random binary data.
- The Gaussian random noise model.
- Signal design to minimize the Prob[error].
- System simulation to verify performance.

10.6.1 Modeling Random Data

A source that produces random binary data is typically modeled by using a pseudo-random number generator (PRNG) that produces random numbers that lie uniformly within the interval $[0, 1)$, which we term the *uniform PRNG*. The closed interval, which is denoted with a square bracket on the left, indicates that 0 is a value in the set that can occur, while the open interval with the parenthesis on the right indicates that 1 is not in the set, although any non-negative number < 1 is possible.

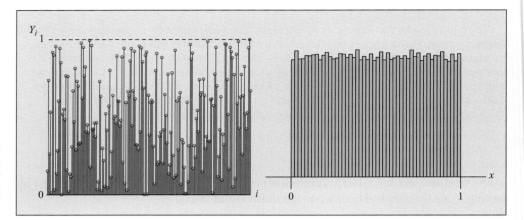

Figure 10.14

Y_i for $i = 0, 1, 2, ..., n_Y - 1$ ($n_Y = 200$) and histogram with 50 bins formed by 10^5 random numbers produced by the uniform PRNG. Each bin contains approximately 2,000 counts.

EXAMPLE 10.8

Histogram of uniform PRNG

Let Y_i denote the random numbers generated by the Uniform PRNG. We form bins that cover this interval using equally sized non-overlapping segments producing 50 bins:

$$\text{bin } 1 \rightarrow [0, 0.02)$$
$$\text{bin } 2 \rightarrow [0.02, 0.04)$$
$$\text{bin } 3 \rightarrow [0.04, 0.06)$$

and so on until

$$\text{bin } 50 \rightarrow [0.98, 1.0)$$

Note that the half-open intervals $[.\,,.)$ ensure that each Y_i value falls into only one bin and all Y_i values in $[0, 1)$ produce counts.

Fig. 10.14 shows 200 Y_i values and the histogram formed by 10^5 random numbers. When random numbers are said to be uniformly distributed over $[0, 1)$, we expect the bins to yield counts that are approximately equal for the bins that cover $[0, 1)$, and the figure verifies this. With 50 bins and 10^5 random numbers, each bin contains approximately 2,000 counts.

EXAMPLE 10.9

Simulating random data bits

To simulate random bits, we use the uniform PRNG that generates a pseudo-random number that is uniformly distributed over $[0, 1)$. We divide this interval into two equal non-overlapping parts. The convenient division into $[0, 0.5)$ and $[0.5, 1)$ is shown in Figure 10.15. The importance of dividing an interval using half-open intervals is two-fold:

- The entire interval is covered (that is, all numbers are accounted for).
- Any random number falls into only one interval.

If $0 \le Y_i < 0.5 \rightarrow D_t = 0$ and if $0.5 \le Y_i < 1 \rightarrow D_t = 1$.

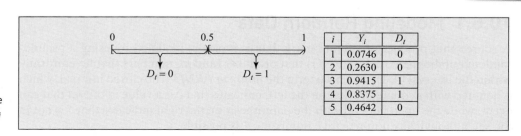

Figure 10.15

Using the uniform PRNG to simulate binary data. Each random number Y_i yields a random data bit.

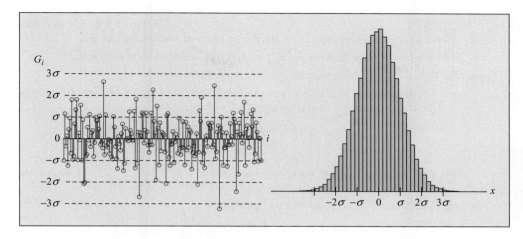

Figure 10.16

G_i for $i = 0, 1, 2, \ldots, n_G - 1$ ($n_G = 200$) and histogram with 51 bins formed by 10^5 random numbers produced by the Gaussian PRNG.

10.6.2 Modeling Random Noise

Noise is typically considered to be a *Gaussian random process* (that is, it obeys the famous *bell-shaped curve* with values that tend to be concentrated about its mean value). Noise samples are simulated with a Gaussian PRNG that produces random numbers having a Gaussian distribution. The most common is the *standardized Gaussian random number* denoted G_i, which has a zero mean ($\mu_G = 0$) and unit variance ($\sigma_G^2 = 1$).

These G_i random numbers simulate zero-mean Gaussian random noise N_i having variance σ_N^2 simply by multiplying G_i by the non-random value σ_N, as

$$N_i = \sigma_N \, G_i \qquad (10.43)$$

Hence, N_i produces a realization (a random number) that has a zero mean and variance σ_N^2.

Histogram of Gaussian random numbers · EXAMPLE 10.10

The standardized Gaussian random numbers G_i have zero mean and unit variance. Fig. 10.16 shows 200 G_i values and the histogram formed by 10^5 G_i values. The bell-shaped form of the histogram distribution is evident only when the number of random numbers is large (10^5 in this example). Note that Gaussian random numbers tend to fall close to the mean value ($=0$) and almost all fall within $[-3\sigma_G, 3\sigma_G]$ ($= [-3, 3]$).

Simulating data transmission · EXAMPLE 10.11

A simulation was performed using the following steps.

1. The basic signal sequence s_i for $i = 0, 1, 2, \ldots, n_x - 1$ was specified and its energy $\mathcal{E}_s$ computed. The signals used in the simulation include the following.

$$s = 1 \quad (n_x = 1, \quad \mathcal{E}_s = 1)$$

$$s_i = \sin(2\pi i/16) \text{ for } i = 0, 1, 2, \ldots, n_x - 1 \quad (n_x = 16, \quad \mathcal{E}_s = 8)$$

$$s_i = 1 \text{ for } i = 0, 1, 2, \ldots, n_x - 1 \quad (n_x = 16, \quad \mathcal{E}_s = 16)$$

2. Complementary signals $s1_i$ ($= s_i$) and $s0_i$ ($= -s_i$) were formed for $i = 0, 1, 2, \ldots, n_x - 1$.

3. The number of trials was specified. Each trial had a specified value for σ_N.

4. For each trial, a large number n_t (10^6) of data transmissions were performed.

5. The following were computed for each transmission:

   (a) The transmitted bit D_t (0/1) was randomly chosen using the uniform PRNG and the corresponding transmitted signal sT_i was formed as

   $$sT_i = s1_i = s_i \quad \text{for } D_t = 1$$

   $$= s0_i = -s_i \quad \text{for } D_t = 0$$

   (b) The Gaussian random noise sequence N_i for $i = 0, 1, 2, \ldots, n_x - 1$ having the specified σ_N was generated using a Gaussian PRNG and Eq. (10.43).

   (c) The received signal sequence was formed

   $$X_i = sT_i + N_i \quad \text{for } i = 0, 1, 2, \ldots, n_x - 1$$

   (d) The X_i was applied to the matched processor tuned to s_i to produce output V.

   (e) The V value was applied to the threshold detector with $\tau = 0$ to determine D_r as

   $$\text{If } V \leq 0 \rightarrow D_r = 0, \text{ otherwise } D_r = 1$$

   (f) If $D_r \neq D_t$, the error count N_e was incremented.

6. At the end of each trial, the probability of error was estimated as

$$\text{Prob[error]} = N_e/n_t$$

This simulation produced the following results.

Figure 10.17 shows Prob[error] decreases as $\mathcal{E}_s/\sigma_N^2$ increases. Note the following values from the graph.

$$\text{Prob[error]} = 0.37 \quad \text{for} \quad \mathcal{E}_s/\sigma_N^2 = 0.1 \quad \text{(the first value in the curve)}$$

and

$$\text{Prob[error]} = 0.16 \quad \text{for} \quad \mathcal{E}_s/\sigma_N^2 = 1$$

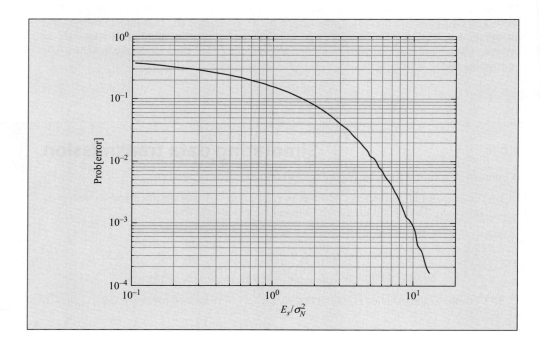

Figure 10.17

Probability of error versus $\mathcal{E}_s/\sigma_N^2$.

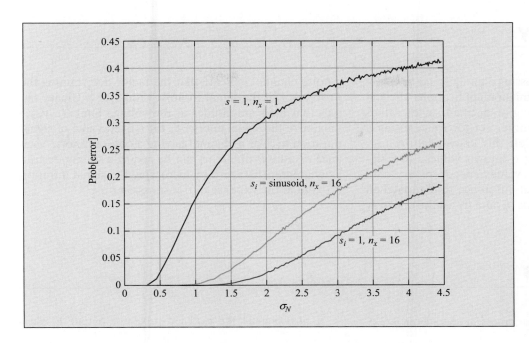

Figure 10.18

Comparing probability of error for different signal durations and types as a function of σ_N.

It does not matter whether $\mathcal{E}_s$ is increased by increasing amplitude or by increasing n_x. The factor that determines the Prob[error] is solely the SNR $= \mathcal{E}_s/\sigma_N^2$.

Figure 10.18 compares the Prob[error] versus σ_N for three signal types having the same amplitude.

1. Single sample $s = 1$ having $\mathcal{E}_s = s^2 = 1$.
2. Sinusoidal sequence $s_i = \sin(2\pi i/16)$ for $i = 0, 1, 2, \ldots, n_x - 1$ with $n_x = 16$, having $\mathcal{E}_s = n_x/2 = 8$.
3. Binary sequence $s_i = 1$ for $0 \le i \le 15$, having $\mathcal{E}_s = n_x = 16$

We note that at the same σ_N value, the Prob[error] decreases as $\mathcal{E}_s$ increases, as expected. As σ_N increases and begins to swamp the signal, the Prob[error] $\to 0.5$. This is the probability of error that is expected if we simply guess what data value was transmitted (without even detecting X_i), because we would be correct half the time.

In practice, the requirements for the error probability depend on the application. One successful system may have Prob[error] $< 10^{-3}$ or less than 1 error in 1,000 transmissions on average, while critical systems typically require Prob[error] $< 10^{-6}$. Practical data transmission systems tolerate the achievable Prob[error] by employing error-correcting codes, as discussed in Chapter 13, to correct rarely occurring errors.

10.7 Research Challenges

10.7.1 Optimum Coding Choice for Noise Tolerance

Practical considerations dictate the design of signal sequence s_i for $i = 0, 1, 2, \ldots, n_x - 1$ for a particular application. These include cost, channel characteristics, allowable energy limits, and acceptable error performance. Analysis, experimentation, and simulation are required to determine which scheme achieves the best results.

10.8 Summary

This chapter described processing data signals in the presence of noise. In data transmission applications, the signal values are known, having a specified set of values for transmitting a 0 and another set for transmitting a 1. Matched processors exploit this knowledge by tuning their coefficients to these known values. Two signals having complementary values encode binary data values to produce a maximal difference in the matched processor output. Simulations of data transmissions over a noise channel indicate that the factor determining the Prob[error] is the signal-to-noise ratio $\mathcal{E}_s/\sigma_N^2$. Binary signals that have amplitudes at the voltage limits ($\pm V_{\max}$) maximize the signal energy $\mathcal{E}_s$ for a fixed value of signal duration n_x. For a channel having a constant noise variance σ_N^2, the Prob[error] can be reduced to any desired level greater than zero by increasing the signal duration n_x—but at the expense of transmission time.

10.9 Problems

10.1 **Linear processor.** A linear processor operates on $n_x = 5$ samples with coefficients

$$c_0 = 1, c_1 = 2, c_2 = 3, c_3 = 2, c_4 = 1$$

The observed signal sequence X_i equals

$$X_0 = 2, X_1 = -3, X_2 = 1, X_3 = 0, X_4 = -1$$

Compute output V produced by the linear processor.

10.2 **Linear processor with $c_i = 0$.** A linear processor operates on $n_x = 5$ samples with coefficients

$$c_0 = 0.5, c_1 = 2, c_2 = 0, c_3 = -3, c_4 = -1$$

Rank the importance of each of the input signal sequence values X_i for $0 \le i \le 4$, as determined by the coefficient magnitudes.

10.3 **Matched processor.** Consider the signal sequence

$$s_0 = 1, s_1 = 2, s_2 = 3, s_3 = 4, s_4 = 5$$

1. Compute the signal energy $\mathcal{E}_s$.
2. Design the matched processor.
3. Compute the matched processor output V when $X_i = s_i$ for $0 \le i \le 4$.
4. Specify a different signal sequence s_i' for $0 \le i \le 4$ having the same $\mathcal{E}_s$.
5. Compute the matched processor output V' when $X_i = s_i'$.

10.4 **Designing a maximum energy signal.** A transmission channel is constrained by allowing signals that have magnitudes $|s_i| \le 2$ V.

1. Design a valid signal sequence s_i for $0 \le i \le 4$ that has the maximum $\mathcal{E}_s$.
2. Compute the signal energy $\mathcal{E}_s$.
3. Design the matched processor.

4. Compute the matched processor output V when $X_i = s_i$.

10.5 **Complementary signals.** Let $s_i = i + 1$ for $0 \le i \le 4$.

1. Form the complementary pair $s1_i$ and $s0_i$. Compute the matched filter output $\mathcal{V}$ when each signal is observed.
2. Specify a different signal sequence s_i' for $0 \le i \le 4$ having the same $\mathcal{E}_s$.
3. Compute the matched processor output V' when $X_i = s_i'$.

10.6 **Sinusoidal signal energy.** Compute the energy of the sinusoidal signal

$$s_i = 2\sin(2\pi i/8) \text{ for } 0 \le i \le 31$$

10.7 **Determining σ_N^2 from observing a time sequence.** You observe a sequence containing a large number of Gaussian noise samples and notice that almost all samples fall within the range $[-120, 120]$ mV. What are reasonable values for σ_N and σ_N^2?

10.8 **Signal-to-noise ratio of sinusoidal signal.** Let the sinusoidal signal be given by

$$s_i = 2\sin(2\pi i/8) \text{ for } 0 \le i \le 31$$

The detected signal is

$$X_i = s_i + N_i \text{ for } 0 \le i \le 31$$

where N_i is Gaussian noise having $\sigma_N = 0.1$. What are the values of the signal-to-noise ratio and SNR_{dB}?

10.9 **Simulating random bits.** If Y_i is a random number produced by the uniform PRNG and *floor* is a function that rounds a number value down to its integer value, explain why $D = floor(2 * Y_i)$ produces values 0 and 1 that are equally likely.

10.10 Matlab Projects

10.1 General linear processor. Extend Example 16.47 to implement a general linear processor having a specified coefficient sequence. Verify using the values

$$c_i = i + 1 \quad \text{for } 0 \le i \le 4$$

Enter the matched input sequence and compute the output V.

10.2 Sinusoidal signal energy. Extent Example 16.48 to verify that the energy of a sinusoid sequence having amplitude A, duration n_x, and frequency $f_k = k/n_x$ equals

$$\mathcal{E}_s = \frac{n_x A^2}{2}$$

Over what values of k is this formula valid? (Hint: Why is the signal energy equal to zero for $k = n_x/2$?.)

10.3 Complementary signals. Extend Example 16.49 to implement a matched processor having coefficient set

$$c_i = i + 1 \quad \text{for } 0 \le i \le 4$$

Generate the corresponding pair of complementary signals and compute outputs $V|_{X=s1}$ and $V|_{X=s0}$.

10.4 Gaussian noise having specified σ_N^2. Modify Example 16.50 to generate Gaussian noise N_i for $0 \le i \le 99$ having $\sigma_N^2 = 16$. Compute the sample variance of the generated sequence.

10.5 Histogram of Gaussian noise. Modify Example 16.51 to generate Gaussian noise N_i for $i = 0, 1, 2, \ldots, n_x - 1$ having $\sigma_N^2 = 16$ and plot the 50-bin histograms for $n_x = 100, 10^4,$ and 10^6.

10.6 Sinusoidal signal with Gaussian noise. Modify Example 16.52 to generate and display Gaussian noise N_i for $i = 0, 1, 2, \ldots, n_x - 1$ having $\sigma_N^2 = 16$ that adds to a two-period sinusoid of duration n_x, for $n_x = 10$ and 100. Compute and display the SNR $\mathcal{E}_s/\sigma_N^2$ on the plot.

10.7 Matched processor output when signals contain noise. Modify Example 16.53 by applying to a matched processor your favorite signal s_i for $0 \le i \le 9$ that is corrupted by Gaussian noise having $\sigma_N^2 = 4$.

Compare the results when the signal duration is increased to $2n_x$ by simply repeating the signal twice and adding $2n_x$ Gaussian noise samples having $\sigma_N^2 = 4$.

10.8 Generating random bits. Use Matlab's PRNG *rand()* to generate random binary data (1/0) that can be used for simulating random bits in a communication system.

10.9 Estimating probability of error of a sinusoidal signal. Modify Example 16.55 by specifying a 2-period sinusoidal signal with $n_x = 16$ and amplitude 2. Compute and plot the probability of error as σ_N^2 varies between 0 and $2\mathcal{E}_s$, in $0.1\mathcal{E}_s$ steps.

10.10 Estimating probability of error for a maximum-energy signal. Modify Example 16.55 to specify a maximum-energy signal having maximum amplitude 2 and $n_x = 16$. Compute and plot the probability of error as σ_N^2 varies between 0 and $2\mathcal{E}_s$, in $0.1\mathcal{E}_s$ steps.

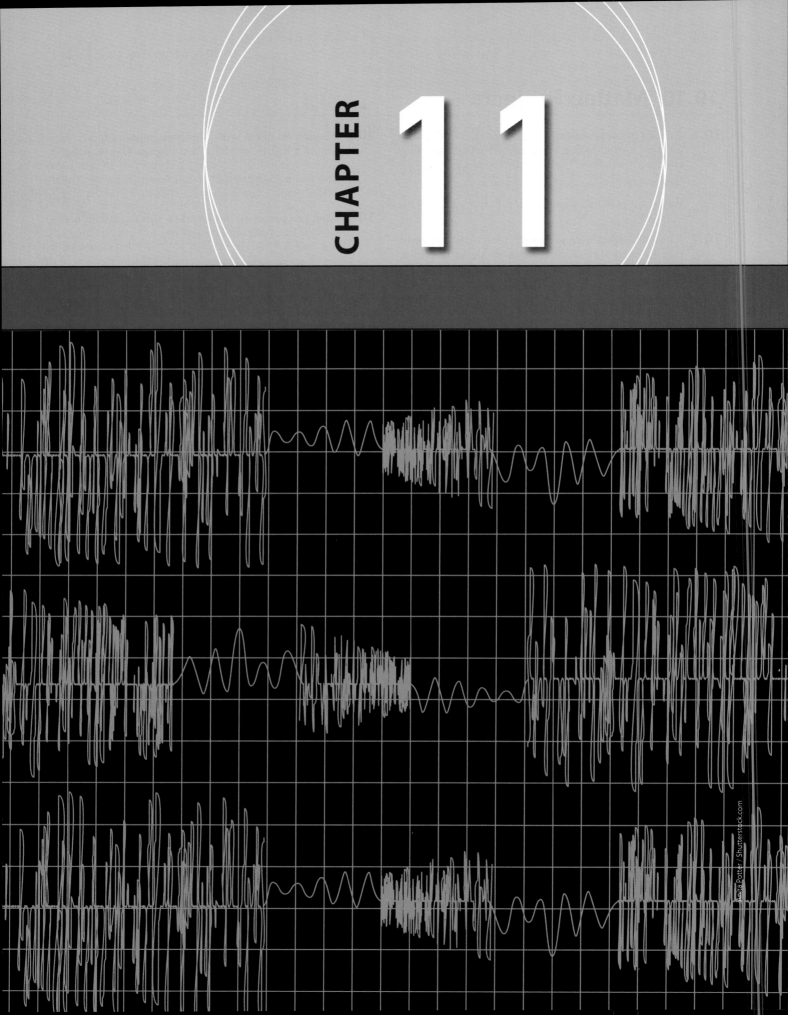

DESIGNING SIGNALS FOR MULTIPLE-ACCESS SYSTEMS

The previous chapter described transmitting a single data bit over a noisy channel by encoding the bit value using a pair of complementary signals. Some data transmission applications need to service multiple users over the same channel simultaneously. An analogous situation arises when one user wants to transmit multiple data bits at one signal transmission time. This chapter considers multiple orthogonal signals that can be transmitted simultaneously over a channel and decoded by separate receivers using matched processors. The concept of orthogonality is described in the time, frequency, and code domains.

11.1 INTRODUCTION

The main topics in this chapter are covered in the following sequence.

Multiple simultaneous user systems—Typical data communication systems serve more than one user simultaneously. Wi-Fi systems connect multiple smartphones and laptops to the Internet, and multiple users talk over a cell-phone network simultaneously. This multi-user system also allows a single user to transmit multiple bits simultaneously.

Orthogonal data signals—The conventional processor that operates on received signals forms the sum of weighted signal values. This processing suggests the *orthogonality condition* that prevents signals having the same energy from interfering with each other. Orthogonal signal pairs then transmit the binary data produced by multiple users or multiple bits produced by one user over a channel simultaneously.

Orthogonal signal types—Signals can be orthogonal in time, in frequency, and in code word. Each system has its advantages and disadvantages with each application choosing the type that meets its specifications.

Processing orthogonal signals in noise—When a matched processor operates on orthogonal signals that are corrupted with noise, the reliability of detecting the correct (transmitted) signal value is determined solely by the signal-to-noise ratio (SNR).

Probability of error—A simulation procedure is described to estimate the probability of error by performing multiple transmissions in the presence of noise and counting the number of errors.

11.2 MULTIPLE SIMULTANEOUS USER SYSTEMS

Some digital communication systems need to transmit more than one bit value per signal or to service multiple users, allowing each to transmit and receive data or to *access* the system simultaneously. This additional consideration leads to the need for orthogonal signals, which can be transmitted over a data channel at the same time without interfering with each other. Figure 11.1 shows the extension from a single signal pair to multiple orthogonal signal pairs described in this chapter. Orthogonal signals permit two or more users to communicate over the same channel simultaneously or to allow one user to send multiple data bits in one signal transmission time.

To illustrate the multi-user problem, imagine you are at a party where there is loud music playing and other people are talking while you are trying to carry on a conversation with a friend. There are features in your friend's utterances that allow you to separate his/her voice from the other sounds, even if the other sounds are louder. A similar method for extracting desired signals is being used by another pair of friends in the same party for their conversation. The same approach is employed in simultaneous

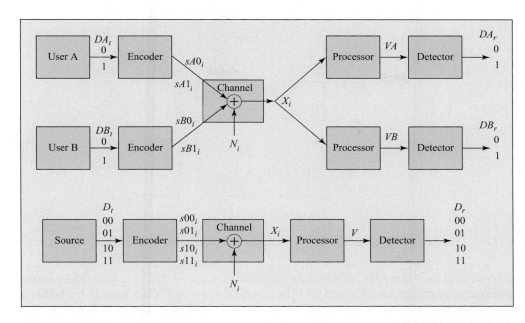

Figure 11.1

Orthogonal signals permit two or more users to communicate over the same channel simultaneously or allow one user to send multiple data bits in one signal transmission time.

digital communications between pairs of users (for example, your cell phone and the local transmission tower). A typical cell contains many phones that are communicating with the tower at any one time. Signals having features that allow them to be extracted in the presence of other signals are called *orthogonal*.

 11.3 ORTHOGONALITY CONDITION

To illustrate orthogonal signals, we consider the signal transmitted by a single user (user A) who designs data sequence sA_i for $i = 0, 1, 2, \ldots, n_x - 1$ to form complementary signals

$$sA0_i = -sA_i \text{ if } DA_t = 0 \quad \text{and} \quad sA1_i = sA_i \text{ if } DA_t = 1 \quad \text{(11.1)}$$

When transmitted over a channel that introduces additive noise N_i for $i = 0, 1, 2, \ldots, n_x - 1$, the signal detected at A's receiver is

$$X_i = sA0_i + N_i = -sA_i + N_i \quad \text{if} \quad sA0_i \text{ is transmitted} \quad \text{(11.2)}$$

$$X_i = sA1_i + N_i = sA_i + N_i \quad \text{if} \quad sA1_i \text{ is transmitted}$$

Under ideal (noiseless) conditions, the received signal X_i is simply one of the transmitted signals. This chapter assumes that the amplitudes do not decrease as the signals pass through the channel. The matched processor tuned to sA_i has coefficients matched to $c_i = sA_i$. When $X_i = sA0_i$ for $i = 0, 1, 2, \ldots, n_x - 1$, A's matched processor produces output as

$$VA_{|X=sA0} = \sum_{i=0}^{n_x-1} c_i X_i \quad \text{(11.3)}$$

$$= \sum_{i=0}^{n_x-1} sA_i \overbrace{X_i}^{=sA0_i=-sA_i}$$

$$= -\sum_{i=0}^{n_x-1} (sA_i)^2 = -\mathcal{E}_{sA}$$

Similarly, when $X_i = sA1_i$, $VA_{|X=sA1} = \mathcal{E}_{sA}$.

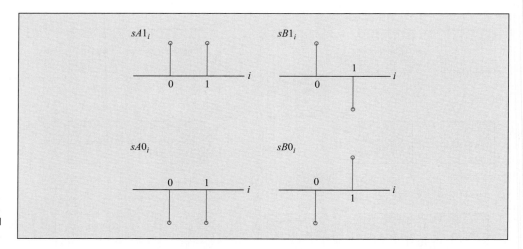

Figure 11.2

Simple orthogonal signal pair. All signal magnitudes are equal to one.

A second user (user B) designs data signal sB_i for $i = 0, 1, 2, \ldots, n_x - 1$ and forms complementary signals $sB0_i$ and $sB1_i$ in a similar manner. When user B transmits a signal, say $sB1_i = sB_i$, over a channel shared by A and B, the output of A's processor tuned to sA_i equals

$$VA_{|X=sB1} = \sum_{i=0}^{n_x-1} sA_i \overbrace{X_i}^{=sB1_i} = \sum_{i=0}^{n_x-1} sA_i \, sB_i \tag{11.4}$$

To counter the large magnitudes $VA = \pm\mathcal{E}_{sA}$ when user A's signals occur, we want VA to be small when user B's signals are detected. We can do this by choosing sB_i such that

$$\sum_{i=0}^{n_x-1} sA_i \, sB_i = 0 \tag{11.5}$$

This is the *orthogonality condition* relating sA_i and sB_i. To prevent the trivial result $sB_i = 0$, all signals are required to have the same energy $\mathcal{E}_s$.

When sA_i and sB_i are orthogonal, then clearly so are $sA1_i$ and $sB1_i$, as well as $sA0_i$ and $sB0_i$, as the following example illustrates.

EXAMPLE 11.1 **Simple orthogonal signals**

Figure 11.2 shows orthogonal signals using two samples ($n_x = 2$) for users A and B with

$$sA_0 = 1 \quad \text{and} \quad sA_1 = 1$$

and

$$sB_0 = 1 \quad \text{and} \quad sB_1 = -1$$

The energy in each signal is

$$\mathcal{E}_{sA} = \sum_{i=0}^{1} sA_i^2 = 1 + 1 = 2$$

$$\mathcal{E}_{sB} = \sum_{i=0}^{1} sB_i^2 = 1 + (-1)^2 = 2$$

Hence, both signals contain the same energy.

Let us verify that sA_i and sB_i are orthogonal signals. The orthogonality condition requires that

$$\sum_{i=0}^{n_x-1} sA_i\, sB_i = 0$$

Inserting the values of $n_x = 2$ and these signals, we get

$$\sum_{i=0}^{1} sA_i\, sB_i = (sA_0\, sB_0) + (sA_1\, sB_1) = (1 \times 1) + (1 \times -1) = 0$$

Hence, sA_i and sB_i are orthogonal.

We now examine the values produced by user A's matched processor. The VA processor output is given by

$$VA = \sum_{i=0}^{n_x-1} sA_i\, X_i = \overbrace{sA_0}^{=1}\, X_0 + \overbrace{sA_1}^{=1}\, X_1 = X_0 + X_1$$

When $X_i = sA1_i = sA_i$,

$$VA_{|X=sA1} = sA_0 + sA_1 = 1 + 1 = 2 \quad (= \mathcal{E}_{sA})$$

When $X_i = sA0_i = -sA_i$,

$$VA_{|X=sA0} = -sA_0 + -sA_1 = -1 - 1 = -2 \quad (= -\mathcal{E}_{sA})$$

When $X_i = sB1_i = sB_i$,

$$VA_{|X=sB1} = sB_0 + sB_1 = 1 + (-1) = 0$$

Similarly, $VA_{|X=sB0} = 0$.

We now examine the values produced by user B's matched processor. The VB processor output is given by

$$VB = \sum_{i=0}^{1} sB_i\, X_i = (\overbrace{1}^{=sB_0} \times X_0) + (\overbrace{-1}^{=sB_1} \times X_1) = X_0 - X_1$$

When $X_i = sA1_i = sA_i$,

$$VB_{|X=sA1} = sA_0 - sA_1 = 1 - 1 = 0$$

Similarly, $VB_{|X=sA0} = 0$.

When $X_i = sB1_i = sB_i$,

$$VB_{|X=sB1} = sB_0 - sB_1 = 1 - (-1) = 2$$

Similarly, $VB_{|X=sB0} = -2$.

Hence, each processor identifies its own signal and ignores the other.

Single-User, Multi-Bit Transmission

The previous discussion considered two users, A and B, transmitting their separate binary data over a single channel. When there is only one user transmitting data over a channel, orthogonal sequences can form complementary signals to transmit *two data bits during one transmission time*. For example, user A designs orthogonal signals $sA0_i$ and $sA1_i$ and forms their complementary pairs. User A then combines two data bits into a *two-bit* packet (called a dibit), DA_t, and transmits each packet using the assignment:

$$sA00_i = -sA0_i \quad \text{if} \quad DA_t = 00 \tag{11.6}$$

$$sA01_i = sA0_i \quad\ \ \text{if} \quad DA_t = 01$$

$$sA10_i = -sA1_i \quad \text{if} \quad DA_t = 10$$

$$sA11_i = sA1_i \quad\ \ \text{if} \quad DA_t = 11$$

Clearly, additional orthogonal signals can service additional users or can transmit additional bits simultaneously over a channel. Current data transmission systems do this.

Practical Orthogonal Signals

What makes this problem interesting and challenging is that, unlike complementary signals that form a unique and obvious pair, there are many ways to form orthogonal signals sA_i and sB_i. Consider a channel time interval

$$T_c = n_x T_s \tag{11.7}$$

over which the data sequences occur. There are three common approaches for forming orthogonal signals with each having its own merits.

Orthogonal in time—in *time-division multiple access (TDMA)* systems, the signals sA_i and sB_i occur during specific, non-overlapping, sub-intervals in T_c. At times other than the specified time slot, a user's signals are zero-valued in order to not interfere with signals of other users.

Orthogonal in frequency—in *frequency-division multiple access (FDMA)* systems, the signals sA_i and sB_i occur during the entire interval T_c but occupy separate frequency bands.

Orthogonal in code—in *code-division multiple access (CDMA)* systems, the signals sA_i and sB_i are *binary*, taking on values equal to either $-V_{max}$ or $+V_{max}$, and occupy the entire interval T_c. CDMA signals have codes that may not be *exactly* orthogonal and that still can be *adequately differentiated* through their *approximately* orthogonal properties.

Theoretically, each technique has the same effectiveness, but each also has practical advantages and disadvantages. The problems and trade-offs encountered are specific to a given application and ultimately determine the appropriate method. We illustrate each technique using a pair of users, A and B, but the methods can be extended to more users with the maximum number of users equal to the maximum number of available orthogonal signals that still operate with sufficient probability of error in noisy channels.

 ## 11.4 SIGNALS ORTHOGONAL IN TIME—TDMA

Signals that are orthogonal in time occupy non-overlapping time slots in the channel data interval $0 \leq i \leq n_x - 1$ using a technique commonly called *time-division multiplexing (TDMA)*. If there are n_u users, each has a slot duration

$$n_d = \frac{n_x}{n_u} \tag{11.8}$$

Typically, n_d is an integer value by choosing n_x to accommodate n_u users. For example, to serve three users, choose $n_x = 15$ to make $n_d = 5$; to serve four, choose $n_x = 16$ to make $n_d = 4$. The system also assigns each user to a specific time slot.

Both signals sA_i and sB_i can have any form—even identical forms—because they occur at different time intervals. Figure 11.3 shows two different signal forms to differentiate the two more clearly.

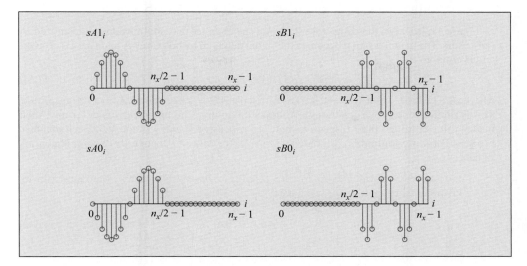

Figure 11.3

TDMA signals.

TDMA signals   EXAMPLE 11.2

Figure 11.3 shows two users, A and B, using signals sA_i and sB_i respectively, that simultaneously access the same channel for transferring data. Each user forms two signals that occupy non-overlapping time intervals within the n_x data interval divided into two sub-intervals. The figure shows signals having duration $n_x = 32$, making

$$n_d = \frac{n_x}{n_u} = \frac{32}{2} = 16$$

User A occupies the first half of the interval with signal sA_i for $0 \le i \le n_d - 1$. Although user A can use any signal, Figure 11.3 shows a single sinusoidal period

$$sA_i = \sin(2\pi f_A i) \text{ for } 0 \le i \le n_d - 1$$
$$= 0 \text{ for } n_d \le i \le n_x - 1$$

with $f_A = 1/n_d$ to form a single period. User A uses $sA0_i = -sA_i$ for transmitting data $DA_t = 0$ and its complementary signal $sA1_i = sA_i$ for transmitting $DA_t = 1$.

User B uses sB_i equal to two periods of a sinusoid during the second half of the interval, as

$$sB_i = 0 \text{ for } 0 \le i \le n_d - 1$$
$$= \sin(2\pi f_B i) \text{ for } n_d \le i \le n_x - 1$$

with $f_B = 2/n_d$ to form two periods. User B uses $sB0_i = -sB_i$ for transmitting data $DB_t = 0$ and its complementary signal $sB1_i = sB_i$ for transmitting $DB_t = 1$.

The energy of the two signals is computed by summing the squares of the non-zero components, producing

$$\mathcal{E}_{sA} = \sum_{i=0}^{n_x-1} sA_i^2 = \sum_{i=0}^{n_d-1} \sin^2(2\pi i/n_d) \quad (= 8 \text{ when } n_d = 16) \tag{11.9}$$

$$\mathcal{E}_{sB} = \sum_{i=0}^{n_x-1} sB_i^2 = \sum_{i=n_d}^{n_x-1} \sin^2(4\pi i/n_d) \quad (= 8 \text{ when } n_d = 16) \tag{11.10}$$

Figure 11.4a shows the signal sequence user A uses to transmit $DA_t = 1$ followed by $DA_t = 0$, and Figure 11.4b shows user B's signals to transmit a 1 followed by a 0. Each n_x interval contains the signals from both users. User A has been assigned the first half of each n_x data interval, and user B has been assigned the second half.

Figure 11.4c shows the composite signal on the channel that contains data transmitted by both users. The channel signal becomes the signal detected by both user A and user B receivers and is equal to

$$X_i = sAT_i + sBT_i \text{ for } i = 0, 1, 2, \ldots, n_x - 1 \tag{11.11}$$

where $sAT_i = sA0_i$ when $DA_t = 0$ is transmitted or $sAT_i = sA1_i$ when $DA_t = 1$ is transmitted. This is similar for sBT_i. Even though X_i looks unfamiliar, the two receivers determine their binary values through their respective matched processor values VA and VB. Each matched processor produces outputs $\mathcal{E}_s$ (for data $= 1$) or $-\mathcal{E}_s$ (for data $= 0$) to identify its complementary signals.

VA: Processor evaluates X_i for $i = 0, 1, 2, \ldots, n_x - 1$. For each n_x time interval with VA processor tuned to sA_i, we have

$$VA = \sum_{i=0}^{n_x-1} sA_i \, X_i \tag{11.12}$$

When $X_i = sA1_i$,

$$VA_{|X=sA1} = \sum_{i=0}^{n_x-1} sA_i^2 = \mathcal{E}_s \tag{11.13}$$

When $X_i = sBT_i$ (either $sB1_i$ or $sB0_i$), the orthogonality condition produces

$$VA_{|X=sBT} = \sum_{i=0}^{n_x-1} sA_i \, sBT_i = 0 \tag{11.14}$$

VB: Processor evaluates X_i for $i = 0, 1, 2, \ldots, n_x - 1$ and for each n_x time interval assuming $X_i = sB1_i (= sB_i)$. When $X_i = sB1_i$,

$$VB_{|X=sB1} = \sum_{i=0}^{n_x-1} sB_i^2 = \mathcal{E}_s \tag{11.15}$$

When $X_i = sA1_i$, the orthogonality condition produces

$$VB_{|X=sA} = \sum_{i=0}^{n_x-1} sB_i \, sA1_i = 0 \tag{11.16}$$

Figure 11.5 shows the outputs of the VA and VB processors with X_i equal to the composite channel signal after each of the two n_x data sequences are detected. A large positive processor value indicates that the signal corresponds to the processor's tuned coefficients. In the case shown for sinusoidal signals $\mathcal{E}_s = n_d/2 = n_x/4$.

The VA processor detects the $sA1_i$ TDMA signal in the first $n_x = 32$ data interval by producing a large-magnitude output ($= n_x/4 = 8$) at the end of the interval when the complete signal is within the processor. In user A's TDMA matched processor having 32 coefficients, half the coefficients are zero in order to ignore user B's TDMA signals. The composite channel signal also contains $sB0_i$ in the first n_x interval, and the VB matched processor detects it by producing a large-magnitude negative value at the end of the interval. The second n_x interval contains $sA1_i$ and $sB1_i$. Both the VA and VB matched processors detect their respective TDMA signals by producing large output magnitudes at the end of the second interval.

Maximum Number of TDMA Users

The number of users in a TDMA system is limited by the number of orthogonal signals available. Let T_c, or equivalently n_x, be the channel time interval that contains the signal sequences for data transmission that is subdivided among the users. The number of users n_u employing strictly TDMA techniques is limited by the sample time T_s. Thus,

$$n_{u,\max} \le \frac{T_c}{T_s} = \frac{n_x \, T_s}{T_s} = n_x \tag{11.17}$$

This number can be increased by increasing n_x, but this slows down the entire system.

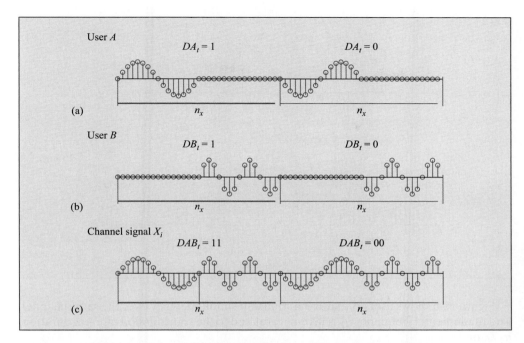

Figure 11.4

TDMA signals. (a) User A transmitting data $DA_t = 1$ followed by $DA_t = 0$. (b) User B transmitting $DB_t = 1$ then $DB_t = 0$. (c) X_i is sum of signals observed on channel.

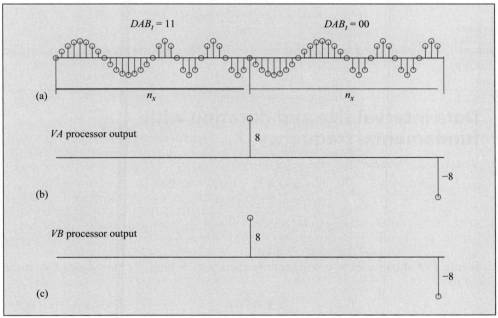

Figure 11.5

Processing TDMA signals. (a) The composite channel signal. (b) The VA matched-processor outputs. (c) The VB matched-processor outputs. The processor outputs are shown after each of the two n_x data sequences are detected and processed.

 ## 11.5 SIGNALS ORTHOGONAL IN FREQUENCY—FDMA

In FDMA systems, signals with predetermined frequencies are assigned to specific users. For a practical system, three considerations occur.

1. Each frequency must produce unique signals, which may present a problem in sampled-data systems.

2. To be orthogonal, the frequencies need to be harmonically related.

3. The code length n_x must encode a complete period of the fundamental frequency.

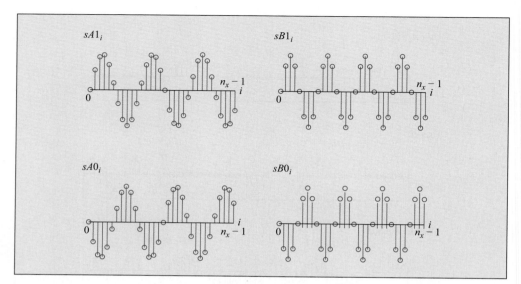

Figure 11.6

FDMA signals with time interval $n_x (= 32)$ and using sampling period T_s. Orthogonal signals for user A is at the third harmonic and for user B at the fourth harmonic.

Figure 11.6 shows users A and B that transmit simultaneously over the same channel. The fundamental frequency f_1 is inversely related to the transmission interval duration, $n_x T_s$, or

$$f_1 = \frac{1}{n_x T_s}$$

where T_s is the sampling period, making $T_c = n_x T_s$ the channel bit signal transmission interval.

EXAMPLE 11.3 Data interval size and duration with fundamental frequency

With $T_s = 10^{-5}$ s and $n_x = 32$, the signal interval is

$$T_c = n_x T_s = 32 \times 10^{-5}\,\text{s} = 0.32\,\text{ms}$$

$$f_1 = \frac{1}{n_x T_s} = \frac{1}{32 \times 10^{-5}\,\text{s}} = 3,125\,\text{Hz}$$

Figure 11.6 shows user A is assigned frequency f_A, which is a harmonic of f_1, to be equal to

$$f_A = 3 f_1 \tag{11.18}$$

and is used to generate

$$sA_i = \sin(2\pi f_A i T_s) \quad \text{for} \quad i = 0, 1, 2, \ldots, n_x - 1$$

User A uses $sA0_i = -sA_i$ for transmitting data $DA_t = 0$ and its complementary signal $sA1_i = sA_i$ for transmitting $DA_t = 1$.

User B is assigned frequency f_B, which is another harmonic of f_1, to be equal to

$$f_B = 4 f_1 \tag{11.19}$$

and is used to generate

$$sB_i = \sin(2\pi f_B i T_s) \quad \text{for} \quad i = 0, 1, 2, \ldots, n_x - 1$$

User B uses $sB0_i = -sB_i$ for transmitting data $DB_t = 0$ and its complementary signal $sB1_i = sB_i$ for transmitting $DB_t = 1$.

When $sA0_i$ is evaluated by the matched processor tuned to sA_i, its output equals

$$VA_{|X=sA0} = \sum_{i=0}^{n_x-1} sA_i \overbrace{X_i}^{=sA0_i} = -\sum_{i=0}^{n_x-1} sA_i^2 = -n_x/2 \qquad (11.20)$$

For example, for $sA0_i = -\sin(6\pi i f_1 T_s)$ for $i = 0, 1, 2, \ldots, n_x - 1$ ($n_x = 32$), we have

$$VA_{|X=sA0} = -\sum_{i=0}^{15} \sin^2(6\pi i f_1 T_s) = -16 \qquad (11.21)$$

When $X_i = sA1_i$ is processed, the VA output equals

$$VA_{|X=sA1} = \sum_{i=0}^{n_x-1} sA_i \overbrace{X_i}^{=sA1_i} = \sum_{i=0}^{n_x-1} sA_i^2 = 16 \qquad (11.22)$$

Similarly, user B uses the same $n_x = 32$ interval but is assigned frequency $f_B = 4f_1$. Thus,

$$sB1_i = \sin(2\pi i f_B T_s) = \sin(8\pi i f_1 T_s) \text{ for } i = 0, 1, 2, \ldots, n_x - 1 \qquad (11.23)$$

transmits a 1 data value and complementary signal $sB0_i$ transmits a 0. Similar matched processor results occur for user B's signals.

To show sA_i and sB_i are orthogonal, we compute the output of the matched processor tuned to sA_i when $X_i = sB1_i (= sB_i)$. Thus,

$$VA_{|X=sB1} = \sum_{i=0}^{n_x-1} sA_i \overbrace{X_i}^{=sB1_i} = \sum_{i=0}^{n_x-1} sA_i \, sB_i = 0 \qquad (11.24)$$

because harmonically related sinusoidal sequences are orthogonal.

We can examine the system operation when both sources transmit their signals over the channel simultaneously. The signal is detected by both matched processors that belong to users A and B. The channel signal is then given by

$$X_i = sAT_i + sBT_i \text{ for } i = 0, 1, 2, \ldots, n_x - 1 \qquad (11.25)$$

where $sAT_i = sA0_i$ when $DA_t = 0$ is transmitted, or $sAT_i = sA1_i$ when $DA_t = 1$ is transmitted. This is similar for sBT_i. Figure 11.7 shows the signals that are produced

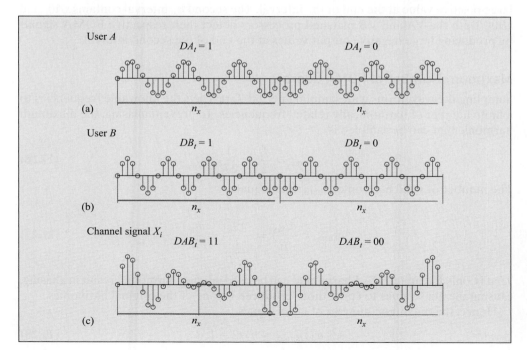

Figure 11.7

FDMA signals from two users A and B and the resulting channel signal. (a) User A transmits $DA_t = 10$. (b) User B transmits $DB_t = 10$. The channel data interval $n_x = 32$. (c) The sum of signals observed on channel ($= X_i$).

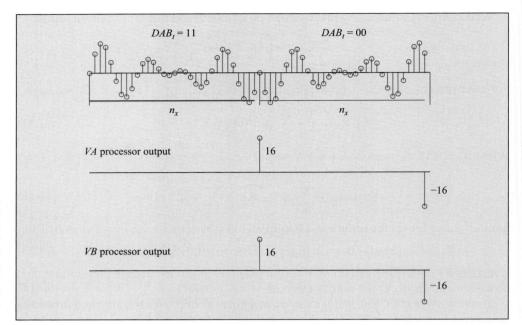

Figure 11.8

Processing FDMA signals. The outputs of the processors with $n_x = 32$ with X_i equal to the summed channel signal (top signal). The processor outputs are shown after each of the two n_x data sequences are detected. A positive value, typically $n_x/2$, indicates that the signal corresponding to that expected by the processor has been observed.

separately by user A producing two-bit data $DA_t = 01$ and user B producing $DB_t = 01$. The resulting composite signal that appears on the channel is shown. Even though X_i looks unfamiliar, the matched processors can determine their transmitted binary values.

Figure 11.8 shows the outputs of the two matched processors with X_i equal to the composite channel signal after each of the two $n_x = 32$ data sequences are detected. A large positive matched-processor output value ($V = n_x/2 = 16$) indicates the presence of the signal corresponding to the processor's expected value.

The VA processor detects the $sA0_i$ FDMA signal in the first $n_x = 32$ data interval by producing a large positive output ($= n_x/2 = -16$) at the end of the interval (when the complete signal is within the processor). The composite channel signal also contains $sB1_i$ in the first n_x interval, and the VB matched processor detects it by producing a large positive value at the end of the interval. The second n_x interval contains $sA0_i$ and $sB0_i$. Both the VA and VB matched processors detect their respective FDMA signals by producing large negative output values at the end of the second interval.

Maximum Number of FDMA Users

Sampling the waveforms with sampling period T_s restricts the allowable frequencies to a finite number of harmonically related frequencies. To prevent aliasing, the maximum harmonic that can be sampled is

$$f_{max} < \frac{f_s}{2} = \frac{1}{2T_s} \tag{11.26}$$

The number of valid harmonics n_{harm} then equals

$$n_{harm} < \frac{f_{max}}{f_1} = \frac{\frac{1}{2T_s}}{\frac{1}{n_x T_s}} = \frac{n_x}{2} \tag{11.27}$$

That is, only the first $n_x/2 - 1$ harmonics can be used. Higher harmonics result in aliasing, causing the alias values to equal those produced by one of the original harmonics.

Hence, the maximum number of users equals

$$n_{users,max} \leq n_x/2 - 1 \tag{11.28}$$

Valid and invalid harmonics   EXAMPLE 11.4

Figure 11.9 illustrates the problem of aliasing that limits the number of harmonics that can be used in sampled-data systems. Each sequence contains $n_x = 16$ points.

With $T_s = 10^{-5}$ s, the interval duration is

$$n_x T_s = 16 \times 10^{-5}\,\text{s} = 1.6 \times 10^{-4}\,\text{s}$$

The fundamental frequency equals

$$f_1 = \frac{1}{n_x T_s} = \frac{1}{1.6 \times 10^{-4}} = 6,250\,\text{Hz}$$

The sampling frequency equals

$$f_s = \frac{1}{T_s} = \frac{1}{10^{-5}\,\text{s}} = 10^5\,\text{Hz}$$

The maximum frequency that limits the harmonic frequency is

$$f_{\max} < \frac{f_s}{2} = 50,000\,\text{Hz}$$

The number of allowed harmonics equals

$$n_{\text{harm}} < \frac{f_{\max}}{f_1} = \frac{50,000}{6,250} = 8 \quad \left(= \frac{n_x}{2}\right)$$

Hence, only the first seven harmonics

$$f_k = k f_1, \quad \text{for } 1 \le k \le \left(\frac{n_x}{2}\right) - 1$$

can be used.

For example, consider three frequencies: the fundamental frequency, the largest valid harmonic, and a higher non-valid harmonic for $\sin(2\pi f i T_s)$ for $i = 0, 1, 2, \ldots, n_x - 1$ ($n_x = 16$). The top curve shows the fundamental frequency

$$f_1 = \frac{1}{16 T_s} = 6,250\,\text{Hz}$$

The second curve shows the highest valid harmonic $f_k = k f_1$ for $k = (n_x/2) - 1 = 7$ or

$$f_7 = 7 f_1 = \frac{7}{16 T_s}$$

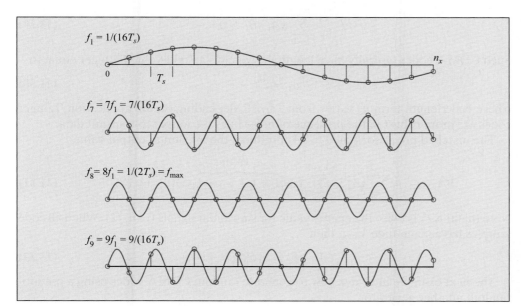

Figure 11.9

Valid and invalid harmonics for FDMA. Valid harmonics are $f_k = k f_1$ for $1 \le k \le (n_x/2) - 1$.

The third curve shows that the $k = 8$ harmonic, $f_8 = 1/(2T_s)$, is not valid because it produces only zero values. Hence, $\mathcal{E}_s = 0$.

The fourth curve shows a higher non-valid harmonic

$$f_9 = 9f_1 = \frac{9}{16T_s}$$

Note that these values are complements of those already used for f_7. Hence, f_9 is not a valid harmonic.

11.6 SIGNALS WITH ORTHOGONAL CODES—CDMA

A third common method of transmitting multiple data signals simultaneously over the same channel is to assign a unique *code* to each signal in a way that each signal can be recognized in the presence of the others, which is called *code division multiple assess (CDMA)*.

The CDMA approach has two features that distinguish it from TDMA and FDMA systems.

- CDMA signals are binary valued and consist solely of the channel limiting values $-V_{\max}$ and $+V_{\max}$. Chapter 10 showed that such binary signals maximize signal energy $\mathcal{E}_s$.

- CDMA signals are formed to be approximately orthogonal (that is, they are sufficiently orthogonal to result in a practical system). The matched processor tuned to a CDMA signal produces a large output when that signal is received and an output that is sufficiently small when any other CDMA signal is received.

Consider user A assigned code sA_i for $i = 0, 1, 2, \ldots, n_x - 1$ and user B assigned code sB_i for $i = 0, 1, 2, \ldots, n_x - 1$, with all code values being either -1 or $+1$ for illustration ($V_{\max} = 1$). To find an orthogonal CDMA code pair, start with a randomly generated sequence sA_i. Then search for a second randomly generated sequence sB_i that is approximately orthogonal to sA_i, as

$$\sum_{i=0}^{n_x-1} sA_i \, sB_i \approx 0 \tag{11.29}$$

Such CDMA codes typically have length n_x, which is an odd-valued integer equal to

$$n_x = 2^{n_c} - 1 \tag{11.30}$$

where code length term n_c varies from 2 to 15, depending on the application. Longer codes are more robust in the presence of noise but take longer to transmit data.

The matched processor tuned to sA_i produces the maximum output value

$$VA_{|X=sA} = \sum_{i=0}^{n_x-1} sA_i^2 = \sum_{i=0}^{n_x-1} 1 = \overbrace{1 + 1 + \cdots + 1}^{n_x \text{ values}} = n_x \quad (= \mathcal{E}_s) \tag{11.31}$$

Note that this $\mathcal{E}_s$ is twice the energy value for sinusoidal signals ($= n_x/2$). When all code samples have magnitude $V_{\max}$, then

$$\mathcal{E}_s = n_x V_{\max}^2 \tag{11.32}$$

The next example indicates how to generate random CDMA codes using a pseudo-random number generator.

Generating random orthogonal codes   EXAMPLE 11.5

Consider a pseudo-random number generator that yields random numbers Y_i that are uniformly distributed over [0,1). The rule that generates the random CDMA sequence sA_i for $i = 0, 1, 2, \ldots, n_x - 1$ with $n_x = 31$ is

$$\text{If } Y_i < 0.5 \rightarrow sA_i = -1$$

$$\text{If } Y_i \geq 0.5 \rightarrow sA_i = 1$$

We generate the complementary codes

$$sA0_i = -sA_i \quad \text{and} \quad sA1_i = sA_i \text{ for } i = 0, 1, 2, \ldots, n_x - 1$$

Figure 11.10 shows the values of $sA0_i$ and $sA1_i$. The matched processor outputs for the sA_i signal give

$$VA_{|X=sA0} = \sum_{i=0}^{30} sA_i \, sA0_i = -\sum_{i=0}^{30} sA_i^2 = -31$$

$$VA_{|X=sA1} = \sum_{i=0}^{30} sA_i \, sA1_i = \sum_{i=0}^{30} sA_i^2 = 31$$

We then select sB_i in a similar random manner, but after each realization of an sB_i code, we check to determine if the orthogonality condition is met, as

$$\left| VA_{|X=sB} \right| = \left| \sum_{i=0}^{n_x-1} sA_i \, sB_1 \right| \leq \epsilon$$

A successful code has $\epsilon \leq 1$. This is a suitably low value for the matched processor output when the other signal is detected.

If this *approximately orthogonal* condition is not satisfied, the search procedure for randomly selecting sB_i is repeated until the condition is met. In practice, this occurs within

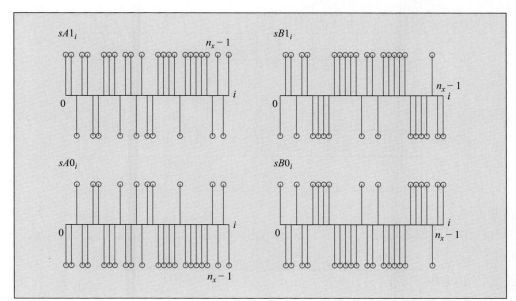

Figure 11.10

CDMA signals for $n_x = 31$. sA_i was chosen randomly to form $sA1_i = sA_i$ and $sA0_i = -sA_i$. sB_i was chosen randomly until $\epsilon \leq 1$ was found. Then, $sB1_i = sB_i$, and $sB0_i = -sB_i$.

ten repetitions for $n_x = 31$. Figure 11.10 shows the values of $sB0_i$ and its complementary code $sB1_i$, which produced the matched processor output

$$\left| VA_{|X=sB} \right| = \left| \sum_{i=0}^{30} sA_i \, sB_i \right| \leq \epsilon$$

with $\epsilon = 1$

The signals that have the maximum magnitude V_{max} are obtained by multiplying the ± 1 values by V_{max} to produce the signal energy $\mathcal{E}_s = n_x V_{max}^2 = 31 V_{max}^2$.

We can examine the system operation when both users A and B transmit their signals over the channel simultaneously. The signal detected by the receivers intended for users A and B equals

$$X_i = sAT_i + sBT_i \text{ for } i = 0, 1, 2, \ldots, n_x - 1 \tag{11.33}$$

where $sAT_i = sA0_i$ when $DA_t = 0$ is transmitted or $sAT_i = sA1_i$ when $DA_t = 1$ is transmitted. This is similar for sBT_i. Figure 11.11 shows the signals that are produced separately by user A transmitting data $DA_t = 10$, and B transmitting $DB_t = 10$. With sA_i and sB_i containing only -1 and $+1$ values, the composite signal X_i can take on values -2, 0, and $+2$. Even though X_i looks unfamiliar, the two matched processors can determine their binary values.

Figure 11.12 shows the outputs of the VA and VB matched processors with X_i equal to the composite channel signal after each of the two n_x data sequences are detected. A large positive processor output indicates that the signal contains the CDMA code corresponding to the processor's tuned value. This output value is typically less than $n_x(= 31)$ because of the interference produced by the other CDMA signal.

The VA processor detects the $sA1_i$ CDMA signal in the first n_x data interval by producing a large positive output at the end of the interval (when the complete signal is within the processor). The composite channel signal also contains $sB1_i$ in the first n_x interval, and the VB matched processor detects it by producing a large positive value at the end of the interval. The second n_x interval contains $sA0_i$ and $sB0_i$. Both the VA and VB matched processors detect their respective CDMA signals by producing large negative output values at the end of the second interval.

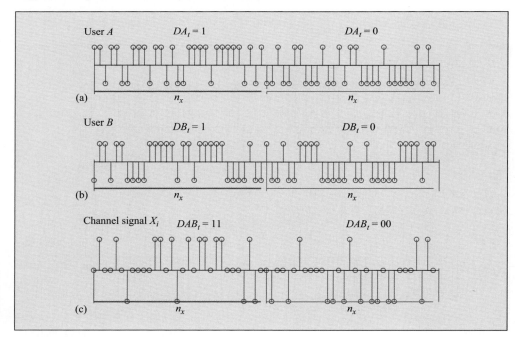

Figure 11.11

CDMA signals from users A and B and the resulting channel signal. (a) User A transmits 10. (b) User B transmits 10. (c) Sum of signals observed on channel (= X_i).

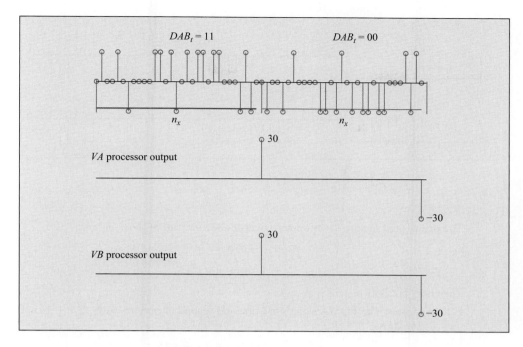

Figure 11.12

Processing CDMA signals. The outputs of the VA and VB processors with $n_x = 31$ with X_i equal to the summed channel signal. The processor outputs are shown after each of the two n_x data sequences are detected. A positive value, typically $\approx n_x$, indicates that the signal expected by the processor has been observed.

Maximum Number of CDMA Users

If additional codes are necessary but those having this optimal quality are no longer available, the non-matching criterion can be relaxed to an acceptable value $|\epsilon| \ll n_x$, which is the signal energy and the value produced by the successful matched processor when $V_{max} = 1$. Adding additional users requires longer codes, with the trade-off being slower system data transmission rates.

CDMA signals are composed of random binary values. Unlike FDMA systems that occur at specific frequency values, CDMA signals are composed of many frequency components. This feature is important for military applications because this technique, termed *spread-spectrum coding*, is difficult to detect. CDMA signals appear as noise and if detected can be difficult to jam using conventional methods.

 # 11.7 DETECTING ORTHOGONAL SIGNALS IN NOISE

This section describes a simulation procedure to estimate the Prob[error] by performing multiple transmissions in the presence of noise and counting the number of errors. This approach follows the methods explained in Section 10.6.

Probability of Error of orthogonal signals EXAMPLE 11.6

This example compares TDMA, FDMA, and CDMA signals of length $n_x = 16$ shown in Figure 11.13.

TDMA signals: The TDMA signals use the maximum allowed amplitude $V_{max} (= 1 \text{ V})$ to maximize signal energy. Signal sA_i is non-zero over the first half of the n_x interval. Thus,

$$sA_i = 1 \text{ for } 0 \leq i \leq 7$$
$$= 0 \text{ for } 8 \leq i \leq 15$$

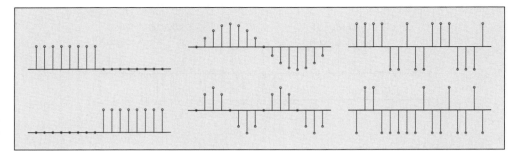

Figure 11.13

TDMA, FDMA, and CDMA orthogonal signals of length $n_x = 16$ and $V_{max} = 1$.

The orthogonal signal sB_i is non-zero over the second half of the n_x interval. Thus,

$$sB_i = 0 \text{ for } 0 \le i \le 7$$

$$= 1 \text{ for } 8 \le i \le 15$$

The energy of both signals is $\mathcal{E}_s = 8$.

FDMA signals: The FDMA signals use first and second harmonics with $f_A = 1/16$ and $f_B = 2/16$. Thus,

$$sA_i = \sin(2\pi f_A i) \text{ for } 0 \le i \le 15$$

and

$$sB_i = \sin(2\pi f_B i) \text{ for } 0 \le i \le 15$$

The energy of both signals is $\mathcal{E}_s = 8$.

CDMA signals: The CDMA signals sA_i and sB_i have maximum V_{max} magnitudes (± 1) that were randomly generated until the orthogonality condition was found with $\epsilon = 0$. Hence,

$$\sum_{i=0}^{15} sA_i \, sB_i = 0$$

with CDMA energy $\mathcal{E}_s = 16$.

A simulation to estimate Prob[error] was performed using the following steps.

1. The basic signal orthogonal sequences sA_i and sB_i for $i = 0, 1, 2, \ldots, n_x - 1$, were specified for TDMA, FDMA, and CDMA, and $\mathcal{E}_s$ was computed.

2. Signals $sA0_i = -sA_i$ and $sA1_i = sA_i$ and $sB0_i = -sB_i$ and $sB1_i = sB_i$ for $i = 0, 1, 2, \ldots, n_x - 1$ were formed.

3. The number of trials was set equal to 101. Each trial had a specified value for noise SD σ_N with

$$\sigma_N = n\sigma_o \text{ for } 0 \le n \le 100$$

where $\sigma_o = 0.1$ to produce σ_N values ranging from 0 to 10 and σ_N^2 from 0 to 100 ($\gg \mathcal{E}_s$).

4. For each trial, a large number n_t ($= 10^5$) of data transmissions were simulated.

5. For each transmission, the following operations were performed.

   (a) Transmitted bits DA_t and DB_t were randomly chosen using the uniform PRNG, and the corresponding transmitted signals were formed as

   $$sAT_i = sA0_i = -sA_i \text{ for } DA_t = 0$$

   $$= sA1_i = sA_i \text{ for } DA_t = 1$$

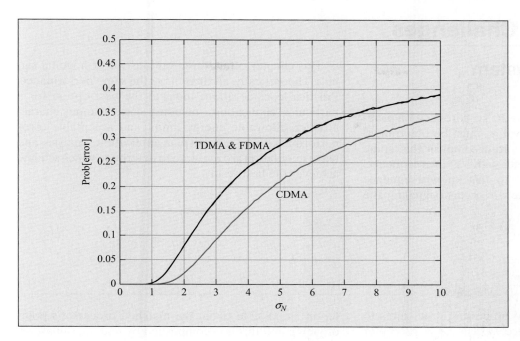

Figure 11.14

Probability of error versus σ_N using TDMA, FDMA, and CDMA orthogonal signals.

and

$$sBT_i = sB0_i = -sB_i \text{ for } DB_t = 0$$

$$= sB1_i = sB_i \text{ for } DB_t = 1$$

(b) The Gaussian random noise sequence N_i for $i = 0, 1, 2, \ldots, n_x - 1$ having the specified σ_N was generated.

(c) The received signal sequence was formed as

$$X_i = sAT_i + sBT_i + N_i \text{ for } i = 0, 1, 2, \ldots, n_x - 1$$

(d) Matched processors used X_i to produce outputs VA and VB.

(e) VA and VB were applied to their respective threshold detectors with $\tau = 0$ to determine DA_r and DB_r.

(f) The error counter N_e was incremented when $DA_r \neq DA_t$ and when $DB_r \neq DB_t$.

6. At the end of each trial, the probability of error was estimated using Prob[error] $= N_e/(2n_t)$. The n_t is multiplied by 2, because two bits are transmitted with each X_i for $i = 0, 1, 2, \ldots, n_x - 1$.

This simulation produced the following results. Figure 11.14 shows Prob[error] versus σ_N. Note that the Prob[error] increases as σ_N increases, as should be expected. The Prob[error]$\rightarrow$ 0.5 for very large σ_N, because when the noise is large, it is the noise that determines the VA and VB values—no matter what signal was transmitted. Hence, half the time the V values produce data values that agree with those that were transmitted, being correct purely by chance.

The Prob[error] values for TDMA and FDMA are equal, because both signals have the same $\mathcal{E}_s$. The Prob[error] for CDMA is smaller, because its signals have $\mathcal{E}_s$ that is twice as large. This increased $\mathcal{E}_s$ for CDMA produces the same Prob[error] for a larger σ_N (by a factor of $\sqrt{2}$) than that for TDMA and FDMA. While FDMA produces a larger Prob[error] than CDMA because of its smaller $\mathcal{E}_s$, FDMA has the advantage of forming its orthogonal signals through direct computations rather than through a search procedure. Commercial systems use a specified set of CDMA codes, called *Kasami codes*.

11.8 Research Challenges

11.8.1 Practical System Considerations

Practical systems must address the synchronization problem: The relevant data interval n_x must be determined from the observed signal X_i. To accomplish this, some systems precede the data signals with a known sequence. A matched processor tuned to this known sequence produces a large output when it occurs, signaling the beginning of the data signals.

This chapter offers three options for orthogonal signals. Theoretically, all three have the same performance, Practical considerations indicate the best option for a particular application, including cost, channel characteristics, allowable energy limits, and acceptable error performance. For a particular application, analysis and experimentation are required to determine which scheme achieves the best result.

11.9 Summary

This chapter described signal processing procedures to detect data signals in multi-access systems. A processor having coefficients tuned to the expected signal was described. Such a matched processor produces maximally different values when the signals encoding binary data 0 and 1 are complementary or negatives of each other. When multiple users access the communication channel simultaneously, each user's signals must be as different as possible (from the matched processor's point of view), and this is accomplished by making signals orthogonal. Three different methods to generate orthogonal signals were described, which included orthogonality in frequency, time, and code. The idea employed for serving several simultaneous users also can be applied to a single user transmitting more than one binary data value per transmission.

11.10 Problems

11.1 **Orthogonal signals.** If $sA1_i$ and $sB1_i$ are orthogonal, show that their complementary signals $sA0_i$ and $sB0_i$ are also orthogonal.

11.2 **Orthogonal signal design in TDMA.** Specify signals sA_i and sB_i for $0 \leq i \leq 7$ that are orthogonal over time. Design matched processors tuned to your signals and explain why they process these signals correctly.

11.3 **Number of orthogonal FDMA signals.** With $n_x = 8$, compute n_{max}, which is the maximum number of unique orthogonal sinusoidal sequences. Show the $n_{max} + 2$ harmonic has values that are identical to one of the complementary signals in the orthogonal set.

11.4 **Orthogonal signal design in FDMA.** Specify signals sA_i and sB_i for $0 \leq i \leq 6$, that are orthogonal in frequency. Let $T_s = 10^{-4}$ s and specify their frequency values in the range $0 \leq f \leq 5,000$ Hz. Design matched processors tuned to your signals and explain why that they process these signals correctly.

11.5 **Orthogonal signal design in CDMA.** Specify signals sA_i and sB_i for $0 \leq i \leq 4$ having values equal to either ±1 that are approximately orthogonal in code. First, flip a coin (tail $\rightarrow -1$, and head $\rightarrow +1$) to specify the values for sA_i. Then do the same repeatedly to generate sB_i until

$$\epsilon = \left| \sum_{i=0}^{4} sA_i \, sB_i \right| \leq 1$$

Design matched processors tuned to your signals and explain why they process these signals correctly.

11.6 **Processing superimposed CDMA signals from users A and B.** Using the codes determined in the previous problem, let X_i equal the sum of the codes that would occur for the four possibilities of transmitted data (00, 01, 10, 11). Design matched processors tuned to your signals and explain why they process these signals correctly.

11.11 Matlab Projects

11.1 **Extending TDMA to three simultaneous users.** Extend Example 16.56 to generate TDMA signals to serve three users simultaneously.

11.2 **Extending FDMA to three simultaneous users.** Extend Example 16.57 to generate FDMA signals to serve three users simultaneously.

11.3 **Extending CDMA to three simultaneous users.** Extend Example 16.58 to generate CDMA signals to serve three users simultaneously.

11.4 **Estimating probability of error.** Using Example 11.6 as a model, modify Example 16.59 to compare performance of TDMA, FMDA, and CDMA orthogonal systems with duration $n_x = 30$ and maximum amplitudes equal to one. Construct CDMA signals to have maximum energy. Compute and plot the TDMA, FMDA, and CDMA Prob[error] as σ_N^2 varies between 0 and $5\mathcal{E}_s$ in steps of $0.5\mathcal{E}_s$.

SOURCE CODING

After completing this chapter, the reader should be able to:

- Compute the information content of a data source using entropy.
- Understand how the data quantity can be reduce without loss of information using variable-length codes.
- Generate a Huffman code to compress the data size of a file.
- Encrypt a file to provide security.

This chapter describes source coding techniques that measure the information content of the symbols produced by a source. In many cases, the quantity of data produced by a source can be quite large, often many megabytes or even gigabytes. This chapter illustrates how data compression reduces the quantity of data without any loss of information. A simple version of encryption is described for securing important data.

 ## 12.1 INTRODUCTION

Previous chapters described how binary data are generated and how the signals encoding them are processed to extract data in the presence of noise. This chapter describes mathematical models of information generation by a source to compute the informational content of data in order to understand how to reduce the data size without affecting the informational content.

The main topics describing source encoding are covered in the following sequence.

Data compression—Data compression reduces the size of data that needs to be transmitted or stored. The quantity of data produced by a source can be quite large, often many megabytes or even gigabytes. *Lossless data compression* occurs without any loss of information, while *lossy compression* causes minor loss in information, which a user would not perceive in practice. This chapter describes lossless compression.

Entropy $\mathcal{E}$—Entropy is a measure of information content. A source produces symbols with each symbol encoded using binary code words. The probabilities of the symbols are used to compute the entropy, having units of bits/symbol. If the entropy value for a source is smaller than the number of bits used in the code words, lossless compression is possible using a Huffman code.

Encryption—Encryption makes information secure by scrambling the binary data in a seemingly random manner. Information is a valuable commodity and should be available only to the intended parties. Data security is a challenge, especially when data are transmitted over a wireless channel, where it can be observed by hackers or sniffed. To keep your important information secure, data are encrypted before they are transmitted or stored. Although practical encryption techniques use advanced mathematical ideas, this chapter presents the fundamental ideas by considering a simple, yet effective, approach.

 ## 12.2 DATA COMPRESSION

Because channel capacity and digital memory size are always limited, *data compression* techniques were developed to limit the quantity of data that needs to be transmitted, while still maintaining most of its informational content. There are two commonly employed approaches to data compression, depending on whether there is any loss of information in the procedure.

1. **Lossless compression** techniques retain all of the information present in the original data. That is, the original data sequence is recovered *exactly* after the coding/decoding operation. These techniques are used for compressing numerical data, such as monetary account information, computer codes, and for document transmissions. Five-fold data compression allows a 10 MB (megabyte) file to be stored as a compressed 2 MB file for faster transmission over a communication channel.

   The Huffman coding procedure is a lossless compression technique that allows the representation of symbols to take on a variable number of bits to save on the *average number* of bits needed to represent the data. Such a code uses a small number of bits to represent the most commonly occurring symbols and a larger number of bits for those that occur less frequently. This idea was employed to encode telegraph messages using Morse code, composed of dots and dashes. The most common English letter "e" is represented by a single dot and a "t" by a single dash, while the less frequently encountered "q" is represented by a dash-dash-dot-dash sequence.

2. **Lossy compression** techniques allow more drastic reductions (hundred-fold) of the data by permitting some loss in information. Such techniques are typically used for encoding data that are to be *perceived*, such as images or sounds. In such cases, using *sufficiently accurate approximations* to the data—rather than the actual data—can lead to hundred-fold compression with little perceptible distortion.

   An interesting feature of lossy compression techniques is that the amount of compression can be varied by specifying the quality of the result. For example, transmitting a sporting event, such as the Olympics with its fast motion and subtlety of movement, demands a high-fidelity reproduction that requires a large number of bits per second. A technique called MPEG-2 encoding specifies a data rate equal to 6 Mbps. At the other extreme, a video conference call does not involve quick movements and can be encoded with the same MPEG-2 encoding at 64 Kbps. Hence, there is almost a factor of 100 on the amount of data that is required to reproduce video information.

12.2.1 Modeling Information

This section develops a mathematical model of information. Information is more than just data, it contains an element that we did not know previously, otherwise it would not be information. To model the novelty of information we use simple probability theory. We can then compute the entropy to quantify information. A pseudorandom number generator is used to simulate such data sources to generate typical data. Finally, we apply the theory to compute the expected compression of a data file.

Figure 12.1 shows our model of a source having the following elements:

■ The source has a vocabulary of m symbols, denoted $X_1, X_2, X_3, \ldots, X_m$, or more concisely X_i for $1 \le i \le m$. By convention, the index starts with 1 rather than 0 used for time sequences.

■ Each symbol has an associated probability of occurrence, with the probability of the source producing symbol X_i denoted as $P[X_i]$. The set of probabilities are denoted concisely as $P[X_i]$ for $1 \le i \le m$. Because the source must produce one of these symbols, these probabilities sum to one:

$$\sum_{i=1}^{m} P[X_i] = 1 \tag{12.1}$$

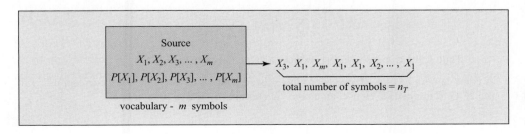

Figure 12.1

A mathematical model of a source.

If the m symbols are equally likely, then the probability of a source producing any particular one equals $1/m$, or

$$P[X_1] = P[X_2] = P[X_3] = \cdots = P[X_m] = \frac{1}{m} \qquad (12.2)$$

■ We also consider the case where the source produces a data file that contains a total of n_T symbols that come from the symbol vocabulary containing m unique symbols.

EXAMPLE 12.1 Binary source

A binary source often models devices that transmit binary data:

- The binary source vocabulary consists of two symbols, making $m = 2$, with symbols

$$X_1 = 0 \quad \text{and} \quad X_2 = 1$$

- Unless there is reason to think otherwise, binary data are typically assumed to be equally probable with

$$P[X_1] = P[X_2] = \frac{1}{2}$$

The sum of these probabilities equals

$$\sum_{i=1}^{2} P[X_i] = P[X_1] + P[X_2] = 1$$

That is, the binary source always produces either a 0 or 1.

Any data file on your computer can be considered to be a sequence of 0's and 1's. Data file sizes vary widely, from a minimum value on many computers being $n_T = 1024$, or a "1 K" file, to image and video files containing many megabytes.

EXAMPLE 12.2 Digital camera source

Consider modeling a digital camera that produces a digital image as a data source:

- The camera vocabulary consists of symbols that represent the 8-bit value of each pixel intensity, composed of RGB colors, yielding $m = 2^8 = 256$ possible symbols, and are denoted by

$$X_i = i - 1 \text{ for } 1 \le i \le 256$$

with $X_1 = 0$, up to $X_{256} = 255$, representing all the possible values that a byte can have.

- Because the camera can take an almost endless variety of pictures, it is reasonable to assume that all symbols are equally likely to occur. This means their probabilities are equal to

$$P[X_1] = P[X_2] = P[X_3] = \cdots = P[X_{256}] = \frac{1}{256}$$

The sum of these probabilities equals

$$\sum_{i=1}^{m} P[X_i] = \sum_{i=1}^{256} \frac{1}{256} = 1$$

That is, the camera always produces a value between 0 and 255.

A color camera having 10^6 pixels produces three symbols (bytes) per pixel, corresponding to the R, G, B intensities. Hence, an image file contains $n_T = 3 \times 10^6$ symbols.

12.2.2 Source Entropy

We can quantify the average information produced by a source by computing its *source entropy* denoted as $\mathcal{H}_S$. To compare different sources using the same units, we compute $\mathcal{H}_S$ in units of *bits per symbol*.

A source that produces the symbols X_1, X_2, X_3, ..., X_m, with the probability of X_i being $P[X_i]$, has a source entropy computed as

$$\mathcal{H}_S = -\sum_{i=1}^{m} P[X_i] \, \log_2 P[X_i] \text{ bits/symbol} \qquad (12.3)$$

The entropy indicates the average number of bits that are required to represent these symbols as a binary code, as illustrated in the following examples.

Entropy of a binary source with equal probabilities EXAMPLE 12.3

Let us consider a source that generates random bits by transmitting symbols X_1 and X_2 that are equally likely. Then their probabilities equal

$$P[X_1] = P[X_2] = 0.5$$

The entropy of this binary source is computed as

$$\mathcal{H}_S = -\sum_{i=1}^{2} P[X_i] \, \log_2 P[X_i]$$

$$= -P[X_1] \, \log_2 P[X_1] - P[X_2] \, \log_2 P[X_2]$$

$$= -0.5 \log_2 0.5 - 0.5 \log_2 0.5$$

Using the results $\log(1/a) = -\log(a)$ and $\log_2(2) = 1$ gives

$$\mathcal{H}_S = -0.5 \log_2 0.5 - 0.5 \log_2 0.5$$

$$= 0.5 \overbrace{\log_2 2}^{=1} + 0.5 \overbrace{\log_2 2}^{=1}$$

$$= 1 \text{ bit/symbol}$$

It is intuitive that when a source produces only two symbols (heads/tails, true/false, ...), then only one bit is needed to indicate which symbol was produced each time a symbol is generated.

The previous example illustrates that a source producing m symbols that are equally probable has probability values

$$P[X_i] = \frac{1}{m} \quad \text{for} \quad 1 \le i \le m \qquad (12.4)$$

The source entropy is simple to compute as

$$\mathcal{H}_S = -\sum_{i=1}^{m} P[X_i] \log_2 P[X_i] = -\sum_{i=1}^{m} \left(\frac{1}{m}\right) \log_2 \left(\frac{1}{m}\right) = \log_2 m \qquad (12.5)$$

EXAMPLE 12.4 Entropy of a binary source with unequal probabilities

Consider a source that produces two symbols having unequal probabilities:

$$P[X_1] = 0.25 \quad \text{and} \quad P[X_2] = 0.75$$

Computing the entropy of this source gives

$$\mathcal{H}_S = -\left(\sum_{i=1}^{2} P[X_i] \, \log_2 P[X_i] \right)$$

$$= -\left(P[X_1] \, \log_2 P[X_1] + P[X_2] \, \log_2 P[X_2] \right)$$

$$= -0.25 \log_2 0.25 - 0.75 \log_2 0.75$$

$$= 0.811 \text{ bits/symbol}$$

A result that is expressed in fractional bits can be interpreted using the following logic: A file containing 1,000 symbols can be encoded using 811 bits.

This example illustrates that a binary source producing bits (symbols) having unequal probabilities has a lower entropy than the source producing two symbols having equal probabilities.

Factoid: A source that produces symbols having equal probabilities generates the most information per symbol and thus produces the greatest entropy.

EXAMPLE 12.5 Source producing one symbol

Consider a source that always produces the same symbol, as in the case of a defective source. If this source is modeled as always producing symbol X_1, then $P[X_1] = 1$. The entropy of this defective source is

$$\mathcal{H}_S = -\sum_{i=1}^{1} P[X_i] \, \log_2 P[X_i] = -(1 \times \overbrace{\log_2 1}^{=0}) = 0 \text{ bits/symbol}$$

That is, a source producing only a single symbol generates zero information.

12.2.3 Effective Probabilities

The previous discussion assumed we know the symbol probabilities. But in many cases these probabilities are not known and must be estimated from the data file.

Consider a data file generated by a source for which we do not know the symbol probabilities. Although a source can produce m symbols, this particular data file may contain only $m_x (\leq m)$ different symbols. Let m_x symbols occur in a data file of size n_T, where $n_T \gg m_x$.

- If the file contains decimal digits, $X_1 = 0$, $X_2 = 1, X_3 = 2, \ldots, X_{10} = 9$, and it is likely that all digits occur, $m_x = m = 10$. For data base file containing telephone numbers or credit card numbers, n_T may be in the millions.

- If the symbols are alphanumeric (ASCII 7-bit) text characters, $X_1, X_2, X_3, \ldots, X_{128}$, and if not all ASCII characters occur $m_x < m = 128$. The data file containing the words in this text has $n_T \approx 10^6$.

- If the symbols are byte values, $X_1 = 00000000$, $X_2 = 0000001$, $X_3 = 00000010$, $\ldots$, $X_{256} = 11111111$, and all symbols occur at least once $m_x = m = 256$. The data file of an instruction manual containing images stored on a thumb drive may have n_T in the billions.

Alphanumberic characters in a book   EXAMPLE 12.6

If the symbols are letters of the alphabet, then $m = 26$, although if we count lowercase, uppercase, spaces, and punctuation as separate symbols, $m_x \approx 100$. If the data file of a book contains 600 pages with 200 words per page and an average of 5 characters per word, the total number of symbols equals

$$n_T = 600 \times 200 \times 5 = 6 \times 10^5$$

A data file contains m_x different symbols with the total number of symbols equal to n_T. If X_i occurs n_i times, the *effective probability* of symbol X_i is computed as

$$P_e[X_i] = \frac{n_i}{n_T} \quad \text{for} \quad 1 \leq i \leq m_x \tag{12.6}$$

Valid symbols that could have been produced by the source but do not appear in the file have zero effective probability values by this method and these are excluded from further consideration.

The book without "e"   EXAMPLE 12.7

The 1939 novel *Gadsby* by Ernest Vincent Wright contains 50,000 words, yet it does not include the letter "e", which is the most probable letter in English. Hence, when sorting the alphanumeric symbols, symbol $X_5 (= e)$, produces count $n_5 = 0$, making $P_e[X_5] = 0$.

12.2.4 Effective Source Entropy

For data files with unknown symbol probabilities, we use the effective probabilities to compute an estimate of the source entropy. The *effective source entropy*, which is denoted as $\hat{\mathcal{H}}_S$, is computed from the effective probabilities as

$$\hat{\mathcal{H}}_S = -\sum_{i=1}^{m_x} P_e[X_i] \, \log_2 P_e[X_i] \text{ bits/symbol} \tag{12.7}$$

Effective source entropy   EXAMPLE 12.8

Consider a binary data file containing 1,000 bits of which 600 are 1's and 400 are 0's. There are two ($m_x = 2$) different symbols, and their effective probabilities are computed as

$$P_e[1] = \frac{600}{1,000} = 0.6$$

$$P_e[0] = \frac{400}{1,000} = 0.4$$

The effective source entropy then gives

$$\hat{\mathcal{H}}_S = -\sum_{i=1}^{2} P_e[X_i] \log_2 P_e[X_i]$$

$$= -P_e[1] \log_2 P_e[1] - P_e[0] \log_2 P_e[0]$$

$$= -(0.6 \times -0.74) - (0.4 \times -1.32))$$

$$= 0.97 \text{ bits/symbol}$$

Note that if the counts of 1's and 0's were equal, $\hat{\mathcal{H}}_S$ would equal 1 bit/symbol. The unequal counts of 1's and 0's reduce the $\hat{\mathcal{H}}_S$ value.

12.2.5 Effective File Entropy

The effective file entropy, which is denoted as $\hat{\mathcal{H}}_f$, represents the minimum number of bits that are needed to encode the file. The $\hat{\mathcal{H}}_f$ value determines whether it is worthwhile to compress a file (that is, whether compression will lead to a significant reduction in data size). To compute $\hat{\mathcal{H}}_f$, we multiply the effective source entropy $\hat{\mathcal{H}}_S$ by the total number of symbols n_T, as

$$\hat{\mathcal{H}}_f = n_T \text{ symbols} \times \hat{\mathcal{H}}_S \text{ bits/symbol} \tag{12.8}$$

Note that this determination of whether a file should be compressed can be made before any compression is even attempted.

EXAMPLE 12.9

Effective file entropy

Consider the data file containing only the following ten symbols

$$A\,B\,A\,D\,A\,E\,C\,D\,B\,A$$

There are five ($m_x = 5$) different symbols (A, B, C, D, and E) in the file of ten symbols ($n_T = 10$). The effective probabilities of the symbol listed in alphabetical order are given by

$$P_e[A] = \frac{n_A}{n_T} = \frac{4}{10} = 0.4$$

$$P_e[B] = \frac{n_B}{n_T} = \frac{2}{10} = 0.2$$

$$P_e[C] = \frac{n_C}{n_T} = \frac{1}{10} = 0.1$$

$$P_e[D] = \frac{n_D}{n_T} = \frac{2}{10} = 0.2$$

$$P_e[E] = \frac{n_E}{n_T} = \frac{1}{10} = 0.1$$

The effective source entropy gives

$$\hat{\mathcal{H}}_S = - \sum_{i=1}^{m_x} P_e[X_i] \log_2 P_e[X_i]$$

$$= -(P_e[A] \log_2 P_e[A] + P_e[B] \log_2 P_e[B] + P_e[C] \log_2 P_e[C]$$

$$+ P_e[D] \log_2 P_e[D] + P_e[E] \log_2 P_e[E])$$

$$= -(0.4 \log_2[0.4] + 0.2 \log_2[0.2] + 0.1 \log_2[0.1]$$

$$+ 0.2 \log_2[0.2] + 0.1 \log_2[0.1])$$

$$= 2.12 \text{ bits/symbol}$$

The effective file entropy computes as

$$\hat{\mathcal{H}}_f = n_T \hat{\mathcal{H}}_S = 21.1 \text{ bits}$$

That is, this file has an information content equal to 21.2 bits. This means that the ten symbols in the file can be encoded using approximately 22 bits. A file with $\hat{\mathcal{H}}_S = 2.12$ bits/symbol containing 100 symbols can be encoded with 212 bits using the data compression procedure described below.

Without compression, fixed-length code words each having three bits ($2^3 = 8$) are needed to provide a unique code for each of the 5 symbols. For example, consider the fixed-length code word assignment

$$A: 000, \ B: 001, \ C: 010, \ D: 011, \ E: 100$$

Clearly, not all possible 3-bit code words are used, but 2-bit code words cannot unambiguously encode these five symbols. Thus, 30 bits are required to encode the file using 3-bit code words. We can then decide whether the savings of 8 bits (=30-22) justifies the effort to compress the file.

12.2.6 Huffman Code

Lossless data compression is achieved by representing symbols with a variable number of bits. Doing so can save on the *average number* of bits needed to represent the data by using a small number of bits for the most commonly occurring symbols and a larger number of bits for those that occur less frequently. The resulting variable-length code generates an average code word size per symbol that is approximately equal to our two entropy measures.

$\mathcal{H}_S$—The source entropy equals the average number of bits required to encode symbols produced by a source with known symbol probabilities.

$\hat{\mathcal{H}}_S$—The effective source entropy of a data file equals the average number of bits required to encode symbols computed from the effective probabilities.

The method of implementing a variable-length code considered here is the *Huffman code*. This procedure employs a *code tree* that consists of *nodes* connected by *branches* that ultimately terminate in *leaves*. Figure 12.2 shows a code tree that describes the four symbols *A, B, C,* and *D*. The node at the top of the tree defines its *root* or starting point. In a *binary tree* the branches originating from each node are designated by the binary values 0 or 1. Hence, only two branches originate from any node. For consistency, the left branch is always designated with a 1 and the right branch with a 0.

Branches terminate in *leaves* with each leaf representing a valid symbol. The sequence of bits in a valid code word specifies a unique path from the root to a leaf. Thus, this sequence of bits (each of which makes a decision at a node) specify the symbol at the

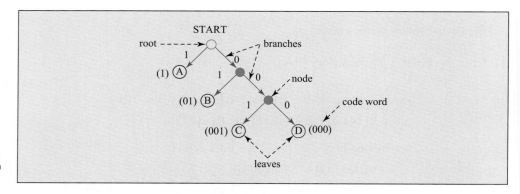

Figure 12.2

A code tree assigns a variable-length code that specifies four symbols.

terminal leaf. The last bit in the code word always terminates at a leaf specifying a symbol.

The tree code in Figure 12.2 indicates that symbol *A* has the shortest code word (a single 1), while symbols *C* and *D* have the 3-bit codes. The corresponding code is called the *Huffman code* and uses the code tree to assign the shortest code words to the most common symbols and longer code words to the ones that occur less often.

EXAMPLE 12.10   Decoding the Huffman code

Using the tree code shown in Figure 12.2, the binary sequence given by

$$00101110001$$

can be decoded into symbols.

Decoding starts at the beginning of the data sequence and the root of the code tree. When a leaf is reached, a symbol has been decoded. The next bit in the data sequence then restarts at the root to search for the next symbol.

Doing this using our binary sequence, we get

$$\underbrace{001}_{C} \quad \underbrace{01}_{B} \quad \underbrace{1}_{A} \quad \underbrace{1}_{A} \quad \underbrace{000}_{D}$$

It is an interesting property of the Huffman code that no valid code word forms the beginning of another code word.

Huffman Coding Procedure

The procedure to form a Huffman code involves the following steps.

1. **Order the symbols by their probabilities.** We form a list of symbols ordered by decreasing probability, using the measure that is available in the problem: $P[X_i]$ if it is known or $P_e[X_i]$ if it is computed from a data file. If X_3 is the most probable symbol, it appears at the top of the list, and if X_2 is the least probable, it appears at the bottom. This is the *first ordering*.

2. **Code bit assignment.** Starting from the bottom of the list, we assign a 0 code bit to the least probable symbol and a 1 code bit to next least probable (that is, the one that is second from the bottom). This code bit forms the rightmost (LSB) bit of the code words that differentiate these two symbols.

3. **Combine symbols by adding their probabilities.** Combine the two least probable symbols into one composite symbol, and the probability of this composite symbol becomes the sum of the constituent probabilities. For example, if the last two symbols on the list are X_2 and X_6, they are combined into the composite symbol

denoted X_2-X_6 and

$$P[X_2\text{-}X_6] = P[X_2] + P[X_6] \qquad (12.9)$$

when the probabilities are known, otherwise, use the effective probabilities

$$P_e[X_2\text{-}X_6] = P_e[X_2] + P_e[X_6] \qquad (12.10)$$

4. **Re-order symbols.** Generate a new ordering of the symbols using the composite symbol as just another symbol having its summed probability. This is the *second ordering*.

5. **Repetition.** We repeat the previous three steps until only two symbols or composite symbols remain on the list. This is the *final ordering*. We assign a 0 code bit to the less probable entry and a 1 code bit to the other. This final assignment completes the formation of the code words.

To decode a Huffman code, a *code table* indicating the correspondence of the code words to the symbols must also be stored (or transmitted) along with the the encoded data.

We construct a Huffman code tree from the root by starting with the final ordering, where the 1 in the code word is assigned to the most probable symbol (or composite symbol). This 1 defines the left branch from the root leading to a node in the second row in the tree. A 0 code bit defines the right branch leading to the other node in the second row.

Each node corresponds to either a symbol or a composite symbol. If the node corresponds to a symbol, it is a leaf node, which designates that symbol. No branches originate from a leaf node. If the node corresponds to a composite symbol, two branches from it lead to nodes that represent the constituent symbols determined from the previous ordering. The tree is formed in this manner until all of the branches end in leaf nodes.

The following examples illustrate the Huffman coding procedure.

Huffman coding the sum of two coins   EXAMPLE 12.11

Consider a coin flip with a head indicating a 1 and a tail a 0. We toss two coins and want to encode their sum. The four possible outcomes and their sums are given by

$$
\begin{array}{ll}
00 & (\text{sum}=0) \\
01 & (\text{sum}=1) \\
10 & (\text{sum}=1) \\
11 & (\text{sum}=2)
\end{array}
$$

For fair coins, each of the four outcomes is equally probable (with probability = 0.25)

Define symbols $X_1 = 0$ (sum=0), $X_2 = 1$ (sum=1), and $X_3 = 2$ (sum=2) with probabilities given by

$$
\begin{array}{ll}
P[X_1] = 1/4 & (\text{one of four possible outcomes}) \\
P[X_2] = 1/2 & (\text{two of four possible outcomes}) \\
P[X_3] = 1/4 & (\text{one of four possible outcomes})
\end{array}
$$

The entropy of the source that produces the symbols that mimic the two-coin-sum outcomes equals

$$
\begin{aligned}
\mathcal{H}_S &= -\sum_{i=1}^{3} P[X_i] \log_2 P[X_i] \\
&= -0.25 \log_2(0.25) - 0.5 \log_2(0.5) - 0.25 \log(0.25) \\
&= 1.5 \text{ bits/symbol}
\end{aligned}
$$

We toss the two coins ten times and observe the following sequence of sum values.

$$2\ 1\ 1\ 2\ 1\ 0\ 2\ 1\ 1\ 0$$

Fixed-length code: We first encode this sequence using fixed-length code words. We have three symbols (0, 1, and 2), so 2-bit code words are needed to provide a unique code for each symbol. Using the arbitrarily assigned code words $X_1 : 00$, $X_2 : 01$, and $X_3 : 10$, the encoded binary sequence appears as

$$\overbrace{10}^{2}\ \overbrace{01}^{1}\ \overbrace{01}^{1}\ \overbrace{10}^{2}\ \overbrace{01}^{1}\ \overbrace{00}^{0}\ \overbrace{10}^{2}\ \overbrace{01}^{1}\ \overbrace{01}^{1}\ \overbrace{00}^{0}$$

Note that these ten realizations are encoded using twenty bits. Because the code word "11" is not used, this fixed-length code is wasteful.

Huffman code: We can encode this same information with a smaller average number of bits using the Huffman code, as shown in Figure 12.3.

In the first ordering, X_1 is assigned the code bit 0 and X_3 a 1. Then these two symbols are combined to form the composite symbol X_1-X_3, with the probability of this composite symbol equal to

$$P[X_1\text{-}X_3] = P[X_1] + P[X_3] = 0.25 + 0.25 + 0.5$$

In the second ordering, because the remaining probabilities are equal, X_1-X_3 is arbitrarily assigned a 0 and X_1 a 1. There were only two symbols in this ordering, so the process is finished.

The resulting Huffman code is given by the code table

sum=0: $X_1 \rightarrow 00$

sum=1: $X_2 \rightarrow 1$

sum=2: $X_3 \rightarrow 01$

The Huffman-encoded version of the ten observations is given by

$$\overbrace{01}^{2}\ \overbrace{1}^{1}\ \overbrace{1}^{1}\ \overbrace{01}^{2}\ \overbrace{1}^{1}\ \overbrace{00}^{0}\ \overbrace{01}^{2}\ \overbrace{1}^{1}\ \overbrace{1}^{1}\ \overbrace{00}^{0}$$

Note that the same ten symbols are encoded in 15 bits, for an average of 1.5 bits/symbol, which equals the source entropy $\mathcal{H}_S$ as computed here.

Figure 12.3

Huffman code for the sum of two coins. (a) Orderings of the symbols with probabilities shown in parentheses and code bits in brackets. (b) Code tree structure.

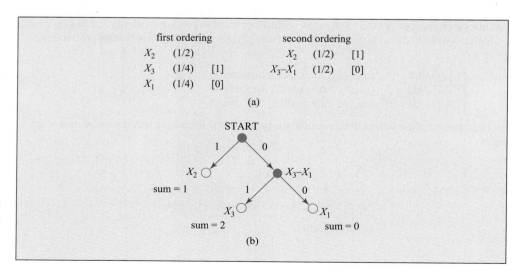

Theoretical Huffman coding of text files

EXAMPLE 12.12

Consider a text file consisting of all capital letters and six punctuation marks (space, comma, period, question mark, semicolon, and colon). These 32 symbols can be encoded with 5-bit code words ($2^5 = 32$). If these symbols were equally probable (with $P[X_i] = 1/32$), the entropy of the source producing them would equal

$$\mathcal{H}_S = \log_2 32 = 5 \text{ bits/symbol}$$

Studies by linguists indicate that the actual probabilities result in the entropy of English text being equal to 4.2 bits/symbol. Hence, compressing text files with a Huffman code using effective probabilities would result in a 16% reduction in the size of a data file.

Huffman coding of data files

EXAMPLE 12.13

The histograms of data in three files types are shown in Figure 12.4.

The first file is the text file of this chapter with the numbers corresponding to ASCII value of each letter. Note that the numbers lie in the range [0, 127], because the most significant bit of an ASCII code is always 0 ($m_x = 128$). The most common symbol value is 32 (the ASCII code for a *space*).

The second file is the postscript file sent to the printer to produce this chapter. Note that the numbers lie in the range [0, 255] ($m_x = 256$). Control characters lie outside the normal ASCII range [0, 127].

The third file includes binary instructions in a typical computer program. Note that these data are distributed almost uniformly over the entire range [0, 255], indicating that almost all of the 8-bit codes appear ($m_x = 256$).

For compressing these data, Huffman coding is more effective when the histogram is more *peaked*, as in the first two data files. These peaks indicate symbols having large effective probability values, which would be assigned short code words. A large number of short code words would result in an efficient data compression. Histograms that are approximately uniformly distributed, as in the third file, result in minimal data reduction.

Huffman Code Application – The Fax Code

EXAMPLE 12.14

The digital facsimile machine (or *fax*) is one of the early success stories in digital technology. Its first patent dates back to 1843 (shortly after the invention of the telegraph) with the first commercial use being in 1865 to transmit newspaper images between Paris, London, and Berlin. A combination of technology, standardization, and the deregulation of the telecommunications network was necessary to make the fax a commercial success and a part of everyday life. Anyone with access to a telephone could connect two fax machines to send or receive messages to or from anywhere in the world.

The resolution of a standard fax machine is a width of 1/8 mm and height of 1/4 mm, as shown in Figure 12.5. When a page of text is inserted into a fax machine, the page is scanned line by line. Text is treated as graphic material and converted into a set of small white or black rectangles, thus producing a binary data sequence.

The typical binary data sequence produced by a fax machine exhibits groupings, or *runs*, of 0's and 1's to indicate white and black regions. For example, the scan in Figure 12.5 indicates that the tops of the E's are encoded with sixteen 1's and the space between the E's as four 0's. The space between the left edge and the first E is encoded with 203 0's. An efficient method of representing such data is to encode the *run lengths* of 1's or 0's, which is denoted as *run-length coding*.

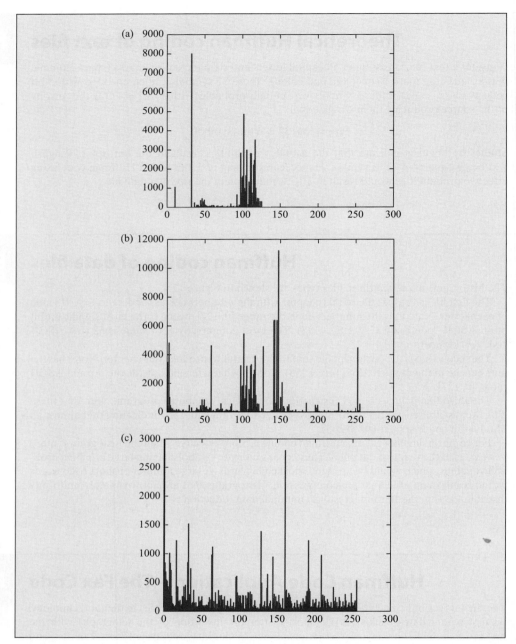

Figure 12.4

Histograms of byte values in three data files. (a) The text file of this chapter, with the numbers corresponding to ASCII value of each letter. (b) The Postscript data file that printed this chapter. (c) The binary version of the assembly code of a typical computer program.

For example, consider the case of n consecutive 1's. This run is represented in fax machine code as

$$n = 64m + r$$

where the value of m is between 0 and 27 and the value of r is between 0 and 63. The value of m is the integer number of 64's that are contained in n, while r is the remainder, computed using the modulo-64 computation as

$$r = [n]_{\text{mod}(64)}$$

In the standard fax code, the term $64m$ is encoded separately as the *make-up* code word and r is encoded as the *terminating* code word. While all code words contain r, most codes contain a $64m$ component followed by an r component. For small values of n, corresponding to short white or black segments, the $64m$ component is zero and only the r component is transmitted.

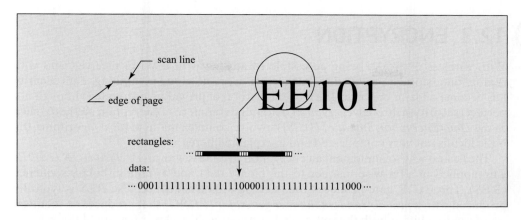

Figure 12.5

Fax sampling operation.

Make-up code words (64m)			Terminating code words (r)		
64m	white	black	r	white	black
64	11011	0000001111	0	00110101	0000110111
128	10010	000011001000	1	000111	010
192	010111	000011001001	2	0111	11
256	0110111	000001011011	3	1000	10
			4	1011	011
1600	010011010	0000001011011	5	1100	0011
1664	011000	0000001100100	6	1110	0010
1728	010011011	0000001100101	7	1111	00011
			8	10011	000101
			9	10100	000100
			10	00111	0000100
			EOL	000000000001	000000000001

Figure 12.6

Part of the standard Huffman code table used in a fax machine.

For example, the make-up (64m) and terminating (r) values for the following n values are computed as:

$n = 100$: $100 = 64 + 36$, so we have $64m = 64$ and $r = 36$.

$n = 8$: Because $8 = 0 + 8$, we have $64m = 0$ and $r = 8$.

$n = 2,000$: Dividing by 64 gives 31.25. Taking m to be the integer part gives $64m = 64 \times 31 = 1,984$ and r is the remainder computed as $r = 2,000 - 1,984 = 16$.

To reduce the number of bits that are transmitted, the values of $64m$ and r are applied to a Huffman code to encode the most common runs with the shortest code words. To obtain the effective probability values for the various runs of 0's and 1's, a wide variety and large number of text documents and line drawings were scanned by manufacturers.

Figure 12.6 shows selected elements of the resulting standard fax code. The end-of-line (*EOL*) is a separate code word that terminates a scan line. The remaining white space on a line does not need to be transmitted, thus saving time. It has been observed that the use of the Huffman code allows a ten-fold reduction, explaining the commercial success of the fax.

12.3 ENCRYPTION

With wireless networks being accessible to anyone with a Wi-Fi receiver, and with remote data storage provided by backup services and cloud computing, data security has become a major concern. The current method of providing adequate (although not perfect) security is through *encryption*. Although commercial encryption methods (such as the *Data Encryption Standard (DES)*) involve complex mathematical algorithms, the basic idea is not very complicated and is explained here.

The easiest way to implement an encryption scheme is with an *Exclusive-OR (ExOR)* gate applied in a bit-wise manner to the binary data and a random binary sequence (RBS). The ExOR gate produces the encrypted data sequence. The RBS is typically produced with a *pseudo-random number generator (PRNG)* that produces an identical set of random bits at both the source and destination. The clever feature of this scheme is that a RBS is applied to the data for encryption at the source and the same RBS is applied to the encrypted data at the destination to yield the original data. Meanwhile, the transmitted data on the channel appears to be a meaningless random collection of bits.

In the usual encryption operation, data are encoded at the source to conceal the values, the encrypted version is transmitted, and the received data are decoded at the destination to reveal the original data. Because large quantities of data are usually transmitted, the coding and decoding operations must be done quickly and efficiently. The ExOR gate is ideal for this purpose, because performing the ExOR operation encrypts the data at the source and then decrypt it at the destination to recover the original data.

To explain the encryption process, we use the symbol $\oplus$ to indicate the bit-by-bit ExOR operation on the data sequence $\mathcal{D}_i$ and random binary sequence RBS_i to produce the encrypted binary sequence $\mathcal{E}_i$ for $1 \le i \le n_x$.

For example, if data $\mathcal{D}_i = 0\ 1\ 0\ 1$ and $\text{RBS}_i = 1\ 1\ 0\ 0$, the bit-wise ExOR output produces the encrypted data $\mathcal{E}_i$ as

$$\mathcal{E}_i = \mathcal{D}_i \oplus \text{RBS}_i$$
$$= 0 \oplus 1 \quad 1 \oplus 1 \quad 0 \oplus 0 \quad 1 \oplus 0$$
$$= 1\ 0\ 0\ 1$$

In practice, the implementation may be with a single ExOR gate to which each bit in $\mathcal{D}_i$ and RBS_i are introduced sequentially or using multiple ExOR gates operating in parallel. For example, eight gates can operate simultaneously on each bit in a data byte.

To accomplish encryption successfully requires the same RBS at both the transmitting and receiving ends. A PRNG uses a seed (the *key*) as the initial value to compute the RBS. The encrypted message $\mathcal{E}$ is then transmitted to the destination, where the receiving party can compute the same RBS employed at the transmitter by using the same key value. To retrieve the original data $\mathcal{D}_i$, the ExOR operation is again performed, but this time on $\mathcal{E}_i$ and RBS_i, as

$$\mathcal{D}_i = \mathcal{E}_i \oplus \text{RBS}_i \quad \text{for} \quad 1 \le i \le n_x \tag{12.11}$$

By the magic of the ExOR operation, this operation reproduces the original data at the destination.

The following example illustrates this simple encryption scheme. We then describe how the random bits are generated with a pseudo-random number generator (PRNG) algorithm and, finally, how the secret value of the key can be (reasonably) securely transmitted over a spy-ridden communication channel.

Encrypting data | EXAMPLE 12.15

Let us consider data that consists of 16 hexadecimal numbers. As a simple illustration, let these be in order, from 0 to F, with each encoded using its equivalent 4-bit value, as shown in Figure 12.7

There are then $4 \times 16 = 64$ bits in this message, forming the binary data sequence $\mathcal{D}_i$ for $1 \leq i \leq 64$. To encrypt this binary sequence, the PRNG produced the RBS_i for $1 \leq i \leq 64$, as shown in Figure 12.7. To form the encrypted binary sequence $\mathcal{E}_i$ use the ExOR operation as

$$\mathcal{E}_i = \mathcal{D}_i \oplus RBS_i \text{ for } 1 \leq i \leq 64$$

The figure shows the encrypted values. The last column gives the hexadecimal equivalents of the encrypted binary values. Note that the hexadecimal values of $\mathcal{E}_i$ do not resemble those corresponding to the original message encoded in $\mathcal{D}_i$.

After $\mathcal{E}_i$ is received at the destination, the received data is decrypted by again applying it to the ExOR gate and the same RBS_i, as

$$\mathcal{D}_i = \mathcal{E}_i \oplus RBS_i \text{ for } 1 \leq i \leq 64$$

with results shown in the figure. Note that the hexadecimal values of the decrypted values are identical to the original data.

Encrypting waveforms | EXAMPLE 12.16

Figure 12.8 shows the encryption of a quantized sinusoidal waveform. Figure 12.8a shows the integer values in the range $[0, 15]$ of a sinusoidal waveform generated for transmission as

$$st_i = \text{round}[7.5(1 + \sin(2\pi i/32)] \text{ for } i = 0, 1, 2, \ldots, n_x - 1 \ (n_x = 33)$$

The corresponding 4-bit code words generated the binary data sequence D_i containing $4 \times 33 = 132$ bits. A RBS sequence of 132 random bits was generated with a PRNG. The source encrypted the data with the ExOR operation as

$$\mathcal{E}_i = RBS_i \oplus \mathcal{D}_i \text{ for } 0 \leq i \leq 131$$

This $\mathcal{E}_i$ was applied to a 4-bit DAC to construct the waveform Eq_i. Figure 12.8b shows that there is no resemblance between Eq_i and st_i, demonstrating the effectiveness of encryption.

At the destination the encrypted binary data were de-encrypted by forming D_i by the ExOR operation using the same RBS_i as in the encryption as

$$\mathcal{D}_i = RBS_i \oplus \mathcal{E}_i \text{ for } 0 \leq i \leq 131$$

This de-encrypted sequence was applied to a 4-bit DAC to construct the received samples sr_i. Figure 12.8c shows that sr_i is identical to st_i, verifying that the data has been retrieved exactly.

Encryption at Source

D	Binary Code	RN (Hex)	RBS	$\varepsilon = D \oplus RBS$	Encrypted Value
0	0000	E	1110	1110	E
1	0001	5	0101	0100	4
2	0010	D	1101	1111	F
3	0011	0	0000	0011	3
4	0100	1	0001	0101	5
5	0101	9	1001	1100	C
6	0110	8	1000	1110	E
7	0111	6	0110	0001	1
8	1000	5	0101	1101	D
9	1001	9	1001	0000	0
A	1010	7	0111	1101	D
B	1011	B	1011	0000	0
C	1100	8	1000	0100	4
D	1101	9	1001	0100	4
E	1110	A	1010	0100	4
F	1111	F	1111	0000	0

Decryption at Destination

ε	Binary Code	RN (Hex)	RBS	$D = \varepsilon \oplus RBS$	Decrypted Value
E	1110	E	1110	0000	0
4	0100	5	0101	001	1
F	1111	D	1101	0010	2
3	0011	0	0000	0011	3
5	0101	1	0001	0100	4
C	0110	9	1001	0101	5
E	1110	8	1000	0110	6
1	0001	6	0110	0111	7
D	1101	5	0101	1000	8
0	0000	9	1001	1001	9
D	1101	7	0111	1010	A
0	0000	B	1011	1011	B
4	0100	8	1000	1100	C
4	0100	9	1001	1101	D
4	0100	A	1010	1110	E
0	0000	F	1111	1111	F

Figure 12.7

Example of encrypting hexadecimal values at the sources and decryption at the destination. RN is a random hexadecimal number that forms the RBS.

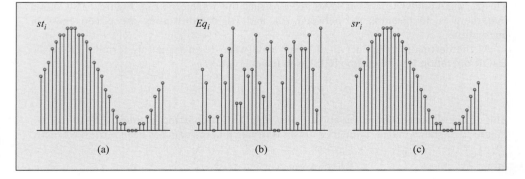

(a)   (b)   (c)

Figure 12.8

Encrypting a quantized waveform sequence: (a) Original. (b) Encrypted. (c) Decrypted.

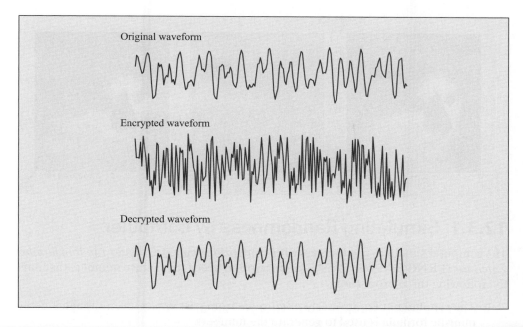

Original waveform

Encrypted waveform

Decrypted waveform

Figure 12.9

Encrypting a speech waveform.

Encrypting speech waveforms

EXAMPLE 12.17

Figure 12.9 shows the encryption of a speech waveform generated by a laptop microphone containing 200 samples. Each sample was expressed as an 8-bit sequence, forming $\mathcal{D}_i$ for $i = 0, 1, 2, \ldots, n_x - 1$ with $n_x = 1,600$. A PRNG generated 1,600 random bits to form RBS_i. The encrypted sequence was formed by the ExOR operation as

$$\mathcal{E}_i = \mathrm{RBS}_i \oplus \mathcal{D}_i \quad \text{for} \quad 0 \leq i \leq 1,599$$

The encrypted values were converted into an audio waveform and displayed in Figure 12.9.

The $\mathcal{E}_i$ was decrypted by forming D_i with the ExOR operation using the same RBS_i as in the encryption as

$$\mathcal{D}_i = \mathrm{RBS}_i \oplus \mathcal{E}_i \quad \text{for} \quad 0 \leq i \leq 1,599$$

The decrypted values were then converted into the audio samples and displayed in Figure 12.9 to show that the decrypted waveform matches the original.

Encrypting images

EXAMPLE 12.18

Figure 12.10 shows the encryption of a color image of a cat. A jpeg image taken with a digital camera was decoded to form a matrix containing 500 rows and 500 columns (250,000 pixels) with a third dimension that encoded 8 bits (256 levels) for red, 8 bits for green, and 8 bits for blue. The total number of data bits was equal to $3 \times 8 \times 2.5 \times 10^5 = 6 \times 10^6$ data bits, forming D_i, for $1 \leq i \leq 6 \times 10^6$. A PRNG generated 6 million random bits, forming RBS_i. The encrypted binary sequence was formed by the ExOR operation as

$$\mathcal{E}_i = \mathrm{RBS}_i \oplus \mathcal{D}_i \quad \text{for} \quad 1 \leq i \leq 6 \times 10^6$$

The encrypted values are displayed in Figure 12.10 to indicate there was no hint of a cat.

The encrypted image was decrypted by the ExOR operation using the same RBS_i as in the encryption as

$$\mathcal{D}_i = \mathrm{RBS}_i \oplus \mathcal{E}_i \quad \text{for} \quad 1 \leq i \leq 6 \times 10^6$$

The decrypted values are displayed in Figure 12.10 to verify that it appears exactly as the original.

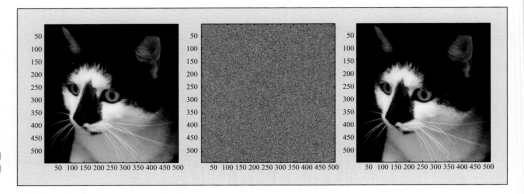

Figure 12.10

Encrypting a color image: Original (left). Encrypted (middle). Decrypted (right).

12.3.1 Simulating Randomness by Computer

The computer simulates randomness with an *algorithm* called a *pseudo-random number generator* (PRNG). A good PRNG for encryption needs to produce numbers that have the following three properties.

1. They should not be obviously *predictable* from past values, even though a deterministic formula is used to generate the numbers.

2. They should be *uniformly distributed* over the interval $[0, n_{max})$.

3. They should be *repeatable*, so that the same sequence of random numbers can be produced at a different time or in a different place.

We first consider generating random integers and then describe a method to generate random bits.

Generating Random Integers

Consider a sequence of n random integers, which is denoted

$$X_1, \ X_2, \ X_3, \ \ldots, \ X_n$$

A common formula used to compute random integers is given by

$$X_i = \mathrm{mod}[\ (\alpha \times X_{i-1} + \beta), \ h] \tag{12.12}$$

where

α is a factor that multiplies the previous integer in the sequence (X_{i-1})

β is a constant that prevents the sequence from degenerating into a set of zero values

h is the divisor in the modulus operation

All values are typically large positive integers. The parenthesis within the *mod* operator helps prevent confusion when the result of the multiplication and addition is a large number that contains commas. (Seeing too many commas makes it difficult to identify the value of h.)

The sequence of random numbers starts with X_1. To compute X_1 we need a starting value X_0, which is called the *seed* (the *encryption key*). This seed starts the sequence and different seeds produce different random sequences. Using the same number for the seed repeats the same random integer sequence. This is a useful property for encryption that needs the same random sequence at the source and destination.

The random integers produced by our PRNG using the modulus operator range from 0 (when the result within the parentheses is a multiple of h) to the largest remainder value $(h - 1)$. Hence, the random numbers lie in the range $0 \leq X_i \leq h - 1$.

Why does the formula in Eq. (12.12) produce random-looking numbers? The reason is that the modulo-h operation discards the quotient (the predictable part of the division) and retains the remainder (which is difficult to predict). It is not a perfect (but reasonable) compromise between true randomness and a repeatable computation.

A successful PRNG should generate billions of different random numbers. This requires α, β, and h to be large integers. The *mod* operator produces integers between 0 and $h-1$. Hence, the value of h limits the number of different random numbers. For example, $h = 1000$ can only produce (at most) one thousand different random numbers in the range [0,999]. Once a random integer X_i is repeated, the PRNG formula begins to repeat the entire sequence.

A pseudo-random sequence of 4-digit integers   EXAMPLE 12.19

Consider generating a sequence of 4-digit random integers. A 4-digit integer falls in the interval $[0, 9{,}999]$ by setting $h = 10{,}000$. Making h a power of 10 (say $h = 10^n$) is particularly convenient for computing Eq. (12.12) because the modulus operation result is simply the n least significant digits in the computation. With $h = 10^4$ the modulus operation produces the 4 least significant digits of the multiplication.

There are no optimum values for α, β, and X_0, other than being large. We arbitrarily set $\alpha = 97{,}531$, $\beta = 54{,}321$, and the seed $X_0 = 6{,}789$. These values generate the sequence of random numbers as

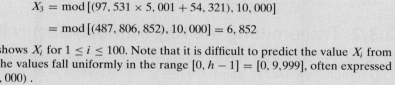

$$X_1 = \mathrm{mod}\,[(97{,}531 \times 6{,}789 + 54{,}321), 10{,}000]$$
$$= \mathrm{mod}\,[(662{,}192{,}280), 10{,}000] = 2{,}280$$

$$X_2 = \mathrm{mod}\,[(97{,}531 \times 2{,}280 + 54{,}321), 10{,}000]$$
$$= \mathrm{mod}\,[(222{,}425{,}001), 10{,}000] = 5{,}001$$

$$X_3 = \mathrm{mod}\,[(97{,}531 \times 5{,}001 + 54{,}321), 10{,}000]$$
$$= \mathrm{mod}\,[(487{,}806{,}852), 10{,}000] = 6{,}852$$

Figure 12.11 shows X_i for $1 \le i \le 100$. Note that it is difficult to predict the value X_i from X_{i-1}, and all of the values fall uniformly in the range $[0, h-1] = [0, 9{,}999]$, often expressed as $[0, h) = [0, 10{,}000)$.

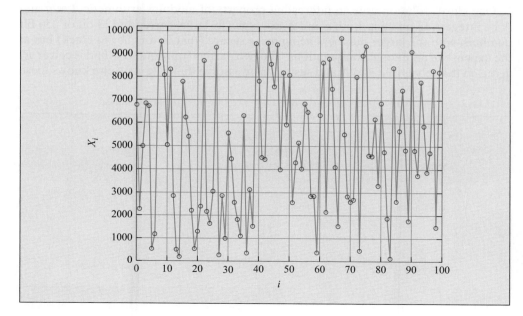

Figure 12.11

Pseudo-random sequence X_i for $1 \le i \le 100$ generated with $\alpha = 97{,}531$, $\beta = 54{,}321$, and $h = 10{,}000$ with $X_0 = 6{,}789$.

Generating Random Bits

The random integers generated by Eq. (12.12) are approximately uniformly distributed over the interval $[0, h)$. We choose an even value for h to form two equal sub-intervals defined as

$$[0, h/2) \text{ and } [h/2, h) \qquad (12.13)$$

The random binary sequence value RBS_i is determined by the sub-interval X_i falls into, as

$$\text{If } 0 \leq X_i < h/2 \;\rightarrow\; \text{RBS}_i = 0 \qquad (12.14)$$

$$\text{If } h/2 \leq X_i < h \;\rightarrow\; \text{RBS}_i = 1$$

EXAMPLE 12.20   **A pseudo-random sequence of random bits**

A sequence of three random bits is generated from the random integers in Example 12.19 using $h = 10,000$. Repeating the sequence of random integers and testing if $0 \leq X_i < h/2 (= 5000)$, we get

$$X_1 = 2,280 \;\rightarrow\; \text{RBS}_1 = 0$$

$$X_2 = 5,001 \;\rightarrow\; \text{RBS}_2 = 1$$

$$X_3 = 6,852 \;\rightarrow\; \text{RBS}_3 = 1$$

12.3.2 Transmitting the Key (Almost) Securely

To generate the same RBS at both the transmitting and receiving ends with the same PRNG formula, we need to transmit only the key and use it as the seed. The problem is how to get the key from the source to the destination in a secure fashion that does not involve a personal meeting. Figure 12.12 shows the process of transmitting the key over an unsecure channel.

One clever solution uses an interesting property of modulus arithmetic. Let a and N be integers. In practice, both a and N are typically large, currently 128-bit or 256-bit numbers, with the larger numbers being more secure (that is, tougher to crack) but at the expense of additional computations. The two parties (transmitter T and receiver R) agree on the values of a and N and suspiciously assume that everyone else knows these

Figure 12.12

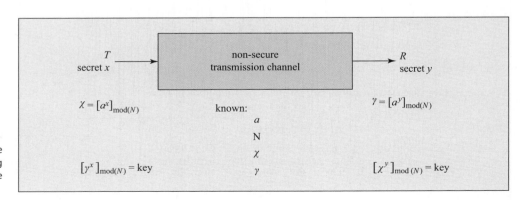

Encrypting information can be done over a non-secure channel by using modulo arithmetic to transmit the value of the *key*.

values as well. T secretly chooses integer x and computes $\mathcal{X}$, the modulus-N value of a^x, denoted as

$$\mathcal{X} = [a^x]_{\text{mod}(N)} \tag{12.15}$$

T transmits $\mathcal{X}$ to R over a channel that everyone sees. R also secretly chooses integer y and computes

$$\mathcal{Y} = [a^y]_{\text{mod}(N)} \tag{12.16}$$

and R transmits $\mathcal{Y}$ to T. Both keep secret x (known only to T) and y (known only to R).

Having received $\mathcal{Y}$ from R, T computes

$$K_T = [\mathcal{Y}^x]_{\text{mod}(N)} \tag{12.17}$$

Having received $\mathcal{X}$ from T, R computes

$$K_R = [\mathcal{X}^y]_{\text{mod}(N)} \tag{12.18}$$

With modulus arithmetic, it turns out that these last two computations produce the same result:

$$K_T = K_R \tag{12.19}$$

Because both T and R arrive at the same number, it can be the seed in the PRNG for generating the RBS for encryption for both parties.

The reason that this simple scheme works is because (even if the values of a, N, $\mathcal{X}$, and $\mathcal{Y}$ are known to all) the problem of determining the value of the key from these quantities is a very difficult and time-consuming computational process, especially if x and y are large integers.

Computing the encryption key   EXAMPLE 12.21

For a simple example, let $N = 23$ and $a = 13$. Secretly, T chooses $x = 7$ and R chooses $y = 10$.

- T computes $\mathcal{X}$ as

$$\mathcal{X} = [a^x]_{\text{mod}(N)} = [13^7]_{\text{mod}(23)} = [62, 748, 517]_{\text{mod}(23)} = 9$$

and sends it to R.

- R computes $\mathcal{Y}$ as

$$\mathcal{Y} = [a^y]_{\text{mod}(N)} = [13^{10}]_{\text{mod}(23)} = [137, 858, 491, 849]_{\text{mod}(23)} = 16$$

and sends it to T.

- T computes the key as

$$K_T = [\mathcal{Y}^x]_{\text{mod}(N)} = [16^7]_{\text{mod}(23)} = [268, 435, 456]_{\text{mod}(23)} = 18$$

- R computes the key as

$$K_R = [\mathcal{X}^y]_{\text{mod}(N)} = [9^{10}]_{\text{mod}(23)} = [3, 486, 784, 401]_{\text{mod}(23)} = 18$$

This key is used by both T and R as the seed in their identical PRNGs to generate the same RBS for encryption (by T) and decryption (by R).

In the previous example, even if we know that $N = 23$, $a = 13$, $\mathcal{X} = 9$, and $\mathcal{Y} = 16$ by intercepting data transmission, to find the key we need to solve for the values of x

and y that obey the formula

$$[16^x]_{\text{mod}(23)} = [9^y]_{\text{mod}(23)} \tag{12.20}$$

The solution is not easy to find because we cannot simply plug in a value for one to find the other. Finding the value of the key from the known quantities requires trying all possible pairs of integers (x, y) to check which pair satisfies the equation

$$[\mathcal{X}^y]_{\text{mod}(N)} = [\mathcal{Y}^x]_{\text{mod}(N)} \tag{12.21}$$

If the pair of integers x and y are each in the range $[1, 100]$, then an exhaustive search needs to examine $10^2 \times 10^2 = 10^4$ combinations. On average, an answer would be found by performing half these computations, because some (x, y) pairs occur sooner in the search than others. However, even the large range $[1, 10^9]$ is small by encryption standards.

EXAMPLE 12.22 Number of calculations for finding key

Assuming you know $\mathcal{X}$ and $\mathcal{Y}$ and N, let us compute the average number of computations needed to find the key value if x and y are 128-bit numbers.

A 128-bit integer spans the range from 0 to

$$2^{128} = 2^8 \times \left(2^{10}\right)^{12} = 256 \times \left(10^3\right)^{12} = 2.56 \times 10^{38}$$

The number of pairs of two 128-bit numbers equals

$$(2.56 \times 10^{38})^2 = 6.55 \times 10^{76}$$

On average, the key could be found by performing half this number or 3.3×10^{76}. If it takes $1\ \mu s$ to perform the computation in Eq. (12.21), it would take

$$3.3 \times 10^{76} \times 10^{-6}\,\text{s} = 3.3 \times 10^{70}\,\text{s}$$

With 32 million seconds per year, the computation would take 10^{62} years!

In addition, evaluating the modulo-N value for 128-bit numbers is an involved computation requiring about 40 digits of precision, which is well beyond the integer range employed by most current computers.

12.4 Research Challenges

12.4.1 Security

With the impending Internet of Things (IoT), security concerns are entering a new dimension. While former issues with securing data have led to increasingly sophisticated encryption methods, computer-controlled homes and automobiles that can be accessed through a smartphone introduce new problems. The competition between hackers hacking and problem solvers responding is shifting to the problem solvers identifying the issues before hackers can exploit them.

12.5 Summary

This chapter introduced source coding concepts of data compression and encryption. The source entropy was used as a measure of information. The source entropy equals the average number of bits that are needed to encode the symbols produced by a source, while the effective file entropy measures the number of bits required to encode a file. The Huffman coding procedure generates variable-length code words as a means for reducing the quantity of data that needs to be stored or transmitted without any information loss. The idea is to assign short code words to the most common symbols and the longest code words to the least common. A code word translation table also needs to be transmitted. Some applications (fax machines) can generate a code and incorporate the table within the system, thus eliminating the need to transmit it. Data security is reasonably maintained through encryption with a pseudo-random number generator that is common to both the transmitter and receiver. Transmitting the seed is performed by employing modular arithmetic, which requires exhaustive computations to crack.

12.6 Problems

12.1 **Modeling a source with a die toss.** A die has six sides and produces a symbol corresponding to the number of dots showing on the top face. A fair die produces symbols that are equiprobable. What are the symbols and their probabilities?

12.2 **Source entropy of the die-toss source.** Compute $\mathcal{H}_S$ for the source that produces symbols X_1 through X_6 with probabilities modeled by a fair die toss.

12.3 **Source entropy of the unfair die-toss source.** Compute $\mathcal{H}_S$ for the source that produces symbols with probabilities modeled by an unfair die toss, in which a 1 appears one quarter of the time and 2 through 6 have equal probabilities.

12.4 **Source entropy of the decimal digit source.** Compute $\mathcal{H}_S$ for the source that produces ten symbols that have equal probabilities.

12.5 **Data sequences having specified $\mathcal{H}_S$.** Compose a representative sequence containing ten or more symbols that would be typical of a source having the following source entropy value.

a. $\mathcal{H}_S = 0$ bits/symbol

b. $\mathcal{H}_S = 1$ bit/symbol

c. $\mathcal{H}_S = 2$ bits/symbol

12.6 **Effective probabilities of symbols in a text file.** A file contains the following text.

```
i need a vacation
```

Include the space as a separate symbol. What are the values for the vocabulary size m_x, total file size n_T, and the effective probabilities of the symbols in this file?

12.7 **Effective source entropy of a text file.** Use the values in Problem 12.6 to compute $\hat{\mathcal{H}}_S$.

12.8 **Huffman code a text file.** Generate a Huffman code for the text file in Problem 12.6. Encode the file with the variable-length code and compare the total bit count to the $\hat{\mathcal{H}}_f$ value.

12.9 **Valid or invalid Huffman code.** Does the following code represent a valid Huffman code? (Hint: draw the tree.)

$$X_1 : 0 \quad X_2 : 01 \quad X_3 : 10 \quad X_4 : 11$$

12.10 **Huffman code for digits.** Generate the Huffman code for the ten digits 0 through 9 that have equal probabilities.

12.11 **Data compression.** Consider the following data file.

AAACAAABAAACAAADAAAE

Implement a Huffman code to compress this file. What is the average number of bits per symbol?

12.12 **Huffman tree encoding.** Find the binary sequence generated by the symbol sequence

$$X_1 \ X_3 \ X_2 \ X_5 \ X_4 \ X_6$$

using the Huffman code tree shown in Figure 12.13.

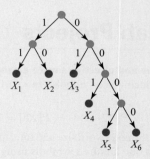

Figure 12.13

Huffman code tree for Problem 12.12

12.13 Huffman tree decoding. What symbols are represented by the binary sequence

$$01110010001000010$$

using the Huffman code tree shown in Figure 12.14.

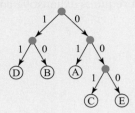

Figure 12.14

Huffman code tree for Problem 12.13

12.14 Huffman code tree for symbols having unequal probabilities. A coin flip produces a 0 if a tail appears and a 1 for a head. A source produces symbols that have the same probabilities as the sums that are observed when three fair coins are flipped. For example, the outcome HTH produces 101, which sums to 2. Determine the Huffman code tree for this source.

12.15 Source encoding. A source generates six symbols with $P[X_1] = 0.05$, $P[X_2] = 0.47$, $P[X_3] = 0.07$, $P[X_4] = 0.20$, $P[X_5] = 0.11$, and $P[X_6] = 0.10$.

a. Compute the source entropy $\mathcal{H}_S$.

b. Generate the Huffman code tree.

c. Encode the ten-symbol sequence

$$X_3 \quad X_1 \quad X_3 \quad X_5 \quad X_6 \quad X_1 \quad X_6 \quad X_3 \quad X_1 \quad X_3$$

d. Compute the average number of bits per symbol used for encoding the sequence.

e. Compare the values of the source entropy and the average bits/symbol computed.

f. Why does the $\mathcal{H}_S$ value differ significantly from the average bits/symbol?

12.16 Huffman code for a symbol file. Consider the data file containing the ten symbols

$$X_3 \quad X_3 \quad X_5 \quad X_1 \quad X_2 \quad X_5 \quad X_1 \quad X_1 \quad X_2 \quad X_4$$

a. Specify a fixed-length code to encode the symbols and compute the total number of bits that encode the file.

b. Compute the effective source entropy of the file $\hat{\mathcal{H}}_S$.

c. Generate a Huffman code to encode the file.

d. Encode the file using this code and compute the total number of bits that encode the file.

12.17 Encryption. Let

$$RBS_i = 10011001$$

Use RBS_i to encrypt the data

$$\mathcal{D}_i = 11001011$$

to form the encrypted binary sequence $\mathcal{E}_i$ and to decrypt it.

12.18 Pseudo-random number algorithm. Letting $X_0 = 123$, generate X_1 and X_2 using the PRNG formula

$$X_i = \text{mod}[\ (\alpha \times X_{i-1} + \beta),\ h]$$

with $\alpha = 766$, $\beta = 369$, $h = 1,000$.

12.19 Random numbers to random bits. Assume the pseudo-random number algorithm in Problem 12.18 generated the eight random numbers

$$979, 122, 475, 986, 730, 286, 66, 899$$

Describe how you would generate eight random bits from this sequence and generate the random bit sequence.

12.20 Encryption key. Compute the value of the encryption key using the values: $n = 10$, $a = 13$, $x = 5$, $y = 6$.

12.7 Matlab Projects

12.1 Source entropy and effective source entropy. Consider a source producing six symbols ($m = 6$) having equal probabilities

$$P[X1] = P[X2] = \cdots = P[X6] = 1/6$$

a. Determine the code word size and total number of bits that encode one hundred symbols using fixed-length code words.

b. Describe your method using Matlab PRNG to generate random symbols that simulate this source.

c. Generate 100 such random symbols. Compute the effective probabilities $P_e[X_i]$ and the effective source entropy $\hat{\mathcal{H}}_S$.

d. Compare the $\hat{\mathcal{H}}_S$ value to the source entropy $\mathcal{H}_S$.

12.2 Data compression. Continue Project 12.1 to do the following.

a. Form a Huffman code tree from the effective probabilities.

b. Encode the 100 symbols using the variable-length code words. Count the number of bits in the encoded file and compare the the effective entropy.

c. Ignoring the code table for the Huffman code, what is the savings in bit count in performing the Huffman code compared to using fixed-length code words.

12.3   **Effective entropy of microphone speech.** Using the Matlab script in Examples 16.9 and 16.60 as guides, acquire speech data from the microphone on your laptop, compute the histogram of the speech samples, and compute the source entropy of the speech source.

12.4   **Encryption of microphone speech.** Using the Matlab script in Example 16.62 as a guide, encrypt speech data from the microphone on your laptop. Play the original speech, followed by the encrypted speech and the de-encrypted speech.

12.5   **Encryption of JPEG image.** Using the Matlab script in Example 16.63 as a guide, encrypt a jpeg image file. Display the original image along with the encrypted and de-encrypted images.

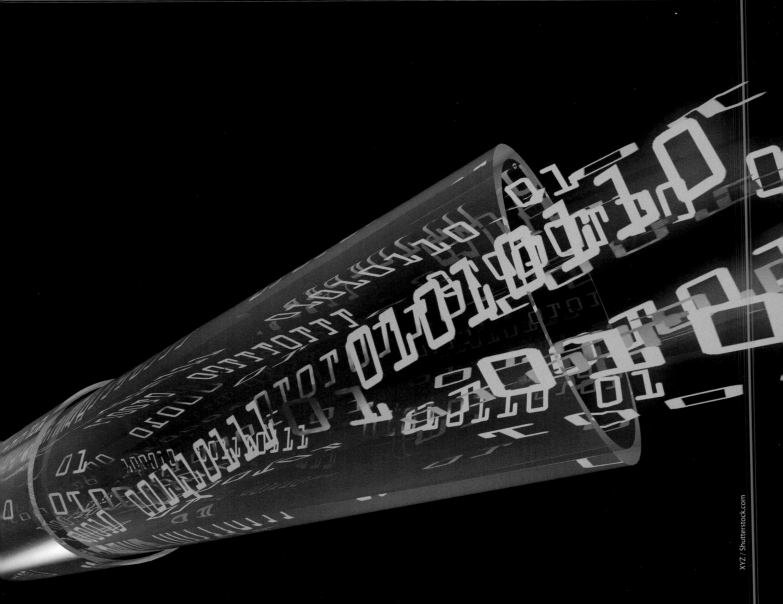

CHANNEL CODING

LEARNING OBJECTIVES

After completing this chapter, the reader should be able to:

- Understand how to model errors in digital data.

- Detect and correct single errors in digital data under a variety of error conditions.

- Understand the trade-off between error-correction capability and data quantity.

- Compute the data carrying capacity of a data transmission channel having a specified bandwidth and signal-to-noise ratio.

This chapter describes channel coding techniques for reliable storage or transmission over a communication channel. Several schemes are suggested to detect and even correct errors in data. The simplifying *single error assumption* (SEA) is employed to illustrate the basic ideas. Commercial systems correct more than one error but are mathematically more complicated and are topics for future study. All error-correction methods use some form of redundancy by adding bits whose values are computed from the original data. This redundancy results in an increased data size that is measured with the *data increase factor* (DIF). These data are transmitted over a channel having a limited *channel capacity*, which measures the rate data can be reliably transmitted over a noisy channel. For reliable data transmission, the channel capacity must be greater than the data rate produced by the source.

13.1 INTRODUCTION

This chapter introduces the problem of transmitting data from a source to destination accurately and reliably. To insure successful operation, a data transmission system must confront the following issues.

Errors—Data errors are inevitable because the Prob[error] > 0 for practical data transmission systems. Because the magnitude of Gaussian noise is theoretically unlimited, there is always the possibility of error occurrence. Error-detection codes determine when errors occur, and error-correction codes restore the correct data. But these codes come with the cost of increased data size and computation times. Errors in most properly operating systems are rare. A practical system usually begins by working flawlessly. As the system degrades, errors begin to occur. It is at this time that error correction is useful: not only does it allow the system to continue to operate as intended, but detecting the occurrence of errors indicates that the system is beginning to fail so that the system can be repaired or replaced.

Error correction—Although practical error correction techniques use advanced mathematical ideas, this chapter presents the fundamental ideas by considering these operations under the simplifying constraint of single errors.

Data rate $\mathcal{D}_S$—For data transmission, it is important to measure how many bits a source produces per second. The source data rate $\mathcal{D}_S$ is expressed in bits per second and is convenient for comparing data sources.

Channel capacity $\mathcal{C}$—Binary data are transmitted over a channel having a capacity measured in bits per second that is limited by noise. For reliable data transmission, the channel capacity $\mathcal{C}$ must be greater than the data rate $\mathcal{D}_S$ of the source.

13.2 DATA STRUCTURES

Figure 13.1 shows the different data structures discussed in this chapter. These include

Data: Binary data are produced by a source, which can be single-bit or multiple-bit units.

Figure 13.1

Data are produced by a source, form code words for error detection and correction, and packed into data packets for transmission. The packet interval is the total count of bits in a packet.

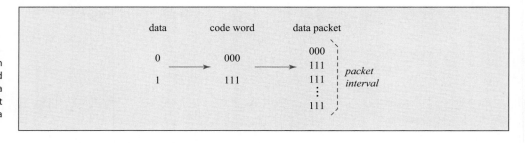

Code word: Code words add redundant bits (that is, bits computed from the data values) to form a fixed-length data structure to facilitate error detection and/or error correction.

Data packet: Data packets are blocks of code words that contain additional redundant code words for error correction. Data packets form the basic data structures for transmission over a communication channel or for storage in a digital memory. Data packet sizes range from a single code word to thousands of code words.

Typical communication systems employ standard data packets with code words that are bytes (containing 8 bits) plus an appended parity bit. Data packets terminate in a checksum byte with its own parity bit. Such data packets conform to computer data structures to simplify processing and storage tasks. For sources that produce different fixed-length or variable-length code words, the transmitter *packs* these binary data into standard data packets and the receiver *unpacks* these standard data packets to form the original code words.

13.3 CODING FOR ERROR CORRECTION

Error correction is the process of adding redundant bits to *data* produced by a source to form *code words*. The code words are grouped into *data packets* in order to detect and correct errors during transmission over a channel or storage in a digital memory. Other terms for this topic are *channel encoding* and *forward error correction*. While this field is mathematically rich, this chapter presents the basic ideas by constraining the number of errors that can occur.

The basic simplifying assumption used in this book is the *single error assumption* (*SEA*): *No more than one error can occur during a data packet interval*, as shown in Figure 13.2. Data packet intervals range in size from the smallest data packet that occurs for error correction (a 3-bit code word) to large data packets that can be formed by grouping a large number of code words into a single block.

The validity of the SEA for specific channel determines the data packet interval: If errors are common, the packet interval must be small (3 bits). If errors are extremely rare, data packets can be large, with the maximum size being determined by transmission factors other than errors. Channel conditions can change, and if multiple errors cause the SEA to be no longer valid for a particular packet interval, the packet interval must be reduced. The shortest packet interval is a 3-bit code word that accomplishes error correction (shown next).

Unlike analog waveforms, digital data can be expanded to include additional bits that help the system detect and even correct errors when they occur. To understand the trade-offs in providing error detection and correction, we compute a *data increase factor* (DIF) that measures the cost of achieving these capabilities.

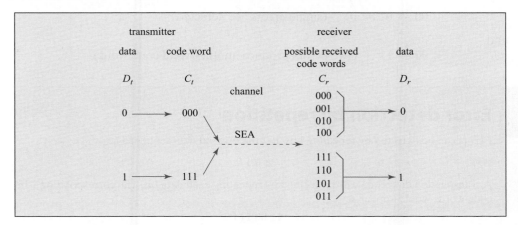

Figure 13.2

Signal error assumption (SEA) where no more than one error can occur during a data packet interval. The simplest data packet is shown, consisting of a single 3-bit code word.

13.3.1 Redundancy and Data Increase Factor

Errors occasionally occur when data are transmitted. To make the system robust, it is desirable to detect when an error has occurred, and if the system requirements demand it, to correct the error. Codes can be designed to detect errors in the received data by adding *redundant* bits to data to form *transmission code words*. Such bits add nothing to the information but serve to confirm it. Redundant bits have values that can be computed from the original data. This computation is also performed by the receiver upon reception of the code word and the result is compared to the observed values of the transmitted redundant bits included in the code word. If the computed and transmitted values of the redundant bits agree, there is no error. Otherwise, error detection or correction can be performed, depending on the system design.

We often use redundancy in our conversations: A *yes* answer is often accompanied by a nod of the head, and a verbal instruction to turn right is often clarified by a hand gesture. When the visible gesture is inconsistent with the verbal information, the recipient detects that an error has occurred. In a similar manner, transmitted code words should contain a mechanism by which the receiver is able to detect that an error has occurred. This is the purpose of error-detection codes.

The basic error recovery mechanism at the detector is ***redundancy***—either by repetition, by including additional bits, and/or by designing specific bit patterns. A little redundancy provides ***error detection***, while additional redundancy provides ***error correction***. However, these additional data take longer to transmit and require more memory to store.

More robust error-correction codes produce more data. We define the *data increase factor* (DIF) as

$$\text{DIF} = \frac{\text{final size of data packet}}{\text{original data size}} \tag{13.1}$$

We now describe error detection and error correction methods.

13.3.2 Bit-Level Error Correction

Different coding systems employ special forms of error detection and correction. Each form is designed to treat the types of errors each system encounters. This section starts with the simplest systems and increases the system complexity to show how the considerations change.

Bit-Level Error Detection

Bit-level error detection can be achieved by simply repeating the bits (that is, transmitting each data bit twice). This results in DIF = 2.

When does it work? As long as there is no more than one error per 2 bits (SEA), so

$$00 \rightarrow 01 \; or \; 10 \quad \text{(single errors are detected)}$$

and

$$00 \rightarrow 11 \quad \text{(two errors produce another valid code word)}$$

EXAMPLE 13.1

Error detection by repetition

Errors can be detected by repeating bits. Let the original data sequence be

$$1\,0\,1\,0$$

A transmission data code word is formed by repeating each data bit. The corresponding 2-bit code words would appear as

$$11\,00\,11\,00$$

Repeating the data, of course, does not add any new information, so these second bits in each 2-bit code word are redundant. These redundant bits allow an error in the transmission of a code word to be detected. If the received data appeared as

11 01 11 00

it would be readily apparent to the receiver that an error had occurred in the second code word. However, the receiver could not determine if the true value of the code word is either 11 or 00. That determination is the function of error-correcting codes.

The trade-off for this error-detection capability is that the size of the original data has been doubled in forming the transmission code words. Thus, DIF = 2.

13.3.3 Bit-Level Error Correction

Single errors can be corrected by adding additional redundant bits to the code words. Upon reception, these redundant values are computed from the data and compared against the values that were received. With the assistance of error detection, which indicates the code word containing the error, error correction corrects the error.

Bit-level error correction is achieved by repeating the bits three times. This results in DIF = 3. The error-correction mechanism uses the majority count to determine the correct bit value.

When does it work? As long as there is only one error per three bits, the error can be corrected. If there are two errors, the error can be *detected*, but the *correction* will be wrong. Figure 13.3 shows an interpretation of an error correction code that illustrates an idea proposed by a Bell Labs engineer Richard Hamming in the 1950s. Representing an error-correction code in a three-dimensional space, we can see how far (called a *Hamming distance*) from a valid code a single error will take us. A single error-correction technique has a Hamming distance equal to one and still keeps the code within the vicinity of a valid code word.

Error correction by repetition   EXAMPLE 13.2

Errors can be corrected by repeating bits. A simple error-correction method is to form a 3-bit code word from each original data bit. Let the original data sequence be represented by

1 0 1 0

A 3-bit code word representation of these data would appear as

111 000 111 000

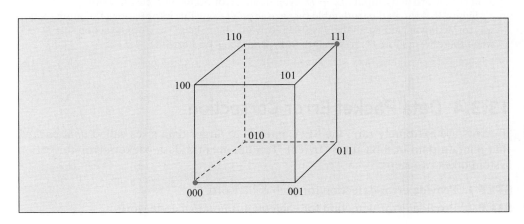

Figure 13.3

Representing an error correction code in a three-dimensional space. A single error still keeps the code within the vicinity of a valid code word (either 000 or 111).

Clearly, repeating the data does not add any new information, so these additional two bits are redundant. These redundant bits allow a single error in the transmission code word to be corrected. If the received data appeared as

$$111 \quad 010 \quad 111 \quad 000$$

a majority count performed on each 3-bit code word involves noting the values of most of the bits in the received code word. It would be readily apparent to the receiver that an error had occurred in the second code word and that the true value should be 000. Correcting binary data is especially simple, requiring only flipping the erroneous bit value from a 1 to a 0 in this case, to produce the corrected sequence. Hence,

$$111 \quad 000 \quad 111 \quad 000$$

The trade-off for this error-correction capability is that the size of the data has been tripled in forming the transmission code words. Thus, DIF = 3.

Error Correction Logic Circuit

A repeated-bit transmitted triplet permits error correction. Let each binary value D_t be transmitted as a triplet of identical values DA_t, DB_t, and DC_t with

$$DA_t = DB_t = DC_t = D_t \tag{13.2}$$

That is, if $D_t = 0$, the transmitted values will be $DA_t\ DB_t\ DC_t = 000$, and if $D_t = 1$, the transmitted values will be $DA_t\ DB_t\ DC_t = 111$.

These transmitted values travel over a communication channel that introduces noise that can cause an error at the receiver. Under SEA, only one error at most can occur in the transmitted data triple. If $DA_t\ DB_t\ DC_t$ are transmitted, the receiver detects the triplet, forming the received data as $DA_r\ DB_r\ DC_r$, which can only have one of the following eight values

$$000 \quad 001 \quad 010 \quad 011 \quad 100 \quad 101 \quad 110 \quad 111$$

Since $DA_t\ DB_t\ DC_t$ had identical values, transmission errors occurred in the second through seventh triples, in which non-identical binary values were detected.

An error-correction logic circuit examines $DA_r\ DB_r\ DC_r$ and produces $D_r = 0$ if $D_t = 0$ and $D_r = 1$ if $D_t = 1$ when SEA holds.

EXAMPLE 13.3

Error-correction logic circuit

Assume SEA holds and only a single error can occur in $DA_r\ DB_r\ DC_r$. Let D_t be transmitted as a code word of three identical values. An error-correction circuit examines each received 3-bit code word and outputs $D_r = D_t$ even if one error occurs in the code word.

Figure 13.4 shows the truth table that was formed using this information. The input section was formed by including all the possible received 3-bit values. The output section sets $D_r = 0$ when there were more 0's received than 1's, and $D_r = 1$ otherwise.

13.3.4 Data Packet Error Correction

If errors are extremely rare, the SEA extends to larger data sizes called *data packets* that contain data in 8-bit units (bytes). To implement a data packet error-correction system takes two steps:

STEP 1. Provide error detection for each code word.

STEP 2. Provide error detection for each *bit position* in a code word.

DA_r	DB_r	DC_r	D_r
0	0	0	0
0	0	1	0
0	1	0	0
0	1	1	1
1	0	0	0
1	0	1	1
1	1	0	1
1	1	1	1

Figure 13.4

Truth table for error correction with received data DA_r, DB_r, DC_r as inputs when transmitted data $DA_t = DB_t = DC_t = D_t$. Output D_r equals D_t under SEA.

This combination identifies the erroneous code word and the bit position of the error. This information is all that is needed to correct single errors in the data packet merely by flipping the erroneous bit. The allowable data packet size is related to the validity of SEA: If SEA is valid over a large number of bits, the data packet can be large. As a channel degrades, however, multiple errors begin to be encountered in each data packet and is sensed by the error detection logic. However, with our scheme, two errors cannot be corrected, although more sophisticated error-correction techniques can correct such multiple errors. When error-detection logic shows more than one error per packet, the SEA is no longer valid, necessitating the reduction in the data packet size. These smaller packets pay the cost of an increase in the DIF, as shown next.

Code Word-Level Error Detection

Code word-level error detection is achieved by adding an extra bit, called a *parity bit*, to each byte. The parity bit value is such that the count of 1's in the byte and the parity bit is always either an even or odd number for all code words. When the count is even, it is called *even parity*, and when odd, it is *odd parity*. For interpreting the count, zero is considered to be an even number. The choice of which parity to use is predetermined— known to both the transmitter and receiver—and depends on the failure mechanism that causes errors, as explained next. This error-detection scheme is successful as long as the SEA holds. If the data size is n bits, the code word size is $n + 1$, producing

$$\text{DIF} = \frac{n+1}{n} \qquad (13.3)$$

For the common byte-sized data $n = 8$ and the DIF $= 9/8 = 1.125$.

Parity bit for code-word error detection   EXAMPLE 13.4

Let the original data sequence consist of 2-bit data values (called dibits). An example of four dibits is

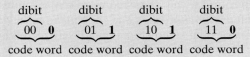

An *even-parity* system forms code words by appending a parity bit to each dibit such that each code word contains an even number of 1's. Forming 3-bit even-parity code words from the dibits produces 3-bit code words, as

dibit   dibit   dibit   dibit
00 **0**   01 **1**   10 **1**   11 **0**
code word code word code word code word

Since the parity bits do not add any new information, they are redundant. These parity bits allow an error in the transmission of each 3-bit code word to be detected.

To illustrate the error-detection procedure, consider receiving code words equal to

$$010 \quad 011 \quad 101 \quad 110$$

It is readily apparent to the receiver that an error occurred in the first received code word because its parity is not even. However, the receiver could not determine the bit position of the error. If the error occurred in the first bit position, the correct code word is 110, and if the error occurred in the third bit position, the correct word is 011.

This error-detection capability is achieved at the expense of data size, which has been increased by one bit in forming the 3-bit code words. Thus,

$$\text{DIF} = \frac{2+1}{2} = 1.5$$

EXAMPLE 13.5 — Choice of parity

The choice of even or odd parity is determined by the failure mode of the system. For example, the standard computer memory is structured to contain byte-sized data, each with an additional ninth bit for parity to form 9-bit code words. Assume the particular technology used to implement the computer memory fails by having a faulty memory location produces all 0's, as

data byte
00000000 0
9-bit code word

These nine 0's produce a valid even-parity code word, so the bad memory location would not be detected when using even parity. In this case, the memory should employ odd parity so the all-0's value would be detected as a failure.

Now consider a memory technology that fails by producing all 1's, as

data byte
11111111 1
9-bit code word

These nine 1's produce a valid odd-parity code word. In this case, the memory should employ even parity so the all-1's value would be detected as a failure.

Correcting Data Packet Errors

If adding parity bits to bytes detects code-word errors, adding a parity bit to detect errors in each *bit position* would provide sufficient information to correct errors in a data packet. This bit-position parity is implemented by appending a *longitudinal redundancy check code word (LRC)* to the end of a data packet. The LRC value is computed from the data in the packet, so it is redundant. Because each data byte appends a parity bit (for code word error detection), providing a parity bit for each bit position in the code word would locate the position within the code word that contains the error. For data packets, SEA is interpreted as *no more than one error in the entire data packet, including the LRC*. Once the erroneous bit is detected, it can be corrected by merely flipping its value.

1.

code word 0	*code word 1*	*code word 2*
10101011 **1**	11110000 **0**	10111010 **1**
data byte 0	*data byte 1*	*data byte 2*

2.

10101011	**1**	*code word 0*
11110000	**0**	*code word 1*
10111010	**1**	*code word 2*

3.

10101011	1	*code word 0*
11110000	0	*code word 1*
10111010	1	*code word 2*
11100001		*LRC byte*

4.

10101011	1	*code word 0*
11110000	0	*code word 1*
10111010	1	*code word 2*
11100001	**0**	*LRC code word*

Figure 13.5

Steps in forming a data packet with even parity and LRC.

Forming a data packet with a LRC code word

EXAMPLE 13.6

Consider a system that performs code word error correction on data bytes with even-parity bits that form 9-bit code words. A data packet with three data bytes, even parity, and an LRC is formed using the following steps, as shown in Figure 13.5

1. Append even-parity bits to the three data bytes to form code words.
2. Stack the code words so the bit positions align in columns.
3. Add a LRC byte whose bit values form even parity in each data-bit column.
4. Form an LRC code word by appending an even parity bit to the LRC byte. The LRC parity bit also forms the even parity for the entire data packet.

The original data consisted of three bytes (24 bits). Code-word parity bits increased the data packet to 27 bits and adding the 9-bit LRC produced a data packet containing 36 bits. The data increase factor is

$$\text{DIF} = \frac{36}{24} = 1.5$$

The DIF decreases when the data packet contains more data. A data packet that contains m 8-bit bytes has m 9-bit code words. The LRC code word increases the packet size by 9 bits. Thus,

$$\text{DIF} = \frac{9(m+1)}{8m} \tag{13.4}$$

The previous example had $m = 3$, giving DIF $= 36/24 = 1.5$.

As the data packet size m increases, the additional LRC code word becomes less significant and the DIF approaches $9/8 = 1.125$, which is the value for parity-bit error detection. This small DIF makes the addition of an LRC code word an efficient and preferred method for transmitting data packets over cellular systems and the Internet.

(a) Received data packet

b_7	b_6	b_5	b_4	b_3	b_2	b_1	b_0	P_{odd}		CW#
1	0	1	0	1	1	1	1	1		CW0
1	1	0	1	1	1	1	0	0		CW1
0	0	1	1	0	0	1	0	0		CW2
0	1	0	0	1	0	1	0	0		CW3
1	1	1	0	0	1	1	0	0		LRC_{odd}

(b) Processed data packet

b_7	b_6	b_5	b_4	b_3	b_2	b_1	b_0	P_{odd}	P'_{odd}	CW#
1	0	1	0	1	1	1	1	1	1	CW0
1	1	0	1	1	1	1	0	0	1	CW1
0	0	1	1	0	0	1	0	0	0	CW2
0	1	0	0	1	0	1	0	0	0	CW3
1	1	1	0	0	1	1	0	0	0	LRC_{odd}
1	1	1	1	0	1	1	0			LRC'_{odd}

Figure 13.6

Data-packet error correction in Example 13.7. (a) Received data packet contains one error. (b) Processed data packet. Error clues shown in red text indicate b_4 in CW1 is an error.

EXAMPLE 13.7

Correcting a single data packet error

A data packet contains four data bytes within four 9-bit code words CW0, CW1, CW2, and CW3 and a LRC_{odd} that forms an odd parity in each column. Each code word in the data packet has its own odd-parity bit denoted as P_{odd}. The data packet was transmitted over a noisy channel. The received data packet shown in Figure 13.6a contains one error.

Find and correct the error using the following steps.

1. The receiver computes its own parity values P'_{odd} using *only the data bytes*. That is, the parity bits in the received data packet (P_{odd}) are not included in the receiver's parity calculations.

2. The receiver compares its computed P'_{odd} to the received P_{odd}. Differing values indicate the code word that contains an error. Figure 13.6b shows an error occurred in CW1.

3. The receiver computes its own odd-parity LRC code word, which is denoted as LRC'_{odd}, to form odd parity using *only the data columns in the data packet*. The bits in the received data packet LRC_{odd} are not included in the receiver's LRC'_{odd} calculations.

4. The receiver compares LRC_{odd} to LRC'_{odd}. Differing bit values indicate that a transmission error occurred in that bit position. The difference in data column b_4 indicates an error in that code word column.

5. The differing P'_{odd} and LRC'_{odd} values point to b_4 in CW1 as being the error. It is a simple matter for the receiver to change the received $b_4 = 1$ in CW1 in Figure 13.6b to the correct value $b_4 = 0$.

EXAMPLE 13.8

Error correction not possible with two errors

Figure 13.7a shows an error-free data packet that contains three code words CW0, CW1, and CW2 using even parity and an LRC that uses even-parity in the column bits.

Two errors are introduced in the data packet (a SEA violation) at b_3 and P_e of CW0, as shown Figure 13.7b. The figure shows the processing performed in the receiver that corrects

(a) Error-free data packet

b_7	b_6	b_5	b_4	b_3	b_2	b_1	b_0	P_e	CW#
1	0	1	0	1	1	1	1	0	CW0
1	1	0	1	1	1	1	0	0	CW1
0	0	1	0	1	0	1	0	1	CW2
0	1	0	1	1	0	1	1	1	LRC_e

(b) Processed data packet containing two error

b_7	b_6	b_5	b_4	b_3	b_2	b_1	b_0	P_e	P'_e	CW#
1	0	1	0	0	1	1	1	1	1	CW0
1	1	0	1	1	1	1	0	0	0	CW1
0	0	1	0	1	0	1	0	1	1	CW2
0	1	0	1	1	0	1	1	1	1	LRC_e
0	1	0	1	0	0	1	1			LRC'_e

Figure 13.7

Data-packet error correction in Example 13.8. (a) Error-free data packet. (b) Processed data packet containing errors in b_3 and P_e of CW0 prevents error correction although LRC difference indicates error detection.

single errors. The following steps illustrate that two errors prevent error correction.

1. The receiver computes P'_e, compares it to P_e, and finds no disagreements because CW0 contains errors in b_3 and P_e.

2. The receiver computes LRC'_e and compares it to LRC_e. The differing values in the b_3 column indicate an error in that code word position.

3. The receiver determines that an error occurred in the b_3 column but cannot determine the code word that contains the error. Thus, *error correction* is not possible, but *error detection* is still possible.

Different systems treat uncorrectable errors in data packets in different ways. If the data must be correct (such as in banking transactions) the receiver sends a message back to the transmitter that the data packet contains an error and to re-transmit the data packet. In time-critical (real-time) systems, such as streaming video, re-transmitted data packets would arrive too late to be useful. In this case, these systems often substitute the previous correct data packet for the erroneous one. This causes a stuttering video playback if there are many erroneous data packets.

 # 13.4 DATA RATE

The *data rate* measures the amount of data a source generates in one second. Let $\mathcal{R}$ denote the *source symbol rate* equal to the number of symbols a source produces every second. For example, if a source transmits ten symbols per second, $\mathcal{R} = 10$ symbols/second.

A source that has a vocabulary of m symbols produces the equivalent of $\log_2 m$ bits of data when it transmits a symbol (or $\log_2 m$ bits/symbol). For example, the source that has $m = 4$ (symbols X_1, X_2, X_3, and X_4) can encode two bits ($\log_2 4 = 2$) of data with each symbol transmission.

The data rates of different sources are compared using common units of *bits/second* (*bps*). Multiplying the symbol rate $\mathcal{R}$ (symbols/second) and $\log_2 m$ (bits/symbol) we obtain the *source data rate*, which is denoted as $\mathcal{D}_S$ with units bps, as

$$\mathcal{D}_S = \mathcal{R} \log_2 m \text{ bps} \tag{13.5}$$

EXAMPLE 13.9 Smartphone in a Wi-Fi network

A smartphone transmits binary data using two signals ($s0$ and $s1$) that are considered symbols ($m = 2$) at a source symbol rate equal to

$$\mathcal{R} = 10^6 \text{ symbols/second}$$

Its data rate is computed as

$$\mathcal{D}_S = 10^6 \overbrace{\log_2 2}^{=1} = 10^6 \text{ bps}$$

EXAMPLE 13.10 Digital audio source

The microphone converts your speech into a waveform that is limited to have a maximum frequency $f_{max} = 3.5\,\text{kHz}$. The ADC in your smartphone samples this waveform using sampling rate $f_s = 8\,\text{kHz}$. Each sample is a symbol, and the source symbol rate equals

$$\mathcal{R} = 8{,}000 \text{ symbols/second}$$

The ADC quantizes each sample with 8-bit code words. Thus, the ADC has a vocabulary of different symbols having a size equal to

$$m = 2^8 = 256$$

The data rate your smartphone produces when you talk on the phone is equal to

$$\mathcal{D}_S = 8{,}000 \overbrace{\log_2 256}^{=8} = 64{,}000 \text{ bps}$$

13.5 CHANNEL CAPACITY

The *channel capacity* measures the number of bits per second that can be transmitted reliably over a channel. The quality of a channel for transmitting data depends on two parameters.

> **Bandwidth $\mathcal{B}$** is the frequency range from the low frequency limit f_L to the high frequency limit f_H (measured in Hz) that the data transmission signals can occupy with $\mathcal{B} = f_H - f_L$. As $\mathcal{B}$ increases, the duration of the data signals decreases. Thus, it takes less time to transmit data.

> **Signal-to-noise ratio (SNR)** is the ratio of the energy of signals propagating on a channel $\mathcal{E}_s$ to noise on the channel measured by its variance σ_N^2. Chapters 10 and 11 describe the significance of $\mathcal{E}_s/\sigma_N^2$ in extracting data values from detected signals. Noise present on the channel limits the amount of data that can be reliably transmitted over a channel.

A formula developed by Claude Shannon (of entropy fame) and Ralph Hartley is called the *Shannon-Hartley theorem* and computes the channel capacity, which is denoted as $\mathcal{C}$ in units of *bits per second (bps)* as

$$\mathcal{C} = \mathcal{B} \log_2 (1 + \text{SNR}) \text{ bps} \tag{13.6}$$

Let us examine the effects of $\mathcal{B}$ and SNR separately.

This formula states that the channel capacity $\mathcal{C}$ varies linearly with bandwidth. Thus, doubling $\mathcal{B}$ doubles $\mathcal{C}$. Reducing $\mathcal{B}$ increases the duration of the data signals that can propagate on the channel and thereby reduces $\mathcal{C}$.

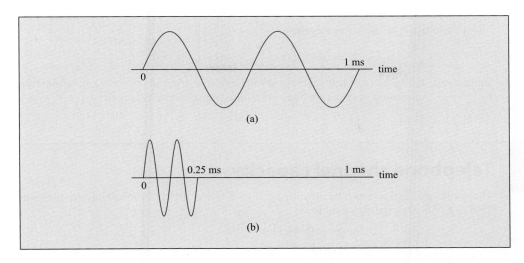

Figure 13.8

Data signals at two bandwidth values. (a) $\mathcal{B} = f_H = 2$ kHz. (b) $\mathcal{B} = f_H = 8$ kHz.

Effect of $\mathcal{B}$ on the channel capacity   EXAMPLE 13.11

To simplify, let the lower frequency limit $f_L = 0$, which makes $\mathcal{B} = f_H$. Assume a data signal consists of two periods of a sinusoidal waveform and has a frequency equal to $\mathcal{B} = f_H$, as shown in Figure 13.8.

When $\mathcal{B} = f_H = 2$ kHz, the two-period data signal has a duration equal to 1 ms.

When $\mathcal{B} = f_H = 8$ kHz, the data signal has a duration equal to 0.25 ms. Thus, four data signals can be transmitted when $\mathcal{B} = 8$ kHz in the time that one data signal can be transmitted when $\mathcal{B} = 2$ kHz. Thus, C varies linearly with $\mathcal{B}$.

The C value also varies as the logarithm of the signal-to-noise ratio. For data signals the SNR $= \mathcal{E}_s / \sigma_N^2$ where $\mathcal{E}_s$ is the energy in the signal and σ_N^2 is the noise variance. Substituting these values into the capacity equation gives

$$C = \mathcal{B} \log_2 \left(1 + \frac{\mathcal{E}_s}{\sigma_N^2} \right) \text{ bps} \qquad (13.7)$$

Effect of noise on the channel capacity   EXAMPLE 13.12

To interpret the effect of noise on channel capacity, consider a channel having a fixed bandwidth $\mathcal{B}$ and vary the noise variance σ_N^2.

When $\sigma_N^2 = 0$ (true only for an ideal channel) the capacity equation gives

$$C = \mathcal{B} \log_2 \left(1 + \frac{\mathcal{E}_s}{0} \right) \rightarrow \mathcal{B} \log_2 (\infty) = \infty$$

The infinite capacity occurs because, when there is no noise, we can differentiate *any two signals* no matter how close they are to each other. For example, we can make a quantizer step size Δ extremely small and still differentiate two adjacent levels. Recall that the number of quantizer bits are related to Δ as

$$\Delta = \frac{V_{\max}}{2^b} \rightarrow b = \log_2 \left(\frac{V_{\max}}{\Delta} \right)$$

If Δ is very small, the number of bits b that can be transmitted becomes very large. Thus, a large capacity is achieved.

At the other extreme, $\sigma_N^2 = \infty$ gives

$$C = \mathcal{B} \log_2 \left(1 + \frac{\mathcal{E}_s}{\infty} \right) \rightarrow \mathcal{B} \log_2 (1) = 0$$

In this case, the noise is so large that it overwhelms any signal and prevents any information from being transmitted over the channel. Thus, the channel capacity is zero.

EXAMPLE 13.13 — Telephone channel capacity

The telephone channel has a bandwidth equal to $\mathcal{B} = 3 \times 10^3$ Hz. A telephone channel with SNR = 1, 000 has capacity equal to

$$C = \mathcal{B} \log_2 (1 + \text{SNR})$$
$$= 3 \times 10^3 \times \log_2 (1 + 1,000)$$
$$= 3 \times 10^3 \times 9.97 = 29,900 \text{ bps}$$

An approximate value is obtained without a calculator by recalling $1,000 \approx 2^{10}$. Then $\log_2 (1,001) \approx 10$, making $C = 30$ kbps.

For a practical data transmission system, the channel must reliably transmit all the data produced by the source. Thus, the *channel capacity must be greater than or equal to the source data rate*, as

$$C \geq \mathcal{D}_S \tag{13.8}$$

The capacity equation is useful for computing the effect of bandwidth $\mathcal{B}$ and SNR on channel performance. However, its values are slightly larger than achievable in practice. As a practical measure of system performance, we substitute for the C value calculated in Eq. (13.6) with the *data transmission capability* routinely achieved by commercial data channels. Figure 13.9 lists some common data transmission systems, the technology they use, and their data transmission capability.

To accommodate the two common units used for data rates, we use the convention that the rate expressed in bits per second is written as *bps* and in bytes per second as *Bps*.

EXAMPLE 13.14 — Watching a video download

A video source produces data at the rate equal to $D_V = 10^4$ bytes per second (Bps) and your smartphone must display it at that rate. Let the size of the data file that stores the video be denoted by n_f.

Assume a video segment is stored in a video file of size $n_f = 1$ MB (10^6 bytes). Thus, the video segment lasts a time denoted by T_V and equal to

$$T_V = \frac{n_f}{D_V} = \frac{10^6 \text{ B}}{10^4 \text{ Bps}} = 100 \text{ s}$$

Figure 13.9

Data transmission systems, their technology, and their data transmission capabilities.

System	Technology	Data Transmission Capability
WiFi	Short-range radio wave	1 Mbps (10^6 bps)
Cellular 4G LTE	Long-range radio wave	100 Mbps (10^8 bps)
Ethernet	Long copper cable	100 Mbps (10^8 bps)
USB 2.0	Short copper cable	480 Mbps (0.48×10^9 bps)
USB 3.0	Short copper cable	5 Gbps (5×10^9 bps)
Fiber-optic	Glass fiber	1 Gbps (10^9 bps)

Let the channel capacity (data transmission capability) of your smartphone connection equal $C = 7.5 \times 10^3$ Bps. Because $C < D_V$, some buffering is required on the smartphone to produce continuous playback without pauses. The data transfer time over the channel is denoted as T_D and equals

$$T_D = \frac{n_f}{C} = \frac{10^6 \, \text{B}}{7.5 \times 10^5 \, \text{Bps}} = 133 \, \text{s}$$

There are two approaches that can be used for your smartphone to display a continuous video.

One approach is to first download the entire video file (in 133 seconds) to your smartphone and play the 100-second video.

The alternative approach is to download a sufficient amount of data to start the video and transmit data while the video is playing. Ideally, the video ends just after the last byte in the file is transmitted. With the video lasting T_V seconds, the video can start at time after transmission starts denoted as T_{VS} and equal to

$$T_{VS} = T_D - T_V = 133 - 100 = 33 \ \text{seconds}$$

During this time the quantity of data transferred to your smartphone equals

$$33 \, \text{s} \times 7.5 \times 10^3 \, \text{Bps} = 0.25 \, \text{MB}$$

During the one hundred seconds the video is playing on the smartphone, the remaining data is transferred and is equal to

$$100 \, \text{s} \times 7.5 \times 10^3 \, \text{Bps} = 0.75 \, \text{MB}$$

thus completing the 1 MB data transfer.

13.5.1 Capacity Loss with Transmission Range

As your smartphone moves away from the cell-phone company transmitting antenna the signal strength $\mathcal{E}_s$ detected at your smartphone drops (your bars become smaller). This section examines the effect of the signal energy ($\mathcal{E}_s$) attenuation on the channel capacity C.

Signal Energy Decay with Transmission Tange

Imagine shining a flashlight onto a wall, as shown in Figure 13.10. As the flashlight (transmitter) range from the wall increases, the beam spot becomes larger and dimmer, because the transmitter power (which is fixed) spreads over a larger area. The intensity

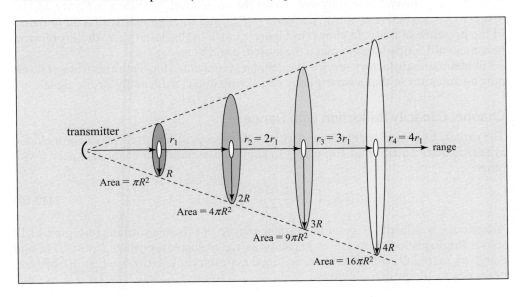

Figure 13.10

Transmission intensity loss with range. The smartphone antenna converts intensity into signal energy and is indicated as a small circle along the transmitter beam axis.

of the transmitted signal at the receiver (your smartphone) equals the transmitter power divided by the area. The smartphone antenna converts the intensity into the detected signal energy $\mathcal{E}_s$.

The decay of intensity with range is governed by the *inverse square law*, which states that the intensity is reduced by the square of the range to the transmitter. To show this, let R_o denote the radius of the beam at range r_o and $A(r_o)$ denote the beam area range $r = r_o$. Thus,

$$A(r_o) = \pi\, R_o^2 \tag{13.9}$$

When the range doubles to $r = 2r_o$, the radius doubles and the area $A(2r_o)$ equals

$$A(2r_o) = \pi\,(2R_o)^2 = 4\pi\, R_o^2 = 4A(r_o) \tag{13.10}$$

In general, when the range increases by factor α, the area increases by α^2. The area at range $r = \alpha r_o$ equals

$$A(\alpha r_o) = \alpha^2 A(r_o) \tag{13.11}$$

Because the signal intensity at your smartphone decreases as $A(\alpha r_o)$ increases, the signal energy detected by your smartphone $\mathcal{E}_s$ also decreases as

$$\mathcal{E}_s(\alpha r_o) = \frac{\mathcal{E}_s(r_o)}{\alpha^2} \tag{13.12}$$

A calibration procedure is required to employ this formula. When the receiver is located at a known range $r = r_o$ from the transmitter, the system is calibrated by measuring the signal energy detected by the smartphone $\mathcal{E}_s(r_o)$.

Figure 13.11 shows the consequences of the decay of $\mathcal{E}_s(\alpha r_o)$ with range $r = \alpha r_o$. At close range (small α), $\mathcal{E}_s(\alpha r_o)$ is large. As range increases (α increases), $\mathcal{E}_s(\alpha r_o)$ reduces, as in Eq. (13.12). Problems arise when the signal energy falls below a minimum value, which is denoted as $\mathcal{E}_{\min}$, because errors become too common.

The maximum range of satisfactory service (that is, the Prob[error] is sufficiently small) is denoted as $r_{\max}$ and equals

$$r_{\max} = \alpha_{\max}\, r_o \tag{13.13}$$

The minimum signal energy value equals

$$\mathcal{E}_{\min} = \frac{\mathcal{E}_s(r_o)}{\alpha_{\max}^2} \tag{13.14}$$

For range $r < r_{\max}$ (that is, $\alpha < \alpha_{\max}$) satisfactory service is provided to the users, because $\mathcal{E}_s(\alpha r_o) > \mathcal{E}_{\min}$ produces signals well above the noise level, as shown in Figure 13.11a. When $r > r_{\max}$ (that is, $\alpha > \alpha_{\max}$), $\mathcal{E}_s(\alpha r_o) < \mathcal{E}_{\min}$ and the signals are difficult to detect in the presence of noise, as shown in Figure 13.11b. In this latter case, the Prob[error] becomes sufficiently large to make the system unreliable.

To maintain satisfactory service, the cellular system must locate the network of transmitting antennas so that a smartphone can always detect sufficiently strong signals.

Channel Capacity Reduction with Range

The variation of signal energy $\mathcal{E}_s(\alpha r_o)$ with range expressed in Eq. (13.12) can be included in the capacity equation in Eq. (13.7) to show the variation in channel capacity with range, as

$$C(\alpha r_o) = B\log_2\left(1 + \frac{\mathcal{E}_s(\alpha r_o)}{\sigma_N^2}\right) = B\log_2\left(1 + \frac{\mathcal{E}_s(r_o)}{\alpha^2 \sigma_N^2}\right) \tag{13.15}$$

Assume the bandwidth B and noise variance σ_N^2 do not change with range. Figure 13.12 shows the expected decrease in capacity as range increases when $B = 10^4$ Hz and $\mathcal{E}_s(r_o)/\sigma_N^2 = 10^4$ at $r_o = 1$ m. At $r = r_o = 1$ m, the capacity equals $C(r_o) = 1.3 \times 10^5$ bps and decreases to 10 bps at $r = 4{,}000$ m.

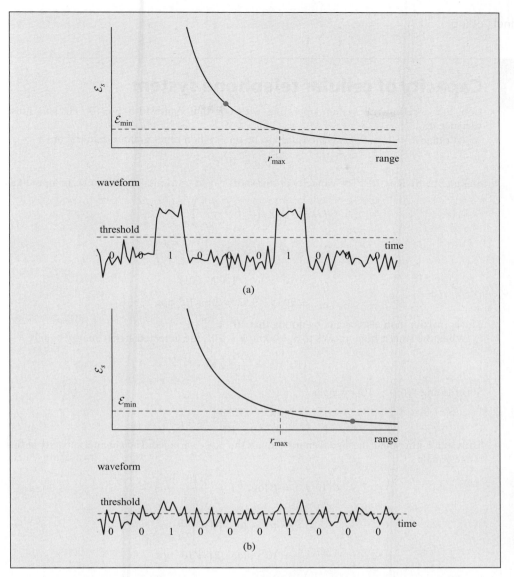

(a)

(b)

Figure 13.11

Signal energy decay through the inverse square law. (a) Digital signals at $r < r_{max}$ are detected reliably. (b) Digital signals at $r > r_{max}$ are unreliably detected in the presence of noise.

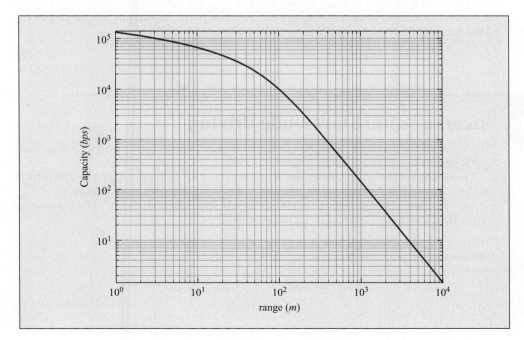

Figure 13.12

Capacity as a function of range. $\mathcal{B} = 10^4$ Hz, and $\mathcal{E}_s(r_o) = 10^4 \sigma_N^2$ with $r_o = 1m$.

EXAMPLE 13.15 Capacity of cellular telephone system

Consider a cell-phone system operating with constant bandwidth $\mathcal{B} = 10^5$ Hz and noise variance σ_N^2.

At range $r_o = 1$ m, $\mathcal{E}_s(r_o)$ is measured to be one million times greater than σ_N^2, or

$$\mathcal{E}_s(r_o) = 10^6 \, \sigma_N^2$$

making the SNR $= 10^6$. The capacity of this short-range transmission channel is computed as

$$
\begin{aligned}
\mathcal{C}(r_o) &= \mathcal{B} \log_2 \left(1 + \frac{\mathcal{E}_s(r_o)}{\sigma_N^2} \right) \\
&= 10^5 \times \log_2 \left(1 + \frac{10^6 \times \sigma_N^2}{\sigma_N^2} \right) \\
&= 10^5 \times \log_2 (1,000,001) \\
&= 10^5 \times 19.93 \approx 20 \times 10^5 \text{ bps}
\end{aligned}
$$

The approximation also occurs by noting that $10^6 = 2^{20}$.

When the smartphone moves to $r = 1$ km, $\alpha = 10^3$, the detected signal energy equals

$$\mathcal{E}_s(10^3 \, r_o) = \frac{\mathcal{E}_s(r_o)}{(10^3)^2} = \frac{\mathcal{E}_s(r_o)}{10^6}$$

Substituting $\mathcal{E}_s(r_o) = 10^6 \, \sigma_N^2$ gives

$$\mathcal{E}_s(10^3 \, r_o) = \frac{10^6 \, \sigma_N^2}{10^6} = \sigma_N^2$$

Thus, at a 1-km range, the signal energy equals the noise variance. The channel capacity at this range equals

$$
\begin{aligned}
\mathcal{C}(10^3 \, r_o) &= \mathcal{B} \log_2 \left(1 + \frac{\mathcal{E}_s(10^3 \text{ m})}{\sigma_N^2} \right) \\
&= 10^5 \times \log_2 \left(1 + \frac{\sigma_N^2}{\sigma_N^2} \right) \\
&= 10^5 \times \log_2 (2) = 10^5 \text{ bps}
\end{aligned}
$$

Note that, even though the signal energy decreased by one million, the channel capacity has decreased by a factor of twenty. This is a consequence of the logarithmic relationship of the SNR in computing the channel capacity.

EXAMPLE 13.16 Inverse-square law while driving

This example examines the reduction of signal-strength bars on your smartphone as your journey starts close to a cell-phone transmission tower and you drive away.

The signal energy at your smartphone is measured to be $\mathcal{E}_s(r_o) = 10^4 \sigma_N^2$ at $r_o = 0.5$ km from the tower, providing the calibration value and a maximum signal-strength indication.

At $r = 25$ km, your smartphone begins to lose the signal (the number of bars decreases to zero). Thus, the maximum reliable operating range is

$$r_{max} = 25 \text{ km} = 50 r_o \;\rightarrow\; \alpha_{max} = 50$$

The minimum signal energy equals

$$\mathcal{E}_{min} = \mathcal{E}_s(\alpha_{max} r_o) = \frac{\mathcal{E}_s(r_o)}{\alpha_{max}^2} = \frac{10^4 \sigma_N^2}{2500} = 4\sigma_N^2$$

EXAMPLE 13.17

Increasing antenna power

A more powerful antenna is added to the system described in Example 13.16 and increases $\mathcal{E}_s(r_o)$ by a factor of 100 from $10^4\sigma_N^2$ to $10^6\sigma_N^2$. The new maximum operating range is

$$r'_{max} = \alpha'_{max}r_o$$

The minimum energy required for proper operation is

$$\mathcal{E}_s(\alpha'_{max}r_o) = \mathcal{E}_{min}$$

Substituting $\mathcal{E}_s(\alpha'_{max}r_o) = \dfrac{\mathcal{E}_s(r_o)}{(\alpha'_{max})^2} = \dfrac{10^6\sigma_N^2}{(\alpha'_{max})^2}$ and $\mathcal{E}_{min} = 4\sigma_N^2$ from the previous example gives

$$\frac{10^6\sigma_N^2}{(\alpha'_{max})^2} = 4\sigma_N^2$$

Solving for α' yields

$$\alpha'_{max} = \sqrt{250{,}000} = 500$$

The new maximum range is

$$r'_{max} = \alpha'_{max}r_o = 250 \text{ km}$$

Thus, increasing the antenna power by a factor of 100 increases the operating range from 25 km to 250 km (a factor of 10).

This result can be generalized: Increasing the transmitter power by a factor increases the operating range by the square root of the factor.

EXAMPLE 13.18

Space probe channel capacity

A space probe has the bandwidth $\mathcal{B} = 10^4$ Hz that does not vary with range. The channel capacity measured at range r_o is $C(r_o) = 10^6$ bps.

The SNR of the detected signals at r_o is computed from

$$\frac{C(r_o)}{\mathcal{B}} = \frac{10^6 \text{ bps}}{10^4 \text{ Hz}} = 100 = \log_2\left(1 + \frac{\mathcal{E}_s(r_o)}{\sigma_N^2}\right)$$

Because $\mathcal{E}_s(r_o)/\sigma_N^2 \gg 1$, we use the approximation $\log_2(1 + x) \approx \log_2(x)$ to give

$$\log_2\left(\frac{\mathcal{E}_s(r_o)}{\sigma_N^2}\right) = 100$$

Raising both sides to the power of 2 yields the SNR at r_o as

$$\frac{\mathcal{E}_s(r_o)}{\sigma_N^2} = 2^{100}$$

The probe ventures out to range $r = 1024r_o$ (making $\alpha = 2^{10}$). If σ_N^2 remains the same, the new SNR of the signals detected on Earth is found by computing the decay in the signal energy, as

$$\mathcal{E}_s(\alpha r_o) = \frac{\mathcal{E}_s(r_o)}{\alpha^2} = \frac{\mathcal{E}_s(r_o)}{2^{20}}$$

The new SNR equals

$$\frac{\mathcal{E}_s(1024r_o)}{\sigma_N^2} = \frac{\mathcal{E}_s(r_o)/\sigma_N^2}{2^{20}} = \frac{2^{100}}{2^{20}} = 2^{80}$$

If $\mathcal{B}$ remains the same, the capacity at $r = 1024r_o$ is

$$C(1024r_o) = \mathcal{B}\log_2(\text{SNR}) = 10^4\log_2(2^{80}) = 10^4 \times 80 = 8 \times 10^5 \text{ bps}$$

Note: Even though the range increased by a factor of more than 1,000, the capacity dropped by only 20%.

13.6 Research Challenges

13.6.1 5G Mobile Wireless System

Wireless data rates of up to 1 Gbps for smartphones are being investigated using millimeter wavelengths for communications. At these short wavelengths, antenna arrays are small enough to fit in a smartphone and produce a *steerable beam pattern*. Instead of current cellphone towers that transmit horizontally in all directions, *5GmmWave* transmitting antennas scan like a radar.

When the system senses your smartphone, it steers a beam in your direction to focus on your smartphone for both transmitting and receiving data. In turn, the antenna array within the smartphone steers its signals in the direction of the antenna. This focussing increases the transmission and reception signal energies, improving their signal-to-noise ratios, thereby increasing the data transmission rate up to 1 Gbps.

13.7 Summary

This chapter described several schemes to detect and correct errors in data. The simplifying single error assumption (SEA) was employed to illustrate the basic ideas. All error-correction methods use some form of redundancy by adding bits whose values are computed from the original data. This redundancy results in an increased data size that is measured with the data increase factor (DIF).

Sources that produce audio or video data exhibit a specific data rate. Data are transmitted over a channel having a capacity that measures the rate at which data can be reliably transmitted. For real-time operation, the channel capacity must exceed the source data rate. The channel capacity is determined by its bandwidth and the signal-to-noise ratio of the data signals. The inverse square law described the decrease in signal energy with range. The resulting variation of the channel capacity with range allows cell-phone systems to determine antenna locations in a network.

13.8 Problems

In the problems below, the rate expressed in bits per second is written as bps and the rate bytes per second as Bps.

13.1 Error correction by repetition. Design a bit-level code that corrects up to 2 errors per code word. What is the data increase factor for your code?

13.2 Data-packet error correction. A data packet contains the 5-digit number 24680. Each code word contains a single digit encoded using binary-coded-decimal (BCD) codes that form quadbit (4-bit) data plus an even-parity bit. Generate a data packet that includes a LRC code word. What is the data increase factor for your data packet?

13.3 Data rate of texting. You enter text on a smartphone keypad at a rate of ten characters per second and an 8-bit ASCII code word is generated for each character. What is your data rate in bps and Bps?

13.4 Data rate of a video source. A digital video is a sequence of digital frames with each forming a 480×640 pixel image and with each pixel encoded with 24-bit color. The video displays 15 frames/sec. What is the data rate produced by this video source?

13.5 Digital camera source. My 10 Megapixel camera has three (R, G, and B) sensors in each pixel and each sensor outputs 256 levels of intensity. The camera allows me to take multiple pictures at a rate of four per second when I keep the shutter button depressed.

    a) What is the maximum data rate from the camera into the camera memory card when I keep the shutter button depressed?

    b) If I want to store 300 images, what size memory card should I purchase for my new camera? (Specify size as an integer number of gigabytes, e.g., X Gb memory.)

13.6 Data rate in the telephone system. If the audio waveform sampling period in the smartphone is $T_s = 0.1$ ms and each sample is quantized to 256 levels, what is the data rate produced by your smartphone?

13.7 Time to transfer 1 TB disk data. A external hard disk drive has a 1 terabyte (TB) capacity. How long does it take to transfer 1 TB through a USB 2.0 port having a 480 Mbps transmission rate? How long does it take if you upgrade to a USB 3.0 port having a 5 Gbps transmission rate?

13.8 Video download to your smartphone. A 10 MB video file requires a data rate $D_V = 6$ Mbps for continuous playback. The file is transmitted over a channel having transmission data rate equal to 1 Mbps. At what point in the data transfer should your smartphone begin to play the video to produce continuous playback and with minimal delay?

13.9 **Data transmission between your smartphone and antenna tower.** A powerful antenna tower transmits data *down* to your smartphone at a rate equal to 300 Mbps, while your less-powerful smartphone transmits data back *up* to the tower at 75 Mbps. Assume the bandwidths and noise levels are the same in each transmission direction. How are the signal energies at the tower and cellphone related? (Hint: Use the channel capacity equation and ignore the $+1$ in the $\log_2$ term.)

13.10 **Minimum signal-to-noise ratio.** The data rate produced by your smartphone is $\mathcal{D} = 2 \times 10^5$ bps. Your smartphone transmits data over the frequency band from 100 kHz to 200 kHz. What is the minimum SNR required to transmit your data in real time?

13.11 **Channel capacity reduction with range.** The signal energy detected by your smartphone at range $r_o = 0.5$ km

equals $\mathcal{E}_s(r_o) = 10^6 \sigma_N^2$. The bandwidth $\mathcal{B} = 100$ kHz and is constant with range.

a) What is the value of $\mathcal{C}(r_o)$?

b) As the smartphone distance from the antenna increases, at what range does $\mathcal{E}_s(\alpha r_o)$ become less than $100\sigma_N^2$?

c) What is the channel capacity at that range?

13.12 **Channel capacity of space probe.** A space probe transmitter operates with $\mathcal{B} = 10^5$ Hz and $\mathcal{E}_s(r_o)/\sigma_N^2 = 10^6$ at $r_o = 1$ km. The probe transmits images that each contain 1 Mb of data. If an image takes approximately one hour to transmit, what is the range of the probe from the antenna on Earth?

13.9 Matlab Projects

13.1 **Simulating a data channel with specified error probability.** Following Example 16.65, simulate a data channel with specified error probability for examining the performance of error correction codes.

13.2 **Simulating receiver error correction using bit-repetition.** Generalize Example 16.66 to simulate an n_b-bit repetition error correction code system and measure the data bit error probability.

1. Measure the data bit error probability as a function of n_b for $0 \le n_b \le 11$ when the channel error probability $= 0.25$.

2. Determine the reliability of your error probability value by repeating the program ten times and recording the minimum and maximum observed values.

13.3 **Forming a data packet.** Generalize Example 16.67 to simulate a transmitter that forms a 3×3 (three tribits) data block, appends parity bits and an LRC code word, and saves the 4×4 data packet on the disk. Verify that

the parity bits and LRC code word are correct for your parity conventions.

13.4 **Simulating a channel that generates random errors in a data packet.** Generalize Example 16.68 to simulate a channel that produces random errors to a 4×4 data packet. Save the data packet on the disk.

13.5 **Correcting an error in the data section of a data packet.** Generalize Example 16.69 to correct a single error in the data section of the 4×4 data packet.

13.6 **Correcting an error anywhere in a data packet.** Generalize Example 16.69 to correct a single error that occurs anywhere in a 4×4 data packet including the parity bits and LRC code word.

13.7 **Channel Capacity.** Following Example 16.70 plot the channel capacity of a space probe as a function of range 1 km $\le r \le 10^6$ km. Probe specifications include $\mathcal{B} = 10^5$ Hz and $\mathcal{E}_s(r_o)/\sigma_N^2 = 10^6$ at $r_o = 1$ km.

CHAPTER

14

Maksim Kabakou / Shutterstock.com

SYMBOLOGY

LEARNING OBJECTIVES

After completing this chapter, the reader should be able to:

- Appreciate the issues regarding machine-readable codes.
- Recognize the utility and limitations of bar codes.
- Understand how codes are embedded in data files.

This chapter describes methods to encode digital data in machine-readable form. Magnetic sensing decodes magnetic stripes on ATM and credit cards. Techniques that allow robust scanning of various bar codes are described. Current bar-code symbols contain not only data but also features that determine code location and orientation. Optical sensing is used in the most common scanners and is described for one-dimensional and two-dimensional symbols. Steganography is a technique for embedding secret codes into data files that are difficult to detect. Early scanning systems were expensive, situated in controlled settings, and used by trained people. Current scanning of digital data is performed by smartphone cameras under a variety of conditions and is used by nearly everyone.

14.1 INTRODUCTION

This chapter describes systems that facilitate the transfer of data from products and users to a computer system. To insure successful operation, machine-readable systems must confront the following issues.

Machine-readable codes—Digital data are encoded in bar codes. Such codes are designed to allow reliable recognition of the data under a variety of scanning conditions. Early scanning systems were expensive, situated in controlled settings, and used by trained people. Current scanning of 2D symbols is performed by smartphone cameras, under a variety of conditions, and by nearly everyone.

Machine-unreadable codes—Symbols that humans read easily (using human reasoning) help distinguish human users from malicious computer programs.

Steganography—Some applications embed secret codes within large data files, making them difficult to detect. One example is a digital watermark that assigns a serial number to each music or video data file to deter piracy of software. The presence of such secret data within the file is difficult to perceive by merely looking at the file contents.

14.2 CREDIT CARD CODES

You have observed your credit card being swiped through a card reader, either by a cashier at a store or by yourself at a gas station pump, to read data from the magnetic stripe. The magnetic stripe is a thin layer of magnetic material that stores magnetic patterns on its surface. Data are recorded on the magnetic stripe using a magnetic head that is similar to one used in an audio tape recorder by storing small magnetic spots (domains) having either a north pole N or south pole S. Magnetic poles are formed by driving current I into a magnetic head to produce magnetic field B. Current flow in one direction forms N and in the opposite direction forms S, as shown in Figure 14.1.

Figure 14.1

Writing and reading data with magnetic domains as the credit card moves in direction of arrow. Magnetic poles are formed on the magnetic stripe by driving current I into a magnetic head in the appropriate direction through the head. The magnetic head senses changes in the magnetic field to produce a voltage $+V$ when $N \rightarrow S$, or $-V$ when $S \rightarrow N$.

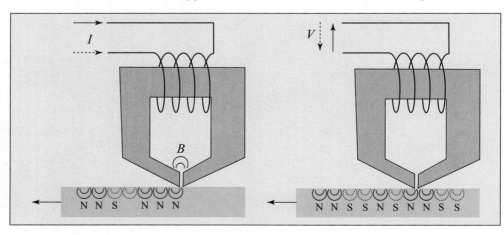

Scanning a Magnetic Stripe

The voltage produced by the magnetic poles passing under the read head is related to the change in the magnetic pole ΔB (equal to either plus or minus $2B$) divided by the time between changes ΔT, as

$$V = \frac{\Delta B}{\Delta T} \tag{14.1}$$

The magnitude of the signal from the magnetic stripe is

$$|V| = \frac{2B}{\Delta T} \tag{14.2}$$

Note that $|V|$ increases as ΔT decreases. To provide a strong signal from a reader, credit-card users are instructed to *remove the card quickly* in order to reduce ΔT, which in turn increases $|V|$. Knowing this, I have always tried to be patient with the careful cashier who does exactly the wrong thing by sliding a credit card ever so slowly through the card reader.

Reading Magnetic Stripe Data

The data signal extracted from the credit-card magnetic stripe provides the information needed to read the recorded data. The challenge is that the signal varies with the card speed, which depends on how quickly the card is pushed or pulled through the reader. In fact, the speed can vary between 3 to 125 inches per second. Furthermore, the speed does not even need to be constant during the time the card is read. To allow the data to be read successfully over such a wide variety of speeds, the data are encoded using a clever *self-clocking* code (also known as the *Manchester code*) that is employed in hard disk drives. The idea is that transitions in the data waveform *always* occur with an almost regular period. These regularly occurring transitions provide the timing necessary to decode the signal into data. Along with these expected periodic transitions, additional transitions can also occur at the mid-period times as well. These mid-period transitions encode the data: the occurrence of an additional transition encodes a 1, while no mid-period transition encodes a 0, as shown Figure 14.2.

For proper operation, the card reader must determine the value of the period and mid-period. To assist in this determination, the beginning of the magnetic stripe produces a waveform that contains only a full-period waveform as shown at the top of Figure 14.2. This waveform allows the card reader to determine the average transition period, which is denoted T_p. The estimate of the next transition time is denoted $\hat{T}_n$ and equals the previous T_p value. The reader uses the value of $\hat{T}_n$ to update the data waveform timing. If a transition (in either direction) does not occur at approximately $\hat{T}_n$, something is wrong, and the message *Please insert the card again* appears on the card reader.

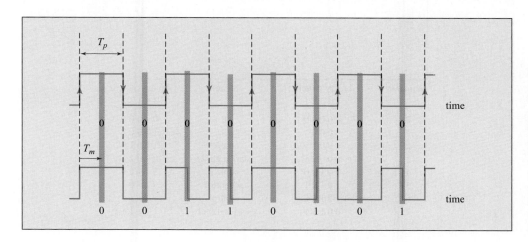

Figure 14.2

Magnetic stripe signals have periodic transitions having period T_p. Binary data are determined by the presence or absence of an additional transition at mid-period time T_m, with a mid-period transition encoding a 1, and no mid-period transition encoding a 0. Top signal shows an all 0 signal that occurs at the beginning of the stripe signal to establish timing.

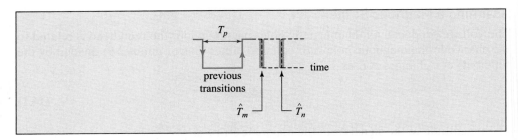

Figure 14.3

From the two previous T_p transitions, the mid-period time $\hat{T}_m = T_p/2$ and the time of the next transition $\hat{T}_n = T_p$ are estimated.

A credit card is typically pulled through the reader at a non-constant speed, which causes T_p to vary slightly. When the card swipe slows down, T_p increases. Because there are 200 transitions per inch (8 transitions per mm) on the magnetic stripe, the card speed does not vary significantly from transition to transition, which causes $\hat{T}_n$ to be close to the previous T_p value. Because $\hat{T}_n$ depends on only the previous T_p value, the $\hat{T}_n$ values track the speed of the card swipe.

Figure 14.3 shows the mid-period time estimate is denoted $\hat{T}_m$ and is estimated from the previous T_p value divided by two. The waveform around $\hat{T}_m$ is monitored for a transition. No transition at time $\hat{T}_m$ indicates a 0 has been encoded and the presence of a transition (in either direction) indicates a 1.

Decoding Magnetic Stripe Data

The data transmitted in credit-card systems use the data packet error correction method similar to that described in Chapter 13. Most credit-card transactions are accomplished by reading data encoded on a magnetic stripe located on the back of the credit card. The magnetic stripe contains three tracks of information, as shown in Figure 14.4. Track 1 stores up to 79 letters or numbers (alphanumeric characters) and track 2 stores up to 40 numbers. These two tracks specify the owner's name, account number, and expiration date. Track 3 stores up to an additional 105 numbers needed by some systems. Tracks 2 and 3 encode digits using 5-bit code words to encode the digits 0 to 9, a separator, a start sentinel, an end sentinel, and odd parity for error detection.

The data on a magnetic stripe begins with the *start sentinel*, which is a predetermined code word that allows the card reader to verify that it is reading the data correctly.

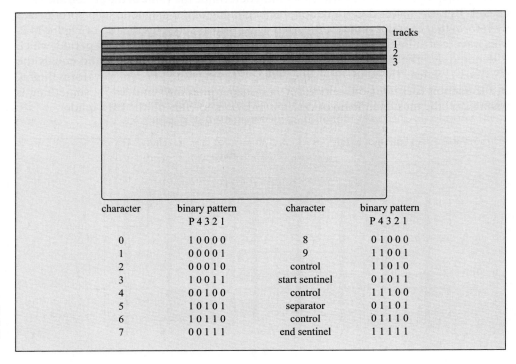

Figure 14.4

The magnetic stripe on the back of a credit card contains three data tracks. Five-bit code words are used in tracks 2 and 3.

character	binary pattern P 4 3 2 1	character	binary pattern P 4 3 2 1
0	1 0 0 0 0	8	0 1 0 0 0
1	0 0 0 0 1	9	1 1 0 0 1
2	0 0 0 1 0	control	1 1 0 1 0
3	1 0 0 1 1	start sentinel	0 1 0 1 1
4	0 0 1 0 0	control	1 1 1 0 0
5	1 0 1 0 1	separator	0 1 1 0 1
6	1 0 1 1 0	control	0 1 1 1 0
7	0 0 1 1 1	end sentinel	1 1 1 1 1

The data are terminated with the *end sentinel*. A *separator* delimits different fields of data (i.e., the account number from the date of expiration). These 5-bit code words are stacked to form a data packet that terminates with a longitudinal redundancy check (LRC) code word as described in Chapter 13.

14.3 BAR CODES

Bar codes were introduced in the early 1970s to implement reliable machine-readable data systems and have quickly spread. They appear on mail, on products as the familiar Universal Product Code (UPC) symbol, and in magazine advertisements and billboards. Bar codes are used for controlling inventory in warehouses, for monitoring books and patrons in libraries, and for identifying patients in hospitals. One major advantage of bar codes over other machine-readable identification devices, such as radio-frequency identification (RFID) tags, is that bar codes can be transmitted electronically and displayed on a smartphone display, such as an airline boarding pass.

To implement a successful and practical bar-code system, the following conflicting requirements must be satisfied.

- The code must have a high density of information.
- The code must be read reliably and quickly.
- The printing process and scanner must be inexpensive.

Trade-offs routinely occur in the design of a bar-code system. For example, some bar codes are designed to have a lower density of information in order to facilitate scanning. Bar codes have evolved through the following four generations.

1. Simple numerical data that are read by machine (US Postal Service codes).

2. Symbols that encode numerical pointers to a data base, or so-called *license plate codes* (UPC symbols).

3. Symbols that contain complete assembly or contextual information, but require specialized readers (PDF codes).

4. Symbols that contain both pointers to web pages and information that can be read with smartphone cameras (QR codes).

Code designers are called *symbologists* and must account for sensor technology, potential users, and applications to implement practical bar-code systems. For example, they must be clever in their designs when they are asked to replace expensive commercial scanners used by trained operators with smartphone cameras operated by the general population.

Bar codes are read by sensing the light reflected from a printed pattern that is black-and-white, while modern bar codes also exploit color coding. Commercial bar-code readers used by package delivery companies are high-speed and read hundreds of codes per second that appear on envelopes. Bar codes allow automation by designing robot delivery systems that automatically locate and read codes on cartons.

We first describe commercial systems that read codes on mail envelopes and products and proceed to smartphone camera bar-code readers.

14.3.1 U.S. Postal Service Bar Code

To speed mail delivery, the United States Postal Service (USPS) implemented the Zoning Improvement Plan (ZIP) code in 1963. A 5-digit ZIP code is sufficient to direct mail to a specific post office anywhere in the United States. The bar code was developed to allow the ZIP code to be read by machine at high speeds, up to ten pieces of mail per second.

Current USPS bar codes appear in several variations.

- The standard 5-number ZIP code, such as 06520, to direct mail to a post office.

- The ZIP+4 code appends a 4-digit number, such as 06520-8284, to sort mail along the delivery route.

- The ZIP+4+2 code, which also includes a 2-digit point-of-delivery location number that is useful when the 9-digit number corresponds to a large apartment complex.

EXAMPLE **14.1**

My ZIP code at Yale

The post office that receives mail directed to Yale's School of Engineering and Applied Science has the designation 06520. Mail addressed to me contains the ZIP-plus-4 number 06520-8284. The last four digits allow letters arriving at post office number 06520 to be grouped together with other mail directed to my building to permit efficient delivery.

The USPS currently uses two bar code systems. The first one is the *early USPS* and uses two bars (short and tall) to encode digits. The current one is the *recent USPS* and uses four bars (short, tall, ascending and descending). Figure 14.5 shows both code types that appeared on mail sent to me. The early USPS code is simpler and encodes only the ZIP or Zip+4 code, and we describe its features next. The recent USPS code is more sophisticated and encodes the ZIP code, level of service (First Class, Overnight, Bulk Mail, . . .), and other options. The recent code also takes advantage of advances in high-resolution printing and scanning technology, as well as improvements in digital technology.

The early USPS bar code uses the 5-bit code words shown in Figure 14.6. While a 4-bit code would suffice for digits, the 5-bit code words implement a clever error-detection code. Each digit is encoded as two 1s (tall bars) and three 0s (short bars) and any other detected pattern of short and tall bars indicates the occurrence of an error.

Figure 14.7 shows how is the code is arranged in to a bar code for a 5-digit ZIP code. An additional digit is appended to the ZIP code for error correction, as described next. The set of six 5-bit code words is preceded by a tall start bar and terminated with a tall stop bar to aid in scanning the bar code.

USPS Bar Code Scanner

The early USPS bar code appears on an envelope as a series of tall and short bars. These are read with an array of optical sensors as the letter moves quickly past the

Figure 14.5

Scanned copies of early USPS (top) and recent USPS (bottom) bar codes. The early code was produced by a dot-matrix printer, while the bottom code used a laser printer. (I have two offices at Yale, which explains the different addresses.)

265   T5   P1
Dr. Roman Kuc
Electrical Engineering
Yale University
P.O. Box 208267
New Haven, CT 06520–8267

Mr. Roman Kuc
Yale University
P.O. Box 208284
New Haven, CT 06520–8284

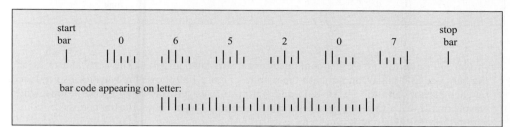

Figure 14.6

The Early-USPS bar code.

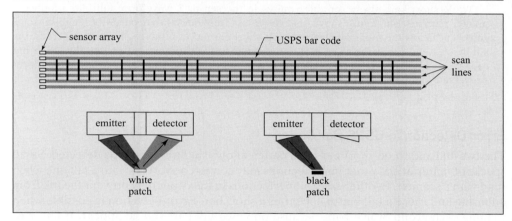

Figure 14.7

Early USPS bar code representing ZIP code 06520.

Figure 14.8

An array of optical sensors scans the USPS bar code. The sensor differentiates between black and white patches forming the bar code.

sensor, as shown in Figure 14.8. The emitter shines a small light spot on the code. The black ink forming each bar absorbs the light and produces no reflection, while the white spaces between the bars reflect light back to the detector. As the bar code passes by the array, each sensor registers the presence or absence of reflected light along a narrow horizontal stripe across the bar code. This scanning system automatically aligns the bar code by determining which sensors detect the first tall bar, which is called the *start bar*. Tall *start* and *stop* bars are always present in the bar code to aid in aligning the bar code within the scanner. The sensors above and below the bar code detect reflected light without interruption. These sensors determine the location and the height of the bar code on an envelope.

After alignment, a single pair of sensors differentiates tall from short bars. The *lower* sensor scans the short bars, while the *higher* sensor scans the tall bars. The lower sensor provides timing information, because a short bar is always present in the code. When the lower sensor detects a short bar, the higher sensor is checked to determine if it too detects a black bar. If so, a tall bar is present, otherwise it is a short bar.

Reading up-side-down USPS bar codes EXAMPLE 14.2

The USPS bar code is typically printed horizontally on an envelope. But what if the envelope is up-side-down in the scanner? The USPS code contains sufficient information to determine its orientation.

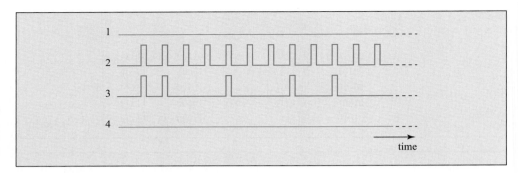

Figure 14.9

Reading a USPS bar code with a vertical array of four sensors determines the code is up-side-down.

Consider a sensor array that contains four sensors labeled 1 to 4. The vertical array scans the bar code as it passes underneath the scanner. Figure 14.9 shows the waveforms produced by each sensor. The top (labeled 1) and bottom (4) waveforms show no variations, while waveforms 2 and 3 show variations typical of a bar code. This indicates the bar code is contained within the vertical extent of the array.

While short bars always produce pulses in the sensor 2 waveform, tall bars produce a smaller number. The figure shows that these less-frequent tall-bar pulses occur in sensor 3 and occur below the more regular, periodic, short bar pulses produced by sensor 2. Thus, the tall bars extend below the short-bars and indicate to the scanning system that the USPS bar code is up-side-down. Being up-side-down, it is being scanned from right to left and requires a reverse-order correction to determine the correct ZIP code.

Error Detection in USPS Bar Codes

The two-tall and three-short bar-code patten allows single errors to be detected. Small specks of material and other imperfections may convert a short bar into a tall one when read with a scanner. In other cases, imperfections in the paper may prevent the ink from adhering and cause a tall bar to appear as a short bar. Error detection is possible when reading each group of five bars, because only two tall bars should appear. If fewer or more appear, the scanner recognizes that an error has occurred.

Error Correction in USPS Bar Codes

The USPS bar code appends an additional digit at the end of the bar code. This extra digit is called a *checksum digit* (CSD) and enables the correction of single errors analogous to the LRC code word described in Chapter 13. To determine the CSD value the digits in the ZIP code (or the ZIP+4 or ZIP+4+2 codes) are added. The CSD value is set to take the ZIP sum to the next multiple of ten. Viewed as a modulo-10 operation, the value of the CSD is computed from

$$[\text{ ZIP sum } + \text{ CSD }]_{\text{mod}(10)} = 0 \tag{14.3}$$

This operation is illustrated in the following example.

EXAMPLE 14.3 **USPS bar code checksum digit**

To determine the CSD value for the ZIP code 06520, we add the digits in the ZIP code to give

$$0 + 6 + 5 + 2 + 0 = 13$$

The CSD is set to the number that makes this sum plus the CSD have a modulo-10 value equal to zero, as

$$[0 + 6 + 5 + 2 + 0 + \text{CSD}]_{\text{mod}(10)} = [13 + \text{CSD}]_{\text{mod}(10)} = 0$$

By inspection (that is, by trial-and-error using different CSD values—there are only 10), CSD = 7. The complete bar code was shown previously in Figure 14.7.

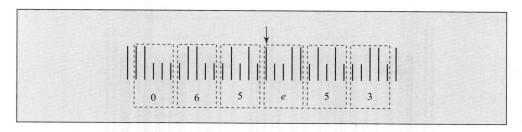

Figure 14.10

Correcting an error in the USPS bar code. The arrow indicates the erroneous bar.

The combination of the error detection and the checksum digit allows single errors in the bar code to be corrected. In this case, SEA means one error in the postal bar code symbol. The error correction procedure is illustrated in the following example.

Error correction in a USPS bar code   EXAMPLE 14.4

The USPS bar code shown in Figure 14.10 contains an error. To find the error, decode as many digit code words as possible. *First, remove the start and stop tall bars.* Collect the remaining bars in groups of five. Correctly encoded digits contain two tall bars and three short bars, so any other configuration indicates an error with the erroneous configuration being decoded as an *e*. Applying this procedure to the bar code shown in Figure 14.10 gives the sequence

$$0, \ 6, \ 5, \ e, \ 5, \ \text{and } 3$$

with 3 being the CSD value. The checksum formula gives

$$[0 + 6 + 5 + e + 5 + 3]_{\text{mod}(10)} = [19 + e]_{\text{mod}(10)} = 0$$

Thus, by inspection (that is, by *trial-and-error*) we find $e = 1$. This converts the erroneous bar code into the correct ZIP code 06515. The arrow in Figure 14.10 marks the location of the erroneous tall bar that should have been a short bar.

14.3.2 Universal Product Code

Stores use a clever system of identification tags called the *universal product code (UPC) symbols*, to automatically identify products. A standard UPC symbol is shown in Figure 14.11 and consists of 12 digits that represent a product identification number. This number accesses the price look-up (PLU) data file in the store computer, which transmits the price of the item. These digits are encoded in white and black bars. To allow correct interpretation, the UPC symbol is designed to have the following format.

- A guard pattern forms the 3-unit bar-space-bar (101) sequence that defines the left edge.

- Six digits are encoded on the left side of the symbol:

    1. The first digit specifies the industry type, such as 0 for grocery and 3 for pharmaceuticals.

    2. The remaining five digits encode the product manufacturer indentity.

- A center guard pattern forms the 5-unit sequence (01010) that separates the two halves of the symbol.

- Six digits encode the information on the left side of the symbol in the following manner:

    1. Five digits encode the item identity.

    2. One digit is used for a checksum value, which is analogous to the CSD in the USPS bar code.

Figure 14.11

Universal Product Code (UPC) symbol.

- A guard pattern forms the 3-unit bar-space-bar (101) sequence that defines the right edge.

A UPC symbol encodes each digit using seven bars with each bar being either a void or a black bar having the same thickness. Figure 14.12 shows each code word consists of two pairs of regions with each region being either black (joining black bars) or light (joining void spaces). The code words on the left side are the complements (or logical NOTs) of those on the right. Thus, a code word on the left side of the symbol has an odd number of black bars, while the code word for same digit on the right side has an even number. This coding allows the scanner to determine the pose of the UPC symbol (normal or inverted), as the next example illustrates.

EXAMPLE 14.5 UPC symbol pose

Figure 14.13 shows an edge of a UPC symbol. Is the symbol right-side-up or up-side-down?

After removing the 1-0-1 guard band that occurs on both edges, the count of black bars in the first 7-segment unit is four (an even number). Thus, the code word comes from the right side of the UPC symbol and makes the symbol up-side-down.

UPC Scanner

Unlike the USPS scanner, which moves the bar code across the sensor array, a UPC scanner contains rotating prisms and mirrors to quickly move a laser light spot across the UPC symbol. Even though the product containing the UPC symbol is typically in motion during the scanning process, its speed is much slower than that of the laser spot, so the symbol can be considered stationary during the scan. The laser in a commercial UPC scanner forms a web pattern, as shown in Figure 14.14. The detector produces a waveform that contains a specific pattern (a *signature*) that allows the scanner to recognize the symbol and decode it.

One problem in scanning the UPC symbol is the time variation in the signature caused by a variable scan angle shown in Figure 14.15. A light spot moving at right angles to

(a)

(b)

Figure 14.12

The digit code words in the UPC symbol form complementary pairs on the left and right sides of the symbol. (a) The number of black bars in left-side code words is odd, while on the right side it is even. (b) A typical UPC symbol of a fictitious product.

Figure 14.13

Is this UPC symbol segment right-side-up or up-side-down?

the bars (shown on the left side of the figure) produces the shortest signal. An oblique scan (shown on the right side) elongates the signal duration. To interpret the symbol's signature correctly the amount of time stretching that occurs must be determined. The clever engineering solution to this problem is to insert a bar pattern (1-0-1) that has a known spacing on both sides of the UPC symbol. The width of these bars provides the timing interval that is used to interpret the signals from the remaining bars on the symbol.

Figure 14.14

A UPC scanner directs a laser to form a web pattern across the UPC symbol for acquiring data regardless of symbol orientation.

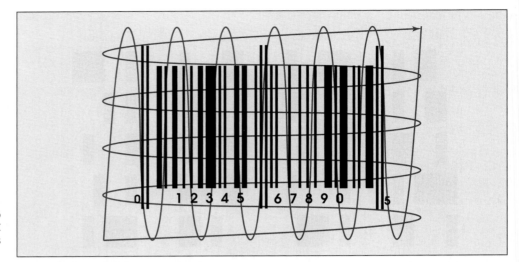

Figure 14.15

Scanning a UPC symbol produces a signal having a variable duration that depends on the scan angle. A black bar produces a 1 and a white space produces a 0.

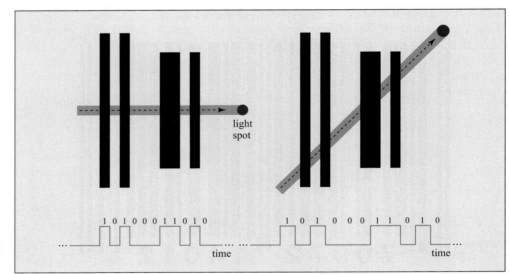

EXAMPLE 14.6 Laser spot speed

A UPC symbol has end guard bars that are 1 mm thick and a scanner laser spot has a speed of 100 m per second. When the laser passes directly across the UPC symbol, as shown on the left side of Figure 14.15, the duration (denoted by T_x) of the pulse produced by crossing over the first black bar equals the bar thickness divided by the laser speed, as

$$T_x = \frac{1 \text{ mm}}{100 \text{ m/s}} = \frac{10^{-3} \text{ m}}{10^2 \text{ m/s}} = 10^{-5} \text{ s} = 10 \ \mu s$$

The scanner reads the rest of the symbol by examining the detector waveform every $T_x = 10 \ \mu s$ for the presence of a black bar or a space between bars.

When the laser scans the symbol at a 45° angle, as shown on the right side of Figure 14.15, the spot takes longer to cross the black bar, producing a longer pulse. The path length through the black bar when the angle is 45° is

$$1 \text{ mm}/ \cos(45°) = 1.414 \text{ mm}$$

making the initial pulse duration equal to

$$T_x = \frac{1.414 \text{ mm}}{100 \text{ m/s}} = \frac{1.414 \times 10^{-3} \text{ m}}{10^2 \text{ m/s}} = 14.14 \ \mu s$$

The scanner reads the rest of the symbol by examining the waveform using a period equal to $T_x = 14.14 \ \mu s$.

Error Detection in UPC Symbols

For error correction, each digit is represented as a 7-bar code with each bar being either black or white. The number of black bars used in the 7-bar code words on the left half is odd. Thus, all left-side digits all produce code words having odd parity. For example, the digit 3 appearing on the left side of the symbol consists of five black bars (five 1s) in its 7-bar (7-bit) code word. The number of black bars on the right half of the symbol is an even number. Thus, all right-side digits produce even-parity code words. For example, the digit 3 appearing on the right side of the symbol consists of two black bars (two 1's). Any deviation from these patterns would signal an error.

Error Correction in UPC Symbols

The checksum digit (CSD) is analogous to that in the USPS bar code that can correct single errors. In this case, SEA means one error in the UPC symbol. The following formulas compute the value of the checksum digit. If d_1 represents the first digit of the code, d_2 the second, and up to d_{11} represents the eleventh digit, the error-correction method computes the sum

$$c = 210 - 3(d_1 + d_3 + d_5 + d_7 + d_9 + d_{11}) - (d_2 + d_4 + d_6 + d_8 + d_{10}) \qquad \textbf{(14.4)}$$

The CSD equals the modulo-10 value of the sum

$$\text{CSD} = [c]_{\text{mod}(10)} \qquad \textbf{(14.5)}$$

Rather than simply computing the sum of the digits, as was done in the USPS bar code, separating the even- and odd-indexed digits also detects other common problems, such as an accidental interchange of digits by the printer of the label. This is illustrated in the next example.

Check digit calculation EXAMPLE 14.7

The UPC bar code shown in Figure 14.11 contains the eleven digits

$$d_1 = 0 \quad d_2 = 1 \quad d_3 = 2 \quad d_4 = 3 \quad d_5 = 4 \quad d_6 = 5$$

$$d_7 = 6 \quad d_8 = 7 \quad d_9 = 8 \quad d_{10} = 9 \quad d_{11} = 0$$

The checksum formula produces the value

$$210 - 3(0 + 2 + 4 + 6 + 8 + 0) - (1 + 3 + 5 + 7 + 9) = 210 - 3(20) - 25 = 125$$

The checksum digit equals

$$\text{CSD} = [125]_{\text{mod}(10)} = 5$$

To demonstrate the significance of the formula for CSD, let two of the digits be mistakenly interchanged by the printer. That is, instead of

012345678905

the mistaken label appears as

$$012435678905$$

Clearly, the sum of the digits remains the same, so this error is not detectable by simply computing the sum of all the digits. However, in evaluating the CSD formula, we get

$$210 - 3(0 + 2 + 3 + 6 + 8 + 0) - (1 + 4 + 5 + 7 + 9) = 210 - 3(19) - 26$$
$$= 127$$

and

$$[127]_{\mod(10)} = 7$$

This value does not match the CSD printed on the label (5) and indicates an error occurrence.

EXAMPLE 14.8 — Correcting UPC errors

A scanner produces the following eleven-digit sequence

$$0\,7\,0\,9\,e\,2\,1\,4\,0\,1\,2\,7$$

where e corresponds to a error that occurs on the left side of the UPC symbol.

The correct value of e can be found from the checksum formula. Using the observed CSD = 7, the checksum formula gives

$$[210 - 3(0 + 0 + e + 1 + 0 + 2) - (7 + 9 + 2 + 4 + 1)]_{\mod(10)} = 7$$

Doing the arithmetic, we get

$$[210 - 3(e + 3) - 23]_{\mod(10)} = [210 - 3e - 32]_{\mod(10)}$$
$$= [178 - 3e]_{\mod(10)}$$
$$= 7 \quad (\text{CSD on UPC symbol})$$

We need to find the value of e such that $178 - 3e$ ends in a 7. By trial-and-error, we find that $e = 7$ gives

$$[178 - 21]_{\mod(10)} = [157]_{\mod(10)} = 7$$

Thus, the erroneous digit should be a 7.

14.3.3 Two-Dimensional Bar Codes

An extension to the previous one-dimensional bar codes is the two-dimensional (2D) bar codes that contain much more data by exploiting both the horizontal and vertical dimensions. 2D bar codes have become increasingly popular, because smartphones with high-resolution cameras permit scans by merely taking their pictures. This section briefly describes two generations of 2D bar codes that have evolved with the technology.

PDF Symbols

Portable data file (PDF) symbols were designed to contain more data rather than simply identifying the product as with the UPC symbol. The PDF symbol is meant to be read

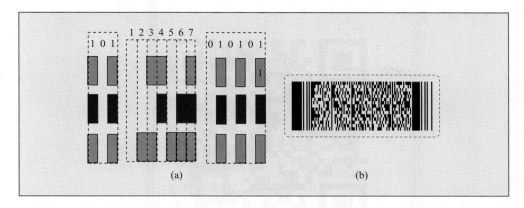

Figure 14.16

PDF symbols. (a) Constructing a PDF symbol by stacking UPC symbols that contain left-side digits 0 9 8, left guard band 101 and right guard band 010101. (b) Commercial PDF symbol. (Nik Niklz / Shutterstock.com)

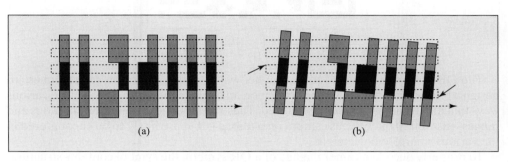

Figure 14.17

A PDF symbol is read by a series of horizontal laser scans that are densely spaced in the vertical direction. (a) In the ideal case the symbol is aligned with the scans. (b) Skewed symbol scan (blue arrows indicate scans that cross boundaries and produce non-repeating code words).

using specialized scanners. The code structure can be thought of as a series of short UPC symbols stacked on top of one another. Figure 14.16(a) shows a simple example where the left-side UPC code previously shown in Figure 14.12 encodes a single digit in each row. The symbol encodes 0 in row 1, 9 in row 2, and 8 in row 3. The UPC coding scheme allows these codes to be scanned from left-to-right or right-to-left and still provide sufficient information in the scanner waveform to decode the digits correctly. A special right-side guard band (010101) identifies the right side of the code in the scanner waveform. Commercial PDF codes are more complex and can contain thousands of alphanumeric characters. An example of a commercial PDF symbol is shown in Figure 14.16b.

This stacked structure requires scanners to perform a sequence of dense horizontal scans across the symbol, which is called a *raster scan*. Figure 14.17 shows a laser raster scan that forms many scan lines across each digit code word in the symbol. Multiple passes occur in each row and the duplicate data can be recognized as such and discarded to yield the codes for each row. The set of row data is combined to form the entire data set represented in the PDF symbol.

Several horizontal scans occur within each row in the symbol to produce the same code words in successive scans. Such repeating code words indicate a valid scan of a row and are reduced to the single code word. Horizontal scans that cross row boundaries produce different code words in adjacent scans that do not repeat the sufficient number of times. Isolated and non-repeating code words are identified as coming from two different rows and are eliminated.

QR Codes

Figure 14.18 shows an example of a *quick response (QR) code,* which is a popular 2D bar code that appears in many advertisements and business cards. The QR code contains information that directs your smartphone browser to a web page where the product details can be viewed. The QR code is finding new applications in manufacturing and architecture to replace (bulky and difficult to maintain) paper drawings and instructions with electronic versions that exist in digital databases that can be accessed when and where needed.

Figure 14.18

Example of QR code that includes the URL of my Web page:
http://pantheon.yale.edu/~ kuc.
(Code generated by Kaywa, see
http://qrcode.kaywa.com)

The QR code was designed to be scanned with digital cameras in smartphones. To
design a reliable smartphone camera scanner, an engineer needs to consider the various
ways that an QR image would be acquired. These include the different orientations and
ranges—as well as the various camera resolutions—that are likely to be encountered in
the various smartphone versions.

To properly decode a camera image of a QR symbol, the symbol contains structural
features that are helpful.

- Each informational element in the symbol must occupy at least several pixels
  in the camera image for robust detection. If an individual QR block occupies
  less than a pixel in the image, it may fall between two pixels and be missed
  altogether.

- The symbol must include simple-to-recognize features that indicate the symbol
  orientation. Camera software can align the image so that it can be scanned horizontally (within the smartphone memory) as in the conventional manner employed
  by raster scanners.

EXAMPLE 14.9 Reading a QR code with a digital camera

Figure 14.19 shows the considerations with reading a QR symbol with a digital camera. To
decode the camera image of a QR symbol, there must be at least one pixel that is situated
within a white region and another in an adjacent blue region. Clearly, this minimum resolution
requires careful registration of the QR image onto the camera pixel array—both in position
and tilt—as shown in the top of the figure. Such a fortuitous arrangement is an unreasonable
expectation under normal operation. The high-resolution cameras in current smartphones
make such accurate registration unnecessary. If we double the number of pixels to allow four
pixels to lie within each region—as shown in the bottom of the figure—even in the worst case
there always will be at least one pixel in each region.

Figure 14.20 shows the problems encountered with decoding an image of a QR symbol taken
with a digital camera. The pixel information must be sufficient to align and orient the QR code
by processing the image. This transformation is simplified by including additional pixels—
either by taking the picture at closer range or using a higher-resolution camera. Experience
with smartphone cameras indicates that reliable decoding results when a QR symbol occupies
at least one quarter of the image.

Figure 14.19

Reading a QR code with a digital camera. QR symbol regions are shown in blue and white, and camera pixel locations in black. Left-side images show ideal registration of camera to symbol. Other images show problems with registration and camera tilt.

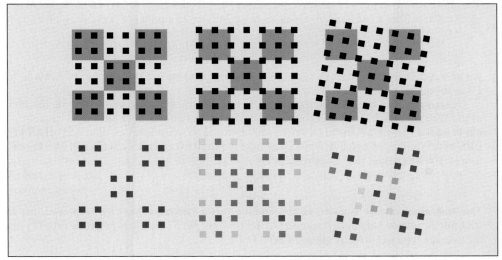

Figure 14.20

Decoding a digital image of a QR symbol.

 # 14.4 MACHINE UNREADABLE CODES

On the other end of the spectrum, how does one make a symbol that *cannot* be read by a computer? One reason is to distinguish a human user from a web robot or *bot*—a program that searches the Web and accesses Web pages much faster than human users can. Malicious bots include those that ticket re-sellers use to buy up good seats at concerts—to be resold later at inflated prices. To prevent such activity, Web sites use text and symbols that humans read easily (using human reasoning), that are called *CAPCHAs*, in order to distinguish human users from computer programs. CAPCHAs are a type of *Turing test* used to distinguish humans from computers that use artificial intelligence (AI).

CAPCHAs are fully automated programs that may contain visual features (or audio features for visually impaired users). The CAPCHA generation process consists of the following steps.

1. A random character generator uses a pseudo-random number generator (PRNG) to form a random string of six keyboard characters.

2. A graphics program forms a random background that humans readily ignore but that confuse AI pattern-recognition programs.

Figure 14.21

Example of a simple automated CAPCHA.

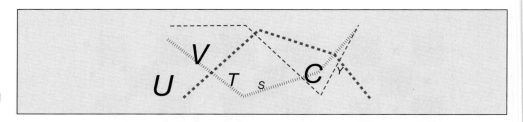

3. A program randomizes the location and appearance of the characters by using a PRNG to generate random offsets and font sizes.

Figure 14.21 shows an example of a simple CAPCHA using upper-case letters that was generated automatically.

 EXAMPLE 14.10

Hacking a CAPCHA

Assume that AI programs are ineffective when trying to recognize the letters in a CAPCHA. The alternative is to use a brute-force technique (that is guessing). A CAPCHA that contains six characters from a 26-character alphabet can generate the number of possible combinations equal to

$$26^6 = 3 \times 10^8$$

On average, a search produces a successful result by guessing half this number of patterns, or 1.5×10^8 queries.

Guessing over an internet connection involves transmitting data packets from your machine to the Web site. Each letter is encoded as a data byte and enclosed within a data packet that has at least a 10-byte overhead (4-byte destination and source addresses, data size byte, and a LRC byte) yielding 16 bytes (128 bits). The time required to transmit each CAPCHA response using a 100 Mbps data rate is approximately

$$\frac{128 \text{ bits}}{10^8 \text{ bits/s}} \approx 1 \ \mu s$$

This transmission time is short, so the limiting factor becomes the 1 ms response time to transmit an acknowledgment response (ACK) from the Web site. If this response time is 1 ms, the average time to correctly guess a CAPCHA is

$$1.5 \times 10^8 \text{ queries} \times 10^{-3} \text{ seconds/query} = 1.5 \times 10^5 \text{ s} = 41 \text{ hrs}$$

Clearly, this would slow down any ticket-scalper.

▶ 14.5 STEGANOGRAPHY

Encrypted data files draw attention: *This encrypted file contains important (and possibly useful) data.* Encrypted files are usually easy to spot because they occur over special secure connections, such as Web sites that begin with "https" (the "s" indicates *secure*) and between financial institutions.

Another form of transmitting small amounts of data secretly is called *steganography*, which includes (or *embeds*) the data within a large data file (such as image, audio, or video) that does not call attention to itself like an encrypted file does. The idea is that by simply looking at a file's contents, it is difficult to determine if it contains embedded data. This form of transmitting secret data is actually thousands of years old: Ancient rulers were known to shave the head of a loyal servant, tattoo a message on the bare scalp, and let the servant's hair grow back before sending him to the destination. One can only imagine the process at the receiving end (or if the servant was captured by a suspecting enemy).

This section provides two examples of steganography: digital watermarks that indicate ownership of printed material that is distributed illegally, and binary data embedded within audio or image data files.

14.5.1 Digital Watermarks

You have written an original article that you want to distribute to a limited number of clients who pay for your service. You discover that one of your clients has made an unauthorized copy of your article and has distributed it widely. You can determine which client did this through a *digital watermark*.

One digital watermarking technique involves changing the text copy almost imperceptibly by shifting particular lines slightly in the vertical and horizontal directions. This can be done easily with a standard layout program and current laser printers. Each document has a different set of lines that are shifted. The shifted lines are determined by a unique binary code that is assigned to each client. Thus, each client receives a uniquely (and secretly) coded document that can be traced back to the offending client.

Digital watermark   EXAMPLE 14.11

Figure 14.22 shows example of a digital watermark. The original text is shown in an unshifted version. The title is a landmark used for registering shifted versions. The watermarked copy sent to client 1010 shifts the lines that correspond to the 1's in the binary code assigned to each client. That is, client 1010 is sent a copy in which lines 1 and 3 have been shifted in the vertical and horizontal directions. Viewing the watermarked copy alone makes it difficult to detect the shifts. To illustrate the shifts visually, the original and watermarked copies are superimposed in Figure 14.22. Note that the shifted lines appear darker because the superimposed text prints with more ink. Even electronic photocopies of the watermarked text maintain the shifts - and the *identity* of the dishonest client. (Very clever!)

How large a binary code is necessary? The title is used to register the original and shifted versions. This leaves the remaining lines on the page for the coded shifts. A typical newsletter prints seven lines per inch. A 9-inch column allows for 63 lines for the code or $2^{63} (= 8 \times 10^{18})$ unique code words—a client base that would satisfy most businesses!

Even though such a digital watermark is subject to attack in an attempt to remove it, the code is still robust. The watermark can be extracted even if the document is photocopied or faxed. My devious colleagues tell me that the way to remove the watermark is to scan the text page, use an optical character recognition program to extract the text from the image, and reprint the document. However, this may involve more expense than simply subscribing to your service in the first place.

As a historical note, mapmakers also had this problem with unauthorized copies. Their solution was to include a fictitious street in the map, establishing their authorship.

14.5.2 Embedding Secret Codes in Data

If the data file size is large, secret data can be embedded with minimal perceptual distortion. For example, consider a data file containing quantized samples of a sinusoidal waveform. If the quantization step size is Δ, changing the least significant bit (LSB) affects the waveform amplitude by (at most) a single Δ. If Δ is sufficiently small, the change is imperceptible. Using this method, we can insert secret binary data (say 1010) into the first four sample values by setting their LSB's equal to these to secret values. The receiving party would know to extract the LSBs from the data to discover the secret data. A casual viewer would not notice any difference when viewing the normally decoded sinusoidal waveform.

Original Document

Important Information

This information is very sensitive and for our clients only.

Unauthorized copying of this material is prohibited.

If sent to non-subscribers, your identity is discoverable.

Then our lawyers will be contacting you.

Version sent to client 1010

Important Information

This information is very sensitive and for our clients only.

Unauthorized copying of this material is prohibited.

If sent to non-subscribers, your identity is discoverable.

Then our lawyers will be contacting you.

Superposition of Original and Version sent to client 1010

Important Information

This information is very sensitive and for our clients only.

Unauthorized copying of this material is prohibited.

If sent to non-subscribers, your identity is discoverable.

Then our lawyers will be contacting you.

Figure 14.22

Example of a digital watermark.

EXAMPLE 14.12   Adding secret data to the LSBs

Let the secret code be 1010. We insert this code into the existing binary data in a file with one code bit replacing the existing LSB of each binary value. Let the first four waveform samples be represented by 8-bit sequences

$$01001101 \quad 01011001 \quad 00001100 \quad 11111111$$

We first set all of the LSBs in the data to 0, as

$$01001100 \quad 01011000 \quad 00001100 \quad 11111110$$

and add the secret data, as

$$01001100 + 1 \quad 01011000 + 0 \quad 00001100 + 1 \quad 11111110 + 0$$

to produce the binary values that embed the code. Thus,

$$01001101 \quad 01011000 \quad 00001101 \quad 11111110$$

Note that adding the secret code leaves some 8-bit data values unchanged, such as in the first byte. Values can change by no more than $\pm$LSB, thus increasing or decreasing the reconstructed amplitude by Δ. If Δ is small, the changes are imperceptible.

To extract the code, we merely extract the LSBs. Hence,

$$01001101 \rightarrow 1 \quad 01011000 \rightarrow 0 \quad 00001101 \rightarrow 1 \quad 11111110 \rightarrow 0$$

To illustrate the perceptual quality of steganography, we inserted the ASCII code for 'steganography' ('s' = 01110011, 't' = 01110100, 'e' = 01100101, etc.) into samples of a sinusoid. Figure 14.23 shows the results for samples that are encoded with 6 bits and encoded with 8 bits. The bottom curves show differences in more detail, with dashed curves showing original values. The Δ value for 6-bit quantization is larger than that for 8-bit quantization by a factor of 4, so the changes of $\pm\Delta$ in the waveform are more evident for 6-bit quantization. For a 16-bit quantization (CD-quality audio), Δ is so small that the figure would not display the differences produced by the code, making it imperceptible.

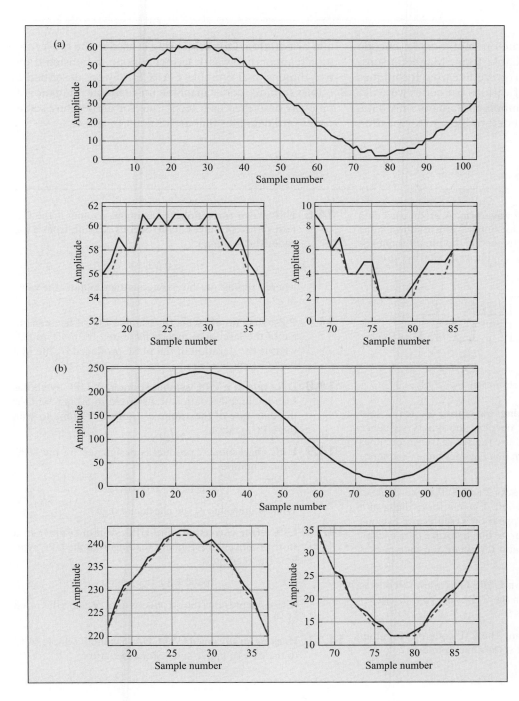

Figure 14.23

Steganography. The ASCII code for "steganography" was inserted into the least-significant bit of the samples of a sinusoidal waveform. (a) Waveform samples encoded with 6 bits. (b) Waveform samples encoded with 8 bits. Bottom curves show detail of differences, with dashed curves showing original values.

14.6 Research Challenges

14.6.1 Active Identification

The symbols described in this chapter were passive, requiring a scanner to probe the symbol and interpret the acquired waveform. With automated systems and the advent of the Internet of Things (IoT), bar codes are being replaced by smart tags that recognize that the data they contain is needed and respond by transmitting it wirelessly. RF-ID tags and wireless smart cards are the first versions of the devices to come.

14.7 Summary

This chapter described common methods to encode digital data in machine-readable form. Magnetic sensing decodes magnetic stripes on credit cards. Techniques that allow robust scanning of various bar codes were described. Current bar code symbols contain not only data, but also include features that determine symbol location and orientation. Optical sensing is used in the most common scanners and was described for one-dimensional and two-dimensional symbols. CAPCHAs were described as efforts to make codes machine un-readable. Steganography was described as a technique for embedding secret codes into data files that are difficult to detect.

14.8 Problems

14.1 Credit card self-clocking waveform. A credit card data stripe produces the self-clocking waveform shown in the top half of Figure 14.24 to encode the binary sequence 00000. What is the binary sequence encoded by the other waveform?

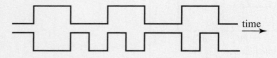

Figure 14.24

Self-clocking waveform in Problem 14.1.

14.2 Generating a self-clocking waveform. Sketch a self-clocking waveform that encodes the binary sequence 00000. Below this sketch, use the above transition times to sketch in the waveform that encodes binary sequence 10101.

14.3 Credit Card Code on Tracks 2 and 3. The two-digit sequence 09 is very important. Encode these digits in a data block as they may appear on a credit card. Encode each code word on a separate line to form a data packet. Add the sentinel codes and terminate the data packet with a LRC code word.

14.4 Interpreting up-side-down USPS bar codes. Decode as many digits as possible in the ZIP code from the waveforms shown in Figure 14.9.

14.5 USPS bar code checksum. The ZIP code of Mountain View, CA is 94043. What is the checksum digit?

14.6 USPS error correction. The bar code scanner in the U.S. Post Office produces an erroneous reading from a bar code. The reading is

$$6061e8$$

where e represents the erroneous digit. What is the value of e?

14.7 Pulse time for 60° scan. In Example 14.6, if the scan angle of the laser spot path is increased from 45° to 60°, what is the duration of the pulse produced by the first bar? Give answer in μs.

14.8 Data rate of a UPC symbol scanner. A UPC symbol is scanned and the data is transmitted in 50 ms. What is the data rate of the scanner? (Ignore the time to scan the UPC symbol.)

14.9 UPC checksum. A product is designated by the UPC symbol number

$$07231000042$$

What is the value of the checksum digit?

14.10 UPC error correction. The UPC symbol scanner in a store produces an erroneous reading from the UPC symbol. The reading is

$$07073e053359$$

where e represents the erroneous digit. What is the value of e?

14.11 Hacking an advanced CAPCHA. Extend Example 14.10 to the case of upper-and lower-case letters.

14.9 Matlab Projects

14.1 **Processing a jpeg image of USPS bar code.** Using Example 16.71 as a guide, take a picture of a USPS bar code and write a Matlab script that decodes it into a binary sequence.

14.2 **Processing a jpeg image of QR symbol.** Using Example 16.71 as a guide, take a picture of a QR bar code and write a Matlab script that decodes the top row into a binary sequence.

14.3 **Generating a CAPCHA.** Using Example 16.72 as a guide, write a Matlab script to automatically generate eight random lower-case letters, and red, blue, and black background curves that contain eight random line segments each.

14.4 **Generating an advanced CAPCHA.** Using Example 16.72 as a guide, write a Matlab script to automatically generate six random upper and lower-case letters, and red, blue, and black background curves that contain four random line segments each.

DATA NETWORKS

LEARNING OBJECTIVES

After completing this chapter, the reader should be able to:

- Understand how data packets are transmitted asynchronously over a data channel.
- Recognize the structure of data packets for robust communication over the Internet.
- Appreciate the differences between circuit switching and packet switching.
- Probe the topology of the Internet.
- Describe packet collisions that occur in wired and wireless networks.
- Comprehend the possible utility of cloud computing.

The *Internet* is a network of interconnected computers called *router nodes* that direct data from sources to specified destinations. Many sources generate data at irregular and random time instants, and such data transfer upon demand is called *asynchronous data transmission*. Data of various sizes form data packets that are transmitted over a channel. We describe how data packets are structured to ensure that data are received reliably and without error. The network interconnections provide robust operation by automatically by-passing individual routers that fail. Problems associated with packet collisions are discussed. Cloud computing resources are described with examples of the flexibility they provide.

15.1 INTRODUCTION

The Internet copes with data communication that varies widely in timing, quantity, and delays, as illustrated in Figure 15.1. This chapter describes how a networks manages to operate successfully by considering the following issues.

Internet packets and protocols—Data packets that travel between different transmitters to different receiver experience variable delays. Data packets contain source and destination addresses for proper routing, a data size value for smaller packets, and a data segment number to re-sort data that get delayed.

Asynchronous data transmission—Data packets occurring at random times contain more than just data in order to avoid confusing the network by incorporating features that indicate their presence on the communication channel. The simplest packets using these techniques travel from a specific transmitter to a specific receiver and have a specific data size, such as the remote control (transmitter) that switches channels on your TV (receiver).

Probing Internet topology—To prevent erroneous data packets from endlessly circulating within the network, data packets contain a timer code word that expires if a problem occurs. This timer feature is used to probe the Internet to find the number of computers between a source and destination, as well as the travel time of the data packet.

Detecting collisions—Practical limitations cause data packets to collide. Techniques are described to detect collisions in wired and wireless data communication systems.

Figure 15.1

Data within a network travel from various sources to one or more destinations with data packets being transmitted and received at random times and with different delays. C are components (sensors) and D are devices (smartphones).

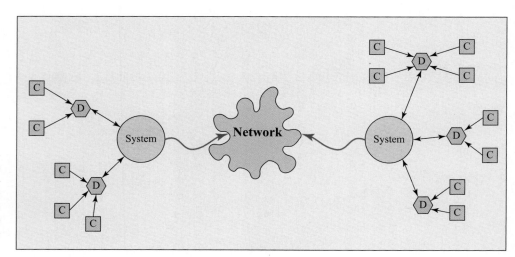

15.2 EVOLUTION OF DATA NETWORKS

So, where are digital networks heading, and why? We have seen computers grow smaller and faster, consume less energy, and communicate information using graphic displays on tablet or smartphone screens in place of computer monitors. Connections to the Internet, which used to be solely by cable, are now increasingly wireless. It all started with email, then cell phones that merged with Blackberries, and currently with ubiquitous tablets and smartphones.

Connectivity describes the study and practice of how to interconnect these devices so that they work together seamlessly. While the programming issues are beyond the scope of this book and usually in the domain of computer science, the electrical engineering issues include models and protocols. The topics covered in the previous chapters permit a meaningful overview of these connectivity issues.

Protocols are rules governing operation. Internet protocol (IP) was established to allow computers to communicate with each other. In data transmission, solving data communication problems between users in complex networks usually means that additional information is attached to the data packet and is sufficient for the network to solve problems. In data transmission, this additional information is stripped away and never seen by the application requesting the data, such as a video player.

32-bit to 64-bit addresses EXAMPLE 15.1

The Internet started getting crowded, with many additional Web sites appearing daily, and IP addresses were running out. The solution was to expand the address size from 4 bytes (= 32 bits) to 16 bytes (= 128 bits). With 32-bit addressing, under Internet Protocol Version 4 (IPv4), the number of unique addresses is

$$2^{32} = 2^2 \times \overbrace{2^{10}}^{=10^3} \times \overbrace{2^{10}}^{=10^3} \times \overbrace{2^{10}}^{=10^3} = 4 \times 10^9$$

or nine billion separate addresses. The 128-bit addressing under IPv6 provides

$$2^{128} = \left(2^{32}\right)^4 = (4 \times 2^{30})^4 \approx (4 \times 10^9)^4 = 256 \times 10^{36}$$

separate addresses, enough to have one for each conceivable device: computers, smartphones, televisions, GPS, automobiles, tablets, and appliances. In the envisioned Internet of things (IoT), each device incorporates a small Web-Interface-computer (WIC), each having its own address that has the ability to receive and send data.

15.2.1 Packet Switching and Circuit Switching

Historically, there have been two main methods for transmitting data from a source to a destination on a system level or, equivalently, from a transmitter to a receiver when describing operations on a circuit level.

- *Packet switching (the postal model).* In this model, the complete data file is broken up and packed into envelopes (data packets) that can be accommodated by the postal system. In today's electronic transfers, the address of the destination and the return address of the source are *Internet protocol (IP) addresses*. When you type a Web name (such as www.yale.edu), which is technically called a *Universal Resource Locator (URL)*, this URL is looked up by a *Domain Name*

Server (DNS) in a local computer called a *router* to find the corresponding IP address of the destination, in this case (130.132.35.53). This IP address was assigned by an Internet administration group when Yale registered both the name and a particular computer on the Yale campus, which is identified by its unique *Media Access Control (MAC)* number specified by the computer manufacturer when the computer was built. This process produces a unique IP address assigned to www.yale.edu.

■ *Circuit switching (the telephone operator model).* In early telephone days, a telephone connection was made by an operator physically connecting a wire from your phone to that of the person being called. More recent telephone systems made the connection with electrically-operated switches. In both cases, the result is a dedicated wire connection between the source and destination. Circuit switching is used for special (and costly) applications where data transmission time is critical.

The rest of the chapter describes the more common packet-switching method. Sufficiently fast Internet speeds make these two methods less distinguishable. Special services demanding high-speed data transfer in the future will likely provide a hybrid solution.

15.2.2 Internet Transmission Protocols

While data on the Internet are transmitted in the form of data packets, different applications have different requirements. This section describes the two most common *protocols* or methods for data transfer that represent a trade-off between reliability and speed.

1. *Transmission control protocol/Internet protocol (TCP/IP)* guarantees that the transmitted data are correct—no matter how long it takes. The term *TCP* indicates the structure of data packets, while *IP* specifies the addressing and data-packet sizes to enable communication between two computers. Data integrity at the data receiver is ensured by including error detection elements (parity bits and a checksum character) within a data packet. If no error is detected, the destination transmits a short *acknowledgment data packet (ACK)* back to the source, as Figure 15.2 illustrates. If the source receives no ACK within a specified time delay, it assumes the data packet was lost and re-transmits the data packet. If a data packet error is detected at the destination—or along the path to the destination—a short *not-acknowledge data packet (NAK)* is sent back to the source requesting re-transmission.

Figure 15.2

Data packets are transmitted and acknowledged.

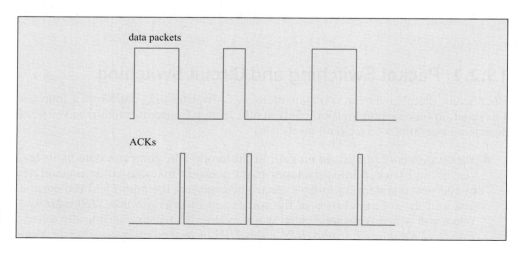

Sending ACKs and NAKs (with re-transmissions) slows the data transfer rate, but data are transferred successfully. TCP/IP is the current method used for Web surfing and downloading documents on the Internet. While most data packet transfers are handled in a timely manner, transmission delays become noticeable when the data traffic becomes congested. Under severe congestion, Web pages fail to load at first request.

2. *Universal datagram protocol/Internet protocol (UDP/IP)* transmits data from source to destination with minimal delay by eliminating the re-transmission of erroneous packets. UDP/IP is employed for applications requiring real-time operation, such as telephone calls over the Internet, called *voice over Internet protocol (VoIP)*, or for video streaming. If a received data packet contains an error under TCP/IP, the time delay involved in the destination sending a NAK and receiving a re-transmitted correct packet would cause annoying pauses in the audio or video playback. To eliminate such pauses, UDP/IP uses its *best effort* to transmit the data. This means not transmitting acknowledgment (ACK or NAK) data packets and no data packet re-transmissions.

In this case, the destination needs to accommodate received data packets that contain errors. Different systems have different approaches: some replace an erroneous data packet with the previous good data packet (common in video streaming), while others use empty data packets or *average* data packets (common in VoIP). When you make a call using your smartphone, the data are usually transmitted using UDP/IP. If you listen carefully, you will hear short pauses caused by missing or erroneous data packets. As long as there is less than a 5% error rate, the audio does not sound too objectionable. However, in times of high Internet congestion, delays and errors cause the audio to have an annoying choppy quality, sometimes making a telephone conversation frustratingly unintelligible.

As an alternative to UDP/IP, if you are willing to accept a significant delay in order to get good sound quality, you could transfer an audio or video file using standard TCP/IP, wait until all (or most of) the data are received, and then start playing it on your computer. However, this approach would have too long of a delay to allow for normal telephone conversations using VoIP.

 # 15.3 ASYNCHRONOUS DATA TRANSMISSION

Transmitting data between two digital systems requires special considerations. For example, data packets typically occur at random times: when a mouse is clicked, when a key on the keyboard is pressed, or when the next frame of video is required. In *asynchronous data transmission*, data packets are transmitted when they become available.

To illustrate transmission of randomly occurring data, we describe the transmission of a single data byte that is enclosed within a time waveform called a *data character* that contains additional timing information sent by a transmitting circuit (transmitter) to a receiving circuit (receiver). The transmitter and receiver both use the same timing interval T_B that is set when specifying the transmission rate (*setting the baud rate*). Other systems use *self-clocking codes* (described in Chapter 14) that are less efficient.

The transmitter controls the signals on the channel and the receiver monitors the channel for the arrival of new data. In asynchronous data transmission, the transmission channel is idle for much of the time and has an *idle-channel* value supplied by the transmitter that indicates the transmitter is connected. For example, the idle-channel value being 5 V indicates that the transmitter is working, while a 0 V level does not.

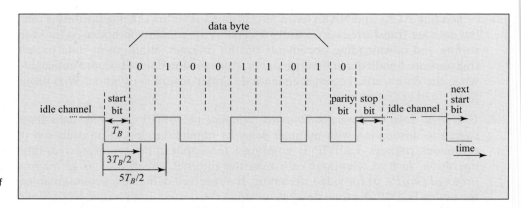

Figure 15.3

Asynchronous character format of data byte 01001101.

The asynchronous data character contains the following four components shown in Figure 15.3.

1. **Start bit.** The beginning of each data character sent by the transmitter is indicated with a start bit whose value produces a change from the idle-channel value (for example, from 5 V to 0 V). The receiver uses this start-bit transition time to start timing the data character. The start-bit duration is T_B seconds (s) and is used by the transmitter and receiver.

2. **Data.** The data bits are transmitted one at a time with each lasting T_B s. The receiver determines the value of each data bit by examining the data character waveform value in the middle of its T_B interval. That is, the middle of the first data bit interval occurs at time $3T_B/2$ as measured from the start-bit transition time as shown in Figure 15.3. The middle of the second data bit occurs at time $5T_B/2$, and so on.

3. **Parity bit.** The data character appends a parity bit to the data for error detection. The parity bit is optional and not included for reliable transmitter-receiver connections to increase the data transmission rate. For systems using even parity, the value of the parity bit is such that the number of 1's in the data plus the parity bit is even.

4. **Stop bit(s).** The data character is terminated with one or more stop bits—each have the idle-channel value and last for T_B seconds. Stop bits provide a delay between character transmissions. The transmitter inserts a sufficient number of stop bits to allow the receiver to process the received data before it transmits the next character.

EXAMPLE 15.2

Asynchronous transmission

A data source transmits data characters asynchronously over a transmission channel having a capacity equal to $C = 10^4$ bps. The transmission time for each bit over the channel is computed as

$$\frac{1}{C} = \frac{1}{10^4 \text{ bps}} = 0.10 \text{ ms/bit}$$

which makes $T_B = 0.1$ ms.

A data character contains a start bit, eight data bits, a parity bit, and one stop bit, which is eleven bits for each byte of data. The data character duration is denoted T_{char} and equals

$$T_{char} = 11 \ T_B = 1.1 \text{ ms}$$

The source transmits eight bits of data during each T_{char}. The source data rate is computed as

$$\mathcal{D}_S = \frac{8 \text{ bits}}{T_{char}} = \frac{8 \text{ bits}}{1.1 \text{ ms}} = 7,270 \text{ bps}$$

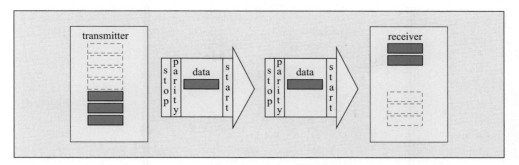

Figure 15.4

Asynchronous transmission divides large data files into smaller units that are transmitted as data characters.

Data characters typically have short durations as longer durations increase the probability of incurring errors. Large data files are broken up into smaller and more manageable units that are transmitted as a series of manageable data characters, as shown in Figure 15.4.

15.4 INTERNET DATA PACKETS

Communicating over the Internet often involves the transmission of large quantities of data, such as downloads or videos. In such cases, the data is broken up into small segments and transmitted as a sequence of separate data packets. To illustrate the data exchange process, consider using the Internet to request funds from your bank account at an automated teller machine or to validate a credit card purchase. Information is sent in the form of these data packets into a network using a data format that allows the packet to travel through the network and arrive at the desired destination. Figure 15.5 shows a simplified data packet that contains six *fields* of essential information.

1. The *address* field contains the routing information about the destination and the source. If a response is required, such as an acknowledgement or error condition, the destination knows the return address.

2. The *time-to-live (TTL)* field contains a byte value that limits the number of times the packet can be relayed through intermediate nodes. The TTL value equals 255 in the packet transmitted by the source. Each time the packet is relayed by a computer node the TTL is decremented. If the packet TTL count reaches zero at a particular node, that node deletes the packet and sends a message back to the source indicating the packet has been deleted. The TTL prevents packets from circulating endlessly through the Internet due to some malfunction. Otherwise, errant data packets would clog the Internet and even bring it to a halt. This feature is also used for determining the structure of the Internet in the next section.

3. The *data length specifier* field indicates the number of bytes in the data field. Data size is an important consideration for the operation of the network. Text data could be transmitted one byte at a time, but this would be inefficient because the overhead in the packet structure must be transmitted with each byte. Data sizes between 500 and 1,500 bytes are found to be transmitted most efficiently.

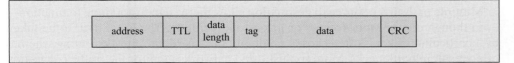

Figure 15.5

Simplified structure of an Internet data packet.

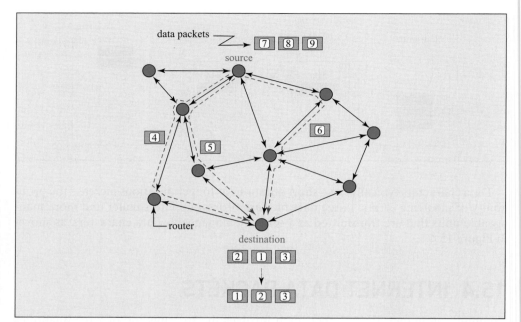

Figure 15.6

A network consists of many switching computer nodes or *routers*. These routers form multiple links with their neighbors. Data packets can travel over any available path (indicated in dashed line) that links the source with the destination, causing packets to arrive at random times. The tag in the data packet allows the destination to arrange the packets in the proper order.

4. The *tag* field is a byte that indicates the packet number in a sequence. The Internet is a dynamic, changing environment consisting of many switching routers between the source and the destination, as shown in Figure 15.6. Transmitting a large data file is usually done by partitioning it into multiple smaller data packets. Each individual packet can travel over different paths, using a different set of routers to travel from source to destination. Figure 15.6 shows alternate paths as dashed lines. Such flexibility maintains the integrity and reliability of the network. With such a flexible network, however, one of the packets may encounter a slow path, as when passing through a router that is handling a large quantity of data. This packet can arrive at the destination after a packet that was transmitted at a later time. In such a case, the packets must be put into the correct order at the destination, and this is the function of the *tag* field. The tag is a single byte with values from 0 to 255. The tag forms a modulo-256 counter, because having more than 256 packets in the network at one time is rare and the destination keeps track of tag overflow values.

5. The *data* field contains the bytes that are being transmitted. Parity bits are not appended to the data bytes to conserve data packet size.

6. The *cyclic redundancy check (CRC)* field performs the error-detection function in the data packet. It is a one-byte value that is computed at the source by summing the bits that are 1's in the data portion of the packet. The CRC is the modulo-256 value of this sum. As the packet passes through each node, the router computes this sum and compares this value to the CRC value stored in the packet. If the two are equal, the probability is high that the data packet contains no errors.

   Data packets transmitted over a network occasionally experience *burst errors* that effect a large number of bits in the data packet. Burst errors occur when disturbances are comparable in duration to that of the data packet, such as lighting strikes or power surges produced by heavy equipment. The CRC is better suited than parity bits for detecting burst errors.

Multiple paths from source to destination allow the system to continue to function—even though one or more routers or links stop operating. The exact path to be taken by a particular packet is often chosen randomly, because deterministic routing schemes often result in bottlenecks. More sophisticated networks monitor the traffic congestion and select the fastest paths available. However, such monitoring requires additional circuits and complicates the network.

15.4.1 Probing the Internet

We can probe the structure of the Internet and determine data transfer times using a program called *Tracert*[1] (pronounced *trace route*). The *Tracert* (or Traceroute in Macs) command is followed by entering a Web address. For example, entering the command *tracert www.yale.edu* (on a PC) displays the number of nodes and data-packet travel times between your computer and the destination (in this case) www.yale.edu.

Tracert operates in the following manner.

1. It repeatedly transmits data packets to the URL until the destination is reached.

2. Tracert instructs your computer to set the first data packet TTL value to one and start a timer. Then the first node that this data packet encounters decrements it, making TTL = 0. That zero-valued TTL causes the first node to transmit a "NAK" that includes the node IP address back to the source (your computer). When your computer receives the "NAK", it stops the timer to determine the round-trip travel time to the first node. The minimum time recorded is 3 ms and time resolution is 1 ms on a PC.

3. Each transmission is repeated, and the travel times are reported three times to display any variations in travel times caused by congestion. Reported times to the same node can vary by a factor of two or more.

4. Tracert increments the TTL value in each subsequent data-packet transmission, causing the NAK to be sent by successive nodes. On a PC Tracert limits the maximum TTL value to access 30 nodes.

5. The process stops when the destination is reached and tracert receives an ACK from the destination.

Tracert results   **EXAMPLE 15.3**

Running the tracert program while at Yale and specifying the destination as Illinois Institute of Technology (www.iit.edu—my alma mater) produces the results shown in Figure 15.7. The Yale computer system has seven internal nodes from my computer's Wi-Fi connection to node 8, which is the router that connects Yale's network to the *Internet Service Provider (ISP)*. The node IP addresses indicate that the route enters the IIT computer system (with first-byte IP address 216) at node 11. The last entry (node 13) is the computer that hosts the IIT Web home page and tracert indicates 31 ms lapsed before the round-trip acknowledgements from the that page were received.

A distance calculator on the Web gives the distance between New Haven, CT, and Chicago, IL, to be 1223.6 km. The data speed over the Internet is denoted c_d and can be computed from the round trip travel time as

$$c_d = \frac{2 \times 1223.6 \text{ km}}{31 \text{ ms}} = 8 \times 10^7 \text{ m/s}$$

Recalling that the speed of light $c = 3 \times 10^8$ m/s, the data packet speed over the Internet is approximately 1/4 the speed of light!

Executing tracert from Yale to Oxford (www.oxford.ac.uk) shows 17 nodes and a 100 ms delay caused by crossing the Atlantic Ocean.

---

[1] In Windows, *tracert* is accessed in the *cmd.exe window* (entered by typing "cmd" in the start text box). Macs access *Traceroute* in the *Terminal* window found the *Applications/Utilities* folder.

```
>tracert www.iit.edu
Tracing route to www.iit.edu [216.47.143.249]:
 1    3 ms    3 ms    2 ms       node at Yale
 2    3 ms    3 ms    2 ms       node at Yale
 3    3 ms    2 ms    2 ms       node at Yale
 4    3 ms    3 ms    2 ms       node at Yale
 5    4 ms    3 ms    3 ms       node at Yale
 6    8 ms    8 ms    9 ms       node at Yale
 7    9 ms    8 ms    9 ms       node at Yale
 8    9 ms    8 ms    9 ms       node of ISP [192.5.89.237]
 9   30 ms   31 ms   30 ms       [192.5.89.18]
10   31 ms   31 ms   31 ms       [64.57.28.204]
11   31 ms   31 ms   32 ms       [216.47.159.162]
12   32 ms   31 ms   31 ms       [216.47.159.5]
13   31 ms   32 ms   31 ms       www.iit.edu [216.47.143.249]
```

Figure 15.7

Results produced by "tracert www.iit.edu" for the data-packet path from Yale to Illinois Institute of Technology.

15.5 DETECTING PACKET COLLISIONS

Computers that are connected to a network either through a cable called an *ethernet cable* or (increasingly more often) through a wireless connection. Fundamental and practical limitations lead to data-transmission interference called *data packet collisions*.

Wired systems: Because signals travel over a pair of wires at approximately the speed of light (equal to one foot per nanosecond), data transmitted by one system is observed by a distant system after a significant time delay. The distant system sensing an idle channel begins to transmit its own packet, which collides with the arriving data packet transmitted the first system.

Wireless systems: Because wireless routers are positioned in the center of an accessible area, two systems on opposite sides of the area can sense the router signals but not those produced by each other. This *hidden terminal* problem occurs when both systems, sensing an idle channel, begin to transmit wireless data packets that collide at the router.

The following sections describe the solutions to packet collisions in wired and wireless systems.

15.5.1 Packet Collisions in Wired Systems

This section discusses the technique that allows two or more computers to exchange information reliably over a pair of wires that implements the communication channel. We use the simple example of two computers to illustrate the basic idea, but the approach extends to many computers.

Figure 15.8 shows two computers (C1 and C2) connected to a pair of wires over which data transmission occurs. The computers use the *data line wire* (labeled DL) for signalling and the other wire (GND) to establish a common ground voltage reference. DL is connected to supply voltage V_s through resistor R (the *pull-up resistor*) to set the idle-channel level of DL at V_s and to limit currents in the communication process. Each computer reads the value of DL with a *sensing gate* (G) that outputs a logic level based upon the DL voltage as

$$G = 0 \quad \text{if} \quad DL = V_s \tag{15.1}$$
$$G = 1 \quad \text{if} \quad DL = 0$$

C1 uses sensing gate G1 and C2 uses G2.

Each computer contains a switch that is normally open between DL and GND. C1 has switch S1 and C2 has switch S2. If both switches are open, $DL = V_s$. When either computer closes its switch, $DL = 0$. A data value D is communicated on DL using

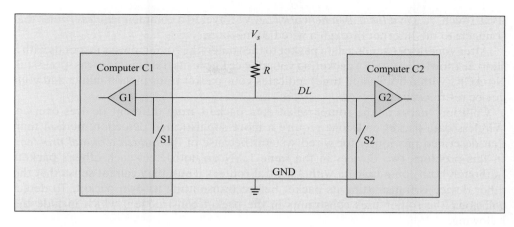

Figure 15.8

Two computers (C1 and C2) transmit data packets over the DL line using switches (S1 and S2) while reading DL with sensing gates (G1 and G2).

negative logic as

$$D = 0 \quad \text{if} \quad DL = V_s \tag{15.2}$$
$$D = 1 \quad \text{if} \quad DL = 0$$

With negative logic, a computer transmits $D = 1$ by closing its switch to connect the data line to ground, thus making DL = 0.

C1 transmits $D = 1$ by closing S1 to make DL = 0 and verifies the DL value by reading the G1 output: If $G1 = 1$ (because it reads DL = 0) the operation was successful. C1 transmits a data packet on DL by using the asynchronous method described previously.

Now let's observe what C2 is doing during C1's transmission. When DL = 0, C2 reads $G2 = 1$. Now this is the clever part: *Because C2 did not close S2, $G2 = 1$ indicates to C2 that another computer (C1 in this simple case) is transmitting data.* C2 acquires the data packet by reading G2. If C1 is transmitting data to C2, C1 transmits the address of C2 in the data packet to inform C2 the data is directed to it.

A packet collision occurs if C2 closes S2 at the same time as C1 closes S1. This is not a problem initially because both C1 and C2 are transmitting $D = 1$ and read their gates to verify they are doing so correctly. A packet collision is noted when the values transmitted by C1 and C2 are different. For example, C1 transmits $D = 1$ (by closing S1 to make DL = 0) and C2 transmits $D = 0$ at the same time (by leaving S2 open). When C2 reads G2 to verify the transmitted value it gets the wrong value, because DL = 0 makes $G2 = 1$. Thus, C2 senses a packet collision, stops transmitting immediately and continues to read G2 to determine if the data packet contains its address. Note that C1's data transmission is still valid. When C2 senses an idle channel (DL = 0 for an extended time), it can begin to transmit its data packet.

Because transmission delays may prevent immediate collision detection, both C1 and C2 may sense packet collisions. In this case, both stop transmitting immediately. Then each computer uses its uniform PRNG to generate a random time delay before starting to transmit again. The computer having the smaller delay starts transmitting its data packet first. The other computer has a larger delay, senses the channel is active, and delays its transmission until it senses an idle channel. This random-delay assignment allows priority to be given to premium customers by adjusting their PRNG to generate smaller random delays.

15.5.2 Packet Collisions in Wireless Networks

Smartphones and most laptops can access wireless networks. Such networks are provided without cost by many merchants as an incentive to visit their establishments. Wireless data packets are transmitted with radio waves that travel through the shared medium of air. The radio signals produced by all active devices (laptop, smartphone, iPod Touch, . . .) in a *local area network (LAN)* travel to a common *wireless router* that connects to the Internet through a wired connection.

After your device sends a data packet to the router, the router always responds with a short acknowledge (ACK) packet, so your device knows its transmission was successful. No ACK (within a specified time) indicates your packet transmission failed, and your device re-transmits it.

Collisions occur at the router when data packets from different devices overlap. Wireless data-packet collisions require a more sophisticated detection method than that described previously for wired systems because of the *hidden terminal problem*. In this problem, two devices in the same LAN do not detect each other's packets (although both communicate with a central router). Thus, they cannot sense that the other device is transmitting its packet before transmitting its own packet. To detect collisions, the router uses constraints in the packet construction, which include the following.

- A known start-bit sequence called a *preamble* occurs at the beginning of each data packet. It is analogous to a long start bit. If a packet begins with a bit sequence that is different than the expected preamble, the router assumes a collision has occurred.

- Parity or checksum errors computed by the router indicate interfering packets, and no ACK(s) are sent.

- A maximum size is imposed on data packets. Let my device start sending a data packet, and the router detects its correct preamble. But before the data-packet transmission finishes, another (hidden) device starts transmitting a data packet that interferes with my packet. This collision would not have occurred if my data packet were shorter and had finished before the other device started transmitting. Thus, shorter data packets reduce the probability of collisions.

EXAMPLE 15.4 Wireless packet collision detection

Figure 15.9 shows a properly operating two-user system with data packets and ACKs transmitted at separate times. The Wi-Fi router assigns a unique IP address to each device and acknowledges each data packet it receives by broadcasting an ACK. Every wireless device in the LAN receives the ACKs and decodes only packets addressed to its IP address.

Each device checks for the presence of radio transmissions to ensure the channel is idle before transmitting data, which does not work when one device is hidden from the other. When the router senses simultaneous packet transmissions, it determines a collision has occurred through error detection and broadcasts a *NAK* to all users, as shown in Figure 15.9 as a negative-going signal. Devices that recently transmitted data packets (and have not received ACKs) wait a random amount of time before they re-transmit their data packet.

EXAMPLE 15.5 Modeling wireless packet collisions

The probability of data-packet collisions is reduced by using shorter data packets. Figure 15.10 shows two data packets having the same duration that start at random times during a constant data transmission time interval. A collision occurs if the data packets overlap. The collision probability increases as the data packet becomes a significant fraction of the data transmission interval. Shorter packets have lower collision probabilities, but each packet transmits less data. Engineers implement systems that maximize the data throughput from a set of sources to a set of destinations.

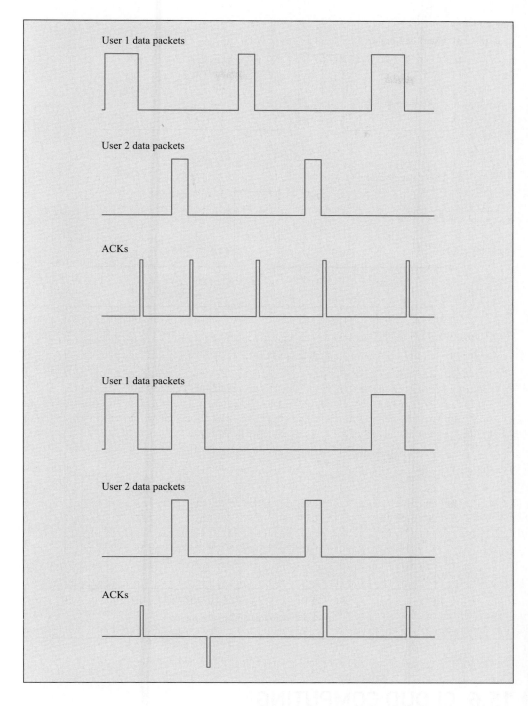

Figure 15.9

Data packet collisions. (a) Properly operating two-user system with data and *ACKs* transmitted at separate times. (b) When the router senses a data packet collision, it broadcasts a *NAK* (shown as a negative pulse).

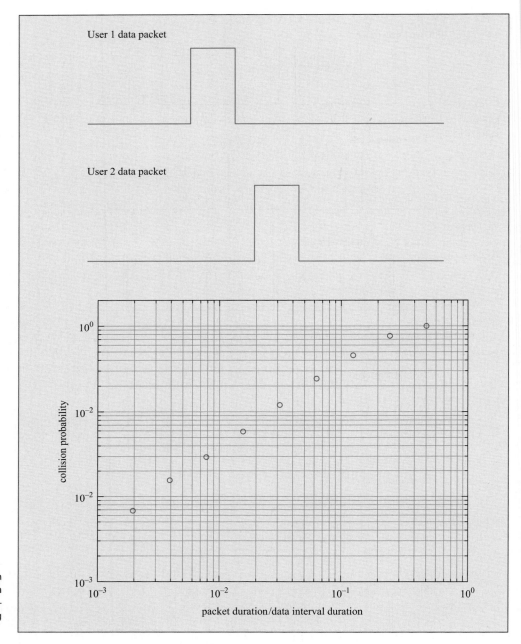

Figure 15.10

Packets of same size and duration starting at random times during a constant data interval duration. Collision probability is reduced by using shorter data packets.

 15.6 CLOUD COMPUTING

A computer-cloud installation is a collection of hardware (CPUs and memories) and software (operating systems and applications) resources that takes advantage of the inter-connectivity and communication speed provided by the Internet. In a manner similar to the electrical service to your home, cloud computing is viewed as a utility that can be used when it is needed and in the quantity required to get the job done according to your schedule. When you buy a PC or laptop, you specify the computing resources, CPU speed, and memory size you need. A computer cloud offers more flexibility to accommodate changing computational needs.

The driving force behind cloud computing is the efficiency in costs that can be achieved.

Acquisition: You do not use your computer at all times. Cloud computing can use "your computer" for another job when you don't need it. The most eager customers for cloud computing are new companies that can easily expand their computer power as their business grows: They can start small (cheaply) to get the business going and acquire additional cloud computing resources as and when they are needed.

Power/cooling: Because a typical cloud installation can have as many as 10^4 servers, power can be more effectively managed by turning off entire sections if they are not used. Similar to multiple-core CPUs, server sections can be cycled to reduce cooling costs.

Maintenance: A cloud facility typically will purchase or lease all of its servers from a single manufacturer. Identical devices are easy to maintain and swap out if a problem occurs. Backups and software updates are similarly simplified for the IT staff.

So what is needed? The current thinking is that there are at least four consumer needs that can be served by cloud computing.

Shared peripherals: Specialized hardware is used rarely by a particular user, but when many users are present, the specialized hardware is used more often. Hence, by charging a small usage fee, even the most specialized hardware can be made available. This is an extension of devices like network printers, which can now be color (rather than black-and-white) or network hard drives that can handle temporary needs of terabyte storage.

Shared computers: Fast computers (super-computers) are available on an hourly use basis, but require complicated sign-up procedures. With tablet computers having flexible GUI's but lacking computing horsepower, sharing powerful computers on the cloud using tablet interfaces is an obvious extension to the super-computer model.

Shared software: There are many shrink-wrapped applications that are very specialized and expensive. Rather than each user buying a license for installation on one's computer, a less expensive fee can be charged for using and sharing the application on the cloud.

Flexible procurement: Buying a computer is a bigger task than simply going on-line and making a purchase. A careful analysis of computer power, memory requirements, and availability is needed and is prone to error by either over- or under-specification. Miscalculations can be costly. The cloud offers the possibility to lease the computer that satisfies current needs without large equipment costs. As the needs change, more or less computing power and memory can be easily obtained without actually getting new equipment that requires additional space, cooling, and maintenance.

The needs are typically generated by application programs that require large computation resources, larger storage capacity, or both. A further advantage of the cloud is the capability of easily sharing information or designs for interactive collaborations. For example, an architect and client can view a design with each using their own computer at their own location.

Figure 15.11 shows two extremes for hosting guests in a computer cloud. At one extreme, the cloud facility contains many physical computers (on the order of 10,000) that are pre-configured with various desirable CPU and memory properties. A guest subscriber can lease a suitable computer having the closest-available desired computing

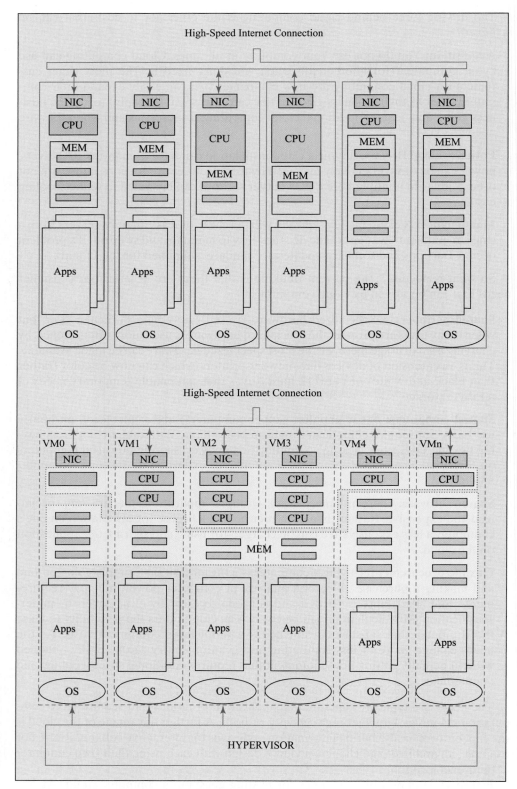

Figure 15.11

Two versions of cloud computing. (a) Dedicated physical computers (solid rectangles) having specified computing power and memory components that are dedicated to guests. (b) Virtual computers (dashed rectangles) configured by a hyperviser for use by guests.

power and memory size along with the applications (CAD, Simulation package) and operating system (OS) for the time that is needed and in a timely fashion. For example, one guest may require a fast computer with a small memory tomorrow for data analysis and sorting, while another guest may require a less powerful computer and a large memory next week for searching a large database when it becomes available. The access to the machine is through a high-speed data network using a *network interface card (NIC)*. When the guest is done, the same machine can be leased to another guest. Assigning and maintaining such a facility is similar to a rental car model.

The other extreme is a share-hardware model that allows a finer resolution in meeting the needs of particular guests. A *hypervisor* is the cloud supervisor that controls massive hardware consisting of CPU and memory elements. The CPU needs of a particular guest are met by combining the requisite number of elemental CPUs. The memory needs are providing by allocating a segment of memory. Operating systems and applications that are commonly used can be accessed from an application database with appropriate licensing. The challenge is to maintain the security of such a complex system by preventing guests from viewing the data of others.

The cloud side has the following benefits.

- Computer resources can be used 24 hours a day by shifting from one guest to another.

- Computer resources can be allocated in server banks, so, if a bank is not being used, it can be turned off, conserving power and providing cooling needs.

- Computer resources from the same vendor are easier to maintain in terms of support personnel and spare parts.

- A facility that is designed to accommodate banks of computers can be powered and cooled more efficiently.

- Common applications can be maintained and updated, so the newest versions are available.

Cloud use by a start-up company guest   EXAMPLE 15.6

When a company is initially formed, venture capitalists prefer to provide funds for salaries and specialized equipment rather than computers. The start-up company can lease a small-sized computer from the cloud for initial designs. As the business grows, additional computing resources can be leased accordingly.

A number of technologies are needed to mature simultaneously to make the cloud feasible.

1. An accessible high-speed network provided by the Internet.

2. Standardized interfaces provided by several popular Web browsers, including Internet Explorer, Firefox, Chrome, and Safari.

3. Security and privacy provided by encryption.

It currently appears that cloud computing is a permanent feature in the computing landscape.

15.7 Research Challenges

15.7.1 Distributed Control

Increasing network data traffic requires trade-offs. While increased *throughput*, (measured in bits/second) is desirable to transmit large amounts of data in reasonable time, the increased use of video and other real-time data applications require that the data arrive at the destination within a maximum time delay denoted as T_{md}, so you can watch videos without pauses. Current smartphones applications strain to maintain the throughput and deliver it within T_{md}. Researchers are investigating local, or *distributed*, control of data packets queues with the number queues numbering in the thousands and packets to be serviced 100 times greater.

15.8 Summary

This chapter described data packet transmission over a network of interconnected computers. Asynchronous data transmission was described to accommodate data packets generated at random time instants. Data of various sizes form data packets that are transmitted over a channel. The chapter described how data packets are structured to ensure that data are received reliably and without error. Problems associated with packet collisions were described. Cloud computing resources were described with examples of the flexibility they provide.

15.9 Problems

15.1 **Asynchronous byte transmission.** A 1 MB data file must be transmitted to another computer using asynchronous data transmission described in Example 15.2. How many bits are transmitted over the channel?

15.2 **Asynchronous data characters.** Construct two serial data characters to transmit bytes 10101010 and 11001100. Assume each character is transmitted in a separate character waveforms each using a start bit, odd parity, and two stop bits. Sketch the two character waveforms of the data and label the character elements. Assume a transmission rate of 1,000 baud (bps), the idle channel level is 5 V, and a logic 1 is 5 V and logic 0 is 0 V.

15.3 **Probing the Internet with Tracert.** This problem asks you to estimate the speed of data transmission over the Internet from data produced by tracert.

   a. Find four Web addresses, each one from a different continent. University Web pages are good addresses to use for this problem.

   b. Find a distance calculator on the web that determines the distance (in km) from your present location to the city hosting the Web address.

   c. Tracert reports three round-trip travel times to each node on the way to the destination. Use the minimum of these three values for only the destination as the time that the data packet took to travel the round trip distance between your location and the destination.

   d. Compute the data travel speed by dividing the round-trip distance by the round-trip travel time.

   e. Compare your results to the speed of light c.

15.4 **Maximum delay for a wired-line telephone call.** Assume an electrical signal travels along a wire at the speed of light. What is the maximum travel delay that occurs for a telephone waveform from NYC to LA (distance = 4,000 km)? What is the minimum delay between any two cities on Earth assuming signals travel along the Earth's surface?

15.5 **Delay for a geostationary satellite telephone call.** Assume an electrical signal travels along a wire at the speed of light. What is the minimum travel delay that occurs for a telephone waveform travelling to and from a geostationary satellite that is orbiting at an altitude of 36,000 km?

15.10 Matlab Projects

15.1 **Generating an asynchronous data packet.** Using Example 16.73 as a guide, design an asynchronous data packet waveform that contains a start bit, 8 data bits, 1 odd parity bit, and 3 stop bits. Compose a Matlab script that displays the waveform that contains the three packets that transmit the ASCII code that corresponds to your first, middle and last name initials.

15.2 **Reading an asynchronous data packet.** Using Example 16.74 as a guide, compose a Matlab script that reads the asynchronous data waveforms generated in Project 15.1.

15.3 **Detecting an error in an asynchronous data packet.** Using Example 16.74 as a guide, compose a Matlab script that performs error detection on the data generated in Project 15.1 by computing the parity in the read data value and comparing to the transmitted parity bit. Verify your code by introducing an error.

15.4 **Modeling wireless data packet collisions.** Example 16.75 simulates wireless data packet collisions to compute the probability of collision occurrence. Small data packets rarely collide, while data packets that are half the duration of the data window always collide. When collision occurs, no data is transmitted. Modify the program to find the data packet size that maximizes data throughput.

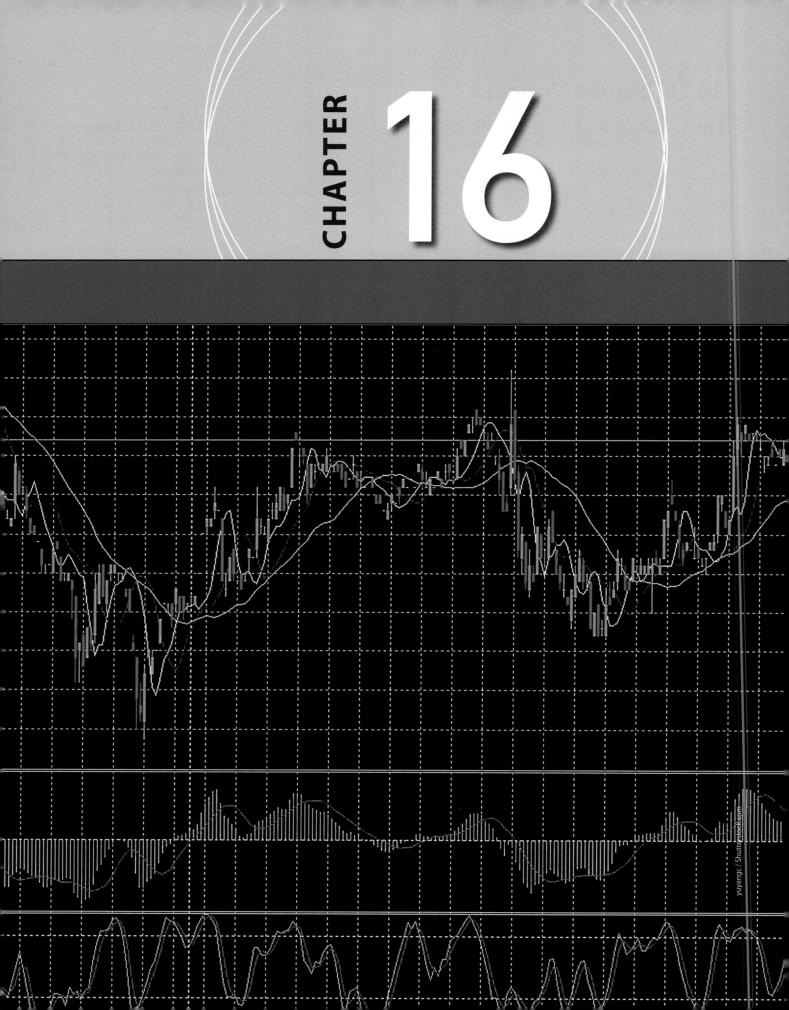

CHAPTER **16**

MATLAB BEST PRACTICES

After completing this chapter, the reader should be able to:

- Compose a Matlab script to analyze and display data effectively.
- Input speech and image data for processing.
- Compute and display the performance of electrical circuits and data coding methods.
- Simulate the operation of a data communication system.

Matlab is a flexible data analysis software package that is used at all levels, from beginning students, graduate students, to signal processing professionals in many fields. This chapter is not meant as a comprehensive Matlab manual but to provide sufficient information to complete the projects assigned in the book. This chapter begins by describing Matlab basics, which are followed by examples that illustrate the theory presented in each chapter. These examples then provide a starting point for the projects assigned for each chapter.

16.1 BASICS

Matlab is an *interpreted language*, as opposed to a *compiled language*. This means that Matlab operates by executing each instruction separately, rather than operating on the entire code to produce an executable version. Hence, Matlab trades off a slower execution time in order to achieve a faster programmer development process. The fast processors in today's computers have reduced this time difference. For quick execution time (not needed in this course), Matlab has tools to convert m programs into executable code by generating the equivalent c-language instructions.

Operating with Matlab is typically done in three windows.

Command window is the communication link between the user and Matlab, with the prompt ≫ indicating that Matlab is ready to receive commands. The user types commands, and Matlab responds by displaying results. The most common command will be the name of an *m-file*, which is a text file that contains Matlab instructions. Matlab has a comprehensive "help" feature: Typing *help* followed by a command name causes Matlab to display documentation that clarifies the use of that command and to provide additional documentation.

Editor window is a text editor in which the user composes a program saved as *filename.m* using Matlab instructions. The editor indicates incorrect syntax and typos as the user types.

Figure window is a graphical display of data that is accomplished easily with the *plot* instruction.

The basic variable in Matlab, say x, is stored as a complex-valued matrix, which is the motivation for the name *Matlab* being the shortened form of *matrix laboratory*. For example, Matlab stores the variable $x = 1$ as a 1×1 matrix $[1 + 0i]$, where $i = \sqrt{-1}$. Doing so simplifies the instructions for matrix operations—which are not necessary but often useful—but requires following the rules of matrix algebra.

One-dimensional arrays can be specified either as a row vector or column vector, depending on the specified (row, column) values. Common array declarations include:

- *Zeros*—specifying x to be a row vector containing ten 0's is done with

  ```
  x = zeros(1,10);
  ```

  The ith element of the row vector is $x(i)$ for $i = 1, 2, 3, \ldots, 10$.

A Matlab instruction typically terminates with a semicolon (;). When it doesn't, Matlab displays the result of the computation in the command window.

Factoid: Note that Matlab indexes an array starting with 1. This is different from the way electrical engineers refer to a time sequence with index beginning with 0, corresponding to $t = 0$. This variation is something we need to consider in comparing analytic results with Matlab programs.

- *Ones*—specifying y to be a column vector containing ten 1's is done with

```
y = ones(10,1);
```

The ith element of the column vector is $y(i)$ for $i = 1, 2, 3, \ldots, 10$.

- *A row vector* is specified with square brackets:

```
x = [ 1 2 3];
```

- *A column vector* is specified by using the enter key to indicate a new row, as in

```
x = [ 1
2
3];
```

An alternate way to specify a column vector is by using a row vector and then transposing with a "prime", as in

```
x = [ 1 2 3]';
```

- *A two-dimensional array* is specified with square brackets and pressing *enter* after each row is complete. Entering spaces is useful for aligning columns. For example, the 2×2 matrix can be specified by

```
x = [ 1 2
      3 4];
```

The $(i, j)^{th}$ element of the matrix x is $x(i, j)$. For example, the element in row 2 and column 1 is $x(2, 1) = 3$.

Matrix operations   EXAMPLE 16.1

Defining x as a row vector of 10 random numbers uniformly distributed over the range [0,1) is accomplished with

```
x = rand(1,10);
```

Then x has one row ($r_x = 1$) and ten columns ($c_x = 10$), making x a $r_x \times c_x = 1 \times 10$ matrix.

Defining y as a column vector of ten random numbers is accomplished with

```
y = rand(10,1);
```

Then $r_y = 10$ and $c_y = 1$, making y a 10×1 matrix.

When multiplying two matrices, say $z = x * y$, the inner dimensions of x and y must agree: If x is an $r_x \times c_x$ matrix that has r_x rows and c_x columns and y is an $r_y \times c_y$ matrix, c_x must equal r_y. The product matrix z is then an $r_x \times c_y$ matrix.

The product $z = x * y$ defined here produces a 1×1 matrix, and $z = y * x$ produces a 10×10 matrix.

Illegal matrix operations   EXAMPLE 16.2

Defining x as

```
x = [ 1 2 ];
```

produces $r_x = 1$ and $c_x = 2$, making x a 1×2 matrix. Defining y similarly, we have

```
y = [ 1 2 ];
```

which produces $r_y = 1$ and $c_y = 2$, making y a 1×2 matrix. The instruction forming the product matrix as

```
z = x*y
```

displays the error message

```
Error using  *
Inner matrix dimensions must agree.
```

which occurs because $c_x \neq r_y$, as required by matrix algebra. Transposing y into a column vector using a y', produces a 2×1 matrix in the instruction

```
z = x*y'
```

that displays $z = 5$, which is a 1×1 matrix.

■ *The colon* (:) is a useful punctuation and is loosely translated as "until". For example, the array $x = 1, 2, 3, \ldots, 10$ is generated with

```
x = [1:10];
```

The array $x = 0, 0.1, 0.2, \ldots, 9.8, 9.9, 10$ is generated by including the increment value 0.1 within the limits in the brackets with

```
y = [0:0.1:10];
```

■ *Plot(x)* is a convenient Matlab instruction that displays the array x graphically in a separate figure window. Matlab does all of the scaling to generate a decent graph. When the graph does not adequately display important information, the axes can by modified using the *axis* instruction

```
axis([xmin xmax ymin ymax])
```

where *xmin (xmax)* are the minimum (maximum) x values in the graph having a Cartesian (x, y) coordinate system, while *ymin (ymax)* are the minimum (maximum) y values.

EXAMPLE 16.3 Moore's law

The Matlab m-file that produced Figure 1.11 illustrates features that Matlab provides to make a plot. The text after % on each line is a comment that indicates the intent of the instruction.

```
clear                                   % clears work space
yr = (1971:2:2021);                     % vector of odd-numbered years
num = 2300*2.^((yr-1971)/1.5);          % number of transistors
semilogy(yr,num,'o')                    % use semilog scale along y axis
axis([1970  2020 2000 1E13])            % specify the graph limits
grid on                                 % display a grid
% add text and labels to graph
text(2013+.5,2300*2.^((2013-1971)/1.5),'\leftarrow 2013 CPU')
xlabel('year')
ylabel('# of transistors in a computer')
```

Note the following in this m-file:

- Comments have been added to most lines and follow the % symbol. It is good practice to add many informative comments to your code, especially when you return to the code later and try to remember what you did.

- The array *num* was generated with the .^ instruction, which applies the exponent operation to each element of the array *yr*.

- *Complex numbers* are designated with a "j" multiplying the imaginary component

```
z = 1+ j*2
```

Matlab displays complex numbers using "i" notation, as in the displayed value

```
z =
   1.0000 + 2.0000i
```

Electrical engineers prefer to use "*i*" to indicate electrical current and index values, so we will use "*j*" to indicate the imaginary component in Matlab scripts. Hence, we should avoid using *j* as an array index in programs that involve complex numbers.

16.2 BASIC INPUT/OUTPUT OPERATIONS

Interacting with Matlab occurs in the following forms.

- *Programs* are typically written into a text file or *script* of file type ".m" using the Matlab editor. Matlab executes the program when the user types the filename in the command window. In the editor, use the "File/New" pull-down menu to open a new file or "File/Open" to open an existing file in your current directory. To execute the file, simply type its file name. A script also can be executed by clicking on the green triangle located on the ribbon in the editor window.

- *The prompt* ≫ appears after completing the execution of an m-file and indicates Matlab is ready for the next instruction. Instructions written on the prompt command line act as if they were instructions appended to the end of the m-file, although they are not permanently stored in the text file. For example, if your program uses variable *x*, typing *x* at the prompt displays its value at the completion of the program.

- *Instructions* can be executed directly on the command line at the prompt ≫. For example, "2+4" followed by return displays "ans=6", and "*plot(x)*" plots whatever *x* is at the current time (very useful).

- *The up-arrow on the keyboard* retrieves the previous instruction entered in the command window. There is a history feature allowing multiple up-arrows to retrieve previous instructions. An *enter* then executes the retrieved instruction.

- *The clear* instruction clears the Matlab memory (the workspace). Matlab remembers variables and their last values during the session. This can cause confusion, especially if you used the same variable name in different programs that have been executed. For this reason, I often start my scripts with the command "clear", which clears the memory.

- *User input* inserts values to Matlab program from the command line. For example, the single value *x* is entered into Matlab by including the instruction

```
x = input('enter x value = ');
```

in the script. Multiple values or an array *x* can be entered as a string by including the following instruction in the script

```
s = input('enter string of values separated by spaces ','s')
d = str2num(s) % converts string to numerical values
```

- *Displaying current variable values* on the command console during script execution is accomplished merely by deleting—or not including—the include space between the and ; terminator. For example,

```
x=1
```

displays the value of x (=1) in the command window when that instruction is encountered in the script, while

```
x=1;
```

does not display its value. Matlab formats the variable values according to its best guess: If x has an integer value, its value displays as an integer, and a floating-point value displays with a decimal point followed by four digits.

Placing the instruction

```
format compact
```

at the beginning of your m-file script reduces the white space in the display.

If x is a single variable and y is an array of ten variables,

```
z=[x y]
```

concatenates x and y to form the z array and displays eleven values using one format, integer if all variables are integers, or floating point if at least one of the variables is a floating point. The displayed values wrap to another line if the command window is too narrow.

Specifying a desired format, in order to save space or to display values on one line, is done by using a string variable defined using single quotes. For example,

```
f = '%4.2f %5d %5d %5d %5d %5d %5d %5d %5d %5d %5d\n';
fprintf(f,x,y)
```

or equivalently

```
fprintf('%4.2f %5d %5d %5d %5d %5d %5d %5d %5d %5d %5d\n',x,y)
```

displays x with %f4.2, (i.e., for values, including the decimal point, with two digits to the right of the decimal point) and each of the ten y values as an integer using five digits and suppressing the leading zeros.

16.3 BASIC PROGRAMMING OPERATIONS

- The *for* loop is useful for repeating commands. (Try *help for* for a fuller explanation.) Setting the array elements $x(1)$ through $x(10)$ to zero using a for loop is done with

```
for i = 1:10
  x(i)= 0;
end
```

The *for* loop always starts with a *for* instruction and terminates with an *end* instruction.

- Add the array elements $x(1)$ through $x(10)$ to $y(1)$ through $y(10)$ and put the sums into $z(1)$ through $z(10)$ using a *for* loop with

```
for i = 1:10
  z(i) = x(i) + y(i);
end
```

- Because Matlab uses matrix algebra and keeps track of the size of the arrays, the same addition is performed more efficiently, in terms of typing effort and execution time, with the command

```
z = x + y;
```

- Matrix multiplication, or outer product, is performed with

```
z = x * y;
```

If the x matrix is $r_x \times c_x$ and the y matrix is $r_y \times c_y$, the resulting z matrix is $r_x \times c_y$. A common sum-of-products occurs with the digital-filter convolution operation

$$y_i = \sum_{j=0}^{n_h-1} h_j x_{i-j}$$

which can be computed simply with

```
y = sum(h .* x);
```

when both h and x are row vectors of the same size. This same result is obtained using $y = h * x$'; but the "sum" provides a reminder of the filtering operation.

Factoid: Whenever using a new function in any signal processing software, ALWAYS apply it to a problem for which you already know the answer. If you get that answer, you are using the new function correctly; otherwise, you need to find your error.

16.4 PROGRAMMING STYLE

Except for the simplest operations, the recommended programming style is to use *user-defined functions* with names that indicate their operation. Composing m-files using your own functions to implement frequently used operations has two major advantages.

1. It reduces tedious programming effort. Once a function is composed, it can be used as a building block in other programs.

2. It makes your program readable. By naming your function by using a *descriptive* name, the sequence of functions makes your program read like a narrative story. This allows you to organize your thought process in program development and also helps you recall your intensions when you revisit your program in the future.

The disadvantage of this approach is that the overhead involved in calling functions slows down the program speed a little. However, it is often better to proceed slowly toward the correct answer than to speed toward the wrong one.

A user-defined function is composed in its own separate m-file having the same name as the function. The first line in the script defining the function must have the form

```
function [output variables] = myFunctionName(input variables)
```

where

function—lets Matlab know it is a function program that can be called in another Matlab program as a standard function.

[output variables]—these are the values that the function returns. Examples include: *[sig, sigLen], [data], . . .*

myFunctionName—is a descriptive name that reminds *you* what the function does. Good examples include *sigGen, dataGen*, while poor ones include *sG* and *dG*.

(input variables)—are the parameter values that the function processes. Examples include *(ampl), (rotAng)*.

The function code in the file named *myFunctionName.m* must reside in the same folder as the programs calling this function.

Regarding variable and function notation, we adopt a common programming style that specifies descriptive names that contain mostly lower-case letters and uses upper-case letters to separate terms. Notation size can often be reduced by shortening or eliminating the vowels, *but only if it makes sense to you.* Examples include:

- A function that generates a sinusoidal waveform would be named *genSine*. The name is descriptive as to its intended function.
- The length of vector *n* would be named *nLen*.
- The size of matrix *a* would be named *aSize*, or *aSz*.

EXAMPLE 16.4

Generating a sinusoidal waveform with a Matlab function

Generating a sinusoidal sequence in a function m-file can be done by specifying its amplitude *amp*, the number of periods, *numPeriods*, and the number of samples per per period *numPerPeriod*. This was composed in the function file *genSine.m* shown in Figure 16.1.

Note that, in the function definition, the name should be sufficient to describe the intended operation. While this function consists of only three lines of code, its name is descriptive in what it accomplishes, and can be used by other m-files that are in the same directory.

It is good practice to compose a test program to verify the function operation *immediately after composing the function*. This was done in the m-file *tstGenSine.m* shown in Figure 16.1.

Function script m-file:

```
%genSine function
function [s,sLen] = genSine(amp, numPeriod, numPerPeriod)
sLen = numPeriod*numPerPeriod;
for i=1:sLen
    s(i) = amp*sin(2*pi*(i-1)/numPerPeriod);
end
```

Script m-file that calls function:

```
%tstGenSine
[x,nx] = genSine(2,4,16);
stem(x)
```

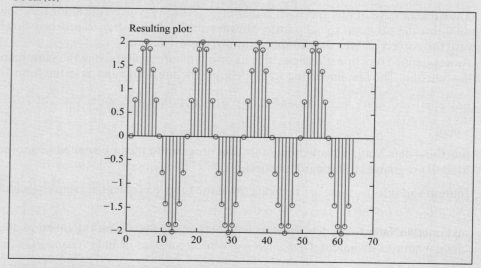

Figure 16.1

Function script m-file, calling m-file, and resulting plot.

16.5 TIPS ON COMPOSING SUCCESSFUL MATLAB PROGRAMS

Complex programs rarely work correctly the first time, so you need a method to find where the problem(s) lie. To fix the problems, the program needs to be *debugged*. The term *debug* comes from an early computer that failed in which technicians found a dead bug. It is doubtful that the bug—dead or alive—caused the malfunction, but the term remains.

The editor catches syntax errors and typos, but after that, Matlab does exactly what you instruct it to do. The following tips for finding coding problems are helpful.

Apply your code to a simple problem. Always apply the program to the simplest, non-trivial, problem for which you know the answer. If Matlab produces the expected result, there is a very good chance that the program operates correctly.

Print intermediate results. Print intermediate results using the following methods.

■ Removing the ; at the end of the instruction, such as,

$$x(i) = 2*i; \quad \rightarrow \quad x(i) = 2*i$$

Removing the semicolon causes Matlab to display the x value each time the instruction is executed.

■ Positioning the cursor over a particular variable in the Editor displays its current value. Also, entering $x1$ in the command line displays the current value of $x1$.

■ Printing selected variables. Let index i and variables $x1$ and $x2$ be intermediate values that are used to compute the result that appears to be wrong. The statement that selects the three variables whose values are printed when Matlab encounters the statement in its program execution is

```
[i x1 x2]
```

Note the absence of ; causes Matlab to display these values. Typically, one of these values will be different than expected because of a typo or otherwise incorrect expression.

For example, if $x = [2\ 3\ 4]$ and $i = 2$, then $[i\ \ x(i)]$ produces

```
ans=
   2.0000    3.0000
```

When you have found the mistake and the program operates successfully, the ; can be added at the end of the statement to prevent time-consuming displays.

Set test points. Set *test points* (also called *break points*) using your Editor *Breakpoints* pull-down menu. A break point is set by right clicking on a line number in your program, which is an appropriate point after a crucial calculation is performed. A small filled circle appears next to the line number. Multiple break points can be set in this manner. Right-clicking on a circle removes the circle and its break point. When a break point is set at a particular line, Matlab suspends operation at that breakpoint and prompts with the debug symbol $K \gg$.

A green arrow will appear next to the circle in your code to indicate the current break point location in the program execution.

Entering *dbcont* in the command line causes the debugger to resume execution until the next break point is encountered. All break points can be removed when the program operates successfully.

Entering *dbquit* in the command line causes the debugger to terminate, producing the normal prompt $\gg$.

Construct a temporary program named *foo.m*. Programmers often use *foo* as the name for a test program that can be deleted without consequence. If a complex Matlab program still exhibits problems after employing the previous tips, save the program m-script as *foo.m*, and then try to isolate the problem by deleting code in *foo.m*. For example, if you have a wrong calculation early in your program, the code after the mistake doesn't matter. So, delete the code after the mistake in *foo.m* to isolate the problem. If you wrote your program as a sequence of function calls, you know the problem is not within a function but in the passing of values between functions.

 # 16.6 MATLAB PROGRAM EXAMPLES

This section gives the Matlab code examples that provide starting points for the projects. Each section refers to the chapter that the scripts are illustrating.

16.6.1 Introduction

EXAMPLE 16.5

Generating noisy signals

The Matlab script shown in Figure 16.2 illustrates the following:

- Inputting constants (A, σ_N).
- Plotting data using plot and stem.
- Matlab's Gaussian pseudo-random number generator.
- Printing string symbols on graph (A, σ_N).

The m-file contains verbose comments that explain the intention of each instruction. Adding such comments is useful for future reference and for tracking down problems.

```
clear                                       % clear all variables
A = input('Amplitude= ') ;                  % Amplitude
sigma_N = input('Noise level (SD)= ');      % noise SD
tau = 0;                                     % threshold
d = [A*ones(1,20) -A*ones(1,20)];           % specify data values
for i=1:10                                   % # of transmissions
    N = sigma_N*randn(1,length(d));          % Gaussian PRNG
    x = d + N;                               % received data
    plot(d)                                  % plot data in blue
    hold on                                  % superimpose plot
    stem(x)                                  % stem plot noisy data
    tau_val = tau*ones(1,length(d));         % define tau array for plot
    plot(tau_val,'k--')                      % plot tau in black dashed line
    hold off
    str = ['A= ' num2str(A)];                % form string array
    text(2,-2,str)                           % display string at (x,y)=(2,-2)
    str = ['\sigma_N= ' num2str(sigma_N)];   % form string array
    text(2,-3,str)                           % display string at (x,y)=(2,-3)
    pause(.2)                                 % pause 0.2 s to display data
end
```

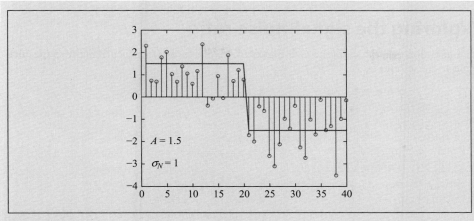

Figure 16.2

Matlab script that simulates noisy data signal.

Thresholding noisy signals EXAMPLE 16.6

The Matlab script shown in Figure 16.3 illustrates data that remains in the workspace and applying the data to threshold detection. Note that the clear statement is not used in order to process the data produced by the program in Example 16.5.

```
r = zeros(size(x));          % define restored data array
plot(x)                      % plot data generated by previous program
hold on                      % superimpose plot
for j = 1:length(x)
    r(j)= A;                 % assign default value
    if x(j) < tau            % apply threshold defined in previous program
        r(j) = -A;
    end
end
stem(r)                      % stem plot restored data
hold off
```

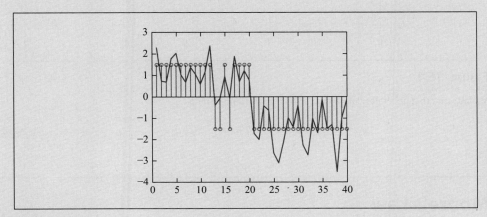

Figure 16.3

Matlab script that restores noisy data signal with threshold detection.

EXAMPLE 16.7 Exploring the signal-noise ratio

The stand-alone Matlab script shown in Figure 16.4 explores the error occurrence for a specified signal-to-noise ratio A/σ_N.

```
clear                                    % clear all variables
SNR = input('SNR= ');
A = 2; % Amplitude
sigma_N = A/SNR;                         % noise SD
tau = 0;                                 % threshold
x = [A*ones(1,20) -A*ones(1,20)];
x = x + sigma_N*randn(1,length(x));      % specify data values
r = zeros(size(x));                      % define restored data array
plot(x)                                  % plot data  generated by previous program
hold on                                  % superimpose plot
for j = 1:length(x)
    r(j)= A;                             % assign default value
    if x(j) < tau                        % apply threshold
        r(j) = -A;
    end
end
stem(r)                                  % stem plot restored data
hold off
```

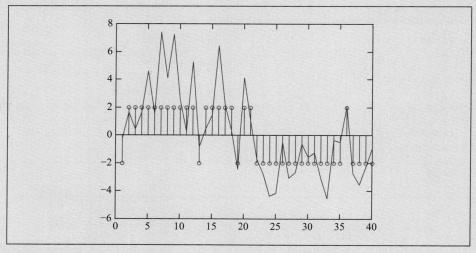

Figure 16.4

Matlab script that generates data with specified signal-to-noise ratio.

EXAMPLE 16.8 Moore's Law

The Matlab script shown in Figure 16.5 generated Figure 1.11. It illustrates semi-logarithmic plots, and adding text and labels to figures.

```
clear                          % clears work space
yr = (1971:2:2021);            % column vector of odd-numbered years
num = 2300*2.^((yr-1971)/1.5); % number of transistors
semilogy(yr,num,'o')           % plot using semilog scale along y axis
axis([1970  2020 2000 1E13])   % specify the axis limits
grid on                        % display a grid
% add text and labels to graph
text(2015-10.5,2300*2.^((2015-1971)/1.5),'2015 CPU \rightarrow')
xlabel('year')
ylabel('# of transistors in a computer')
```

Figure 16.5

Matlab script that generated Figure 1.11.

16.6.2 Sensors and Actuators

Acquiring Microphone Data   EXAMPLE 16.9

The Matlab script shown in Figure 16.6 acquires speech data from a laptop computer microphone, plays it back on the laptop speaker, and plots the speech waveform. The details are given in Example 2.9.

```
clear                                 % clears workspace
recObj = audiorecorder(8000, 8, 1);   % define ADC specs
disp('Start speaking now')            % prompt speaker
recordblocking(recObj, 2);            % record for 2 sec
disp('End of recording');             % indicate end
play(recObj);                         % playback recording.
myRecording = getaudiodata(recObj);   % store data in array
plot(myRecording);                    % Plot the 2-sec waveform.
save('speech.mat','myRecording')      % saves data into file
```

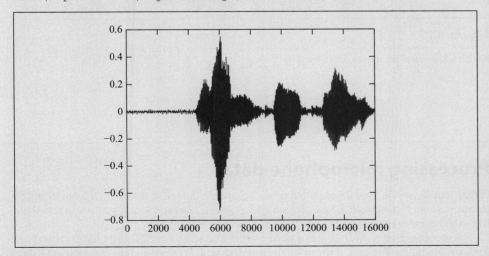

Figure 16.6

Matlab script that acquires microphone speech ("1, 2, 3") data and stores it on disk.

The Matlab command *audiorecorder* acquires data from the microphone input. The code acquires two seconds of audio from a laptop microphone, using $f_S = 8,000$ Hz and 8 bits/sample in the variable *recObj*. The numerical waveform values are then obtained using *getaudiodata(recObj)* and stored in *myRecording*. Speech data are then saved in a file for future processing.

EXAMPLE 16.10   Reading speech data file

Figure 16.7 reads speech data that were generated with the program in Example 16.9. The data in the *myRecording* array is played on the speakers and plotted to confirm it is correct.

```
load('speech.mat','myRecording')  % read data from file into array myData
sound(myRecording,8000)           % play the speech on the speaker
plot(myRecording)                 % plot the speech waveform
```

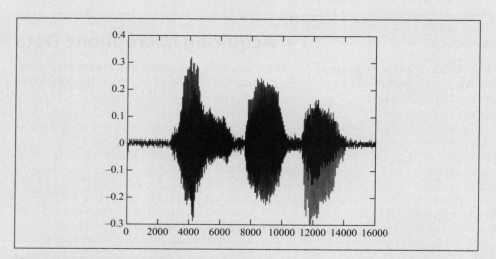

Figure 16.7

Matlab script that reads speech data from disk.

EXAMPLE 16.11   Processing microphone data

Figure 16.8 shows the Matlab script that processes speech data read from the data file stored by the program in Example 16.9. Speech data are extracted from the file by applying a threshold from the beginning and from the end. The details are given in Example 2.9.

The speech segment *myData* is extracted from *myRecording* using threshold detection both from the beginning and end of the original recorded data. Original and processed speech data are plotted on the same figure window using the subplot instruction. Speech data is then saved in a file *speech1.mat* for further processing.

```
load('speech.mat','myRecording')     % read data from file into array myData
sound(myRecording,8000)              % play the speech on the speaker
plot(myRecording)                    % plot the speech waveform
subplot(2,1,1),plot(myRecording);    % Plot the 2-sec waveform.
% Extract the speech waveform
threshold = 0.05;                    % waveform amplitude threshold
n_x = length(myRecording);           % number of samples
for i=400:n_x                        % Start at i=400 to avoid transients
    if myRecording(i) > threshold
        break                        % i = starting point
    end
end
i_strt = max(1,i-400);               % data start - include 400 prior samples
for i=1:n_x
    if myRecording(n_x-i) > threshold % start checking from end
        break                        % i = end point
    end
end
i_end = min(n_x-i+400,n_x);          % data end - include 400 more samples
myData = myRecording(i_strt+1:i_end);
subplot(2,1,2),plot(myData);
axis([1 length(myData) 1.2*min(myData) 1.2*max(myData)])
save('speech1.mat','myData')         % save data into file
```

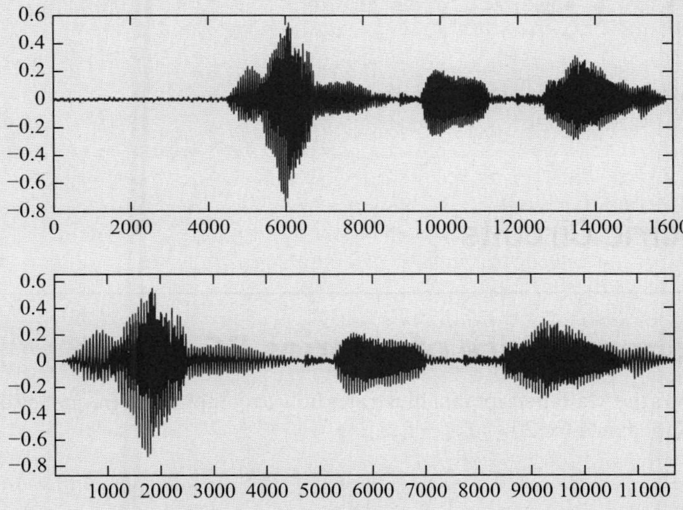

Figure 16.8

Matlab script that reads speech data from disk and extracts speech segment.

Acquiring jpeg image and producing gray scale image

EXAMPLE 16.12

Figure 16.9 shows a Matlab script that acquires image data from a color jpeg picture taken by a digital camera and transferred to disk, and converts the image to gray scale.

```
clear
filename = input('enter filename ', 's');
filename = [filename '.jpg']
Ao = imread(filename);
A = cast(Ao,'single');  % convert from unsigned 8-bit to single (for sqrt)
[R C D] = size(Ao);     % row, column and depth
Gray = zeros(R,C);
for i=1:R
    for j = 1:C
        val = 0;
        for k=1:D
            val = val+ A(i,j,k)^2;      % intensity = sum of square
        end
        Gray(i,j) = floor(sqrt(val)/3); % integer value related to gray level
    end
end
colormap('GRAY')
image(Gray)
axis image % gives plot same scale as image
save('image.mat','Gray') % saves Gray image data into image.mat
```

The following program reads the image stored in image.mat and displays it.

```
clear
load('image.mat');
image(Gray)
axis image  % gives plot same scale as image
```

Figure 16.9

Matlab script that reads JPEG image data from disk and forms gray scale image.

16.6.3 Electric Circuits

EXAMPLE 16.13 **Complex impedance of a series *RC* circuit**

Figure 16.10 shows the Matlab script that illustrates how to compute the frequency dependence of a series RC impedance for $20 \leq f \leq 5 f_x$.

```
clear
C = 1E-6;                                   % C value
R = 1000;                                   % R value
fx = 1/(2*pi*R*C);                          % f_x value
f = (20:10:5*fx);                           % define frequency points
ZR = R*ones(size(f));                       % ZR(f)
ZC = -j./(2*pi.*f*C);                       % ZC(f)   imag uses j
Z = ZR + ZC;                                % Z
Zmag = sqrt(real(Z).^2 + imag(Z).^2);       % magnitude using real and imag components
Zphase = atan(imag(Z)./real(Z));            % phase
subplot(2,1,1), plot(f,Zmag)
grid on
xlabel('f  (Hz)')
ylabel('|Z_R_-_C| (\Omega)')
subplot(2,1,2), plot(f,Zphase)
grid on
xlabel('f  (Hz)')
ylabel('\angle Z_R_-_C (radians)' )
```

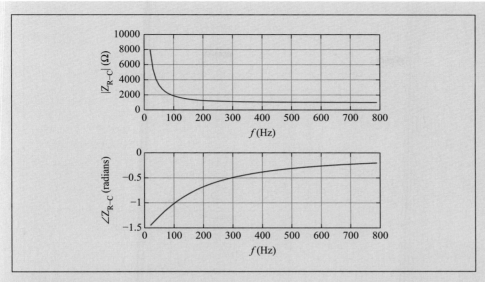

Figure 16.10

Matlab script computes impedance of series *RC* circuit as a function of frequency.

Voltage and current in a series *RC* circuit — EXAMPLE 16.14

Figure 16.11 shows the Matlab script that illustrates how to compute the amplitude and phase relationship between the voltage and current in a series *RC* impedance at $f = f_x$.

```
clear
C = 1E-6;                              % C value
R = 1000;                             % R value
fx = 1/(2*pi*R*C);                   % f_x value
ZR = R;                               % ZR(f)
ZC = -j./(2*pi*fx*C);                % ZC(f) imag uses j
Z = ZR + ZC;                          % Z
Zmag = sqrt(Z * conj(Z));            % magnitude using conjugates
Zphase = atan(imag(Z)/real(Z));      % phase
Tp = 1/fx;                            % period at f_x
t=(0:.01:1)*Tp;                       % 101 points over period
A_in = 10;
v_in = A_in * sin(2*pi*fx.*t);        % input voltage
i_in = (A_in/Zmag)*sin(2*pi*fx.*t - Zphase);    %input current
subplot(2,1,1), plot(t,v_in)
grid on
xlabel('t (s)')
ylabel('v_i_n(t) (V)')
subplot(2,1,2), plot(t,i_in)
grid on
xlabel('t (s)')
ylabel('i_i_n (A)' )
```

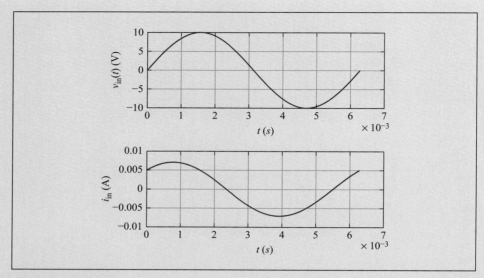

Figure 16.11

Matlab script computes voltage and current in series *RC* circuit at $f = f_x$.

EXAMPLE 16.15 **Complex impedance of a parallel *RC* circuit**

Figure 16.12 shows the Matlab script that illustrates how to compute the frequency dependence of a parallel *RC* impedance for $20 \leq f \leq 5 f_x$. Note the use of the conjugate for computing the magnitude.

```
clear
C = 1E-6;                              % C value
R = 1000;                             % R value
fx = 1/(2*pi*R*C);                   % f_x value
f = (20:10:5*fx);                    % define frequency points
ZR = R*ones(size(f));                % ZR(f)
ZC = -j./(2*pi.*f*C);                % ZC(f) imag uses j
Z = (ZR .* ZC)./(ZR + ZC);           % Z - note .*  and ./
Zmag = sqrt(Z .* conj(Z));           % magnitude
Zphase = atan(imag(Z)./real(Z));     % phase
subplot(2,1,1), plot(f,Zmag)
grid on
xlabel('f  (Hz)')
ylabel('|Z _R_|_|_C| (\Omega)')
subplot(2,1,2), plot(f,Zphase)
grid on
xlabel('f  (Hz)')
ylabel('\angle Z _R_|_|_C (radians)')
```

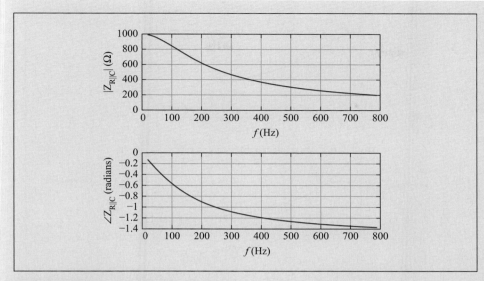

Figure 16.12

Matlab script computes impedance of parallel *RC* circuit as a function of frequency.

Low-pass *RC* filter EXAMPLE 16.16

Figure 16.13 shows the Matlab script that illustrates how to compute $H(f)$ of a series *RC* low-pass filter for $20 \le f \le 5f_x$.

```
clear
C = 1E-6;                              % C value
R = 1000;                              % R value
fx = 1/(2*pi*R*C);                     % f_x value
f = (20:10:5*fx);                      % define frequency points
ZR = R*ones(size(f));                  % ZR(f)
ZC = -i./(2*pi.*f*C);                  % ZC(f)
H = ZC./(ZR + ZC);                     % H - voltage divider
Hmag = sqrt(H .* conj(H));             % magnitude using conjugate
Hphase = atan(imag(H)./real(H));       % phase
subplot(2,1,1), plot(f,Hmag)
grid on
xlabel('f  (Hz)')
ylabel('|H_L_P| (\Omega)')
subplot(2,1,2), plot(f,Hphase)
grid on
xlabel('f  (Hz)')
ylabel('\angle H_L_P (radians)' )
```

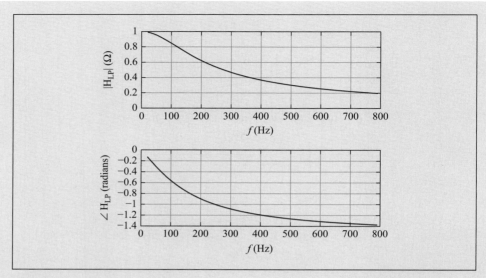

Figure 16.13

Matlab script computes transfer function $H(f)$ of low-pass RC filter as a function of frequency.

EXAMPLE 16.17   **n-stage low-pass RC filter**

Figure 16.14 shows the Matlab script that illustrates how to compute $H(f)$ of a n-stage series RC low-pass filter for $20 \leq f \leq 5f_x$. Note that the *clear* statement is missing and the plots are held on to compare different values of n.

```
n = input('n stage= ');                    % input number of stages
C = 1E-6;                                   % C value
R = 1000;                                   % R value
fx = 1/(2*pi*R*C);                          % f_x value
f = (20:10:5*fx);                           % define frequency points
ZR = R*ones(size(f));                       % ZR(f)
ZC = -j./(2*pi.*f*C);                       % ZC(f)  imag uses j
H = ZC./(ZR + ZC);                          % H - voltage divider
Hmag = (sqrt(H .* conj(H))).^n;             % magnitude of n-stage filter
Hphase = n*atan(imag(H)./real(H));          % phase of n-stage filter
subplot(2,1,1), plot(f,Hmag)
hold on
grid on
xlabel('f  (Hz)')
ylabel('|H_L_P| (\Omega)')
subplot(2,1,2), plot(f,Hphase)
hold on
grid on
xlabel('f  (Hz)')
ylabel('\angle H_L_P (radians)' )
```

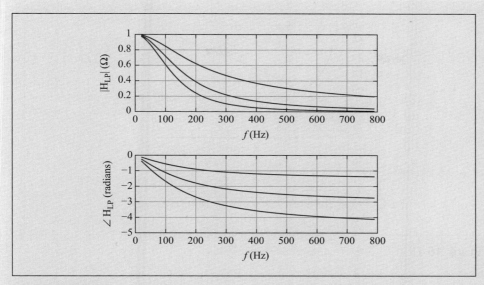

Figure 16.14

Matlab script computes transfer function *H* (*f*) of *n*-stage low-pass *RC* filter as a function of frequency.

Complex impedance of a series *LC* circuit   EXAMPLE 16.18

Figure 16.15 shows the Matlab script that illustrates how to compute the frequency dependence of a series *LC* impedance for $f_o/4 \leq f \leq 4f_o$. Note the use of the conjugate for computing the magnitude.

```
clear
C = 1E-5;                           % C value
L = 1E-3;                           % L value
fo = 1/(2*pi*sqrt(L*C));           % f_o value
f = (fo/4:fo/100:4*fo);            % define frequency points
ZL = i*2*pi.*f*L;                  % ZL(f)
ZC = -j./(2*pi.*f*C);              % ZC(f)  imag uses j
Z = ZC + ZL;                       % Z(f)
Zmag = sqrt(Z .* conj(Z));         % magnitude
Zphase = atan(imag(Z)./real(Z));   % phase
subplot(2,1,1), plot(f,Zmag)
grid on
xlabel('f  (Hz)')
ylabel('|Z_L_C| (\Omega)')
subplot(2,1,2), plot(f,Zphase)
grid on
xlabel('f  (Hz)')
ylabel('\angle Z_L_C (radians)' )
```

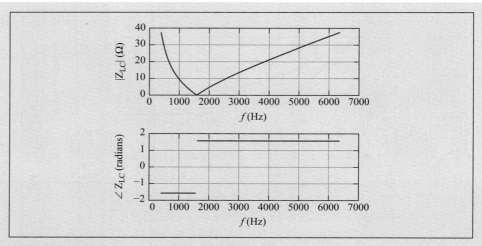

Figure 16.15

Matlab script computes impedance of parallel *LC* circuit as a function of frequency.

16.6.4 Electronics

EXAMPLE **16.19**

Thresholding noisy signals

Figure 16.16 shows the Matlab code that generates Figure 4.5 by applying a threshold to restore data values. The data consist of ten 0's, followed by ten 1's, followed by ten 0's, followed by ten 1's, followed by ten 0's.

```
clear
format compact
p0 = zeros(1,10);
p1 = ones(1,10);
s = [p0 p1 p0 p1 p0];
sd_noise = 0.3;
noise = sd_noise*randn(size(s));
x = s + noise;
subplot(2,1,1),plot(x)
hold on
stem(s,'r')
axis off
tau = 0.5;
threshold = tau*ones(size(x));
plot(threshold,'k--')
axis([1 1.1*length(x) -3*sd_noise 1+3*sd_noise])
hold off
restored = zeros(size(x));
for i=1:length(x)
    if x(i) >= tau
        restored(i) = 1;
    end
end
subplot(2,1,2),plot(restored)
hold on
stem(s,'r')
axis([1 1.1*length(x) -3*sd_noise 1+3*sd_noise])
axis off
hold off
```

Figure 16.16

Matlab script that applies a threshold to noisy data.

Op-amp |*G*| variation with *A* EXAMPLE 16.20

Figure 16.17 shows the Matlab code that generates Figure 4.8.

```
clear
A = (10:10:1e5);
RI = 1e3;
RF = 10e3;
num = A;
denom = A+(RF+RI)/RI;
factor = num ./ denom;
G = abs(-(RF/RI)*factor);
semilogx(A,G,'linewidth',2)
grid on
axis([min(A) max(A) 0 1.1*RF/RI])
hold on
Ginf = (RF/RI)*ones(size(A));
semilogx(A,Ginf,'--')
hold off
xlabel('A')
ylabel('|G|')
```

Figure 16.17

Matlab script that shows op-amp |*G*| variation with *A*.

Op-amp low-pass filter EXAMPLE 16.21

Figure 16.18 shows the Matlab script that generates the gain curves shown in Figure 4.16.

```
clear
f = (1:10:10000);
RI=1E3;
CF= 1E-8;
RF = 1E4;
ZI = RI*ones(size(f));
ZR = RF*ones(size(f));
ZC = -j./(2*pi.*f*CF);   % imag uses j
ZF = (ZR.*ZC)./(ZR + ZC);
G = abs(ZF./ZI);
subplot(2,1,1), plot(f,G,'linewidth',2)
axis([-.05 5000  -.05  13])
grid on
xlabel('f (Hz)')
ylabel('|G|')
subplot(2,1,2), loglog(f,G,'linewidth',2)
axis([50 10000  1 13])
grid on
xlabel('f (Hz)')
ylabel('|G|')
```

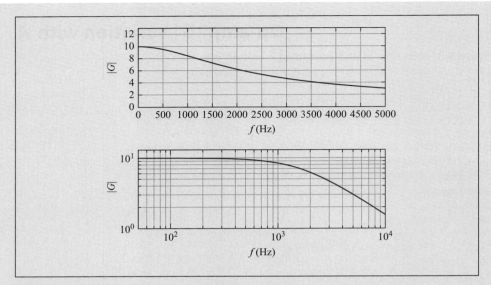

Figure 16.18

Matlab script that generates gain curves for an op-amp low-pass filter.

EXAMPLE 16.22

Diode rectification

Figure 16.19 shows the Matlab script that illustrates rectification performed by an ideal diode.

```
clear
t = (0:512);                     % define time array
s = sin(2*pi.*t/64);             % define sinusoid
plot(t,s,'--')                   % plot sinusoid
hold on
axis off
r = max(0,s);                    % rectification
plot(t,r,'r')
ave = mean(r)*ones(size(t));     % compute average of rectified signal
plot(t,ave)
hold off
```

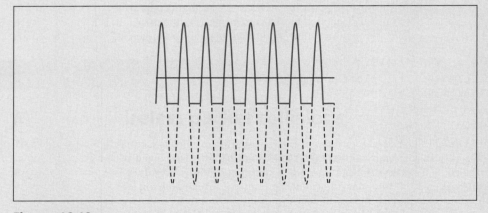

Figure 16.19

Matlab script that illustrates rectification performed by an ideal diode.

16.6.5 Combinational Logic Circuits

The following Matlab programs illustrate how binary values form logic gates and circuits.

AND Gate   **EXAMPLE 16.23**

Figure 16.20 shows the Matlab script that prints the truth table for an AND gate.

```
% Inputs:        A    B  Output: Y = A&B
variables=[ '    A    B    Y']
for i=0:3
    B = mod(i,2);             % least significant bit
    A = mod(floor(i/2),2);    % most significant bit
    Y = A & B ;               % Matlab AND operation
    [A B Y]                   % display truth table row
end
```

Script output:

```
variables =
     A     B     Y
ans =
     0     0     0
ans =
     0     1     0
ans =
     1     0     0
ans =
     1     1     1
```

Figure 16.20

Matlab script that prints the truth table for an AND gate.

OR Gate   **EXAMPLE 16.24**

Figure 16.21 shows the Matlab script that prints the truth table for an OR gate.

```
% Inputs:        A    B    Output Y = A|B
variables=[ '    A    B    Y']
for i=0:3
    B = mod(i,2);             % least significant bit
    A = mod(floor(i/2),2);    % most significant bit
    Y = A | B ;               % Matlab OR operation
    [A B Y]    % display row
end
```

Script output:

```
variables =
     A     B     Y
ans =
     0     0     0
ans =
     0     1     1
ans =
     1     0     1
ans =
     1     1     1
```

Figure 16.21

Matlab script that prints the truth table for an OR gate.

NOT Gate

Figure 16.22 shows the Matlab script that prints the truth table for a NOT gate.

```
% Inputs:          A      B      Output Y = A|B
variables=[ '      A      Y']
for i=0:1
    A = mod(i,2); % least significant bit
    Y = ~A;         % Matlab NOT operation
    [A Y]
end
```

Script output:

```
variables =
      A       Y
ans =
      0       1
ans =
      1       0
```

Figure 16.22

Matlab script that prints the truth table for a NOT gate.

Truth Table Verification

A logic circuit is verified by entering its logic equation and generating the truth table. Figure 16.23 shows the Matlab script that prints the truth table for specified logic equation.

```
% Logic equation Y = A&B&~C | A&B&C | ~A&B&C
 variables=['      A      B      C      Y']
for i=0:7
    A = mod(floor(i/4),2);              % most significant bit
    B = mod(floor(i/2),2);
    C = mod(i,2);                       % least significant bit
    Y = A&B&~C | A&B&C | ~A&B&C;        % logic equation
    [A B C Y]                           % display row
end
```

Script output:

```
variables =
      A       B       C       Y
ans =
      0       0       0       0
ans =
      0       0       1       0
ans =
      0       1       0       0
ans =
      0       1       1       1
ans =
      1       0       0       0
ans =
      1       0       1       0
ans =
      1       1       0       1
ans =
      1       1       1       1
```

Figure 16.23

Matlab script that prints the truth table corresponding to a logic equation.

Figure 16.24 shows the Matlab script that produces a nicer looking truth table by displaying the output as a 2D string array.

```
% Logic equation Y = A&B&~C | A&B&C | ~A&B&C
output=['A' ' ' 'B' ' ' 'C' '    ' 'Y'];
for i=0:7
  A = mod(floor(i/4),2);  % most significant bit
  B = mod(floor(i/2),2);
  C = mod(i,2);           % least significant bit
  Y = A&B&~C | A&B&C | ~A&B&C;
  row = [int2str(A)' ' int2str(B)' ' int2str(C)'   ' int2str(Y)]; % string array
  output = [output,
     row];                % adds another row to the table
end
output                    % display  output
```

Script output:

```
output =
A B C   Y
0 0 0   0
0 0 1   0
0 1 0   0
0 1 1   1
1 0 0   0
1 0 1   0
1 1 0   1
1 1 1   1
```

Figure 16.24

Matlab script that prints a denser truth table corresponding to a logic equation.

16.6.6 Sequential Logic Circuits

Set-Reset Flip-Flop EXAMPLE 16.27

Figure 16.25 shows the Matlab script that implements a set-reset flip-flop that produces output Q in response to set (S) and reset (R) sequence values.

```
% Input Sequence: S, R
% Output: Q0
format compact
S = [ 0 1 0 0 0 0 0 0 1 1 0 0 0]    % specify & display S input timing diagram
R = [ 0 0 0 0 1 1 1 0 0 0 1 0 0]    % specify & display R input timing diagram
Q= [];
Qnext = 0;                          % zero initial condition
for i=1:length(S)
    if S(i) == 1                    % Set = 1
        Qnext = 1;                  % set Qnext
    end
    if R(i) == 1                    % Reset = 1
        Qnext = 0;                  % reset Qnext
    end
    Q = [Q Qnext];
end
Q                                   % display Q time output
```

Figure 16.25

Matlab script that implements a set-reset flip-flop. (*Continued on next page*)

Script output:

```
S =
    0   1   0   0   0   0   0   0   1   1   0   0   0
R =
    0   0   0   0   1   1   1   0   0   0   1   0   0
Q =
    0   1   1   1   0   0   0   0   1   1   0   0   0
```

Figure 16.25

(Continued)

EXAMPLE 16.28

Toggle Flip-Flop

Figure 16.26 shows the Matlab script that implements a Toggle Flip-Flop that produces output Q in response to downward $(1 \rightarrow 0)$ transitions in the T input sequence.

```
% Input Sequence: T
% Output: Y = Q0
clear
format compact
T = [ 0 1 0 1 0 1 1 0 1 0 0]    % specify and display input
Q= [];
Past_T = 0;                      % defines past T value
Qnext = 0;                       % next value: zero initial condition
for i=1:length(T)
    if Past_T==1 && T(i)==0       % 1->0 transition occurs
        Qnext = ~Qnext;          % toggle
    end
    Past_T = T(i);
    Q = [Q Qnext];
end
Q                                % display Q time output
```

Script output:

```
T =
    0    1    0    1    0    1    1    0    1    0    0
Q =
    0    0    1    1    0    0    0    1    1    0    0
```

Figure 16.26

Matlab script that implements a toggle flip-flop.

EXAMPLE 16.29

Mod-4 counter

Figure 16.27 shows the Matlab script that Matlab script that implements mod-4 counter. The count increments on a downward transition on the CLK input.

```
%Ex Mod-4 Counter
% Input Sequence: CLK
% Output: Y0 Y1 = Q0 Q1
clear
format compact
CLK = [ 0 1 0 0 1 0 1 0 1 0  0 ]    % specify and display input
Y0 = [];
Y1 = [];
Q0 = 0;                             %  zero initial condition
Q1 = 0;
Past_CLK = 0;                       %  zero initial conditions
Past_Q0 = 0;                        %  Past Q0 = input to Q1

for i=1:length(CLK)
    if Past_CLK==1 && CLK(i)==0     % 1->0 transition at Q0 input
        Q0 = ~Q0;                   % toggle
    end
    Past_CLK = CLK(i);              % update input
    Y0 = [Y0 Q0];                   % add to output sequence
     if Past_Q0==1 && Q0==0         % 1->0 transition  at Q1 input
        Q1 = ~Q1;                   % toggle
    end
    Past_Q0 = Q0;                   % update Q0 output
    Y1 = [Y1 Q1];                   % add to output sequence
end
Y0       % display Q0 time output
Y1       % display Q1 time output
```

Script output:

```
CLK =
    0    1    0    0    1    0    1    0    1    0    0
Y0 =
    0    0    1    1    1    0    0    1    1    0    0
Y1 =
    0    0    0    0    0    1    1    1    1    0    0
```

Figure 16.27

Matlab script that implements a mod-4 counter.

Mod-6 counter EXAMPLE **16.30**

Figure 16.28 shows the Matlab script that Matlab script that implements mod-6 counter.

```
% Input Sequence: CLK
% Output: Y0 Y1 Y2 = Q0 Q1 Q2
format compact
CLK = [0 1 0 1 0 1 0 1 0 1 0 1 0]    % specify clock input
Y0 = [];
Y1 = [];
```

Figure 16.28

Matlab script that implements a mod-6 counter. (*Continued on next page*)

```
Y2 = [];
Q0 = 0;                    %  zero initial condition
Q1 = 0;
Q2 = 0;
Past_CLK = 0;
Past_Q0 = 0;               % Past Q0 is input to Q1 (zero initial condition)
Past_Q1 = 0;                % Past Q1 is input to Q2 (zero initial condition)
Q2C=1; Q1C=1; Q0C=0;       %  6 pattern to be recognized to clear counter
for i=1:length(CLK)
    if Past_CLK==1 && CLK(i)==0        % 1->0 transition at Q0 input
        Q0 = ~Q0;                      % toggle
    end
    if Past_Q0==1 && Q0==0             % 1->0 transition  at Q1 input
        Q1 = ~Q1;                      % toggle
    end
    if Past_Q1==1 && Q1==0             % 1->0 transition  at Q2 input
        Q2 = ~Q2;                      % toggle
    end
    if Q0==Q0C && Q1==Q1C && Q2==Q2C   % if count =  6 patten
        Q0 = 0;
        Q1 = 0;
        Q2 = 0;
    end
    Past_CLK = CLK(i); % Remember past value
    Past_Q0 = Q0;
    Past_Q1 = Q1;
    Y0 = [Y0 Q0];         % Append to array
    Y1 = [Y1 Q1];
    Y2 = [Y2 Q2];
end
Y0 % display Q0 time output
Y1 % display Q1 time output
Y2 % display Q2 time output
```

Script output:

```
CLK =
    0   1   0   1   0   1   0   1   0   1   0   1   0
Y0 =
    0   0   1   1   0   0   1   1   0   0   1   1   0
Y1 =
    0   0   0   0   1   1   1   1   0   0   0   0   0
Y2 =
    0   0   0   0   0   0   0   0   1   1   1   1   0
```

Figure 16.28

(*Continued*)

16.6.7 Converting Between Analog and Digital Signals

EXAMPLE 16.31 | **Sampling sinusoids**

Figure 16.29 shows a Matlab script that illustrates sampling a sinusoidal waveform.

```
clear
f0 = 6240;                      % arbitrary frequency
dt = 1/(32*f0);                 % 32 samples per period
T = 32*2;                       % 2-period duration
t = (0:T-1)*dt;                 % time array
s_t = sin(2*pi*f0*t);           % original sinusoid waveform
plot(t,s_t)
axis off
hold on
fs = 8*f0;                      % sampling rate
Ts = 1/fs;                      % sample period
n_s = round(max(t)/Ts);         % number of samples
for i=1:n_s
    iTs(i) = (i-1)*Ts;          % sample times
    s(i) = sin(2*pi*f0*iTs(i)); % sample values
end
stem(iTs,s)
hold off
```

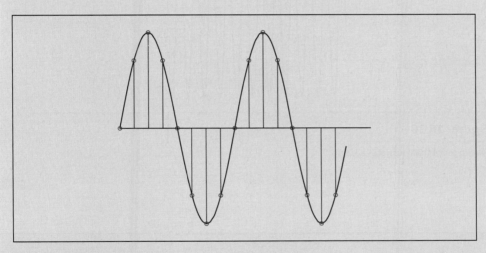

Figure 16.29

Matlab script that illustrates sampling a sinusoidal waveform.

Aliasing using a boxcar DAC EXAMPLE 16.32

Figure 16.30 shows a Matlab script that illustrates aliasing. The reconstructed alias waveform is formed using a boxcar DAC that holds the values over the sample time intervals applied to the analog waveform.

```
clear
f0 = 6240;                   % arbitrary frequency
dt = 1/(32*f0);              % 32 samples per period
T = 32*11;                   % 11-period duration
t = (0:T-1)*dt;              % time array
s_t = sin(2*pi*f0*t);        % original sinusoid waveform
plot(t,s_t)
axis off
hold on
fs = 0.9*f0;                 % sampling rate
Ts = 1/fs;                   % sample period
i = floor(t/Ts);            % sample-and-hold time
s_i = sin(2*pi*f0.*i*Ts);   % sample points show aliased signal
plot(t,s_i,'r','linewidth',2)
hold off
```

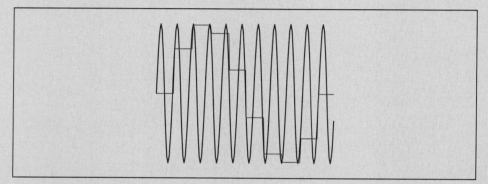

Figure 16.30

Matlab script that illustrates aliasing.

<hr>

EXAMPLE 16.33   Aliasing microphone speech and boxcar DAC

This example uses the microphone speech sample generated in Example 16.9 to illustrate the intelligibility of aliased speech signals. Figure 16.31 shows a Matlab script that illustrates aliasing of microphone speech. The reconstructed alias waveform is formed using a boxcar DAC that holds the values over the time intervals defined by the original waveform.

```
clear
n_skip = input('undersampling factor(1:orig f_s, 2:fs/2, 3:fs/3,...)= ');
load('speech.mat','myRecording')       % read speech data
sound(myRecording,8000)                % play original speech
n1 = length(myRecording)/2 - 200;      % start plot value
n2 = length(myRecording)/2 + 200;      % start plot value
subplot(2,1,1),plot(myRecording(n1:n2)) % plot original speech
axis off
```

Figure 16.31

Matlab script that illustrates aliasing of microphone speech using undersampling factor = 10. (*Continued on following page*)

```
sample = myRecording;                   % define sample array
for i = 1:n_skip:length(myRecording)-1
    sample(i) = myRecording(i);         % sub-sample original
    for j = 0:n_skip-1
        sample(i+j) = sample(i);        % hold sample value
    end
end
sound(sample,8000)                      % play sub-sampled speech
subplot(2,1,2),plot(sample(n1:n2))      % plot sub-sampled waveform
axis off
```

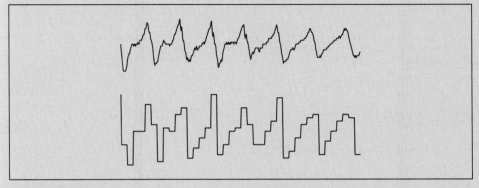

Figure 16.31

(*Continued*)

Quantizing sinusoidal waveform   EXAMPLE 16.34

Figure 16.32 shows a Matlab script that illustrates quantization of a sinusoidal waveform. This
script contains a *while loop* that insures the number of bits specified is 8 or less.

```
clear
nbits = 10;                             % set large default value
while nbits > 8                         % insure valid # of bits
    nbits = input('number of quantization bits (<=8)= ');
end
t = (0:512);
f0 = 1/128;                             % frequency
sine = sin(2*pi*f0.*t );                % generate sinusoid
subplot(2,1,1),plot(sine)               % plot original waveform
axis off
sample = sine;                          % define sample array
max_ampl = 1;                           % maximum amplitude
Delta = 2*max_ampl/(2^nbits-1);         % step size
for i = 1:length(sine)
    sa = sine(i)+1+Delta/2;             % analog value +Delta/2 - 2+Delta/2 V
    sq = floor(sa/Delta);               % quantizer step
    s_DAC = Delta * sq;                 % DAC reconstruction
    sample(i) = s_DAC-1;                % restore zero mean
end
subplot(2,1,2),plot(sample)             % plot quantized waveform
axis off
```

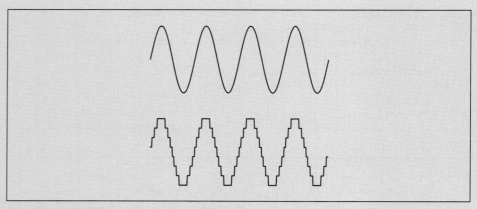

Figure 16.32

Matlab script that illustrates quantization of sinusoid using nbit = 3.

EXAMPLE 16.35 **Quantizing microphone speech**

This example uses the microphone speech sample generated in Example 16.9 to illustrate the intelligibility of quantized speech signals. Figure 16.33 shows a Matlab script that illustrates quantization of microphone speech.

```matlab
clear
nbits = 10;                                 % set large default value
while nbits > 8                             % insure valid # of bits
    nbits = input('number of quantization bits (<=8)= ');
end
load('speech.mat','myRecording')           % read speech data
sound(myRecording,8000)                     % play original speech
n1 = length(myRecording)/2 - 200;          % start plot value
n2 = length(myRecording)/2 + 200;          % start plot value
subplot(2,1,1),plot(myRecording(n1:n2))    % plot original speech
axis off
sample = myRecording;                       % define sample array
speech_ampl = 1;                            % microphone max amplitude
Delta = 2*speech_ampl/(2^nbits-1);         % step size
for i = 1:length(myRecording)
    sa = myRecording(i)+1+Delta/2;         % analog value Delta/2 -> 2+ Delat/2 V
    sq = floor(sa/Delta);                  % quantizer step
    s_DAC = Delta * sq;                    % DAC reconstruction
    sample(i) = s_DAC-1;                   % zero mean for sound player
end
sound(sample,8000)                          % play quantized speech
subplot(2,1,2),plot(sample(n1:n2))         % plot quantized waveform
axis off
```

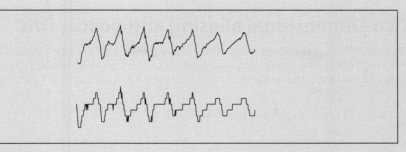

Figure 16.33

Matlab script that illustrates microphone speech quantization using nbit = 6.

Two-Dimensional spacial sinusoids

EXAMPLE 16.36

Figure 16.34 shows the Matlab script that generates two-dimensional spacial sinusoids.

```
clear
R = 256;                            % # of rows
y = (0:R-1);                        % row index
C = 256;                            % # of columns
x = (0:C-1);                        % column index
TwoDSine = zeros(R,C);              % define image array
colormap('GRAY')                    % gray scale image
fR = input('vertical freq= ');
valR = 1+sin(2*pi*fR.*y/R);         % row values
fC = input('horizontal freq= ');
valC = 1+sin(2*pi*fC.*x/C);         % column values
for i = 1:R                         % rows - vertical freq
    for j=1:C                       % columns - horizontal freq
        val = 16*valR(i)*valC(j);   % 0 <= val <= 64
        TwoDSine(i,j) = val;
    end
end
image(TwoDSine)                     % display image
axis image
```

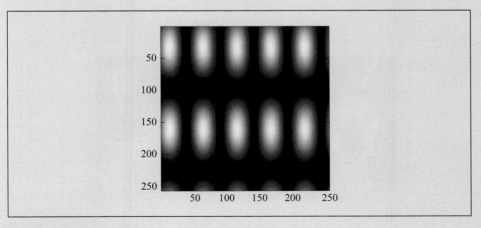

Figure 16.34

Matlab script that generates two-dimensional spacial frequencies using vertical freq = 2 and horizontal freq = 5.

EXAMPLE 16.37 **Two-Dimensional aliasing and boxcar DAC**

Figure 16.35 shows the Matlab script that reads a jpg file and illustrates aliasing by producing a sub-sampled version using samples spaced T_S samples. The reconstructed image is formed using a boxcar DAC that holds the values over the pixel values defined by the original image.

```matlab
clear
filename = input('jpg filename= ', 's');
filename = [filename '.jpg']
Ao = imread(filename);   % read image
A = cast(Ao,'single');   % convert from unsigned 8-bit to single (for sqrt)
[R C D] = size(Ao);      % row, column and depth
GS = zeros(R,C);         % assign gray-scale array
for i=1:R
    for j = 1:C
        val = 0;
        for k=1:D
            val = val+ A(i,j,k)^2; % image intensity
        end
        GS(i,j) = floor(sqrt(val)/3); % gray-scale value
    end
end
colormap('GRAY')         % specify gray-scale image format
subplot(1,2,1),image(GS)
axis image
Ts = input('sampling period (pixels)= ');
spt = floor(Ts/2)+1;     % sample point
B = zeros(R,C);          % assign gray-scale array
for i = 0:floor(R/Ts)-1
    for j = 0:floor(C/Ts)-1
        sample = GS(Ts*i+spt,Ts*j+spt);          % get image sample
        for k = 0:Ts-1
            for m = 0:Ts-1
                B(Ts*i+k+1,Ts*j+m+1) = sample; % hold sample
            end
        end
    end
end
subplot(1,2,2),image(B)
axis image
```

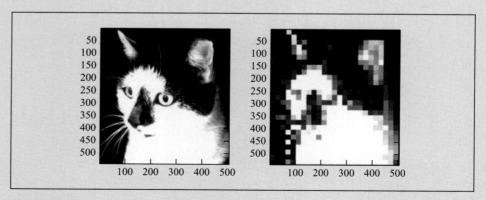

Figure 16.35

Matlab script that generates sub-sampled image using sampling period (pixels) = 20.

Figure 16.36 shows a Matlab script that illustrates pulse width modulation.

```
clear
V = input('voltage level (0-5V)= ');
nbit = input('number of bits= ');
resolution = 2^nbit-1;              % # of intervals in PWM period
dur = round(resolution * V/5);      % time at +5V
p1 = 5*ones(1,dur);                 % high level
m1 = zeros(1,resolution-dur);       % time at 0V
PWM = [p1 m1  p1 m1  p1 m1 p1 m1 ]; % several complete periods
nPWM = length(PWM);
MaxV = 5*ones(size(PWM));
plot(MaxV,'k--')                    % plot Max
text(nPWM+3,5,'5V')
hold on
MinV = zeros(size(PWM));
plot(MinV,'k--')                    % plot Min
text(nPWM+3,0,'0V')
plot(PWM,'linewidth',2)             % plot PWM
AveV = mean(PWM)*ones(size(PWM));   % compute average of PWM
plot(AveV,'r--')                    % plot average
text(nPWM+5,mean(PWM),num2str(mean(PWM)))
axis off
hold off
```

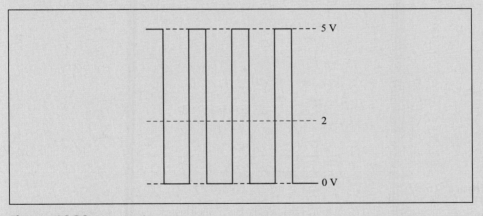

Figure 16.36

Matlab script that illustrates pulse-width modulation with voltage level = 2 V and number of bits = 8.

16.6.8 Digital Signal Processing

Figure 16.37 shows the Matlab function that implements an MA digital filter, and a simple program that tests the function.

```
function y = f_MA_filter(x,h)
mem = zeros(size(h));              % assign and initialize memory array
ny = length(x)+ length(h)-1;       % length of output
y = zeros(1,ny);                   % assign output array
for i = 1:length(x)
    mem(2:length(mem)) = mem(1:length(mem)-1); % shift memory
    mem(1) = x(i);                             % x(i) -> first mem value
    y(i) = sum(h.*mem);                        % sum of inner product
end
for i = length(x)+1:ny
    mem(2:length(mem)) = mem(1:length(mem)-1);  % shift memory
    mem(1) = 0;                                 % 0 -> first mem value
    y(i) = sum(h.*mem);                         % sum of inner product
end
```

This function is tested with the script:

```
x = [ 1 1 1 ];
h = [ 1 1 1 ];
y = f_MA_filter(x,h);
subplot(3,1,1),stem(x)
axis off
subplot(3,1,2),stem(h)
axis off
subplot(3,1,3),stem(y)
axis off
```

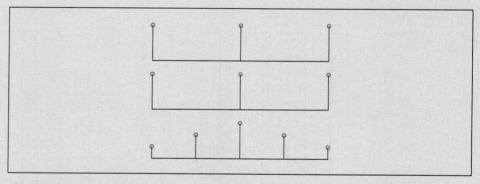

Figure 16.37

Matlab script that generates the output of an MA digital filter when input *x* and *h* are specified.

EXAMPLE 16.40

AR digital filter

Figure 16.38 shows the Matlab function that implements an AR digital filter of order m_y, determined by the number of feedback coefficients. Also shown are the simple programs that test the first-order and second-order AR filter function with a unit-sample sequence to generate the digital filter unit-sample response.

```
function y = f_AR_filter(x,a,ny)
nx = length(x);
mem = zeros(size(a));        % assign and initialize memory array
nm = length(mem);            % memory size
y = zeros(1,ny);             % assign output array
for i = 1:nx
    y(i) = sum(a.*mem) + x(i);  % sum of inner product + input
    mem(2:nm) = mem(1:nm-1);    % shift memory
    mem(1) = y(i);              % y(i) -> first mem value
end
for i = nx+1:ny                 % response when input exhausted
    y(i) = sum(a.*mem);         % sum of inner product
    mem(2:nm) = mem(1:nm-1);    % shift memory
    mem(1) = y(i);              % y(i) -> first mem value
end
```

This script tests this function as a first-order AR filter:

```
x =  1;                % define input-sample sequence
a = 0.9;               % first-order coefficient
ny = 40;               % output sequence size
y = f_AR_filter(x,a,ny);
stem(y)                % plot output (unit-sample response)
```

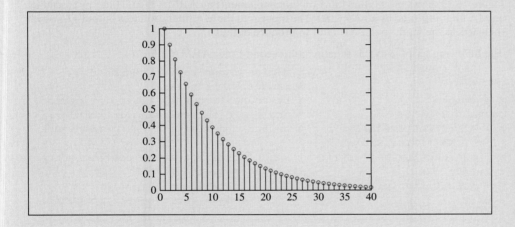

This script tests this function as a second-order AR filter:

```
index = (-3:25);               % specify duration of all sequences
x =  1;                        % define input sequence
r = 0.9;                       % radius (damping) term
f = 1/16;                      % resonant frequency (16-point period)
omega = 2*pi*f;                % angle wrt to real axis in complex plane
a = [2*r*cos(omega) -r^2];     % second-order feedback coefficient array
ny = 40;                       % output sequence size
y = f_AR_filter(x,a,ny);
stem(y)                        % plot output
```

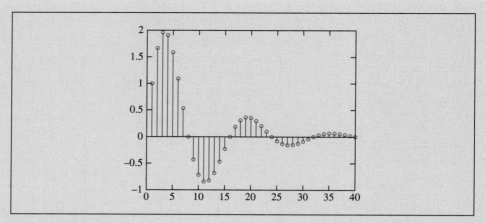

Figure 16.38

Matlab script that generates the output of an AR digital filter when input x and m_y coefficient array, and output y size are specified.

EXAMPLE 16.41 # ARMA digital filter

Figure 16.39 shows the Matlab function that implements an ARMA digital filter by cascading an MA filter followed by an AR filter. The input x_i to the MA filter produces output $y1_i$, which then acts as the input to the AR filter to produce output y_i.

The following script tests this function as a second-order ARMA filter:

```
x = 1;                       % define input-sample sequence
r = 0.9;                     % radius damping term
f = 1/16;                    % resonant frequency (16-point period)
omega = 2*pi*f;              % angle wrt to real axis in complex plane
h = [1 -r*cos(omega)];       % second-order MA filter coefficients
y1 = f_MA_filter(x,h);
a = [2*r*cos(omega) -r^2];   % second-order feedback coefficient array
ny = 40;                     % output sequence size
y = f_AR_filter(y1,a,ny);
stem(y)                       % plot output (unit-sample response)
```

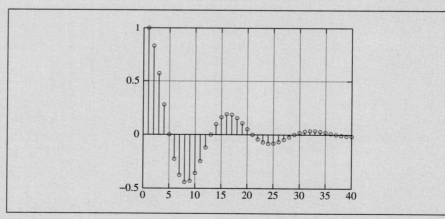

Figure 16.39

Matlab script that implements an ARMA digital filter.

16.6.9 Spectral Analysis

Direct evaluation of the DFT   EXAMPLE 16.42

It is important to note that a complication exists because DFT formulas start with $i = 0$ while Matlab starts an array index at $i = 1$. Accommodating this difference is indicated in the script below.

Figure 16.40 shows the function f_DFTe that evaluates the DFT using the complex exponential. Figure 16.41 shows the function f_DFTcs that evaluates the DFT using sine and cosine functions.

```
function X=f_DFTe(x)
nx = length(x);                         % array size
for k = 1:nx
    s = 0;                              % initialize sum
    for i = 1:nx
        omega = 2*pi*k*(i-1)/nx;    % NOTE (i-1) to start omega_i at 0
        s = s + x(i)*exp(-j*omega);
    end
    X(k) = s;
end
```

This script tests the DFTe function:

```
clear
x = [1 1 zeros(1,30)];    % define 32-element array
X = f_DFTe(x);
stem(abs(X))
```

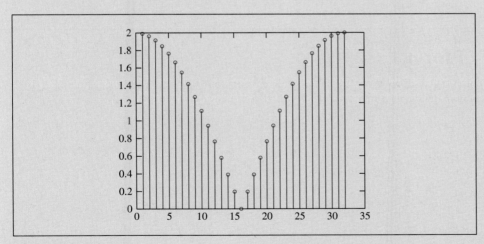

Figure 16.40

Matlab script that implements the DFT using the complex exponential.

```
function X = f_DFTcs(x)
nx = length(x);                        % array size
for k = 1:nx
    c = 0;
    s = 0;                             % initialize sum
    for i = 1:nx
        omega = 2*pi*k*(i-1)/nx;       % NOTE (i-1) to start omega_i at 0
        c = c + x(i)*cos(omega);
        s = s + x(i)*sin(omega);
    end
    X(k) = c + j*s;
end
```

This script tests the DFTcs function

```
clear
x = [1 1 zeros(1,30)];   % define 32-element array
X = f_DFTcs(x);
stem(abs(X))
```

Figure 16.41

Matlab script that implements the DFT using sine and cosine functions.

EXAMPLE 16.43

FFT for $n_x \neq 2^K$

Figure 16.42 shows the Matlab script that computes 100-point DFT using the Matlab *fft* function and displays magnitudes of "actual" frequency components.

```
clear
nx = 100;
x = zeros(1,nx);
x(1) = 1;                   % specify 2-sample sequence
x(2) = 1;
X = fft(x);                 % compute FFT
stem(abs(X(1:nx/2)))        % plot abs(X) over ''actual'' freq range
axis([-.01 nx/2 -.10 2.1])
grid on
xlabel('k' )
ylabel('|X_k|')
```

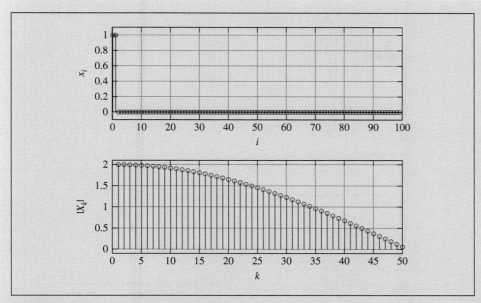

Figure 16.42

Matlab script that tests Matlab's FFT algorithm by computing the 100-point DFT and displaying magnitudes of "actual" frequency components.

Computing power spectrum and log-spectrum with the FFT

EXAMPLE 16.44

Figure 16.43 shows the Matlab script that computes 100-point of a 3-sample rectangular pulse and displays "actual" frequency components of power spectrum and log-spectrum using −10 dB and −60 dB scales.

```
clear
nx = 100;
x = zeros(1,nx);
x(1) = 1;              % specify 3-sample pulse
x(2) = 1;
x(3) = 1;
X = fft(x);            % compute FFT
Sx = abs(X).^2;        % compute power spectrum
subplot(3,1,1),stem(Sx(1:nx/2))    % plot Sx over ''actual'' freq range
Sx_dB = 10*log10(Sx/max(Sx));      % compute log-spectrum
axis([0 nx/2 0 9])
grid on
subplot(3,1,2),plot((Sx_dB(1:nx/2))) % plot log-spectrum using -10dB scale
axis([-.01 nx/2 -10 .1])
grid on
xlabel('k' )
ylabel('Sx_d_B')
subplot(3,1,3),plot((Sx_dB(1:nx/2))) % plot log-spectrum using -60dB scale
axis([-.01 nx/2 -60 .1])
grid on
xlabel('k' )
ylabel('Sx_d_B')
```

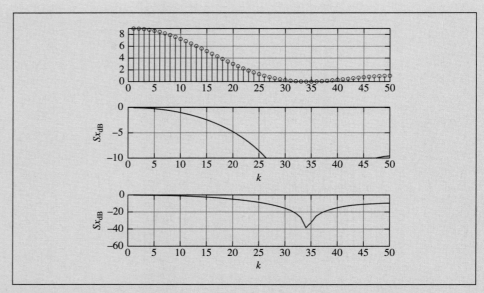

Figure 16.43

Matlab script that computes 100-point FFT and displays "actual" frequency components of power spectrum and log-spectrum using −10 dB and −60 dB scales.

 **EXAMPLE 16.45**

Inverse DFT

Figure 16.44 shows the Matlab script that computes the inverse DFT from specified real and imaginary parts.

```
k=(0:15);                        % specify k array
XMAG = ones(size(k));            % specify magnitude
XPHASE = (-2*pi/length(k))*k;    % specify phase
XR = XMAG.*cos(XPHASE);          % compute real part
XI = XMAG.*sin(XPHASE);          % compute imaginary part
X = XR+j*XI;                     % from complex value
x=ifft(X);                       % compute IDFT using inverse FFT
stem(real(x))                    % plot result
```

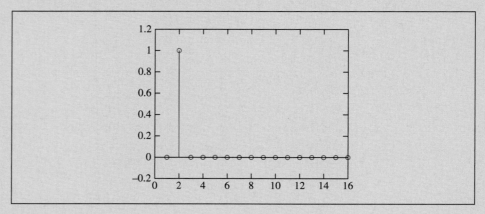

Figure 16.44

Matlab script that computes the inverse DFT from specified real and imaginary parts.

EXAMPLE 16.46

Processing in the frequency domain

Figure 16.45 shows the Matlab script that generated Figure 9.17.

```
clear                  % clear all variables
x = ones(1,10);        % define input sequence
nx = length(x);
h = ones(1,10);        % define MA sys coefficients
nh = length(h);
ny = nx+nh-1;          % length of output
index = (0:ny-1);      % index values for plotting
xp = [x zeros(1,nh-1)]; % pad input w nh-1 zeros
hp = [h zeros(1,nx-1)];% coefficients padded w nx-1 zeros
Xk =fft(xp);           % DFT(xp_i) using FFT
Hk = fft(hp);          % DFT(hp_i) using FFT
Yk = Hk .* Xk;
y = ifft(Yk);          % IDFT(Y_i) using IFFT
subplot(3,2,1),stem(index,xp)        % xp_i
axis([0 ny-1 0 1.1*max(xp)])
subplot(3,2,3),stem(index,hp)        % hp_i
axis([0 ny-1 0 1.1*max(hp)])
subplot(3,2,5),stem(index,y)         % y_i from IDFT
axis([0 ny-1 0 1.1*max(y)])
subplot(3,2,2),stem(index,abs(Xk))   % | DFT(x_i) |
axis([0 ny-1 0 1.1*max(abs(Xk))])
subplot(3,2,4),stem(index, abs(Hk))  % | DFT(h_i) |
axis([0 ny-1 0 1.1*max(abs(Hk))])
subplot(3,2,6),stem(index,abs(Yk))   % |H_k X_k |
axis([0 ny-1 0 1.1*max(abs(Yk))])
```

Figure 16.45

Matlab script that generated Figure 9.17.

16.6.10 Detecting Data Signals in Noise

EXAMPLE 16.47

Linear Processor

Figure 16.46 shows the one-line Matlab instruction that implements the linear processor

$$V = \sum_{i=0}^{n_x-1} c_i\, X_i$$

that operates on a set of user-input values.

```
X = input('enter array [X0 X1 X2]= ');
c = [1 2 3];      % coefficient sequence
V = sum(c.*X)     % linear processor output = sum of inner product
```

Following input produces output:

```
enter array [X0 X1 X2]= [1 2 3]
V =
    14
```

Figure 16.46

Matlab instruction that implements the linear processor.

EXAMPLE 16.48 ## Signal energy of a sinusoidal signal

Figure 16.47 shows the Matlab script that computes the signal energy of a sinusoid of having a specified amplitude and duration that contains two complete periods.

```
format compact
A = input('amplitude= ');
nx = input('n_x containing 2 periods = ');
f =2/nx;               % discrete-time frequency
i=(0:nx-1);            % time index
s = A*sin(2*pi*f.*i);  % signal sequence
Es = sum(s.^2)         % signal energy
```

Script produces output:

```
amplitude= 2
n_x containing 2 periods = 16
Es =
   32.0000
```

Figure 16.47

Matlab script that computes the signal energy of a sinusoid of having a specified amplitude and duration.

EXAMPLE 16.49 ## Complementary signals

Figure 16.48 shows the Matlab script that implements the matched processor that operates on a pair of complementary signals formed from an array of user-input values. Any number of values can be defined, but must begin and end with brackets.

```
format compact
s = input('enter array [X0 X1 X2]= ');
c = s;                  % coefficient matched to input
s1 = s;                 % define s1 signal
V1 = sum(c.*s1)         % matched processor output when s1 observed
s0 = -s;                % define s0 signal
V0 = sum(c.*s0)         % matched processor output when s0 observed
```

Script produces output:

```
V1 =
    30
V0 =
   -30
```

Figure 16.48

Matlab script that implements a matched processor and operates on a pair of complementary signals.

Generating Gaussian noise with specified σ_N EXAMPLE 16.50

Figure 16.49 shows the Matlab script that generates Gaussian noise with a specified σ_N. The $\pm 2\sigma_N$ levels are shown in dashed lines.

```
sigma_N = input('enter noise SD= ');      % specify sigma
N = sigma_N*randn(1,100);                 % generate 100 noise samples
stem(N)
hold on
p2sig = 2*sigma_N*ones(size(N));          % + 2 sigma level
plot(p2sig,'r--')
plot(-p2sig,'r--')                        % - 2 sigma level
axis off
hold off
sigma_N = std(N)                          % compute and display sample SD
```

Script produces output:

```
enter noise SD= .5
sigma_N =
    0.4881
```

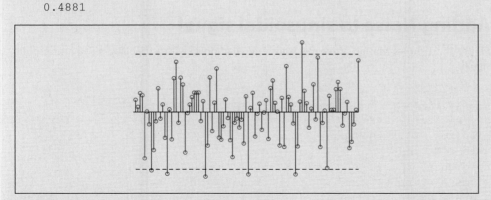

Figure 16.49

Matlab script that generates Gaussian noise with a specified σ_N. The $\pm 2\sigma_N$ levels are shown in dashed lines.

EXAMPLE 16.51

Generating histogram of Gaussian noise with specified σ_N

Figure 16.50 shows the Matlab script that generates a 51-bin histogram of Gaussian noise with specified σ_N from ten thousand Gaussian random numbers. The script also computes the sample standard deviation for comparison to the specified value.

```
sigma_N = input('enter noise SD= ');     % specify sigma
N = sigma_N*randn(1,100000);             % generate 10^5 noise samples
hist(N,51)                               % plot histogram with 51 bins
sigma_N = std(N)                         % compute sample SD
```

Script produces output:

```
enter noise SD= 1
sigma_N =
    0.9994
```

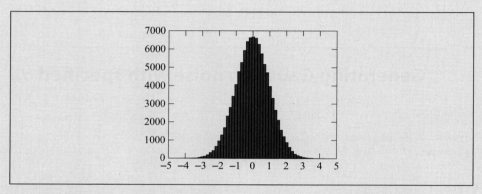

Figure 16.50

Matlab script that generates a 51-bin histogram of Gaussian noise with specified σ_N. Note almost all values fall within $\pm 3\sigma_N$.

EXAMPLE 16.52

Adding Noise to sinusoidal signal

Figure 16.51 shows Matlab script that adds Gaussian noise with a specified σ_N to a sinusoidal signal.

```
fo = 1/256;                          % specify freq
t = (0:512);                         % specify time points
s = sin(2*pi.*t*fo);                 % signal sequence
plot(t,s,'k','linewidth',2)
axis off
hold on
sigma_N = input('sigma_N= ');        % noise SD
X = s + sigma_N*randn(size(s));      % signal plus noise
plot(t,X)
hold off
```

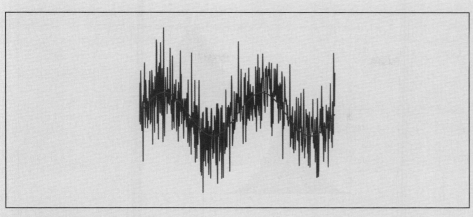

Figure 16.51

Matlab script that adds Gaussian noise with a specified $\sigma_N (= 1)$ to a sinusoidal signal having amplitude $A = 1$.

Matched processor output in the presence of noise   EXAMPLE 16.53

Figure 16.52 shows Matlab script that displays the histograms of matched processor outputs when specified signals are observed in Gaussian noise with a specified σ_N.

```
format compact
s = input('enter array [X0 X1 X2]= ');
Es = sum(s.^2);                    % compute singal energy
c = s;                             % coefficient matched to input
s1 = s;                            % define s1 signal
s0 = -s;                           % define s) signal
sigma_N = input('enter sigma_N= ');
sigma_V = sigma_N*sqrt(Es);        % compute sigma_V
for i=1:10000                      % 10^4 transmissions of 1's and 0s
    X = s1 + sigma_N*randn(size(s));    % X = s1 + N
    V1(i) = sum(c.*X);             % matched processor output when X=s1+N
    X = s0 + sigma_N*randn(size(s));    % X = s0 + N
    V0(i) = sum(c.*X);             %% matched processor output when X=s0+N
end
N1=hist(V1,51);                    % get hist values
N0=hist(V0,51);
subplot(2,1,1),hist(V1,51)
axis([-Es-3*sigma_V Es+3*sigma_V 0 max(N1)]) % use common axes
subplot(2,1,2),hist(V0,51)
axis([-Es-3*sigma_V Es+3*sigma_V 0 max(N0)])
```

Script inputs:

```
enter array [X0 X1 X2]= [1 1 1 1 1]
enter sigma_N= 1
```

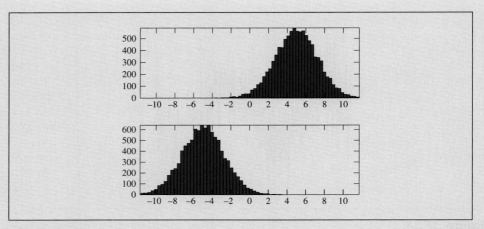

Figure 16.52

Matlab script that displays the histograms of matched processor outputs when specified signals are observed in Gaussian noise with a specified σ_N.

EXAMPLE **16.54**

Generating random bit values

Figure 16.53 shows Matlab script that uses Matlab's uniform random number generator to produce 1 and 0 having equal probability.

```
for i=1:10
    Dt(i) = floor(2*rand());
end
Dt
```

Script output:

```
Dt =
     0     0     1     1     1     1     1     0     0     1
```

Figure 16.53

Matlab script that uses Matlab's uniform PRNG to produce 1 and 0 having equal probability.

EXAMPLE **16.55**

Estimating probability of error of a matched processor

Figure 16.54 shows Matlab script that estimates probability of error for a matched processor and complementary signals.

```
s = input('enter array [X0 X1 X2 ...]= ');
c = s;                               % coefficient matched to input
s1 = s;                              % define s1 signal
s0 = -s;                             % define s) signal
sigma_N = input('enter sigma_N= ');
```

Figure 16.54

Matlab script that estimates probability of error for a matched processor and complementary signals.
(*Continued on following page*)

```
n_err = 0;                              % initialize err count to 0
n_xmit = 10000;                         % # of transmissions
for i=1:n_xmit                          % 10^4 transmissions of 1's and 0s
    [Dt sT] = f_rand_sT(s);             % bit gen and sT select function
    X = sT + sigma_N*randn(size(s));    % X = sT + N
    V = sum(c.*X);                      % matched processor
    Dr = f_threshDetector(V);           % threshold detecto function
    if Dr ~= Dt                         % if Dr not equal to Dt -> error
        n_err = n_err +1;
    end
end
p_err = n_err/n_xmit                    % estimate Prob[error]
```

Functions:

```
function [Dt sT] = f_rand_sT(s)
%   generates random bit Dt and
%   selects corresponding complementary signal
    Dt = floor(2*rand());               % random transmitted bit
    if Dt == 1
        sT = s;
    else
        sT = -s;
    end
end

function Dr = f_threshDetector(V)
% threshold detector
    if V>0                              % threshold detector
        Dr=1;
    else
        Dr=0;
    end
end
```

Script output:

```
enter array [X0 X1 X2 ...]= [1 1 1 1]
enter sigma_N= 1
p_err =
    0.0206
```

Figure 16.54

(Continued)

16.6.11 Designing Signals for Multiple-Access Systems

TDMA orthogonal signals   EXAMPLE 16.56

Figure 16.55 shows the Matlab script that generates two orthogonal TDMA signals of length n_x.

```
n_x = 16;                                    % must be multiple of # of users
sA = [ones(1,n_x/2) zeros(1,n_x/2)];    % sA
sB = [zeros(1,n_x/2) ones(1,n_x/2)];    % sB
subplot(2,1,1), stem(sA)
axis off
subplot(2,1,2), stem(sB)
axis off
OrthCond = sum(sA.*sB)                        %      orthogonality condition
```

Script output:

```
OrthCond =
     0
```

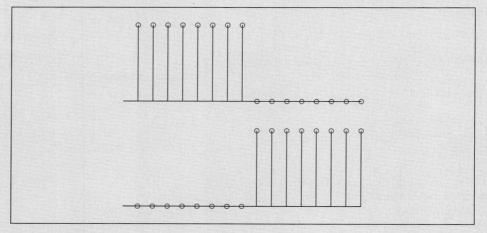

Figure 16.55

Matlab script that generates two orthogonal TDMA signals of length $n_x (= 16)$.

EXAMPLE **16.57** ## FDMA orthogonal signals

Figure 16.56 shows the Matlab script that generates two orthogonal FDMA signals of length n_x.

```
nx = 50;                  % signal duration
i = (0:nx-1);             % time index
fA = 3/nx;                % user A frequency
sA = sin(2*pi*fA.*i);     % user A signal
fB = 4/nx;                % user B frequency
sB = sin(2*pi*fB.*i);     % user B signal
subplot(2,1,1),stem(sA)
axis off
subplot(2,1,2),stem(sB)
axis off
OrthCond = sum(sA.*sB)    % orthogonality condition
```

Script output:

```
OrthCond =
   1.1741e-14
```

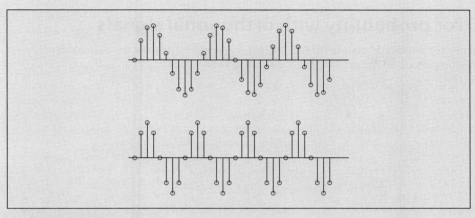

Figure 16.56

Matlab script that generates two orthogonal FDMA signals of length $n_x (= 50)$. Note othogonality condition is a "computed zero" that ≈ 0.

CDMA orthogonal signals EXAMPLE 16.58

Figure 16.57 shows the Matlab script that generates two orthogonal CDMA signals of length n_x.

```
rng('default')                        % initialize PRNG
nx = 20;                              % signal duration
sA = 2*floor(2*rand(1,nx))-1;        % generates random +1s and -1s
epsilon = nx;                         % initial epsilon value
while abs(epsilon) > 1                % search until valid sB code found
    sB = 2*floor(2*rand(1,nx))-1;    % generates random +1s and -1s
    epsilon = sum(sA.*sB);            % test orthogonality condition
end
subplot(2,1,1),stem(sA)
axis off
subplot(2,1,2),stem(sB)
axis off
OrthCond = sum(sA.*sB)                % orthogonality condition
```

Script output:

```
OrthCond =
     0
```

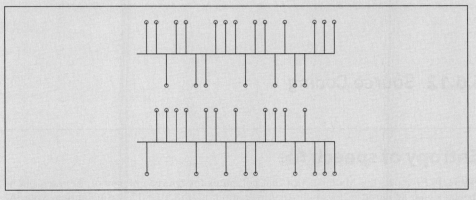

Figure 16.57

Matlab script that generates two orthogonal CDMA signals of length $n_x (= 20)$.

EXAMPLE 16.59 Error probability with orthogonal signals

Figure 16.58 shows Matlab script that estimates probability of error for a matched processor and complementary signals. The functions called in the script are given in Example 16.55.

```
sigma_N = input('enter sigma_N= ');
rng('default')                       % initialize PRNG
nx = 20;                             % signal duration
sA = 2*floor(2*rand(1,nx))-1;        % generates random +1s and -1s
epsilon = nx;                        % initial epsilon value
while abs(epsilon) > 1               % search until valid sB code found
   sB = 2*floor(2*rand(1,nx))-1;     % generates random +1s and -1s
   epsilon = sum(sA.*sB);            % test orthogonality condition
end
cA = sA;                             % coefficients for processor matched to sA
cB = sB;                             % coefficients for processor matched to sB
n_err = 0;                           % initialize err count to 0
n_xmits = 10000;                     % # of transmissions
for i=1:n_xmit                       % 10^4 transmissions of 1's and 0s
   [DAt sAT] = f_rand_sT(sA);        % bit gen and sT select function
   [DBt sBT] = f_rand_sT(sB);        % bit gen and sT select function
   X = sAT + sBT + sigma_N*randn(1,nx); % X = sAT + sBT + N
   VA = sum(cA.*X);                  % matched processor
   DAr = f_threshDetector(VA);       % threshold detecto function
   if DAr ~= DAt                     % if DAr not equal to DAt -> error
       n_err = n_err + 1;
   end
   VB = sum(cB.*X);                  % matched processor
   DBr = f_threshDetector(VB);       % threshold detecto function
   if DBr ~= DBt                     % if DBr not equal to DBt -> error
       n_err = n_err + 1;
   end
end
p_err = n_err/(2*n_xmit)             % estimate Prob[error] 2 bits/xmission
```

Script output:

```
enter sigma_N= 10
p_err =
   0.3256
```

Figure 16.58

Matlab script that estimates the probability of error for a matched processor and orthogonal CDMA signals.

16.6.12 Source Coding

EXAMPLE 16.60 Entropy of speech file

Figure 16.59 shows the Matlab script that computes the source entropy of the microphone speech signal acquired in Example 16.9 by computing the effective probabilities. Note the 8-bit symbols have an entropy of 4.48 bits/symbol due to the long silent periods.

```
load('speech.mat','myRecording') % read data from file into array myData
sound(myRecording,8000)        % play the speech on the speaker
plot(myRecording)              % plot the speech waveform
d = floor(128*(myRecording+1));     % converts samples into bytes
N = hist(d,256);               % generate histogram
n_t = sum(N);                  % total number of samples
PeX = N/n_t;                   % Effective probability
E = 0;
for i = 1:length(N)
    if Px_i(i) > 0             % compute entropy for positive probability
E = E - PeX(i)*log2(PeX(i));
    end
end
disp('Entropy (bits/symbol)=')
E
```

Script output:

```
Entropy (bits/symbol)=
E =
    4.4813
```

Figure 16.59

Matlab script that computes the source entropy of the microphone speech.

Encryption **EXAMPLE 16.61**

Figure 16.60 shows a Matlab script that implements encryption of 4-bit data by generation four random bits at a time. The script performs decryption to retrieve the original data.

```
format compact
key = 6;                         % PRNG seed
rng(key);                        % seed input to Matlab's PRNG
for i =1:16
    RN(i)  =  floor(16*rand()); % 4 random bits
    Data(i) = i-1;              % data 0-15
end
Data                            % disp Data numbers 0 - 15
RN                              % disp Random numbers
E = bitxor(RN,Data)            % bit-wise xor disp encrypted vals
D = bitxor(RN,E)              % bit-wise xordisp decrypted vals
```

Script output:

```
Data =
     0  1  2  3  4  5  6  7  8  9 10 11 12 13 14 15
RN =
    14  5 13  0  1  9  8  6  5  9  7 11  8  9 10 15
E =
    14  4 15  3  5 12 14  1 13  0 13  0  4  4  4  0
D =
     0  1  2  3  4  5  6  7  8  9 10 11 12 13 14 15
```

Figure 16.60

Matlab script that implements encryption and decryption.

EXAMPLE 16.62   **Encrypting microphone speech**

Figure 16.61 shows a Matlab script that encrypts microphone speech.

```matlab
load('speechEx.mat','myData')                    % read data from file into array myData
sound(myData,8000)                               % play the speech (-1,+1) on the speaker
sample1 = myData;
subplot(3,1,1),plot(sample1(2000:2200))          % plot the speech waveform
b = zeros(1,8);
r=b;
E = b;
ADC_sample = myData;
Encrypt = myData;
rng('default')
for i = 1:length(myData)
    ADC_sample(i) = round(128*((myData(i)+1)));  % convert to [0,255]
    for j=1:8
        b(j) = mod(floor(ADC_sample(i)/2^(j-1)),2);
    end
    for j=1:8
        r(j) = floor(2*rand());                  % RBS_i
    end
    for j=1:8
        e(j) = xor(b(j),r(j));                   % exor encryption
    end
    sum = 0;    %DAC
    for j=1:8
        sum = sum + e(j)*2^(j-1);
    end
    Encrypt(i) = sum;
    sample2(i) = sum/255 -1;
end
sound(sample2,8000)                              % play the speech on the speaker
subplot(3,1,2),plot(sample2(2000:2200))         % plot the speech waveform
rng('default')   for i = 1:length(myData)
    for j=1:8
        b(j) = mod(floor(Encrypt(i)/2^(j-1)),2);
    end
    for j=1:8
        r(j) = floor(2*rand());                  % RBS_i
        e(j) = xor(b(j),r(j));                   % exor decryption
    end
    sum = 0;                                     %DAC
    for j=1:8
        sum = sum + e(j)*2^(j-1);
    end
    sample3(i) = sum/255 -1;
end
sound(sample3,8000)                             % play the speech on the speaker
subplot(3,1,3),plot(sample3(2000:2200))         % plot the speech waveform
```

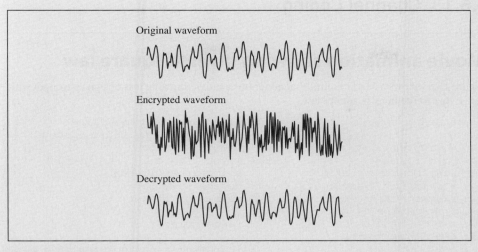

Figure 16.61

Matlab script that encrypts microphone speech.

Encrypting a jpg image EXAMPLE 16.63

Figure 16.62 shows a Matlab script that encrypts a jpg image.

```
str = input('JPEG image filename =','s')   % input file name
str = [str '.jpg'];                         % append jpg type
[a,map]=imread(str);                        % read image data
colormap(map);
subplot(1,3,1),image(a);
axis 'image'
R = uint8 (floor(256.*rand(size(a))));      % generate random bytes
E = bitxor(a,R);                            % encrypt image
subplot(1,3,2),image(E);
axis 'image'
E2 = bitxor(E,R);                           % decrypt image
subplot(1,3,3),image(E2)
axis 'image'
```

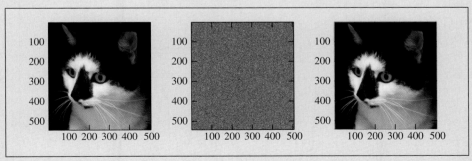

Figure 16.62

Matlab script that encrypts a jpg image and performs decryption.

16.6.13 Channel Coding

EXAMPLE 16.64 **Movie animation showing inverse square law**

Figure 16.63 shows a Matlab script that generates a movie showing loss of signal strength with range due to the inverse square law.

```
clear mex                              % clear any open mex files
close                                  % close any open figure windows
fig = figure;                          % specify figure
movegui(fig)                           % moves figure to main display
set(fig,'DoubleBuffer','on');          % eliminates flicker
mov = avifile('MyMovie.avi')           % define avifile
num_frames = 27;                       % number of frames
sigma_N = 0.01;                        % noise SD
ptr = 0;                               % range pointer
for k = 1:num_frames
    ptr = ptr + 3;                     % increment range pointer
    r = (2:.1:10);                     % range values
    A = 1 ./ r.^2;                     % inverse square law
    subplot(2,1,1),plot(r ,A,'linewidth',2 ) % plot function
    hold on
    plot(r(ptr), A(ptr),'ro','linewidth',3)  %plot specific range point
    threshold = 2*sigma_N*ones(size(r));
    plot(r,threshold,'r--')                   % plot 2sigma_N on graph
    hold off
    i = (1:100);
    s = ones(1,20);                    % signal sequence
    sT = [s -s s -s s];                % 10101 transmitted data sequence
    N = sigma_N*randn(size(sT));       % noise
    X = A(ptr)*sT + N;                 % detected signal
    subplot(2,1,2),plot(i,X)
    axis([min(i) max(i) -max(X) max(X)])
    hold on
    plot(i,zeros(size(i)),'r--')       % zero threshold on data plot
    hold off
    pause(.2)
    F = getframe(gcf);                 % figure into frame
    mov = addframe(mov,F);             % add frame to movie
end
mov = close(mov);                      % close movie file
clear mex                              % clear any open mex files
```

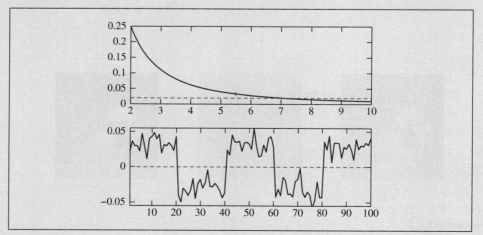

Figure 16.63

Matlab script that generates an animated movie illustrating inverse square law. A single frame is shown.

Simulating a specified error probability   EXAMPLE 16.65

Figure 16.64 shows a Matlab script that simulates a data channel with specified error probability by comparing specified and estimated probabilities.

```
format compact
Pe = input('Enter Prob[err] (.01) ');
nt = round(1000/Pe);        % number of bit transmissions for reliable estimate
ne = 0;                     % error count
for i = 1:nt
    bt = floor(2*rand());   % random transmitted data bit
    br = bt;                % received bit
    if rand() < Pe
        br = ~br;           % bit flipped by error
    end
    if  br ~= bt            % err
        ne = ne+1;
    end
end
'Pe (measured)'
PeM = ne/nt

Command window output:

Enter Prob[err] (.01) .1
ans =
Pe (measured)
PeM =
    0.1110
```

Figure 16.64

Matlab script that simulates a data channel with specified error probability.

Three-bit repetition error correction   EXAMPLE 16.66

Figure 16.65 shows a Matlab script that simulates 3-bit repetition error correction at the receiver.

```
format compact
Pe = input('Enter Prob[err] for each bit (.01) ');
nt = round(1000/Pe);        % number of bit transmissions
ne = 0;                     % error count
br_cw = zeros(1,3);         % 3-bit repetition
for i = 1:nt
    bt = floor(2*rand());   % random  data bit
    for j = 1:3
        br_cw(j) = bt;      % received bit
        if rand() < Pe
            br_cw(j) = ~bt; % bit flipped by error
        end
    end
```

Figure 16.65

Matlab script that simulates 3-bit repetition error correction at the receiver. (*Continued on next page*)

```
        br_sum = sum(br_cw);      % sum received bits
        br = 0;                    % received bit default = 0
        if br_sum >= 2
            br = 1;
        end
        if  br ~= bt              % error
            ne = ne+1;
        end
    end
end
'Pe (measured)'
PeM = ne/nt

Command window output:

Enter Prob[err] (.01) .01
ans =
Pe (measured)
PeM =
    5.0000e-04
```

Figure 16.65

(*Continued*)

EXAMPLE 16.67   Forming a data block code with parity and LRC

Figure 16.66 shows a Matlab script that simulates a transmitter that forms a simple 2×2 data block with parity and a LRC to form a 3×3 block code and saves it on the disk that acts as a transmitter.

```
clear
d = zeros(3,3); % 2x2 data block + parity and LRC
for i = 1:2
    for j = 1:2
        d(i,j) = floor(2*rand());    % random data bit
    end
end
% odd parity in columns
for j = 1:2
    sumd = d(1,j)+d(2,j);
    d(3,j) = ~mod(sumd,2);
end
% even parity in rows including LRC
for i = 1:3
    sumd = d(i,1)+d(i,2);
    d(i,3) = mod(sumd,2);
end
d                       % display data block
save datablock.mat  % saves data to disk (channel)

Command window output:

d =
    0    1    1
    0    0    0
    1    0    1
```

Figure 16.66

Matlab script that forms a simple 2×2 data block to form a 3×3 block code having parity bits and LRC.

EXAMPLE 16.68

Simulating random errors in a block code

Figure 16.67 shows a Matlab script that reads the 3 × 3 block code generated with Example 16.67, introduces random errors, and saves the block code on the disk that acts as a transmission channel.

```
clear
Pe = input('Enter Prob[err] for each bit (.01) ');
nt = round(1000/Pe);     % number of bit transmissions for reliable estimate
load datablock.mat       % read data from disk (channel)
% Insert random errors
[r,c] = size(d);
dr = zeros(size(d));
for i = 1:r
    for j = 1:c
        dr(i,j) = d(i,j);                % received bit
        if rand() < Pe
            dr(i,j) = ~d(i,j);           % bit flipped by error
        end
    end
end
d
dr
save received_data.mat

Command window output:

Enter Prob[err] for each bit (.01) .1
d =
     0     1     1
     0     0     0
     1     0     1
dr =
     0     0     1
     1     0     0
     1     0     1
```

Figure 16.67

Matlab script that introduces random errors into a block code.

EXAMPLE 16.69

Error correction in a block code

Figure 16.68 shows a Matlab script that corrects a single error in the data section of a block code generated with Example 16.68.

```
clear
load received_data.mat  % read d and dr from disk (channel)
% compute p_prime and LRC_prime
% odd parity in columns
d_ec = dr; % error corrected data
% compute p_prime in data and LRC row
```

Figure 16.68

Matlab script that corrects a single error in the data section of a block code. (*Continued on next page*)

```
for j = 1:2
    sumd = dr(1,j)+dr(2,j);
    dr(4,j) = ~mod(sumd,2);
end
% compute  LRC_prime
% even parity in rows including LRC
for i = 1:2
    sumd = dr(i,1)+dr(i,2);
    dr(i,4) = mod(sumd,2);
end
dif = 0;
err_num = 0;
err_row = 0;
err_col = 0;
% compare p and p_prime in each data row
for i=1:2
    if dr(i,3) ~= dr(i,4) % p neq p_prime -> error
        dif =dif +1;
        err_row = i;
    end
end
% compare LRC and LRC_prime in each data column
for j=1:2
    if dr(3,j) ~= dr(4,j) % LRC neq LRC_prime -> error
        dif = dif +1;
        err_col = j;
    end
end
% SEA error correcion
if dif == 0
    disp('no errors found')
end
if dif == 2
    d_ec(err_row,err_col) = ~d_ec(err_row,err_col); % error correction
end
if dif > 2
    disp('2 or more errors - SEA violation')
end
d
dr
d_ec

Command window output:
d =
     0     1     1
     0     0     0
     1     0     1
dr =
     0     1     1     1
     1     0     0     1
     1     0     1     0
     0     0     0     0
d_ec =
     0     1     1
     0     0     0
     1     0     1
```

Figure 16.68

(Continued)

Channel capacity EXAMPLE **16.70**

Figure 16.69 shows the Matlab script that generated Figure 13.12.

```
r = [1:10000];
BW = 10000;
Eso= 10000;
E_s=Eso./(r.*r);
C = BW*log10(1+E_s)/log10(2);
loglog(r,C)
axis([0 max(r) 0 1.05*max(C)])
grid on
xlabel('range (m)')
ylabel('Capacity (bps)')
```

Figure 16.69

Matlab script that generated Figure 13.12.

16.6.14 Symbology

Transforming images of codes EXAMPLE **16.71**

Figure 16.70 shows a Matlab script that transforms jpg images of a USPS bar code and QR symbol into a 2D array.

```
str = input('JPEG image filename =','s');   % input file name
str = [str '.jpg'];                          % append jpg type
[a,map]=imread(str);                         % read image data
colormap(map);
image(a);
axis 'image'
```

Figure 16.70

Matlab script that transforms a jpg image of a USPS bar code and and QR symbol into a 2D array.

EXAMPLE 16.72

Automated CAPCHA

Figure 16.71 shows a Matlab script that generates automated CAPCHAs.

```
str = input('JPEG image filename =','s')    % input file name
str = [str '.jpg'];                         % append jpg type
[a,map]=imread(str);                        % read image data
colormap(map);
subplot(1,3,1),image(a);
axis 'image'
R = uint8 (floor(256.*rand(size(a))));      % generate random bytes
E = bitxor(a,R);                            % encrypt image
subplot(1,3,2),image(E);
axis 'image'
E2 = bitxor(E,R);                           % de-encrypt image
subplot(1,3,3),image(E2)
axis 'image'
```

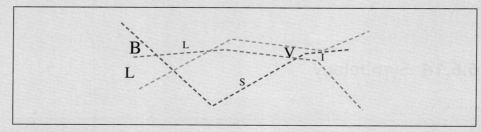

Figure 16.71

Matlab script generates a CAPCHA.

16.6.15 Data Networks

EXAMPLE 16.73

Generating asynchronous waveform of nibble data

Figure 16.72 shows a Matlab script that generates and displays an asynchronous transmission data character waveform containing a start bit, 4 data bits, an even parity bit, and two stop bits. The specified baud rate computes the duration of one data character waveform.

```
clear
baud = 9600;                % baud rate
T_b = 1/baud;               % bit time interval
T_s = T_b/16;               % sample period
nstop = 2;                  % number of stop bits
for i = 1:4
    d(i) = floor(2*rand()); % random data bit
end
```

Figure 16.72

Matlab script that generates an asynchronous data nibble waveform at 9600 baud. (*Continued on following page*)

```
sumd = sum(d);
d(5) = mod(sumd,2);            % even parity bit
d                              % display data
w0 = zeros(1,16);             % zero bit waveform
w1 = ones(1,16);              % one bit waveform
wvfrm = [w1];                 % idle channel bit for illustration only
wvfrm = [wvfrm w0];           % start bit
for i = 1:length(d)
    if d(i) == 0
        wvfrm = [wvfrm w0]; % add 0 bit
    else
        wvfrm = [wvfrm w1]; % add 1 bit
    end
end
for i = 1:nstop
    wvfrm = [wvfrm w1];       % idle channel stop bits
end
plot(wvfrm)
axis off
disp('T_packet (s):')
T_p = length(wvfrm-16)*T_s    % remove first idle channel bit
save Async.mat                % save character on disk

Command Window output:
d =
     0     1     0     1     0
T_packet (s):
T_p =
   9.3750e-04
```

Figure 16.72

(*Continued*)

Reading asynchronous nibble waveform   EXAMPLE 16.74

Figure 16.73 shows a Matlab script that reads an asynchronous transmission data character waveform produced by Example 16.73 and saved in disk file Async.mat.

```
clear
load Async.mat          % read character from disk
d                       % display data (for verification)
for i=2:length(wvfrm)   % search for start bit transition
    if wvfrm(i) ~= wvfrm(i-1)    % find transition
        ptr = i;                 % time of start bit
        break
    end
end
ptr = ptr + 16 + 8;     % position at center of next bit
for i=1:4               % read only data
    dr(i) = wvfrm(ptr);
    ptr = ptr + 16;     % position at center of next bit
end
dr                      % display received data

Command Window output:
d =
    0    1    0    1    0
dr =
    0    1    0    1
```

Figure 16.73

Matlab script that reads an asynchronous data nibble waveform.

EXAMPLE 16.75 Modeling wireless packet collisions

Figure 16.74 shows a Matlab script that simulates wireless data packet collisions using a set of variable packet sizes to compute the probability of collision. Small data packets rarely collide, while data packets that are half the duration of the data window always collide.

```
clear
num_xmits = 10000;          % number of transmissions
max_size = 10;              % maximum data packet size = 2^(max_size-1)
trans_window = 2048;        % transmission window = twice max dp size
PCol = zeros(1,max_size);
for dur = 1:max_size
    dpSz(dur) = 2^(dur);    % data packet size
    dp = ones(1,dpSz(dur)); % data packet
    errCnt = 0;
    for xmit = 1:num_xmits
        P1Time = round(1000*rand());    % start time of user1 packet
        d1 = [zeros(1,P1Time) dp zeros(1,trans_window-P1Time-length(dp))];
        P2Time = round(1000*rand());    % start time of user2 packet
        d2 = [zeros(1,P2Time) dp zeros(1,trans_window-P2Time-length(dp))];
        if sum(d1.*d2) > 0              % packet overlap if sum >0
            errCnt = errCnt+1;
        end
    end
    PCol(dur) = errCnt/num_xmits;
end
```

Figure 16.74

Matlab script that simulates wireless data packet collisions. (*Continued on following page*)

```
disp('Packet size =')
fprintf('%5d %5d %5d %5d %5d %5d %5d %5d %5d %5d\n',dpSz)
disp('Collision Probability =')
fprintf('%5.3f %5.3f %5.3f %5.3f %5.3f %5.3f %5.3f %5.3f %5.3f %5.3f\n',PCol)

Command Window output:
Packet size =
     2     4     8    16    32    64   128   256   512  1024
Collision Probability =
0.003 0.006 0.016 0.032 0.063 0.124 0.243 0.450 0.761 1.000
```

Figure 16.74

(Continued)

16.7 Summary

By the time you reach this point, you should have achieved the following goals.

1. A good grasp of Matlab

2. Knowledge of the common approaches to analyze data

3. An ability to simulate a complete communication system and compute a measure of its capabilities.

4. An understanding of how to present your results effectively

All of these aspects are important. Probably the biggest disappointment I can have with work presented by a student occurs when (s)he has an insightful approach to data analysis, yet produces a final product that looks sloppy, disorganized, and leaves the impression that the student does not understand the problem. I prefer an approach that is well-organized and, while initially falling short of the solution, provides indications where the difficulties lie and what corrective actions need to be applied.

Math Details

 A.1 HEXADECIMAL NOTATION

The basic data unit in a computer is a byte that consists of eight bits. Because $2^8 = 256$, only 256 numbers can be represented. For numbers representing counts, the range of values is from 0 (00000000) to 255 (11111111). Negative numbers are represented using a different more complex notation in which the MSB represents the sign (0 = positive and 1 = negative), but these are not necessary for the material in this text. Larger numbers are represented by using additional bytes.

A number represented in a byte (8-bit) format may have leading zeros, such as

$$53_{10} = 00110101_2 \tag{A.1}$$

Rather than write out all these bits, engineers use hexadecimal notation to represent a 4-bit number ranging between 0 and 15 by a single character. The characters are the ten digits to represent 0 through 9 and the letters A through F to represent the values from 10 to 15. Figure A.1 shows the correspondence of the 4-bit patterns with their decimal and hexadecimal equivalents. Hence, a byte value is represented by two hexadecimal, or "*hex*", numbers.

4-bit Pattern	Decimal Value	Hexadecimal Value
0000	0	0
0001	1	1
0010	2	2
0011	3	3
0100	4	4
0101	5	5
0110	6	6
0111	7	7
1000	8	8
1001	9	9
1010	10	A
1011	11	B
1100	12	C
1101	13	D
1110	14	E
1111	15	F

Figure A.1

A series of 4-bit patterns with their decimal and hexadecimal equivalents.

EXAMPLE A.1

Representing a byte value in hex form

Consider the values that can be represented in a single byte. To simplify the interpretation of the 8-bit pattern, a space is inserted between the two 4-bit units (sometimes called *nibbles*):

$$0_{10} = 0000\ 0000_2 = 00_{16} \tag{A.2}$$

$$53_{10} = 0011\ 0101_2 = 35_{16} \tag{A.3}$$

$$255_{10} = 1111\ 1111_2 = FF_{16} \tag{A.4}$$

The base of the number system is typically not included if it is clear which system is being used from its context.

A.2 COMPLEX NUMBERS

Electrical engineers accommodate both amplitude and delay effects produced by systems by using *complex numbers* that have two components called the *real* and *imaginary* parts. Figure A.2 shows a complex number c is a point in the complex plane and can be written as

$$c = a + jb \tag{A.5}$$

where a is the *real part*, b is the *imaginary part*, and $j = \sqrt{-1}$ indicates the imaginary component. Other fields of engineering use i instead of j (as does Matlab). Electrical engineers use j because they reserve i for denoting current.

A complex number can also be written as a vector with the form

$$c = |c|\, e^{j\theta_c} \tag{A.6}$$

where $|c|$ and θ_c are real numbers with $|c|$ being the length of the vector and θ_c the angle with respect to the x axis expressed in radians.

The complex exponential is equal to

$$e^{j\theta} = \cos\theta + j\sin\theta \tag{A.7}$$

When θ increases by $\pm 2k\pi$ radians for integer k, the vector undergoes a complete rotation k times with counter-clockwise rotations for $k > 0$ and clockwise rotations for $k < 0$, and c falls back onto itself as

$$c = |c|\, e^{j(\theta_c \pm 2k\pi)} = |c|\, e^{j\theta_c} \overbrace{e^{\pm j2k\pi}}^{=1} \tag{A.8}$$

These relationships are summarized in the table shown in Figure A.3.

Figure A.2

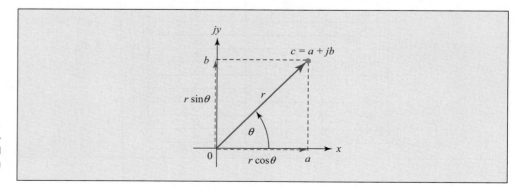

The complex plane displays real-valued numbers along the x axis and imaginary-valued numbers along the jy axis.

$$c = a + jb$$

$$c = re^{j\theta}$$
$$r = \sqrt{a^2 + b^2}$$
$$\theta = \arctan(b/a)$$

$$e^{j\theta} = \cos\theta + j\sin\theta$$

$$c = r\cos\theta + jr\sin\theta$$

$$c = re^{j(\theta \pm 2k\pi)} = re^{j\theta} \text{ for integer } k$$

Figure A.3

Useful relationships for complex numbers.

The *Euler Identities* are given as

$$\cos\theta = \frac{e^{j\theta} + e^{-j\theta}}{2} \tag{A.9}$$

$$\sin\theta = \frac{e^{j\theta} - e^{-j\theta}}{2j}$$

Figure A.4 shows the Euler identities displayed in the complex plane.

Complex Number Arithmetic

Complex numbers are vectors in the complex $(\mathcal{Re}, \mathcal{Im}) = (x, jy)$ plane. Hence, the arithmetic operations are vector operations. Consider complex numbers $a = |a|e^{j\theta_a}$ and $b = |b|e^{j\theta_b}$ that produce complex result $c = |c|e^{j\theta_c}$.

- *Addition.* Adding a to b produces c:

$$c = a + b = \overbrace{(|a|\cos(\theta_a) + |b|\cos(\theta_b))}^{\text{add real parts}} + j\overbrace{(|a|\sin(\theta_a) + |b|\sin(\theta_b))}^{\text{add imaginary parts}} \tag{A.10}$$

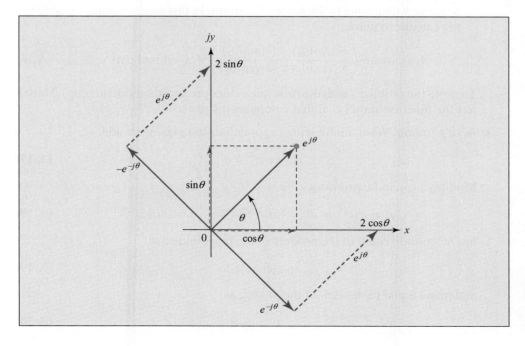

Figure A.4

Euler identities displayed in the complex plane.

■ *Magnitude.* The magnitude of c is computed using Pythagoras' Theorem, as

$$|c| = |a + b| = \sqrt{[|a|\cos(\theta_a) + |b|\cos(\theta_b)]^2 + [|a|\sin(\theta_a) + |b|\sin(\theta_b)]^2} \quad \text{(A.11)}$$

A second method for computing the magnitude of c the complex conjugate, as

$$|c| = \sqrt{c\,c^*} \quad \text{(A.12)}$$

If $c = a + jb$, its complex conjugate is found by negating the imaginary component, producing $c^* = a - jb$. To compute the magnitude let $c = |c|e^{j\theta_c}$. Then $c^* = |c|e^{-j\theta_c}$, and

$$(c) \times (c^*) = \left(|c|e^{j\theta_c}\right) \times \left(|c|e^{-j\theta_c}\right) = |c|^2 \quad \text{(A.13)}$$

■ *Phase.* The phase of a complex number is the angle the vector makes with the positive real axis with positive angles measured in the counter-clockwise direction. The phase equals the arctangent of the ratio of the imaginary to real parts:

$$\theta_c = \arctan\left(\frac{\text{imaginary part}}{\text{real part}}\right) = \arctan\left(\frac{|a|\sin(\theta_a) + |b|\sin(\theta_b)}{|a|\cos(\theta_a) + |b|\cos(\theta_b)}\right) \quad \text{(A.14)}$$

The arctangent function produces values in the range $[-\pi/2, \pi/2)$ or the *principle value*. Be careful with the interpretation of the principle value produced by the arctangent function if the real part is negative.

− A positive result in the division of the arctangent argument occurs when both the real and imaginary parts are positive, producing the correct answer lying in the first quadrant of the (x, jy) plane. A positive division result also occurs when both the real and imaginary parts are negative with c lying in the third quadrant. In this case, $-\pi$ must be added to the calculated phase:

$$\theta_c = \arctan\left(\frac{|a|\sin(\theta_a) + |b|\sin(\theta_b)}{|a|\cos(\theta_a) + |b|\cos(\theta_b)}\right) - \pi \quad \text{if real part} < 0 \quad \text{(A.15)}$$

− A negative result occurs when either the real and imaginary parts are negative. The correct answer lies in the fourth quadrant of the (x, jy) plane in which the imaginary part is negative. A negative result also occurs when the real part is negative with c lying in the second quadrant. In this case, $+\pi$ must be added to the calculated phase:

$$\theta_c = \arctan\left(\frac{|a|\sin(\theta_a) + |b|\sin(\theta_b)}{|a|\cos(\theta_a) + |b|\cos(\theta_b)}\right) + \pi \quad \text{if real part} < 0 \quad \text{(A.16)}$$

To avoid the problems with the principle values produced by arctan(ratio), Matlab has the function atan2(y, x) that computes the phase over $[-\pi, \pi)$.

■ *Multiplication.* When multiplying exponentials the exponents add, as

$$e^{j\alpha} \times e^{j\beta} = e^{j(\alpha+\beta)} \quad \text{(A.17)}$$

Multiplying a and b produces c, so

$$c = |c|e^{j\theta_c} = ab = \left(|a|e^{j\theta_a}\right)\left(|b|e^{j\theta_b}\right) = |a||b|e^{j(\theta_a+\theta_b)} \quad \text{(A.18)}$$

has magnitude equal to the product of the magnitudes, as

$$|c| = |a||b| \quad \text{(A.19)}$$

and phase equal to the sum of the phases, as

$$\theta_c = \theta_a + \theta_b \quad \text{(A.20)}$$

Probability

 ## B.1 INTRODUCTION

Probability is the mathematical tool that helps us describe, interpret, and predict the statistical behavior of random events, specifically those that contribute noise in transmitting data signals. Probability theory is a vast subject that covers many types of randomness. We necessarily limit the topics to those that help us understand how to cope with noise.

Most data-analysis software packages contain a *pseudo-random number generator* (PRNG) that produces pseudo-random numbers. The two most common are random numbers that are uniformly distributed over [0,1) and Gaussian random numbers that are distributed according to the *bell-shaped curve*. This appendix describes each PRNG and their application. The signal-to-noise ratio (SNR) is the important parameter in transmitting data in the presence of noise and is equal to the signal energy over the noise energy. One main result of this appendix shows that deterministic signals add in amplitude and noise signals add in variance. This is the secret to processing signals in the presence of noise for reliable data transmission.

 ## B.2 USING PROBABILITY TO MODEL UNCERTAINTY

Probability is the branch of mathematics that treats randomness. The first practitioners of probability were *probably* gamblers. In the 1600's, a problem involving gambling led two famous mathematicians, Pascal and Fermat, to develop a mathematical theory of probability to describe games of chance. Their method, known as the *relative frequency approach*, is intuitive, still applicable, and used in this text.

Typical data exhibit a known trend that also contains random deviations, collectively called *noise*. Depending on the measurement procedure, these deviations can have one or more causes:

- *Electronic noise*, such as hiss on weak radio stations.

- *Transmission errors* due to computer glitches.

- *Sensor inaccuracies* caused by aging components or weak batteries.

- *Fluctuations caused by dynamic environments*, as when trying to predict hurricane paths.

- *Human mistakes* (typos do happen).

When we call such deviations *random*, we assume that they are not only *unpredictable*, but they also follow laws of probability. Even though the exact values of the deviations cannot be determined, probability provides some information about their *statistics*, such as their average value and average variation.

To apply probability to our data communication problem, the following five elements are helpful.

1. The *experiment* is an activity that produces unpredictable results.

2. The *outcome* indicates the possible results that the experiment can produce.

3. The *random variable* (RV) is a rule that assigns a numerical value to each outcome (because it is easier to deal with the statistics of numbers than more general outcomes). A RV is denoted as an upper-case letter, such as X, which assigns a numerical value to the experimental outcome. In contrast, lower-case letters will refer to non-random (or *deterministic*) values.

    We consider the following two RVs that will be used to generate all other RVs:

    Y describes the RV that is uniformly distributed over the range [0,1).

    G describes the RV that has a normalized Gaussian distribution with zero mean and unit variance.

4. An *event* is a numerical condition of the RV indicated with square brackets. Events include

    - $[X = a]$: the RV X produces realizations (random numbers) equal to the non-random value a.

    - $[X > 0]$: the RV X produces realizations that have positive values.

    - $[a \leq X < b]$: the RV X produces realizations that lie in the interval $[a, b)$.

5. The *probability measure* of an *event*. The probability can take values between 0 and 1 for an experimental outcome.

    - $Prob[.] = 0$ indicates that the event will likely not occur, such as hitting the exact center of a bullseye.

    - $Prob[.] = 1$ indicates that the event will likely always occur.

    - $Prob[.] = 0.5$ indicates that the event is likely to occur half the time an experiment is performed.

 ## B.3 HISTOGRAMS

A *histogram* displays the frequency of occurrence of a set of random numbers that are realized when performing the experiments that produces RV X. Statistical analysis deals with the *average behavior* of RVs, and histograms show the possible values that have occurred.

A histogram is formed by defining a set of bins and then counting the number of times a realization (random number) falls within each bin. Each bin is defined by a numerical interval, such as $[a, b)$. The square bracket denotes a *closed limit* (that is, if $X = a$, that bin is incremented). The parenthesis indicates an *open limit* (that is, if $X = b$, that bin is not incremented, rather the next bin defined by $[b, c)$ is incremented).

Because the values are random, the count in each bin is also random. Our intuition indicates that larger numbers of data points lead to more accurate estimates. Hence, when there are many random numbers, the count within each bin with large counts tend to be accurate and repeatable. The next example illustrates the reliability of bin counts as the quantity of data points increases.

Histogram of uniformly distributed random numbers

The random numbers Y_i generated by the uniform PRNG fall within the interval [0,1). We form bins that cover this interval using equally sized, non-overlapping segments producing 50 bins:

$$\text{bin } 1 \rightarrow [0, 0.02)$$
$$\text{bin } 2 \rightarrow [0.02, 0.04)$$
$$\text{bin } 3 \rightarrow [0.04, 0.06)$$

and so on until

$$\text{bin } 50 \rightarrow [0.98, 1.0)$$

Note that the half-open intervals denoted [,) ensure that each Y value falls into only one bin and all Y values in [0,1) produce counts by falling into some bin.

Figure B.1 shows histograms for

$$Y_i \text{ for } i = 1, 2, \ldots, n$$

when $n = 20$, $n = 1,000$ and $n = 100,000$. When random numbers are said to be uniformly distributed over [0, 1), we expect the bins to yield counts that are approximately equal for the bins that cover [0, 1). Yet, the figure shows that this will happen – and then only approximately – when n is very large. Another way to look at it is that a histogram will be accurate only when the count in a narrow bin is large, say greater than 100.

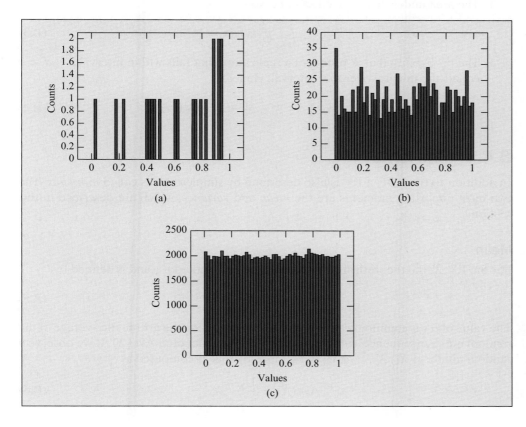

Figure B.1

Histogram with 50 bins displaying the results of uniformly distributed pseudo-random numbers, Y_1 to Y_n. (a) $n = 20$. (b) $n = 1,000$. (c) $n = 100,000$.

> ***Factoid:*** To obtain a representative and repeatable histogram, a large quantity of numbers is needed.

B.4 PROBABILITY DENSITY FUNCTIONS

When n becomes very large (*approaches infinity*) and the bin size is made small, the histogram approximates a function called the *probability density function* (PDF). In the detection problem, the most appropriate event describes the RV producing a value that lies within an interval, such as $[a \leq X < b]$, with the probability of the event being denoted $P[a \leq X < b]$. Indeed, such an interval can be considered to be just one of the bins in a histogram.

The probability density function (PDF), denoted mathematically as $p_X(x)$, is the function describing RV X at point x (a *dummy* variable) that is produced in the limit as both the number of trials n approaches infinity and the bin interval δ centered on x approaches 0. If x is the center of a bin interval and n_x is the number of random numbers that fall into that bin, the PDF is defined by

$$p_X(x) = \lim_{n \to \infty} \lim_{\delta \to 0} \frac{n_x}{n\,\delta} \tag{B.1}$$

When $p_X(x)$ is a function that does not contain discontinuities and the limits are approached correctly (as mathematicians caution), the limit exists and produces the desired PDF.

The PDF has the following three properties.

1. The PDF is greater than or equal to zero for all x. Thus,

$$p_X(x) \geq 0 \text{ for all } x \tag{B.2}$$

2. The area under the curve equals one, so

$$\int_{-\infty}^{\infty} p_X(x)\, dx = 1 \tag{B.3}$$

3. The probability that X produces a realization that falls within interval $a \leq x < b$ equals the integral over that interval. Hence,

$$\text{Prob}\,[a \leq X < b] = \int_{a}^{b} p_X(x)\, dx \tag{B.4}$$

B.4.1 Moments

In addition to the PDF, a RV is also described by simpler values called *moments*. The two most important moments are the *mean* and *variance*, which are described in this section.

Mean

For the RV X, the theoretical value of the mean is denoted μ_X and is defined by

$$\mu_X = \int_{\infty}^{\infty} x\, p_X(x)\, dx \tag{B.5}$$

The value of μ_X is commonly approximated by computing the arithmetic average of the random number sequence, called the *sample average*, denoted Ave(X). If we observe n random numbers $X_1, X_2, \ldots, X_n$, the sample average is computed as

$$\text{Ave}(X) = \frac{1}{n}\sum_{i=1}^{n} X_i \tag{B.6}$$

Ave(X) starts with an upper-case A because, being a function of RVs, it too is a RV whose value depends on the data values actually observed when performing the experiment. The sample average Ave(X) approaches the theoretical non-random value μ_X when n becomes very large, (that is, when we have a lot of data). The larger the n, the closer the value of Ave(X) will likely be to μ_X.

Variance

The other important moment is the variance σ_X^2. Its square root σ_X is the *standard deviation* (SD) that is a measure of the spread of the RVs about the mean value μ_X. The variance is defined by

$$\sigma_X^2 = \int_\infty^\infty (x - \mu_x)^2 \, p_X(x) \, dx \tag{B.7}$$

The sample variance is computed from the square of the individual RVs minus the square of the mean

$$\text{Var}(X) = \text{Ave}(X^2) - (\text{Ave}(X))^2 \tag{B.8}$$

If the RV X has zero mean ($\mu_x = 0$), which is a common occurrence in this book, its sample variance simplifies to

$$\sigma_X^2 = \int_\infty^\infty x^2 \, p_X(x) \, dx \tag{B.9}$$

The sample variance is then

$$\text{Var}(X) = \text{Ave}(X^2) = \frac{1}{n} \sum_{i=1}^n X_i^2 \tag{B.10}$$

As with the sample average and mean, $\text{Var}(X) \to \sigma_X^2$ as n increases.

Factoid: In computing estimates, larger values for n (that is, more data) are likely to produce more accurate results. Alternatively, accurate estimates should not be expected when only small values of n are available.

B.4.2 Uniform PDF

The uniformly distributed RV X that has the PDF:

$$p_X(x) = \begin{cases} \frac{1}{b-a} & \text{for } a \leq x < b \\ 0 & \text{otherwise} \end{cases}$$

Figure B.2 shows the uniform PDF $p_X(x)$ that governs the RV X that produces random numbers that are uniformly distributed over $[a, b)$. The most commonly used form of the uniform RV is Y, which lies in the range $[0, 1)$

The uniform PDF is usually assumed when we know from basic principles that the RV is uniformly distributed between two limits, such as rounding errors.

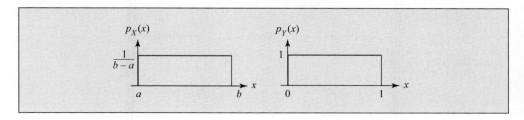

Figure B.2

Uniform PDF $p_X(x)$. Left: General form for RV X lying in the range $[a, b)$. Right: Specific form for the common RV Y in the range $[0, 1)$.

EXAMPLE   B.2

Uniform PDF

The uniform PDF is used to model several important applications in this book:

Uniform pseudorandom number generator (PRNG): Data analysis software packages always include a PRNG that produces random numbers that are uniformly distributed over the interval $[0, 1)$. In this case, $a = 0$ and $b = 1$. We reserve the RV Y to denote the random value produced by this PRNG, and its PDF is denoted $p_Y(x)$.

Rounding error: The residual when rounding a real number to an integer is often modeled as a RV R that is uniformly distributed over $-0.5 \leq x < 0.5$. In this case, $a = -0.5$ and $b = 0.5$. For example, rounding $3.14 \rightarrow 3$ has residual $R = 3 - 3.14 = -0.14$.

Quantization error: When an infinite-precision analog value V_a is quantized to finite-precision V_q having step-size Δ, then the error

$$\epsilon_q = V_q - V_a$$

is modeled as a RV ϵ_q that is uniformly distributed over $0 \leq x < \Delta$. In this case, $a = 0$ and $b = \Delta$. The height is then $1/\Delta$.

Computing the area under the uniform PDF over $-\infty \leq x < \infty$, we need to consider only the interval $[a, b)$ over which $p_X(x)$ is greater than zero or

$$\int_{-\infty}^{\infty} p_X(x)\, dx = \int_a^b \frac{1}{b-a}\, dx = \frac{1}{b-a} x \Big|_a^b = 1 \tag{B.11}$$

The probability of the event $[a \leq x_1 \leq X < x_2 < b]$ is computed as

$$\int_{x_1}^{x_2} p_X(x)\, dx = \int_{x_1}^{x_2} \frac{1}{b-a}\, dx = \frac{1}{b-a} x \Big|_{x_1}^{x_2} = \frac{x_2 - x_1}{b - a} \tag{B.12}$$

EXAMPLE   B.3

Probability of *X* being exactly equal to number

This example shows that $\text{Prob}[X = c] = 0$ for any particular value of c in the range $[a, b)$ by using a limiting argument. Let $\epsilon > 0$ be a small number. The probability that a realization of X produces a value in the range $[c - \epsilon/2, c + \epsilon/2)$ is given by

$$\text{Prob}[c - \epsilon/2 \leq X < c + \epsilon/2] = \int_{c-\epsilon/2}^{c+\epsilon/2} p_X(x)\, dx$$

$$= \int_{c-\epsilon/2}^{c+\epsilon/2} \frac{1}{b-a}\, dx$$

$$= \frac{a + \epsilon/2 - (a - \epsilon/2)}{b - a}$$

$$= \frac{\epsilon}{b - a}$$

In the limit as $\epsilon \rightarrow 0$, the left side equals

$$\lim_{\epsilon \to 0} \text{Prob}[c - \epsilon/2 \leq X < c + \epsilon/2] = \text{Prob}[X = c]$$

and the right side equals

$$\lim_{\epsilon \to 0} \frac{\epsilon}{b - a} = 0$$

Hence, $\text{Prob}[X = c] = 0$. That is, for a RV that is distributed uniformly over an interval, the event that the RV produces a specific value does not make sense. This even, however, does make sense for a RV that takes on one of a set of discrete values, but such a RV is described by the *probability mass function*, described below.

Computing μ_Y and σ_Y^2   ⬛ EXAMPLE   B.4

From the definition of μ_Y, we have

$$\mu_Y = \int_{\infty}^{\infty} x p_Y(x)\, dx$$

$$= \left. \frac{x^2}{2} \right|_0^1 = \frac{1}{2}$$

Let the Y produce the ten ($= n$) realization of $Y_1, Y_2, \ldots, Y_{10}$:

$$0.774, 0.787, 0.699, 0.350, 0.301, 0.562, 0.824, 0.070, 0.918, 0.372$$

Computing the average by summing and dividing by 10, we find the sample average equals

$$\text{Ave}(Y) = \frac{1}{10} \sum_{i=1}^{10} Y_i = 0.566$$

This result is *approximately equal to* 0.5. As n is increased, $\text{Ave}(Y)$ would tend to be closer to μ_Y.

Computing σ_Y^2 from the definition

$$\sigma_Y^2 = \int_{\infty}^{\infty} (x - \mu_x)^2\, p_Y(x)\, dx$$

$$= \int_0^1 (x - 0.5)^2\, dx = \int_0^1 (x^2 - x + 0.25)\, dx$$

$$= \left. \frac{x^3}{3} - \frac{x^2}{2} + 0.25x \right|_0^1 = \frac{1}{12}$$

The SD σ_Y then equals

$$\sigma_Y = \sqrt{\sigma_Y^2} = \sqrt{\frac{1}{12}} = 0.3$$

The random numbers Y_i fall uniformly over the interval $[0,1]$. Using the previous realizations, the sample variance equals

$$\text{Var}(Y) = 0.0696$$

This result is approximately equal to the expected 0.0833. As n is increased , $\text{Var}(Y)$ would tend to be closer to σ_Y^2.

B.4.3 Gaussian PDF

The *Gaussian* probability density function, also known as the familiar *bell-shaped curve*. The Gaussian PDF is important because it occurs in physical processes, such as thermal noise. We model thermal noise N_i as a Gaussian random number for the following reasons.

■ Noise is *random* because we cannot predict the values that will occur and we will never see the exactly same noise sequence again.

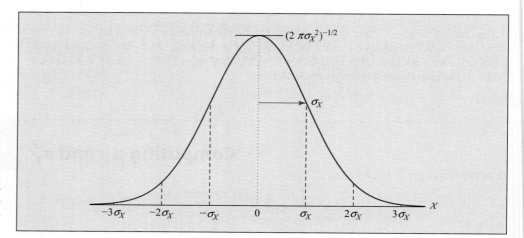

Figure B.3

Gaussian probability density function $p_X(x)$, also known as the bell-shaped curve. The common zero-mean case is shown.

■ Noise is *Gaussian* because each noise sample can be considered to be the resulting sum of many independent contributing random factors.

The bell-shaped curve describing RV X is specified by the mean μ_X and SD σ_X. Figure B.3 shows the general form of a Gaussian PDF that produces RV X is given by

$$p_X(x) = \frac{1}{\sqrt{2\pi}\sigma_x} e^{-\frac{(x - \mu_X)^2}{2\sigma_X^2}} \tag{B.13}$$

One important result for the Gaussian RV X is that 95% of the realizations fall in the range

$$[\mu_x - 1.96\sigma_X, \mu_x + 1.96\sigma_X]$$

This result is used for determining the ability of detectors to differentiate signal from noise.

Our main goal for the Gaussian PDF is to describe thermal noise, denoted by the RV N, which has zero mean $\mu_N = 0$ and SD σ_N, with PDF

$$p_N(x) = \frac{1}{\sqrt{2\pi}\,\sigma_N} e^{-\frac{x^2}{2\sigma_N^2}} \tag{B.14}$$

 ## B.5 PROBABILITY MASS FUNCTION

The previous PDF describes a RV that produce a realization that can have any value within an interval, like [0,1) or $(-\infty, \infty)$. When the RV X produces values that are members of a set of discrete values which is denoted as $\{x_i\}$ we need an alternate description. This case occurs in communication problems when a set of specific values is transmitted. Then X can take on a particular value x_i with probability greater than zero. The assignment of these probability values is described by the *probability mass function* (PMF), which is denoted $p_X(x_i)$ and defined by

$$p_X(x_i) = \text{Prob}[X = x_i] \tag{B.15}$$

The PMF is a set of values defined at a set of discrete values defined by x_i, for $i = 1, 2, 3, \ldots, n$, as shown in Figure B.4.

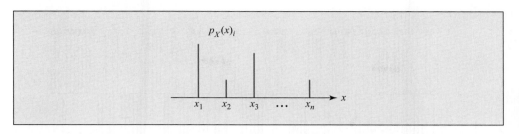

Probability mass function $p_X(x_i)$

The coin-flip experiment   EXAMPLE   B.5

A coin is flipped and produces either a head or tail outcome. Consider assigning the RV X the value $+1$ when the flip outcome is a head and -1 when it is a tail. Intuition tells us that a fair coin will produce each of the two possible outcomes (-1 and $+1$) with equal probability. Hence, the PMF is given by

$$p_X(1) = 0.5$$

$$p_X(-1) = 0.5$$

$$p_X(x) = 0 \text{ for } x \neq -1, 1$$

Figure B.5 shows this PMF.

The PMF has the following three properties.

1. $p_X(x_i) \geq 0$ for all x_i.

2. Since one of the allowable values must occur at each trial, it follows that

$$\sum_{x_i} p_X(x_i) = 1 \tag{B.16}$$

3. Given two non-random constants a and b, where $a \leq b$, then the probability of the event that $a \leq X < b$ equals

$$\text{Prob}[a \leq X < b] = \sum_{a \leq x_i < b} p_X(x_i) \tag{B.17}$$

where the sum is computed for all x_i such that $a \leq x_i < b$. Note that, if $x_i = b$ is an allowed value of X, the events $[a \leq X < b]$ and $[a \leq X \leq b]$ have different probabilities.

Let us proceed as before and display the results of multiple trials of the coin-flip experiment in a histogram. A typical 100-trial series is shown in Figure B.6a. The observed outcomes indicate that runs of six or more of the same outcome are not that unusual. The 100-bin histogram appears in Figure B.6b. The histogram shows a deviation in the actual counts from the expected number of 50 occurrences of each, which is a "statistical fluctuation" due to the small number of trials.

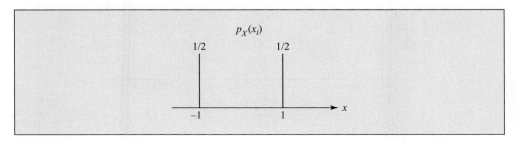

Probability mass function $p_X(x_i)$ for the coin flip experiment.

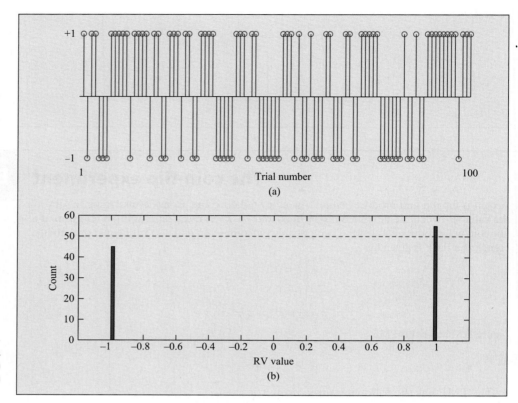

Figure B.6

Coin flip experiment using 100 trials. (a) Observed RV values. (b) 100-bin histogram. Dashed line indicates a count of 50.

Increasing the number of coin-flip trials to a million produces the 100-bin histogram shown in Figure B.7a. It shows that each possible value occurs approximately 1/2 of the time. As before, we reduce the bin size by a factor of ten to produce the 1000-bin histogram shown in Figure B.7b. The narrower bin produces a histogram with values concentrated about the two possible integer values of -1 and $+1$.

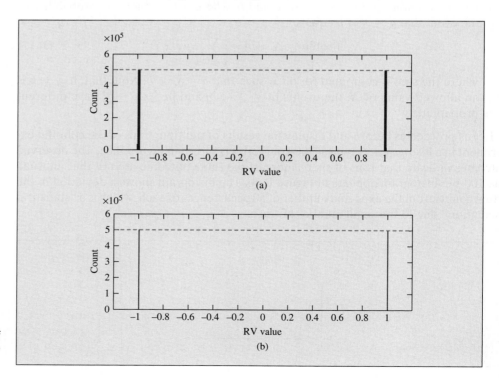

Figure B.7

Histograms from one million trials of coin-flip experiment. (a) 100-bin histogram. (b)1000-bin histogram.

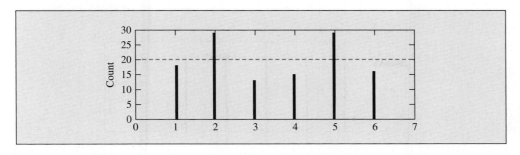

Continuing to increase the number of trials (n_t) to infinity and to reduce the bin size to zero, the histogram form converges to the PMF defined at each possible discrete value of X that can occur. For the coin-flip experiment, the PMF has non-zero values only for $x = -1$ and $x = +1$. If the number of times the random variable is observed to be -1 ($X = -1$) is denoted by n_{-1} and the number of times $X = +1$ by n_{+1} in n_t trials, the PMF is computed at these two points as

$$\text{Prob}[X = -1] = \lim_{n_t \to \infty} \frac{n_{-1}}{n_t} \quad \left(= \frac{1}{2} \text{ for a fair coin} \right) \tag{B.18}$$

and

$$\text{Prob}[X = +1] = \lim_{n_t \to \infty} \frac{n_{+1}}{n_t} \quad \left(= \frac{1}{2} \text{ for a fair coin} \right) \tag{B.19}$$

The PMF values at all other values of x equal zero.

PMF describing a die-toss experiment   EXAMPLE   B.6

In a die-toss experiment, a die is tossed and the outcome is the top face. The RV X is the count of dots on the top face. Then, X can only exhibit the integer values from 1 to 6. Multiple trials of the die-toss experiment produces the data displayed in a histogram. The 100-bin histogram produced from observation is a typical 120-trial experiment is shown in Figure B.8. Intuition tells us that each of the possible outcomes (integer values one through six) is equally likely to occur. The histogram shows a variation in the actual counts from the expected number of $20 (= 120/6)$ occurrences. These are statistical fluctuations due to the small number of trials.

Increasing the number of trials to a 1.2 million produces the 100-bin histogram shown in Figure B.9a. It shows that each possible value occurs approximately 1/6 of the time. As before, we reduce the bin size by a factor of 10 to produce the histogram shown in Figure B.9b. This narrower bin produces a histogram with values concentrated about the possible integer values of one through six. Continuing to increase the number of trials to infinity and to reduce the bin size to zero, the histogram converges to the PMF, which equals 1/6 at the integer values of one through six.

B.6 PSEUDO-RANDOM NUMBER GENERATORS

Most data analysis software packages contain a *pseudo-random number generator* (PRNG) that generates pseudo-random numbers. A PRNG replaces the actual experiment of flipping a coin or tossing a die with a computer program, called a *simulation*. Because a computer program follows a fixed set of instructions, known as the *algorithm*, the results cannot be truly random. The random numbers a PRNG produces, however, do exhibit reasonable *unpredictability*. This section describes the two most common PRNGs.

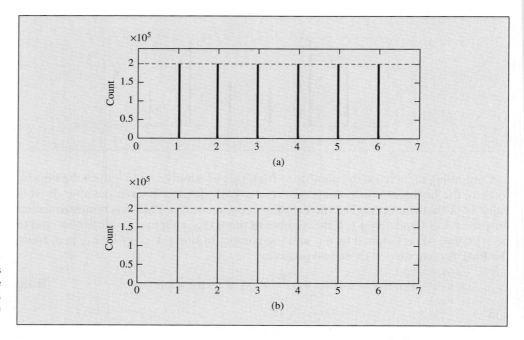

Figure B.9

Histograms from 1.2 million trials of die-toss experiment. Dashed line indicates count equal to 200,000. (a) 100-bin histogram. (b) 1000-bin histogram.

B.6.1 Uniform PRNG

The RV Y produces realizations that are uniformly distributed over the range $[0,1)$. The uniform PRNG generates random numbers that are uniformly distributed over the same range.

EXAMPLE B.7

Simulating a fair coin flip

Let us examine how a computer can simulate the flip of a fair coin. We can simulate the behavior of a coin flip by using the PRNG Y_i values lying in the $[0,1)$ interval by dividing the interval into two equal sub-intervals, and assigning a head outcome to the first sub-interval and a tail to the second. Although any partitioning that produces two equal lengths, Figure B.10 shows the simplest partition of $[0,1)$ into two equal sub-intervals

$$[0, 1) \rightarrow \overbrace{[0, 0.5)}^{\text{head}} \overbrace{[0.5, 1)}^{\text{tail}}$$

This partition into two half-open intervals has the following desirable properties.

- It covers the entire $[0,1)$ interval. Hence, any random number will fall into *some* interval. This prevents the case when a possible Y value does not fall into any interval.
- It covers the entire $[0,1)$ interval with no overlaps. Hence, every possible value falls into *only one* of the sub-intervals. This prevents any ambiguity when $Y = 0.5$.

Figure B.10

Simulating a fair coin flip with the uniform PRNG.

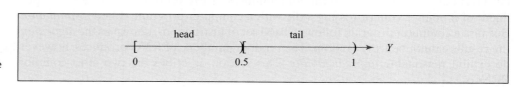

Uniform PRNG   EXAMPLE   B.8

Let us illustrate the four elements to model a problem using probability by observing the values produced by the Matlab uniform PRNG *rand()*.

1. The experiment is making a call to the Matlab PRNG *rand()* that produces a random number.
2. The outcome is random value in the range [0,1).
3. The RV Y corresponds to a numerical value that is produced.
4. To determine the PDF $p_Y(x)$, we perform the experiment n times, producing RVs

$$Y_1, Y_2, Y_3, Y_4, Y_5, \ldots, Y_n$$

Figure B.11 shows typical values for $n = 100$. Note all of the values fall within the range [0, 1), as expected.

To generate histograms, we set the bin size very small ($\delta = 0.01$) and perform many experiments by making n large. Figure B.12 shows histograms that are observed for $n = 100$ and $n = 10^6$.

Note that the more accurate histogram generated with $n = 10^6$ random numbers exhibits counts that are approximately equal over the interval [0,1). A finite number of randomly generated numbers never produce exactly equal bin counts (except by extremely rare accident). Because the bin size $\delta > 0$ and n is finite, the value of $p_Y(x)$ is approximated as

$$p_Y(x) \approx \frac{n_{a_i}}{n\delta}$$

For $n = 100$, the counts vary from 0 to 5 for bins lying within [0,1), producing unreliable estimates for $0 \le x < 1$ ranging from $p_Y(x = 0.02) = 0$ to

$$p_Y(x = 3.8) \approx \frac{5}{100 \times 0.01} = 5$$

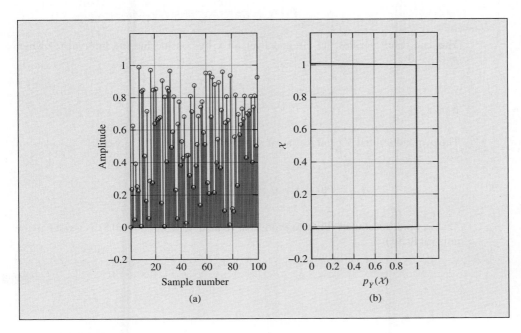

Figure B.11

One hundred values of the uniformly distributed random numbers produced by the Matlab PRNG *rand()*.

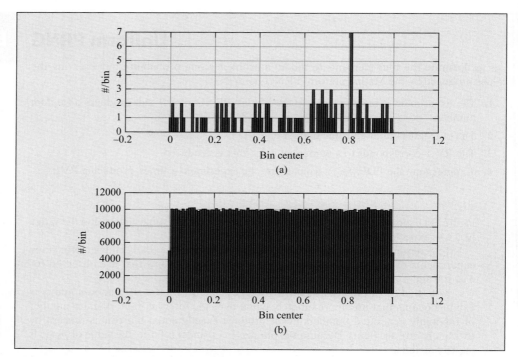

Figure B.12

Histograms for the Matlab PRNG rand() with $\delta = 0.01$. Top: $n = 100$. Bottom $n = 10^6$.

For $n = 10^6$, all counts are approximately 10^4 for bins lying within $[0,1)$, producing constant estimates for $0 \leq x < 1$ equal to

$$p_Y(x) \approx \frac{10^4}{10^6 \times 0.01} = 1$$

This last result verifies the integral over all x by considering the interval for which $p_Y(x) > 0$:

$$\int_{-\infty}^{\infty} p_Y(x)\, dx = \int_0^1 \overbrace{p_Y(x)}^{=1}\, dx = 1$$

The probability that Y produces a value in the interval $[0,0.5)$ is given by

$$P[0 \leq Y < 0.5] = \int_0^{0.5} \overbrace{p_Y(x)}^{=1}\, dx = 0.5$$

That is, rand() is equally likely to produce a value in the interval $[0,0.5)$ as a value in the interval $[0.5,1)$.

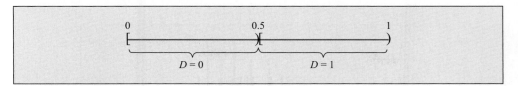

Figure B.13

Using the uniform PRNG to simulate binary data.

Generating random binary data | EXAMPLE B.9

As a model of generating random binary data, use the uniform PRNG that generates pseudo-random numbers that are uniformly distributed in the range $[0, 1)$. This interval is divided into two equal non-overlapping parts with one part corresponding to a $D = 0$ and the other a $D = 1$. The arbitrary (but convenient) division into $[0, 0.5)$ and $[0.5, 1)$ is shown in Figure B.13. The random number produced by the uniform PRNG being less than 0.5 makes $D = 0$, otherwise $D = 1$.

B.6.2 Gaussian PRNG

Most signal processing software contains a PRNG that produces the *standardized* Gaussian RV G having with zero mean ($\mu_G = 0$) and unit variance ($\sigma_G^2 = 1$). The corresponding Gaussian PDF is

$$p_G(x) = \frac{1}{\sqrt{2\pi}} e^{-\frac{x^2}{2}} \tag{B.20}$$

Figure B.14 shows typical values produced by a Gaussian PRNG for $n_x = 100$. The samples do not exhibit discrete values but vary continuously. Hence, a Gaussian PDF is appropriate. We note that almost all the values fall within $[-3, 3]$. This verifies the

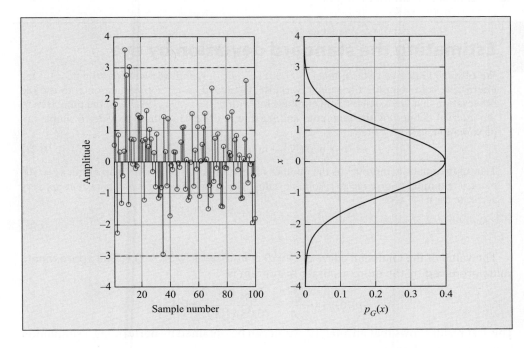

Figure B.14

One hundred values of the standardized RV G_i, $\mu_G = 0$ and $\sigma_G = 1$, produced by the Matlab PRNG *randn()* and the Gaussian curve. Note that 99.7% of the random numbers fall between $\pm 3\sigma_G$.

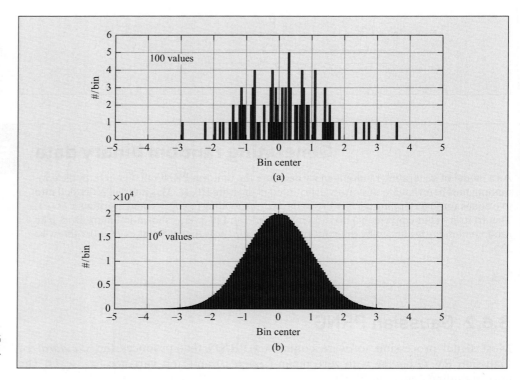

Figure B.15

Histograms for the Matlab PRNG *randn()* with $\delta = 0.05$. (a) $n = 100$. (b) $n = 10^6$.

observation that 99.7% of the Gaussian values fall within $[-3\sigma, 3\sigma]$ and 95.4% of the Gaussian values fall within $[-2\sigma, 2\sigma]$.

To generate a histogram, we set the bin size very small ($\delta = 0.05$) and perform many experiments. Figure B.15 shows histograms for $n = 100$ and $n = 10^6$.

Note that the more accurate histogram generated with $n = 10^6$ random numbers exhibits the familiar bell-shaped curve. Note that Gaussian random numbers almost all fall within $[-3\sigma_G, 3\sigma_G] = [-3, 3]$. This feature is useful for estimating the SD from a set of Gaussian random numbers.

EXAMPLE B.10   Estimating the standard deviation by eye

We observe Gaussian data N_i for $i = 0, 1, 2, \ldots, n_x - 1$, and we want to estimate σ_N^2. One alternative is to compute the sample variance. A second easier approximation is to use the observation that Gaussian random numbers fall within $[-3\sigma_N, 3\sigma_N]$ 99.7% of the time. Hence, the interval defined by the minimum and maximum observed values provides a simple approximation, as

$$\max(N_i) - \min(N_i) \approx 6\sigma_N \qquad \text{(B.21)}$$

This approximation improves as the number of random data points n increases with $n_x = 100$ being a minimum to provide a reasonable value. The approximation is *ball park*, but not very accurate for $n_x < 100$.

The value of the Gaussian curve at $G = 0$ is $p_G(0) = 1/\sqrt{2\pi} = 0.4$. The approximate value produced by the more accurate histogram is

$$p_G(0) \approx \frac{n_0}{n_x \delta} = \frac{2 \times 10^4}{(0.05)(10^6)} = 0.4$$

This last result shows that an analytic result can be obtained by performing a simulation (when n_x is very large).

B.7 RANDOM NUMBER ARITHMETIC

A processor manipulates detected data $X_i = s_i + N_i$ for $i = 0, 1, 2, \ldots, n_x - 1$ containing random noise by computing the random output value V with

$$V = \sum_{i=0}^{n_x-1} s_i X_i = \overbrace{\sum_{i=0}^{n_x-1} s_i^2}^{\text{non-random}} + \overbrace{\sum_{i=0}^{n_x-1} s_i N_i}^{\text{random}} \tag{B.22}$$

The value V is a RV that contains both a non-random component and a random component that is modeled as a Gaussian random variable. The goal here is to determine the mean of V (μ_V), its variance σ_V^2, and its PDF $p_V(x)$.

This processor operation involves three arithmetic manipulations that we consider separately.

1. Adding a non-random constant value to a RV.

2. Multiplying a RV by a non-random constant.

3. Adding independent RVs.

B.7.1 Adding a Constant to a RV

Consider adding a non-random constant to a random variable as shown in Figure B.16. Let G_i be a Gaussian RV and add a non-random constant a to each realization to produce RV $Y1_i$. Then

$$Y1_i = a + G_i \tag{B.23}$$

The mean of $Y1_i$ has been shifted from 0 to a. Hence,

$$\mu_{Y1} = a + \overbrace{\mu_G}^{=0} = a \tag{B.24}$$

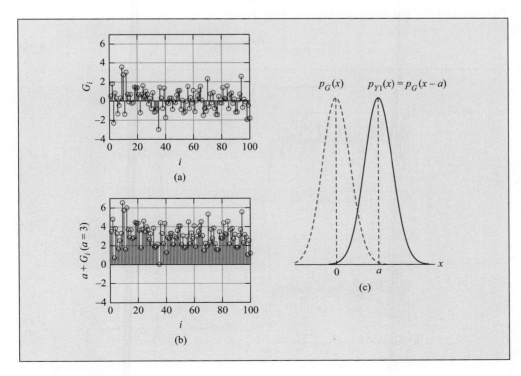

Figure B.16

Probability density functions $p_G(x)$ and $p_{Y1}(x)$ when $Y1_i = a + G_i$. The results for $a = 3$ are shown.

The form of $p_{Y1}(x)$ is the same as that of $p_G(x)$, except for a shift. Hence, the variance, which is a measure of the deviation from the mean, is unchanged:

$$\sigma_{Y1}^2 = \sigma_G^2 \tag{B.25}$$

The form of the PDF of $Y1_i$ is then

$$p_{Y1}(x) = p_G(x - a) = \frac{1}{\sqrt{2\pi}\sigma_G} e^{-\frac{(x-a)^2}{2\sigma_G^2}} \tag{B.26}$$

EXAMPLE B.11 — Non-random signal plus Gaussian random noise

The signal s is a non-random constant. Gaussian noise N having mean $\mu_N = 0$ and variance σ_N^2 is added to s to produce the detected signal X, as

$$X = s + N$$

The detected signal X is random (because of N) and has mean, variance, and PDF computed as

$$\mu_X = s + \mu_N = s$$

$$\sigma_X^2 = \sigma_N^2$$

The PDF of X is Gaussian (because N is Gaussian) with form specified by its mean and variance as

$$p_X(x) = \frac{1}{\sqrt{2\pi}\sigma_N} e^{-\frac{(x-s)^2}{2\sigma_N^2}}$$

B.7.2 Multiplying a RV by a Constant

Consider multiplying a random number by a constant a, as shown in Figure B.17. Let G_i be a Gaussian RV that is multiplied by the constant a to produce RV $Y2_i$. Then

$$Y2_i = aG_i \tag{B.27}$$

The mean of $Y1_i$ has been not been shifted, hence

$$\mu_{Y2} = a \overbrace{\mu_G}^{=0} = 0 \tag{B.28}$$

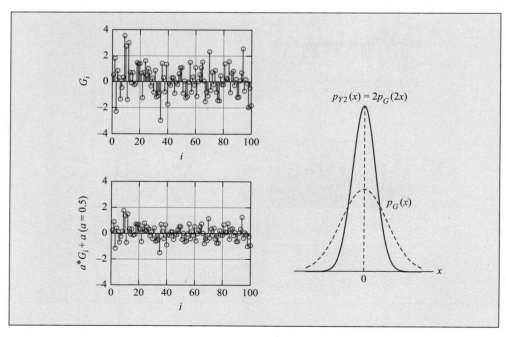

Figure B.17

Probability density function $p_G(x)$ and $p_{Y2}(x)$ when $Y2_i = 0.5G_i$.

The form of $p_{Y2}(x)$ is a stretched (when $a > 1$) or contracted (when $a < 1$) version of $p_G(x)$. To maintain the area under the curve to equal 1 the amplitude of $p_{Y2}(x)$ is scaled by $1/a$, which is a reduced (for $a > 1$) or increased (for $a < 1$) version of $p_G(x)$. The variance, which is a measure of the deviation from the mean, is computed as

$$\sigma_{Y2}^2 = a^2 \sigma_G^2 \tag{B.29}$$

The PDF of $Y2_i$ has a Gaussian form that is specified by its mean and variance as

$$p_{Y2}(x) = \frac{1}{\sqrt{2\pi}\, a\sigma_G}\, e^{-\frac{x^2}{2(a\sigma_G)^2}} \tag{B.30}$$

Multiplying a detected sample value by a non-random constant   EXAMPLE B.12

The detected signal is

$$X = s + N$$

The noise N is a Gaussian RV with mean $\mu_G = 0$ and variance σ_N^2. If we multiply X by 2, we get

$$2X = 2s + 2N$$

The mean, variance, and PDF of $2X$ equal

$$\mu_{2X} = 2s + 2\mu_N = 2s$$

$$\sigma_{2X}^2 = 4\sigma_X^2 = 4\sigma_N^2$$

$$p_{2X}(x) = \frac{1}{\sqrt{2\pi}\,(2\sigma_N)} e^{-\frac{(x-2s)^2}{2(4\sigma_N^2)}} = \frac{1}{\sqrt{8\pi}\,\sigma_N} e^{-\frac{(x-2s)^2}{8\sigma_N^2}}$$

B.7.3 Adding Independent RVs

The RVs X_1 and X_2 are said to be *independent* if knowing the value produced by X_1 tells us nothing about the value produced by X_2. Each time a PRNG produces two random numbers (say X_1 and X_2), the values are assumed to be independent because neither value can be easily predicted from the other. Independence simplifies the calculations involving sums of RVs.

Central limit theorem   EXAMPLE B.13

The *central limit theorem* (CLT) states that, under conditions that typically hold in practice, the sum of n independent RVs approaches a Gaussian distribution as n becomes large, *no matter what the distributions of the RVs themselves*. We apply the CLT to argue that thermal noise has a Gaussian distribution.

Consider a resistor R at room temperature that contains a huge number of thermally excited electrons moving inside it. If there are n electrons and I_k is the random current due to the motion of the kth electron, the total instantaneous (random) current I is the sum of the individual currents. The random instantaneous voltage V measured across the terminals of resistor R is then given by

$$V = RI = R\sum_{k=1}^{n} I_k$$

The current I is the sum of many electrons that move randomly and independently. Hence, the central limit theorem indicates that the distribution of V approaches the Gaussian distribution with the variance σ_N^2 being the noise power. *No matter what the distribution of the original RVs (the individual currents I_k), the random voltage V can be accurately described by the Gaussian distribution by the central limit theorem.*

> **Factoid:** Your SAT score resulted from the averaging effect of many different and mostly independent influences, such as your study habits, memory, stress level, and so on. The CLT is the reason that the histograms of SAT scores typically exhibit a bell-shaped curve.

EXAMPLE B.14

Sum of two independent RVs

Let $X1$ and $X2$ be independent Gaussian RVs having zero means ($\mu_{X1} = \mu_{X2} = 0$), variances σ_{X1}^2 and σ_{X2}^2, and $Y3$ equal their sum. Thus,

$$Y3 = X1 + X2 \tag{B.31}$$

The mean of $Y3$ is the sum of the means of $X1$ and $X2$, as

$$\mu_{Y3} = \mu_{X1} + \mu_{X2} = 0 \tag{B.32}$$

Because $X1$ and $X2$ are independent RVs, the variance of $Y3$ equals the sum of their variances, as

$$\sigma_{Y3}^2 = \sigma_{X1}^2 + \sigma_{X2}^2 \tag{B.33}$$

Because $X1$ and $X2$ are Gaussian, the PDF of $Y3$ is also Gaussian and specified by its mean and variance values, as

$$p_{Y3}(x) = \frac{1}{\sqrt{2\pi\sigma_{Y3}^2}} e^{-\frac{x^2}{2\sigma_{Y3}^2}} = \frac{1}{\sqrt{2\pi(\sigma_{X1}^2 + \sigma_{X2}^2)}} e^{-\frac{x^2}{2(\sigma_{X1}^2 + \sigma_{X2}^2)}} \tag{B.34}$$

Note that the σ_{Y3} was brought into the square root and became σ_{Y3}^2. The PDFs of $X1$, $X2$, and $Y3$ are shown in Figure B.18.

Figure B.18

The Gaussian PDFs $p_{X1}(x)$, $p_{X2}(x)$, and $p_{Y3}(x)$ when $Y3 = X1 + X2$, where X1 and X2 are independent RVs.

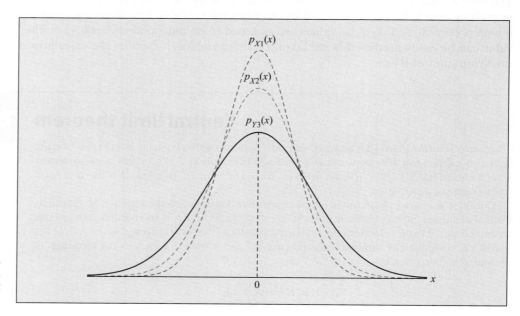

ASCII Code

ASCII (American Standard Code for Information Interchange)

ASCII character	Keyboard	Decimal	Binary pattern 7 6 5 4 3 2 1 0	ASCII character	Keyboard	Decimal	Binary pattern 7 6 5 4 3 2 1 0
NULL	ctrl @	0	0 0 0 0 0 0 0 0	SP	space	32	0 0 1 0 0 0 0 0
SOH	ctrl A	1	0 0 0 0 0 0 0 1	!	!	33	0 0 1 0 0 0 0 1
STX	ctrl B	2	0 0 0 0 0 0 1 0	``	``	34	0 0 1 0 0 0 1 0
ETX	ctrl C	3	0 0 0 0 0 0 1 1	#	#	35	0 0 1 0 0 0 1 1
EOF	ctrl D	4	0 0 0 0 0 1 0 0	$	$	36	0 0 1 0 0 1 0 0
ENQ	ctrl E	5	0 0 0 0 0 1 0 1	%	%	37	0 0 1 0 0 1 0 1
ACK	ctrl F	6	0 0 0 0 0 1 1 0	&	&	38	0 0 1 0 0 1 1 0
BEL	ctrl G	7	0 0 0 0 0 1 1 1	'	'	39	0 0 1 0 0 1 1 1
BS	ctrl H	8	0 0 0 0 1 0 0 0	(	(	40	0 0 1 0 1 0 0 0
HT	ctrl I	9	0 0 0 0 1 0 0 1	)	)	41	0 0 1 0 1 0 0 1
LF	ctrl J	10	0 0 0 0 1 0 1 0	*	*	42	0 0 1 0 1 0 1 0
VT	ctrl K	11	0 0 0 0 1 0 1 1	+	+	43	0 0 1 0 1 0 1 1
FF	ctrl L	12	0 0 0 0 1 1 0 0	,	,	44	0 0 1 0 1 1 0 0
CR	ctrl M	13	0 0 0 0 1 1 0 1	-	-	45	0 0 1 0 1 1 0 1
SO	ctrl N	14	0 0 0 0 1 1 1 0	.	.	46	0 0 1 0 1 1 1 0
SI	ctrl O	15	0 0 0 0 1 1 1 1	/	/	47	0 0 1 0 1 1 1 1
DLE	ctrl P	16	0 0 0 1 0 0 0 0	0	0	48	0 0 1 1 0 0 0 0
DC1	ctrl Q	17	0 0 0 1 0 0 0 1	1	1	49	0 0 1 1 0 0 0 1
DC2	ctrl R	18	0 0 0 1 0 0 1 0	2	2	50	0 0 1 1 0 0 1 0
DC3	ctrl S	19	0 0 0 1 0 0 1 1	3	3	51	0 0 1 1 0 0 1 1
DC4	ctrl T	20	0 0 0 1 0 1 0 0	4	4	52	0 0 1 1 0 1 0 0
NAK	ctrl U	21	0 0 0 1 0 1 0 1	5	5	53	0 0 1 1 0 1 0 1
SYN	ctrl V	22	0 0 0 1 0 1 1 0	6	6	54	0 0 1 1 0 1 1 0
ETB	ctrl W	23	0 0 0 1 0 1 1 1	7	7	55	0 0 1 1 0 1 1 1
CAN	ctrl X	24	0 0 0 1 1 0 0 0	8	8	56	0 0 1 1 1 0 0 0
EM	ctrl Y	25	0 0 0 1 1 0 0 1	9	9	57	0 0 1 1 1 0 0 1
SUB	ctrl Z	26	0 0 0 1 1 0 1 0	:	:	58	0 0 1 1 1 0 1 0
ESC	ctrl [	27	0 0 0 1 1 0 1 1	;	;	59	0 0 1 1 1 0 1 1
FS	ctrl /	28	0 0 0 1 1 1 0 0	<	<	60	0 0 1 1 1 1 0 0
GS	ctr]	29	0 0 0 1 1 1 0 1	=	=	61	0 0 1 1 1 1 0 1
RS	ctrl ~	30	0 0 0 1 1 1 1 0	>	>	62	0 0 1 1 1 1 1 0
US	ctrl	31	0 0 0 1 1 1 1 1	?	?	63	0 0 1 1 1 1 1 1

α—(Greek alpha) range factor or parameter in a PRNG.

β—(Greek beta) parameter in a PRNG.

δ—(Greek lower-case delta) histogram bin width.

Δ—(Greek capital delta) quantizer step size.

Δ_{PWM}—voltage step size produced by PWM waveform.

μ—(Greek mu) one millionth.

Ω—(Greek sigma) Ohm, unit of resistance.

σ—(Greek sigma) statistical standard deviation (SD).

σ^2—statistical variance.

σ_N—noise SD.

σ_N^2—noise variance.

θ—(Greek theta) angle with respect to x axis in complex plane.

A—Ampere (Amp), unit of electrical current.

ACK—acknowledge packet sent by destination in TCP/IP.

ADC—analog-to-digital converter.

AR—autorecursive, as in digital filter.

ARMA—autorecursive-moving average, as in digital filter.

ASCII—American Standards Committee for Information Interchange.

Asynchronous transmission—randomly-time occurring data transfer.

ATM—asynchronous transfer mode, or automatic teller machine.

b—bit (binary digit).

B—byte (8-bit data unit).

$\mathcal{B}$—bandwidth in channel capacity equation.

Bandwidth—interval of frequencies present in a waveform, or measure of frequency transmission quality of a channel.

BCD—binary coded decimal, 4-bit code words used for integers 0-9.

Bit—binary digit, a 0 or 1.

Bps—bytes per second.

bps—bits per second.

Byte—8-bit data unit.

C—capacitor or Coulomb (unit of electrical charge).

$\mathcal{C}$—channel capacity.

CAPCHA—machine-unreadable figure.

CD—compact disk used for storing data in a digital format.

CDMA—code division multiple access.

Character—asynchronous transmission waveform enclosing a code word.

CLR—clear input of T-FF.

Code word—set of bits that represent a symbol.

Combinational logic—logic whose output depends only on the current input values.

Combinatorial logic—another term for combinational logic.

CPU—central processing unit in a computer.

CRC—cyclic redundancy check—Internet code word used for data packet error correction.

CSD—check sum digit, used for error detection.

$\mathcal{D}_S$—source data rate.

$\mathcal{D}_V$—video source data rate.

DAC—digital-to-analog converter.

data packet—collection of code words transmitted as a unit.

dB—decibel, equal to ten times the base-10 logarithm of a power ratio.

Dibit—data unit formed by two bits.

Diode—electronic device that passes current in only one direction.

DFT—discrete Fourier transform.

$\mathcal{E}$—electrical energy.

$\mathcal{E}_{min}$—minimum signal energy for reliable data communication.

$\mathcal{E}_s$—signal energy.

EEPROM—electrically erasable programmable read-only memory, used to store digital data in thumb drives and smartphones.

Encryption—data coding that allows access by only intended recipients.

EOL—end-of-line character.

Ethernet—communication channel over dedicated cable.

ExOR—exclusive-OR logic gate.

F—Farad, unit of capacitance.

f_a—alias frequency, caused by sampling waveforms too coarsely in time.

FDMA—frequency division multiple access—form of orthogonal signals.

FET—field-effect transistor.

FFT—fast Fourier transform—fast algorithm that evaluates DFT.

F_{PWM}—fundamental frequency of a PWM waveform.

f_o—frequency of analog sinusoidal waveform.

f_s—sampling frequency.

f_x—filter crossover frequency at which real and imaginary impedance components are equal.

giga-—prefix meaning one billion, or 10^9.

GHz—gigaHertz, or 10^9 Hz.

GPS—global positioning system.

$\mathcal{H}_S$—source entropy in units of bits/symbol.

$\hat{\mathcal{H}}_S$—effective entropy computed using effective probabilities.

Hz—Hertz, or cycles per second.

i—index of time-domain samples.

I/O—input/output, or a connection that can both receive and transmit data.

IC—integrated circuit.

Im—imaginary part of an imaginary number.

Impedance—generalized resistance that includes frequency dependent effects caused by capacitors and inductors.

IR—infrared.

JPEG—Joint Photographers Experts Group.

jpg—image file using JPEG standard.

k—index of frequency-domain samples.

kB—kilobyte, or $2^{10} = 1,024$ bytes.

kbps—thousand bits per second.

KCL—Kirchoff's current law.

kilo-—prefix meaning one thousand.

kHz—kiloHertz, or 10^3 Hz.

km—kilometer, or 10^3 m.

KVL—Kirchoff's voltage law.

L—inductor.

LAN—local-area network.

LCD—liquid crystal display.

LED—light-emitting diode.

LRC—longitudinal redundancy check—code word appended data packet for error correction.

LSB—least significant bit, usually the rightmost bit in a sequence.

LTE—long term evolution.

m—meter or number of symbols in a source vocabulary.

MA—moving average, as in digital filter.

magnitude—absolute value of amplitude.

MB—Megabyte, or 2^{20} bytes.

Mbps—Megabits per second (a data rate).

MCU—microcontroller unit.

mega-—prefix meaning one million, as in 1 megaHertz, or 1 MHz.

MHz—megaHertz, or 10^6 Hz.

micro-—prefix meaning one one-millionth, abbreviated μ, as in 1 microsecond, 1 μs.

milli-—prefix meaning one one-thousandth, or 10^{-3}.

mm—millimeter, or 10^{-3} m.

MPEG—Motion Picture Experts Group.

ms—milliseconds.

MSB—most significant bit, usually the leftmost bit in a code word.

MUX—digital multiplexer.

mV—millivolt, or 10^{-3} V.

mW—milliwatt, or 10^{-3} W.

n—number of countable objects.

n_d—number of non-zero samples in TDMA window of size n_x.

n_t—number of trials in a simulation.

n_T—number of symbols produced by a source.

n_x—number of sample points in a time sequence.

N—random noise value.

n_{steps}—number of steps in a quantizer.

nano-—prefix meaning one one-thousand-millionth, or 10^{-9}.

ns—nanosecond, or 10^{-9} s.

outcome—the result of performing an experiment.

op amp—operational amplifier.

$\mathcal{P}$—electrical power.

$P[X]$—probability of symbol X.

$P_e[X]$—effective probability of symbol X determined from its relative frequency of occurrence.

Parity—forming an even or odd count of 1s in a data character included for error detection.

PDF—probability density function.

pico-—prefix meaning one one-million-millionth, abbreviated p, as in 1 picosecond, 1 ps.

pixel—picture element.

PMF—probability mass function.

PRNG—pseudo-random number generator.

protocol—set of rules for data transfer.

pseudo-random numbers—random numbers generated by an algorithm.

PWM—pulse width modulation.

Q—flip-flop output.

QR symbol—quick response symbol.

Quantizer—device that converts continuous values into discrete values.

R—resistor.

R—signal transmission rate measured in symbols/second.

RAM—random access memory.

Random—unpredictable and obeying the laws of probability.

Re—real part of an imaginary number.

Redundant—computed from available data.

r_o—range at which a measurement is taken (calibration range).

ROM—read-only memory.

RMS—root-mean-square value (square root of the average of the squared values).

RS-FF—reset-set flip-flop.

RV—random variable, a rule that assigns a numerical value to an outcome.

s—second.

s_i—time signal value (non-random).

SEA—single error assumption.

Sequential logic—logic whose output depends on the current inputs and past outputs.

Signal—informational time waveform.

S_k—frequency transform value.

SNR—signal-to-noise ratio.

Spectrum—magnitude display of waveform frequency components.

Start bit—first bit in asynchronous transmission character that indicates its beginning.

Stop bit—last bit or bits in asynchronous transmission that separate characters.

Symbol—informational unit generated by a source.

T_B—time interval representing a single bit in a data waveform.

T_C—data signal waveform duration.

TCP/IP—Transmission control protocol/Internet protocol.

TDMA—time division multiple access.

TEC—Thevenin equivalent circuit.

tera-—prefix meaning one thousand billion, or 10^{12}.

T-FF-—toggle flip-flop.

THz—(teraherz) 10^{12} Hz.

T-FF—toggle flip-flop.

TOF—time of flight in a sonar.

T_{OFF}—time duration when PWM waveform is off.

T_{ON}—time duration when PWM waveform is on.

T_{PWM}—time period of PWM waveform.

T_s—sampling period in ADC.

T_x—duration of pulse caused by laser passing over UPC black bar.

TTL—time-to-live counter in an Internet data packet.

UDP/IP—Universal datagram protocol/Internet protocol.

UPC—Universal product code.

USPS—United States Postal Service.

V—volt, unit of electrical potential.

V_{ave}—average of waveform produced by PWM.

VOIP—voice over Internet protocol.

V_{min}—minimum voltage in a quantizer.

V_{max}—maximum voltage in a quantizer or maximum voltage allowed for a signal.

V_S—supply voltage that powers a circuit.

W—watt, unit of power.

WAN—wide-area network.

Wi-Fi—wireless network.

WWW—World Wide Web.

X_{RMS}—value measured with RMS meter.

Z—impedance.